COASTAL MANAGEMENT

COASTAL MANAGEMENT
Global Challenges and Innovations

Edited by

R. R. Krishnamurthy
M.P. Jonathan
Seshachalam Srinivasalu
Bernhard Glaeser

Academic Press is an imprint of Elsevier
125 London Wall, London EC2Y 5AS, United Kingdom
525 B Street, Suite 1650, San Diego, CA 92101, United States
50 Hampshire Street, 5th Floor, Cambridge, MA 02139, United States
The Boulevard, Langford Lane, Kidlington, Oxford OX5 1GB, United Kingdom

© 2019 Elsevier Inc. All rights reserved.

No part of this publication may be reproduced or transmitted in any form or by any means, electronic or mechanical, including photocopying, recording, or any information storage and retrieval system, without permission in writing from the publisher. Details on how to seek permission, further information about the Publisher's permissions policies and our arrangements with organizations such as the Copyright Clearance Center and the Copyright Licensing Agency, can be found at our website: www.elsevier.com/permissions.

This book and the individual contributions contained in it are protected under copyright by the Publisher (other than as may be noted herein).

Notices
Knowledge and best practice in this field are constantly changing. As new research and experience broaden our understanding, changes in research methods, professional practices, or medical treatment may become necessary.

Practitioners and researchers must always rely on their own experience and knowledge in evaluating and using any information, methods, compounds, or experiments described herein. In using such information or methods they should be mindful of their own safety and the safety of others, including parties for whom they have a professional responsibility.

To the fullest extent of the law, neither the Publisher nor the authors, contributors, or editors, assume any liability for any injury and/or damage to persons or property as a matter of products liability, negligence or otherwise, or from any use or operation of any methods, products, instructions, or ideas contained in the material herein.

Library of Congress Cataloging-in-Publication Data
A catalog record for this book is available from the Library of Congress

British Library Cataloguing-in-Publication Data
A catalogue record for this book is available from the British Library

ISBN 978-0-12-810473-6

For information on all Academic Press publications
visit our website at https://www.elsevier.com/books-and-journals

Publisher: Candice Janco
Acquisition Editor: Louisa Hutchins
Editorial Project Manager: Emily Thomson
Production Project Manager: Nilesh Kumar Shah
Cover Designer: Christian Bilbow

Typeset by SPi Global, India

Contents

Contributors

M.S. Achary Environment and Safety Division, Indira Gandhi Centre for Atomic Research, Kalpakkam, India

Dilanthi Amaratunga Global Disaster Resilience Centre, University of Huddersfield, Huddersfield, United Kingdom

Jiménez-Illescas Ángel R Centro de Interdisciplinario de Ciencias Marinas (CICIMAR), Instituto Politécnico Nacional (IPN), La Paz, Mexico

Md Sharif Imam Ibne Amir School of Engineering and Technology, Central Queensland University, QLD, Australia

A. Anandkumar Department of Applied Geology, Faculty of Engineering and Science, Curtin University Malaysia, Miri, Malaysia

S. Arockiaraj National Centre for Coastal Research (NCCR), Ministry of Earth Sciences, Government of India, Chennai, India

Faisal Ashar Global Disaster Resilience Centre, University of Huddersfield, Huddersfield, United Kingdom

Isaac Azuz Adeath CETYS Universidad—Ensenada-Campus Internacional, Ensenada, México

K. Banerjee National Centre for Sustainable Coastal Management (NCSCM), Ministry of Environment, Forest and Climate Change, Anna University Campus, Chennai, India

Mrittika Basu United Nation's University Institute for Advanced Studies (UNU-IAS), Tokyo, Japan

S. Biswas Environment and Safety Division, Indira Gandhi Centre for Atomic Research, Kalpakkam, India

Smitanjali Choudhury Zoological Survey of India, Andaman and Nicobar Regional Centre, Port Blair, India

Alejandra Cortés Ruiz CETYS Universidad—Ensenada-Campus Internacional, Ensenada, México

N.P.I. Das Environment and Safety Division, Indira Gandhi Centre for Atomic Research, Kalpakkam, India

Rajarshi DasGupta Institute for Global Environmental Strategies (IGES), Hayama, Kanagawa, Japan

Dixon T. Gevaña Department of Social Forestry and Forest Governance, College of Forestry and Natural Resources, University of the Philippines, Los Baños, Philippines

Bernhard Glaeser German Society for Human Ecology (DGH) and Berlin Free University (FUB), Berlin, Germany

Cristina Gómez GeoDrone Survey Ltd., United Kingdom; INIA-Forest Research Centre, Department of Forest Dynamics and Management, Madrid, Spain

Billy J. Gregory AICSM/UCEMM, Department of Geography and Environment, School of Geosciences, University of Aberdeen, Aberdeen, Scotland; DroneLite, Forres, Moray, United Kingdom

David R. Green AICSM/UCEMM, Department of Geography and Environment, School of Geosciences, University of Aberdeen, Aberdeen, Scotland, United Kingdom

Jason J. Hagon AICSM/UCEMM, Department of Geography and Environment, School of Geosciences, University of Aberdeen, Aberdeen, Scotland; GeoDrone Survey Ltd., United Kingdom

Richard Haigh Global Disaster Resilience Centre, University of Huddersfield, Huddersfield, United Kingdom

K. Jayakumar Center for Remote Sensing and Geoinformatics, Sathyabama Institute of Science and Technology, Chennai, India

M.P. Jonathan Centro Interdisciplinario de Investigaciones y Estudios sobre Medio Ambiente y Desarrollo (CIIEMAD), Instituto Politécnico Nacional (IPN), Ciudad de Mexico (CDMX), Mexico

R.S. Kankara National Centre for Coastal Research (NCCR), Ministry of Earth Sciences, Government of India, Chennai, India

M. Shah Alam Khan Institute of Water and Flood Management, Bangladesh University of Engineering and Technology, Dhaka, Bangladesh

Bharat Kumar Environment and Safety Division, Indira Gandhi Centre for Atomic Research, Kalpakkam, India

Elango Lakshmanan Department of Geology, Anna University, Chennai, India

Ahana Lakshmi National Centre for Sustainable Coastal Management (NCSCM), Ministry of Environment, Forest and Climate Change, Anna University Campus, Chennai, India

Maxime Le Bail Instituto Politécnico Nacional (IPN)—Centro Interdisciplinario de Investigaciones y Estudios sobre Medio Ambiente y Desarrollo (CIIEMAD), Ciudad de México, México

Espinosa-Carreón T. Leticia Instituto Politécnico Nacional, Centro Interdisciplinario de Investigación para el Desarrollo Integral Regional (CIIDIR) Unidad Sinaloa, Guasave, Mexico

Alfredo Salazar López Escuela Superior de Comercio y Administración (ESCA), Instituto Politecnico Nacional (IPN), Ciudad de Mexico, Mexico

Zayas-Esquer Ma Magdalena Instituto de Investigación en Ambiente y Salud de la Universidad de Occidente, Los Mochis, Mexico

A.K. Mohanty Environment and Safety Division, Indira Gandhi Centre for Atomic Research, Kalpakkam, India

Norma Patricia Muñoz Sevilla Instituto Politécnico Nacional (IPN)—Centro Interdisciplinario de Investigaciones y Estudios sobre Medio Ambiente y Desarrollo (CIIEMAD), Ciudad de México, México

Tomokazu Murakami National Research Institute for Earth Science and Disaster Resilience, Tsukuba, Japan

R. Nagarajan Department of Applied Geology, Faculty of Engineering and Science, Curtin University Malaysia, Miri, Malaysia

Rekha Nianthi K.W.G. Department of Geography, University of Peradeniya, Peradeniya, Sri Lanka

Nimal Sri Rajarathna W.A. Coast Conservation and Coastal Resource Management Department, Colombo, Sri Lanka

Toshinori Ogasawara Iwate University, Morioka, Japan

R.K. Padhi Environment and Safety Division, Indira Gandhi Centre for Atomic Research, Kalpakkam, India

A. Paneerselvam National Centre for Sustainable Coastal Management (NCSCM), Ministry of Environment, Forest and Climate Change, Anna University Campus, Chennai, India

S.N. Panigrahi Environment and Safety Division, Indira Gandhi Centre for Atomic Research, Kalpakkam, India

R.C. Panigrahy Department of Marine Sciences, Berhampur University, Berhampur, India

Omar Mayorga Pérez Centro Interdisciplinario de Investigaciones y Estudios sobre Medio Ambiente y Desarrollo CIIEMAD), Instituto Politécnico Nacional (IPN), Ciudad de Mexico, Mexico

Juan M. Pulhin Department of Social Forestry and Forest Governance, College of Forestry and Natural Resources, University of the Philippines, Los Baños, Philippines

R. Purvaja National Centre for Sustainable Coastal Management (NCSCM), Ministry of Environment, Forest and Climate Change, Anna University Campus, Chennai, India

C. Raghunathan Zoological Survey of India, Andaman and Nicobar Regional Centre, Port Blair, India

R. Raghuraman Zoological Survey of India, Andaman and Nicobar Regional Centre, Port Blair; National Centre for Sustainable Coastal Management (NCSCM), Ministry of Environment, Forest and Climate Change, Anna University Campus, Chennai, India

R. Ramesh National Centre for Sustainable Coastal Management (NCSCM), Ministry of Environment, Forest and Climate Change, Anna University Campus, Chennai, India

B.K. Rawlins Department of Hydrology, University of Zululand, Kwa Dlangezwa, South Africa

I. Retama Centro Interdisciplinario de Investigaciones y Estudios sobre Medio Ambiente y Desarrollo (CIIEMAD), Instituto Politécnico Nacional (IPN), Ciudad de Mexico (CDMX), Mexico

D.M. Rivera Rivera Centro Interdisciplinario de Investigaciones y Estudios sobre Medio Ambiente y Desarrollo (CIIEMAD), Instituto Politécnico Nacional (IPN), Ciudad de Mexico (CDMX), Mexico

María Concepción Martínez Rodríguez Centro Interdisciplinario de Investigaciones y Estudios sobre Medio Ambiente y Desarrollo (CIIEMAD), Instituto Politécnico Nacional (IPN), Ciudad de Mexico, Mexico

P.F. Rodríguez-Espinosa Centro Interdisciplinario de Investigaciones y Estudios sobre Medio Ambiente y Desarrollo (CIIEMAD), Instituto Politécnico Nacional (IPN), Ciudad de Mexico (CDMX), Mexico

G. Sahu Environment and Safety Division, Indira Gandhi Centre for Atomic Research, Kalpakkam, India

M.K. Samantara Environment and Safety Division, Indira Gandhi Centre for Atomic Research, Kalpakkam, India

S.K. Sarkar Department of Marine Sciences, University of Calcutta, Kolkata, India

K.K. Satpathy Environment and Safety Division, Indira Gandhi Centre for Atomic Research, Kalpakkam, India

Rajib Shaw Graduate School of Media and Governance, Keio University, Tokyo; Keio University, Japan

Shinya Shimokawa National Research Institute for Earth Science and Disaster Resilience, Tsukuba, Japan

V.C. Shruti Centro Interdisciplinario de Investigaciones y Estudios sobre Medio Ambiente y Desarrollo (CIIEMAD), Instituto Politécnico Nacional (IPN), Ciudad de Mexico (CDMX), Mexico

Pournima Sridarran Global Disaster Resilience Centre, University of Huddersfield, Huddersfield, United Kingdom

S.B. Sujitha Centro Interdisciplinario de Investigaciones y Estudios sobre Medio Ambiente y Desarrollo (CIIEMAD), Instituto Politécnico Nacional (IPN), Ciudad de Mexico (CDMX), Mexico

Maricel A. Tapia Department of Social Forestry and Forest Governance, College of Forestry and Natural Resources, University of the Philippines, Los Baños, Philippines

Manivannan Vengadesan Department of Geology, Anna University, Chennai, India

E. Vetrimurugan Department of Hydrology, University of Zululand, Kwa Dlangezwa, South Africa

H. Vijith Department of Applied Geology, Faculty of Engineering and Science, Curtin University Malaysia, Miri, Malaysia

Editors Bio

Dr. R.R. Krishnamurthy is a geologist specializing in the field of satellite remote sensing applications in coastal management. He has 27 years of teaching and research experience and has served in different capacities at Anna University, the M.S. Swaminathan Research Foundation, and the University of Madras. He was a YSSP participant in the Advanced Computer Applications (ACA) Program of the International Institute for Applied Systems Analysis (IIASA), Austria, and served as visiting professor at the Inter-Graduate School Program for Sustainable Development and Serviceable Societies (GSS), Kyoto University, Japan. As a recipient of the Environmental Award from the government of Tamil Nadu, he has produced several end-user-oriented, significant research outputs such as a Climate Action Plan (CAP), Climate Disaster Resilience Index (CDRI), and Climate Disaster Recovery Process (CDRP) for Chennai, a Tsunami Vulnerability Atlas for the Chennai coast, teaching tools for high school students on disaster management, etc. As the founding principal of the University of Madras Constituent College in Nemmeli Village, East Coast Road, Kanchipuram District, during 2011–14, he successfully implemented the introduction of disaster education for rural coastal communities to enhance their resilience to disasters. His remarkable research accomplishments include participation and contributions in India's ambitious national programs on the Marine Satellite Remote Sensing Information Service (MARSIS) and the establishment of India's early warning system for tsunami and storm surges. Based on his expertise in this domain, he has closely interacted with the Ministry of Environment and Forests, the National Disaster Management Authority, the Ministry of Science and Technology, and the Ministry of Earth Sciences. His contributions under extension and outreach activities include (i) training of key decision makers in India under the UK-DFID program on coastal management jointly with the Universities of Newcastle and Bath in the United Kingdom and (ii) UNISDR, Geneva and the World Bank Institute, Washington, DC, funded/supported program on building resilience to Indian Ocean tsunami, which aimed to train different target groups in India and the Maldives. In the aftermath of the 2004 Indian Ocean tsunami, he has trained about 1000 school teachers in India and several field practitioners both in India and the Maldives.

Dr. M.P. Jonathan is a proficient expert in the fields of marine geochemistry and trace metal pollution of aquatic environments. He earned a doctorate in the field of geology from the University of Madras in Chennai, India. He is the author of more than 75 research articles on numerous multidisciplinary aspects of coastal pollution in various reputed journals. With a wide vision to generate a database of pollution status in tourist beaches all over the world, he has visited 10 countries to date, and his programs are expanding year by year to other countries. He is also a great teacher in orienting and inspiring young researchers to develop an urge for writing good scientific articles through his "Art of Scientific Writing" course. He has also partaken in various bilateral and mega-scientific research projects between India and Mexico which have dealt with coastal pollution and development/mitigation efforts through mangroves. His research interests include geochemical processes in aquatic systems, environmental geology, tsunami (past and present), GIS applications, and innovative approaches in pollution studies. Presently, he works as a research professor and is also a member (SNI Level 2) of the National Council of Science and Technology (CONACyT) and also attests to his abilities as the coordinator of doctorate program in the Center for Interdisciplinary Studies on Environment and Development (CIIEMAD), National Polytechnic Institute (IPN), Mexico City, Mexico. He is also the chief editor of an International Scientific Journal *Envirogeochim Acta* and is also a member/reviewer of leading journals of Elsevier and Springer Verlag Publishers.

Dr. S. Srinivasalu has more than 30 years of teaching and research experience. He is currently the director of the Institute for Ocean Management at Anna University, Chennai. His main research interest lies in coastal zone management. He has used coastal stratigraphy, sedimentology, and micropaleontology to study the recurrence interval of catastrophic coastal flooding events such as tsunami and large storms. His most significant contributions to the field include characterization of modern tsunami storm deposits of the southeastern coast of India and the western coast of Mexico, assessment of heavy metal concentrations in the coastal zones, mapping of the coastal geomorphology of India, mineral mapping of ICZM sites of India, and mapping of ecologically sensitive areas of the Tamil Nadu coast. He has more than 80 publications in national and international journals. His citation index is 1133, and his h-index is 17. His i10-index is 28. He is an expert member of UNECO-IOC IOTWS programs. He has guided 9 PhD students and more than 60 postgraduate students. He is a member of many academic and management committees in India and abroad. He has visited 12 countries for various collaborative research programs in the field of coastal management.

Dr. Bernhard Glaeser, PhD, is professor of sociology in the Department of Political and Social Sciences, Free University of Berlin (Germany), with emphasis on environment and development and research projects in Europe, East Africa, and Asia. He is the honorary president of the German Society for Human Ecology (DGH), and retired as professor of human ecology at the University of Göteborg (Sweden) and as senior researcher at the Social Science Research Center, Berlin (WZB). Ever since 1996, his research has focused on integrated and sustainable coastal management, with projects in Sweden, Germany, Poland, and Indonesia. He is LOICZ corresponding member (previously, Scientific Steering Committee) and corresponding member of the IMBER (Integrated Marine Biogeochemistry and Ecosystem Research) Human Dimensions Working Group. He has served on multiple international advisory boards and is founder and editor of the book series "Routledge Studies in Environment, Culture, and Society" (RSECS, United Kingdom).

Foreword

Ramesh Ramachandran

National Centre for Sustainable Coastal Management, Ministry of Environment, Forest and Climate Change (MoEF&CC), Government of India, Chennai, India

This book, *Coastal Management: Global Challenges and Innovations*, addresses several important and cutting-edge aspects of global coastal system research. Despite several global research and development initiatives, large gaps in our knowledge exist that clearly indicate the complexities of this narrow zone. The multiple uses of the coast often cause an imbalance between conservation and increasing developmental pressures. It is important to assess the rapid changes along the coast using a multidimensional scientific and social approach.

GIS and remote-sensing technology have undergone by the far the largest transition, from low-resolution satellite data to high-resolution aerial photographs of the highest precision to the current use of drones. This enables close monitoring of coastal features, including subtle changes at finer scales. Besides mapping, numerical modeling is a powerful tool to determine highly vulnerable coastal areas in order to safeguard the life and assets of the coastal community.

The greatest threat facing coasts is the exposure to extreme conditions, making the coastal community highly vulnerable. There is an urgent need to integrate the traditional knowledge of the coastal communities while planning coastal development. Their traditional knowledge should also be used as a tool in the conservation of coastal/marine living and nonliving resources. The best option would be to manage these natural resources through comanagement or ecosystem-based resource management. It is equally important to restore lost ecosystems and provide adequate resources and technology for coastal pollution abatement.

Cutting across all these natural and human-induced challenges to the coast is climate change. It is extremely important to move toward a blue economy by conserving coastal ecosystems by enhancing carbon sequestration and renewable energy resources such as offshore wind and biomass.

These challenges demand wider coastal management approaches that unify sectoral planning through Integrated Coastal Zone Management. The challenges in research are now being overcome by the path-breaking evolution of science and technology, leading to practical solutions that are highlighted in this book. It is my firm belief that the compilation of a reference book based on case study experiences from across the globe under the title *Coastal Management: Global Challenges and Innovations* by RR Krishnamurthy, MP Jonathan, S Srinivasalu, and Berhard Glaser will be immensely beneficial to researchers, field practitioners, policymakers, and the coastal community, working on building the Future Coasts.

Editorial Note

Ramasamy R. Krishnamurthy
Department of Applied Geology, School of Earth and Atmospheric Sciences, University of Madras, Chennai, India

My memory goes back to the early 1990s, when Anna University was establishing the Institute for Ocean Management (IOM) to fulfill the need for a scientific approach toward coastal management, with a handful of young scientists constituting a multidisciplinary team. We started the initial research under the National Programme entitled "Marine Remote Sensing Information Service (MARSIS)," which was coordinated by the Indian government's Department of Space. The launch of the Indian Remote Sensing (IRS) satellite series with different spectral and spatial resolutions, including the availability of satellite data such as NOAA-AVHRR from other countries, was helpful in initiating several pioneering studies with regard to mapping and monitoring India's lengthy coastal zone. The sea surface temperature (SST) and ocean productivity information derived from AVHRR sensor data were used to demarcate the Potential Fishery Zone (PFZ), which was one of the pioneering end user products generated for the benefit of fishermen along the coast.

The scarcity and accuracy of scientific data pertaining to coastal areas presented major challenges during the 1990s, but such conditions have been superseded by more precise technological tools such as high spatial/spectral resolution remote sensing data and spatially enabled decision support systems, including the latest mapping methods using drones, which have proven to be extremely helpful globally in the effective monitoring and management of coastal resources. The majority of the critical coastal habitats such as mangroves, coral reefs, sea grass meadows, etc., are also being monitored using thematic information derived from multisensor remote sensing data. High spatial resolution satellite data, coupled with digital analysis methods, have aided in demarcating even small patches of mangroves and, to some extent, delineating the zonation of major species. A systematically organized scientific database pertaining to all mangrove sites along coastal India using the GIS tool has helped end users and local authorities effectively restore degraded sites and quantify the extent of restoration. A considerable amount of research has been carried out by researchers and institutions in the country using satellite remote sensing and GIS tools during the last three decades, and the quantum of data generated is highly laudable.

India has a contrasting coastal geomorphology consisting of a gently sloping East coast with numerous deltas, lagoons, mangroves, and coral reef resources, and a relatively more steeply sloping West coast with numerous estuaries, sea cliffs, backwaters, etc. The influence of coastal geomorphology and topography on the coastal ecosystems/resources can be witnessed apart from the variations in tidal amplitude and inundation. Several pockets of coastal India suffer from the chronic disaster of shoreline erosion as well as accretion, both of which affect the traditional communities inhabiting the coast. The National Centre for Sustainable Coastal Management (NCSCM) has been mapping the coastal erosion/accretion sites on a larger scale and quantitatively estimating the rate of shoreline change to extrapolate erosion/accretion data for the next 100 years in order to demarcate the coastal hazard line. The outcomes of this study are expected to help local authorities and vulnerable communities develop better preparedness in the future.

Vulnerable to natural hazards, especially the increasing intensity of cyclones, the coastal communities have invariably become more resilient as a result of the experience they gain from every disaster. Mitigating the impact of climate change on the coast is an important focus area for research, which must address sea level changes, saline water intrusion into coastal aquifers, etc. Though tsunami terminology was once unfamiliar, it is now well-known, even at the grassroots level, and the 2004 Indian Ocean tsunami was an unforgettable lesson for all those affected by it. One positive result of this megadisaster is that it emphasized the importance of establishing a dedicated tsunami warning system in India. In improving vulnerable communities' disaster preparedness, it is necessary to take their socio-economics into account, including their livelihood capital, in order to understand their resilience levels. Major challenges still exist in this research domain, especially incorporating the traditional knowledge and local wisdom of coastal communities with the scientific data being generated through the use of technology tools.

In India, there have been considerable attempts during the last two decades toward Integrated Coastal Zone Management (ICZM), which includes (i) training and capacity building for different target groups, (ii) engaging the stakeholders in planning, and (iii) research and development activities concerning different aspects of coastal management by various institutions. About 570 Indian islands, which are important biodiversity hotspots, are to be managed through specific policies, and toward that end, the Ministry of Environment and Forests has produced the Integrated Island Management (IIM) plan separate from Island CRZ. Not only have these efforts generated enormous amounts of scientific data, they have also enhanced knowledge about coastal governance, as exemplified by the shift in Coastal Regulation Zone (CRZ) implementation from a regulatory approach to a management approach between 1991 and 2011. Invariably, all the stakeholders have realized the importance of community participation and coastal governance in coastal management.

Higher education (including research in the field of ICZM) must still address the challenges in developing site-specific approaches and effectively transferring technology to vulnerable communities. My collaborators in India, Germany, and Mexico came forward to receive contributions of chapters from across the world and compile a reference book based on case study experiences. This book, titled *Coastal Management: Global Challenges and Innovations*, will be immensely useful to the researchers, field practitioners, policy makers, and NGOs working in this domain. I am indebted to my editorial colleagues, professors M.P. Jonathan, S. Srinivasalu, and Bernhard Glaser, for their valuable support in completing this ambitious project. My sincere thanks are also due to my colleagues at the Elsevier office, particularly Louisa Hutchins and Emily Thomson, for their valuable help and support from time to time. There were a few unexpected delays in completing this project, due to recovery from flood and cyclone disasters in Chennai, one after the other, during November and December of 2015 and 2016. However, all the chapter contributors for this book were very supportive during those challenging times, and I am very thankful to them.

I am also thankful to Prof. Ramesh Ramachandran for his concise yet informational Foreword.

CHAPTER

1

Global Coasts in the Face of Disasters

Rajib Shaw
Graduate School of Media and Governance, Keio University, Tokyo, Japan

1 INTRODUCTION

Being the most prominent and traditional part of the transportation system, coastal areas have several infrastructures such as ports and fishing harbors. Several areas in the coastal zones are dominated by industries. Some of them can cause significant impacts to the people and communities if they are not resilient, as seen in the case of the nuclear meltdown in Fukushima due to the East Japan earthquake and tsunami. Coastal zones also have rich biodiversity with coastal buffers such as mangroves, which reduces the impacts of hazards and enhances the livelihoods of the people and communities living nearby. It is said that more than 45% of the world's population lives in coastal areas (within 100 km of the coast area). The Low Elevation Coastal Zone (LECZ, less than 10 m elevation) is also considered for its different threats.

Coastal areas are known for their resources as well as vulnerabilities. While people go to the coastal areas for business (as part of their livelihoods) as well as entertainment (coastal resorts), the coastal hazards (typhoon/cyclone/hurricane, storm surge, tsunami) pose significant threats to the population and infrastructure. While the above-mentioned hazards are more visible, there are also the impacts of creeping disasters such as rising sea levels, which are already affecting several small island countries and communities. Coastal erosion is also affecting a significant number of lives and habitats.

2 GLOBAL FRAMEWORK AND COAST

Many global frameworks have mentioned coastal areas in different ways. Sustainable Development Goals (SDGs) mentioned the importance of coasts as follows:

> Secure blue wealth by ensuring a healthy and productive marine environment with all basic provisioning, support, regulation, and cultural services. Provide equitable access to resources, and ensure that neither

https://doi.org/10.1016/B978-0-12-810473-6.00001-7
© 2019 Elsevier Inc. All rights reserved.

pollution nor the harvesting and extraction of animate and inanimate resources impairs the basic functions of the ecosystem. Facilitate the development of sustainable and resilient coastal communities. Harmonize national and regional maritime policies, and encourage cooperation in coastal and global marine spatial planning.

The Paris climate agreement also focused on the needs of the SIDS (Small Island Developing States), and highlighted the importance of mitigation and adaptation as well as damage and loss due to climate change. The Sendai Framework for Disaster Risk Reduction (SFDRR) also focuses on the need for transboundary collaboration for a resilient coastline, and also to promote mainstreaming disaster risk reduction in coastal zone management.

A recent study by the G-7 countries on climate fragility suggested that the complex nature of climate impacts international relations, especially related to the coast. The study mentioned that gradual sea temperature change has caused specific fish varieties to move away from Japan's coast, thereby causing fishermen to move farther out to sea to catch fish, which is making additional resources necessary. That has increased the food price in Japan. Also, competition with neighboring countries has increased, often causing violations of the international fishing line.

Apart from these global frameworks, the international science policy framework and the global environmental research framework have also identified the importance of coastal zone management. The IPCC (Intergovernmental Panel on Climate Change), in its chapter (Chapter 5) on "Coastal System and Low-lying Areas," said:

"The population and assets exposed to coastal risks as well as human pressures on coastal ecosystems will increase significantly in the coming decades due to population growth, economic development, and urbanization." It has also clarified that: "Some low-lying developing countries (e.g., Bangladesh, Vietnam) and small islands are expected to face very high impacts and associated annual damage and adaptation costs of several percentage points of gross domestic product (GDP)."

The Future Earth research program has also emphasized the importance of coastal areas in its KAN (Knowledge Action Network):

> The ocean, including coastal and near-shore areas, thus provide services essential for life on earth and to the history, culture, and livelihoods of people across the globe. However, the ocean is also facing multiple challenges from climate change, overfishing, acidification, deoxygenation, and pollution. Accordingly, the United Nations referred to the importance of a healthy ocean in several of (its) Sustainable Development Goals.

3 DISASTER RISK AND COASTS

For effective disaster management in the coastal areas, there needs to be a combination of predisaster planning and preparedness issues as well as during and postdisaster activities. As a part of predisaster preparedness, coastal buffer and coastal zone planning and management become of utmost importance. This includes the planting and protecting of coastal green belts as well as the protection of the coastline through coastal dykes or other hard infrastructures, based on a carefully performed risk analysis. On the other hand, softer components such as coastal watching and disaster education also become important to enhance the perception and understanding of coastal hazards. The "education for sustainable development (ESD)" in some of the coastal schools in Kesennuma, Japan, has been extremely useful for the safer evacuation of school children and local communities, notably in the case of the East Japan earthquake and tsunami of 2011. Similarly, regular disaster drills involving elementary and junior high school children along with the local communities have helped joint evacuation of the children to the higher ground during the tsunami, resulting in no casualty in those schools in Kamaishi city in Iwate prefecture of Japan.

In the time of disaster, three issues become important, especially in the coastal areas: (i) an early warning system (EWS), (ii) evacuation shelters, and (iii) the human network. For an effective evacuation for coastal hazards, the first and foremost necessity is the timely warning system, including sounding that warning to the people and communities in the coastal towns and villages. The "last mile communication" of the EWS is always a challenge, and it needs possibly a balanced mix of technology and local knowledge. The second important issue is safer places for evacuation, which are often schools as cyclone shelters, built in many coastal areas in different countries. However, in spite of having an EWS and existing shelters, timely evacuation is a human behavior that depends on one's judgment and risk perception. Therefore, a human network of volunteers plays an important role. This means identifying people who can facilitate an early evacuation of specific vulnerable groups such as the aged population, the physically challenged, pregnant mothers, and small children. A combination of these three elements is found to be useful for effective evacuation, not only for developing countries but for developed countries as well.

4 WAY FORWARD

The 21st century is considered to be the century of information age and technologies, where new emerging technologies such as the IoT (Internet of Things), robotics, drones, three-dimensional printing, block chains, and artificial intelligence have a strong role to play in our daily lives. These emerging technologies, along with information technology, can make significant changes in society.

However, this technological development needs a balanced approach of a mechanism of governance, technological development, and education/awareness. For effective coastal zone management in terms of disaster risk reduction, we need an appropriate governance mechanism, which is the demarcation of coastal zones; the planning and development of regulation on the maintenance and utilization of coastal resources; sticking to the national, regional, and global policies and practices, etc. Community governance also becomes crucial when it comes to the management of local resources. Different types of technologies become important to support the implementation of a governance system. Traditional GIS (geographic information systems) and remote sensing have been used for several years to monitor coastal changes over time. Added to these are the emerging technologies and mobile applications. For example, drones can be used for monitoring coastal resources more closely. Mobile phone applications can be used to promote "citizen science" in engaging communities in a community governance system. This will also be linked to the education and awareness of people and communities, including school students. An educative mobile application for coastal schools can make a large difference in engaging school students, teachers, and their parents to be more interested in coastal risks and resources.

The key target is to make ourselves "responsible citizens" so that the current resource use does not compromise the future need for those resources. That is the key principle of "sustainable" development goals, and it would take the collective efforts of governance, education, and technology to make a difference.

CHAPTER

2

☆Special Coastal Management Area Concept Experience in Sri Lanka

*Nimal Sri Rajarathna W.A.**, *Rekha Nianthi K.W.G.*†

*Coast Conservation and Coastal Resource Management Department, Colombo, Sri Lanka

†Department of Geography, University of Peradeniya, Peradeniya, Sri Lanka

Abbreviations

ADB	Asian Development Bank
CC&CRMP	Coast Conservation and Coastal Resource Management Department Plan
CRMP	Coastal Resource Management Project
CBO	Community Base Organization
CCC	Community Coordinating Community
FAO	Food and Agriculture Organization
GTZ	German Technical Cooperation Agency
GEF-RUK	Global Environment Facility-Rakawa, Ussangoda, Kalamatiya
HICZMP	Hambanthota Integrated Coastal Zone Management Project
IUCN	International Union for Conservation of Nature and Natural Resources
MFAR	Ministry of Fisheries and Aquatic Resources
NECDEP	North East Coastal Community Development Project
NGO	Nongovernmental Organization
SCMA	Special Coastal Management Area
UNDP	United Nation Development Program

☆Because this study has been based on original research work conducted by the CC&CRMD, Colombo in Sri Lanka, the study wanted to refer only to the original reports of the CC&CRMD and MFAR (2004, 2004: b, c, d, 2005: a, b, c, d, e, f, 2007), Special Area Management Plans of 1996: Rekawa, 2005, Hikkaduwa, Special Area Management (SAM) Reports and Integrated Resources Management Program Plans (2003). It is also necessary to mention that the Special Area Management (SAM) is referred as Special Coastal Management Area (SCMA) in this chapter.

https://doi.org/10.1016/B978-0-12-810473-6.00002-9

© 2019 Elsevier Inc. All rights reserved.

1 INTRODUCTION

The coastal environment in Sri Lanka is greatly influenced by the island location in the northern part of the Indian Ocean between 5°54′ and 9°52′ north latitude and 79°39′ and 81°53′ east longitude. Sri Lanka has a coastline of approximately 1620 km, including the shorelines of bays and outlets but excluding lagoons. The country has 103 rivers (Atlas of Sri Lanka, 1997), most of which radiate from the middle of the hill country and flow down to the sea. These then form estuaries that are important features of the coastal landscape, providing vital habitats for species of commercial and subsistence use. Table 1 shows that the coastal area contains a variety of terrestrial habitats, including sandy beaches, barrier beaches, sand spits, sand dunes, rocky shores, mangrove stands, and salt marshes. The coral reefs, lagoons, estuaries, and seagrass beds in the coastal waters off Sri Lanka are also equally important (CC&CRMP, 2018).

2 THE EVOLUTION OF THE SPECIAL COASTAL MANAGEMENT AREA IN SRI LANKA

The Special Coastal Management Area concept was first introduced to Sri Lanka in the late 1980s with the pilot testing projects carried out at Rekawa Lagoon and the Hikkaduwa coastal and marine area. It was hoped that this concept would address the issues emerging due to rapid economic development coupled with anthropogenic activities. The experiences and outcomes gained from the pilot testing in the local scenarios encouraged officials to promote and incorporate 23 coastal sites for SCMA planning into the 1997 National Coastal Zone Management Plan to identify and address site-specific issues pertaining to environmental, social, and economic aspects. During the implementation phase of the Coastal Resource Management Project (CRMP) funded by ADB, the CC&CRMD was able to initiate and extend the SCMA process in several sites: Bar Reef in Kalpitiya, Negombo Estuary and Muturajawela Marsh, Lunawa Lagoon, Madu Ganga, Koggala-Habaraduwa, and the Kalametiya and Mawella coastal areas (Fig. 1). In addition to the above SCMA sites,

TABLE 1 Extent of Coastal Resources in Sri Lanka (Hectares)

Types of Coastal Resources	Hectares
Mangroves	11,656
Salt Marches	27,520
Sand Dunes	10,363
Beaches, Barriers, Spits	5732
Lagoons, Estuaries	214,522
Other Water Bodies	13,062
Seagrass beds	37,137

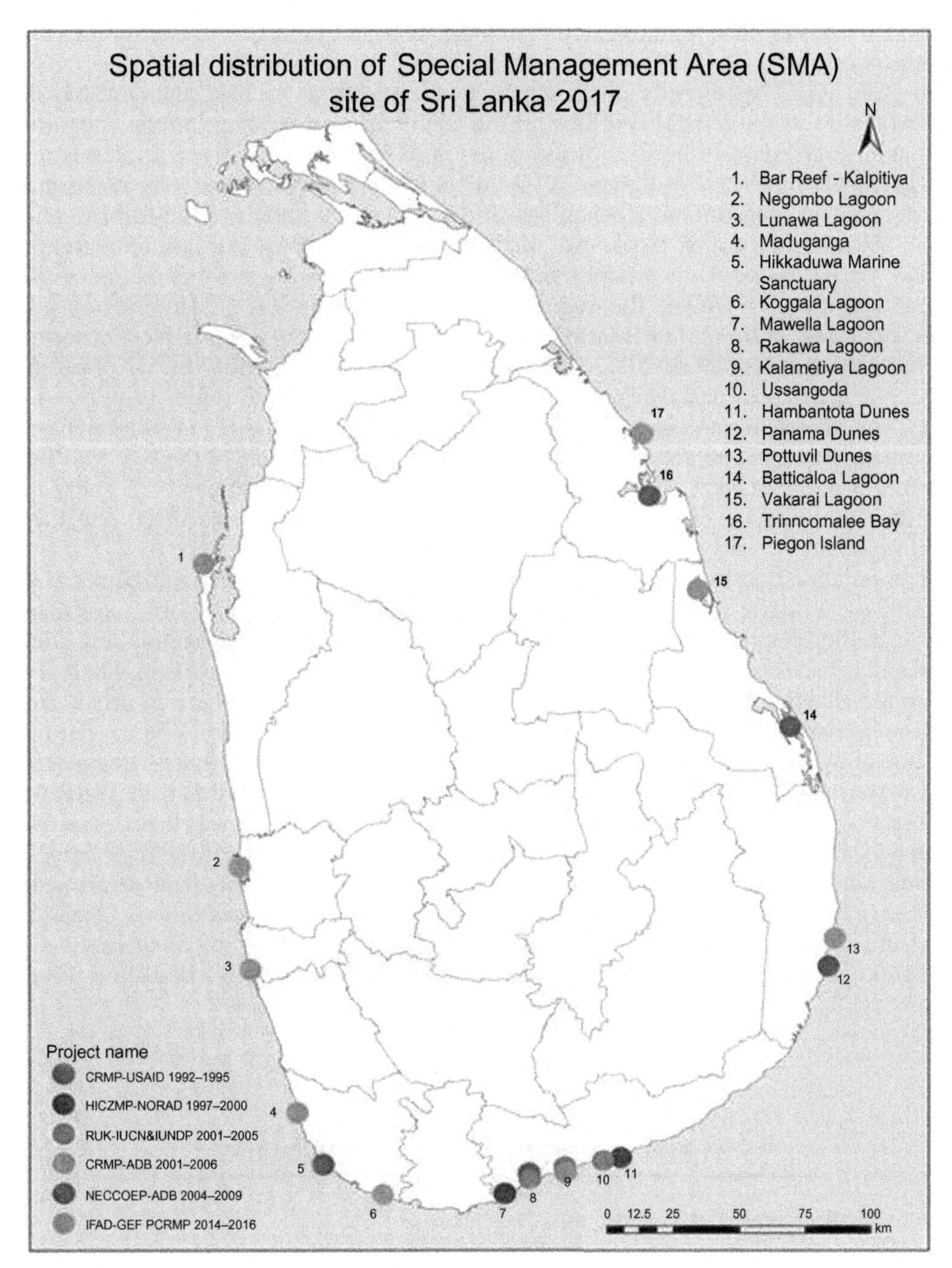

FIG. 1 Distribution of SCMAs in Sri Lanka implemented between 1992 and 2017 (CC&CRMD).

the Revised Coastal Zone Management Plan of 2004 has identified 27 candidate sites for the SCMA process.

The CC&CRMD has already prepared the new revision of its 2018 plan and has listed the SCMA sites in the coastal belt around the island after organizing public consultation meetings in each coastal district. Although the SCMA concept has been implemented by the CC&CRMD since 1992 as a major ICZM management policy, there were no proper legal provisions or institutional mechanisms for the implementation of the SCMA concept. In that absence of proper legal and institutional mechanisms for implementation of SCMAP, the CC&CRMD is mainly reliant on administrative arrangements with the relevant agencies. However, the required legal provisions have been introduced for SCMA planning and implementation with the enactment of the Coast Conservation (Amendment) Act No. 49 of 2011. The previous attempts to identify SCMA sites in the Northern and Eastern coastal segments were carried out mainly through secondary information. The experience and exposure of the CC&CRMD officials due to the country's civil war prevailed in those regions until 2009. However, since the end of the war and the resumption of peace in 2009, opportunities have now been opened to obtain primary information for the purpose of proper identification of SCMAs in the Northern and Eastern coastal regions.

The population density is weighted heavily toward the coastal region, particularly along the southern, western, and northwestern coastal areas. The coastal regions accommodate about 4.6 million people, which comprises about 25% of the island's total population. The municipal and urban councils in the coastal areas cover 285 km^2 and comprise nearly half of such land on the island. This has resulted in the concentration of a large share of urban growth and development activities within the coastal region (CZM Plan, 2004). The coastal region is the hub of industrial production and contains 61.6% of all industrial units. The tourism industry is the fifth-largest revenue source for the island while hotels in the coastal region continue to have around 70% of all hotels registered with the country's Tourist Board. The fishing industry in the marine and brackish waters of Sri Lanka is also one of the most important economic activities, providing more than 80% of the nation's protein. The Asian tsunami in 2004 gave a good lesson for the people about the protection and conservation of the island's coastal resources. That was the worst natural disaster in the history of Sri Lanka, leaving more than 38,000 people dead, 7100 missing, and around 1 million feeling some effect from the disaster.

3 MAIN OBJECTIVES OF THE STUDY

The main objectives of the study are:

i. To highlight the challenges and constraints for the sustainability of the SCMA process for the management of coastal resources in Sri Lanka.

ii. To provide best practices for the implementation of the SCMA process.

The CC&CRMP is the main national government agency in Sri Lanka for the conservation and management of coastal resources within the legally defined[1] coastal zone. The department has to implement the National Coast Conservation and Coastal Resource Management Plan within the coastal zone during 5 years. The management plan has been mandated by the Coast Conservation and Coastal Resource Management Act 57 of 1981, which was amended by Act 64 of 1988 and Act 49 of 2011. The CC&CRM plan was adopted by the department and is designed to ensure the sustainable use of the coastal environment and its resources in the long term while satisfying current national development goals. The management objectives, policies, strategies, and actions have been introduced in the management plan for the required effective and sustainable management of coastal zones in Sri Lanka for 5 years.

4 WHAT IS A SPECIAL COASTAL MANAGEMENT AREA (SCMA)?

> In respect of any area of land within the "coastal zone" or adjacent to the "coastal zone" or comprising both areas from the "coastal zone" and the adjacent area of land declare such area by order published in the Gazette to be a "Special Coastal Management Area" needed to adopt a collaborative approach to planning resource management within the defined geographic area. ***Coast Conservation (Amendment) Act 49 of 2011, Sri Lanka***

A Special Coastal Management Area (SCMA) is a locally based, geographically specific planning process that, in theory, is a highly participatory practice that allows for the comprehensive management of natural resources with the active involvement of the local community as the main stakeholder group. It involves comanagement of resources through which decision-making, responsibility, and authority in respect to natural resource use and management are shared between the government and the local resource users or community. Government institutions and other planning agencies assume the role of facilitator by providing technical and financial assistance to the local community management effort. The local community groups are considered the "custodians" of the resources being managed under the SCMA process, through which sustainable livelihood practices allow for sustainable natural resource use and management within the designated site.

One of the major objectives of an SCMA is to resolve competing demands on natural resource use within a specific geographical boundary by planning the optimal sustainable use of resources. In a broad sense, the SCMA approach seeks to ensure both the economic well-being of the local communities as well as the ecological well-being of the natural ecosystems by the practice of sound natural resource management. The SCMA concept is now considered a key component of coastal zone management policy in Sri Lanka.

[1]The "coastal zone" was legally defined by the Coast Conservation and Coastal Resource Management Act 57 of 1981, which was amended by Act 64 of 1988 and Act 43 of 2011, as the area lying within a limit of 300 m landward of the mean high water line and a limit of 2 km seaward of the mean law water line. In the case of rivers, streams, lagoons, or any other body of water connected to the sea either permanently or periodically, the landward boundary shall extend to a limit of 2 km measured perpendicular to the straight base line drawn between the natural entrance points thereof and shall include the waters of such rivers, streams, and lagoons or any other body of water so connected to the sea. Any other body of water has a farther extended limit of 100 m inland from the zero mean sea level along the periphery."

5 CRITERIA USED FOR RANKING POTENTIAL SCMA SITES FOR IMPLEMENTATION

i. The severity of social, economic, and environmental issues prevailing in the sites.
ii. The relative richness and abundance of coastal ecosystems.
iii. The feasibility of management based on size, location, and legal and institutional factors.
iv. The existing or potential value of economic development in the area.
v. The level of exposure/vulnerability to the impact of climate change.
vi. Vulnerability to coastal disasters, both episodic and chronic.
vii. Significance of archaeological and historic values of the site.

6 SCMA PROCESS

The following steps are adopted for the SCMA process:

i. Identify the SCMA site with the guidelines given by the National CC&CRM plan and agreement.
ii. Compile an environmental profile for the particular SCMA site.
iii. Enter the community with full-time professional facilitators and community organizers.
iv. Conduct planning/training workshops in the SCMA site.
v. Organize resource management core groups.
vi. Draft a management plan for the special management area through community involvement and the determination of indicators for monitoring.
vii. Implement pilot projects as whole planning continues.
viii. Refine the management plan for the SCMA.
ix. Implement the SCMA plan for the relevant site.
x. Monitoring.

The planning and implementation process for the 11 SCMA sites will be relisted and action will be initiated to declare the SCMA sites through a government gazette notification in compliance with the legal provisions of the Coast Conservation (Amendment) Act 49.

7 WHY SHOULD SCMA BE STRENGTHENED?

In Sri Lanka, the "Coastal 2000" policies highlighted the need for a more integrated approach to coastal zone management. It was specifically suggested to use the SCMA process to develop plans and implement actions simultaneously in selected sites. The SCMA process is based on comanagement principles, and is considered an effective and viable approach for integrated coastal resources management. This concept properly acknowledges the complex relationship between coastal and marine uses and the coastal ecosystems. The SCMA process also promotes linkage and harmonization among varied types of coastal activities and the physical processes of nature. The flexibility of policies management system pays proper attention to both coastal resource systems as well as human systems. The main influencing factors behind the requirement of SCMA as a complementary tool for integrated coastal resources management are summarized as follows:

i. The principles adopted in SCMA planning are the concept that has been followed since ancient times in Sri Lanka as well as internationally accepted principles relating to sustainable development.
ii. SCMA is viewed as an effective means of promoting sustainable management of coastal resources within a defined geographic setting and possibly to deal more comprehensively and effectively with complex management issues.
iii. Neither the government nor the market is uniformly successful in enabling individuals to sustain long-term productive use of coastal resources.
iv. The decentralization policies that have been implemented since the late 1980s positively contributed to adopting collaborative management.
v. The recognition of the need to formalize indigenous or traditional sustainable resources management practices within a legalistic and wider governance framework to minimize coastal resource depletion, overexploitation, and user conflicts.
vi. The characteristics of public or state-owned coastal resources and the prevailing status of open access present formidable challenges to manage coastal resources.
vii. Coastal habitats are being rapidly degraded due to both man-induced causes and natural phenomena. Thus, a user-centered management approach is vital.
viii. The effective approaches should be introduced to minimize the poverty and overexploitation of marine and coastal resources.
ix. To facilitate local management interventions to maintain consistency and compliance with national level coastal resource management policies and regulations.
x. A community demand for greater legitimacy and transparency in resource management decision-making.
xi. Increasing user conflicts are parallel to new development activities taking place in coastal regions.
xii. Empowering and building a sense of ownership among civil society, communities, and community-based organizations to enable them to manage coastal resources in a sustainable manner.
xiii. To address gender issues related to coastal resource uses.
xiv. To incorporate a sustainable livelihood perspective to address site-specific coastal environmental issues.
xv. To build resilience and reduce vulnerability among coastal communities against natural coastal hazards.
xvi. Attitudes and perceptions of community on decentralization policies are being persuaded in the recent past in administrative and political fields provide enabling environment for effective and sustainable coastal resources management through SCMA process.

8 WHY SHOULD AN SCMA BE IMPLEMENTED?

The need for integrated management to conserve, develop, and sustain the use of dynamic and resource-rich coastal zones has long been recognized in Sri Lanka. The Coast

Conservation and Coastal Resource Management Plan (CC&CRMP) is the plan adopted by the CC&CRMD for the management of the coastal zone during a five-year period. It is designed to ensure the sustainable use of the coastal environment and its resources in the long term while satisfying current national development goals. It outlines the management objectives of the CC&CRMD for the period under consideration, the policies to be adopted, and the strategies and actions required for effective management of the coastal zone in the face of competition for resource use.

So far, coastal resource management planning has been based on the recognition that the CC&CRMD is only one of many institutions that has jurisdiction over management and conservation of coastal resources within the legally defined coastal zone. The necessary policies, strategies, and actions have to be based on a realistic assessment of the department's capacity to directly manage the development activities affecting coastal resources within the legally defined coastal zone.

Because the department does not have the ability to control development activities out of the coastal zone under the Coast Conservation and Coastal Resource Management Act, the SCMA is a very important mechanism used to manage and conserve ecosystems in an adjacent area of the legally defined coastal zone. The following areas have been identified for the successful implementation of this concept in the coastal region.

- Empowering and building a sense of ownership among civil society, communities, and community-based organizations to enable the management of coastal resources in a sustainable manner.
- Community demand for greater legitimacy and transparency in resource management decision-making.
- Incorporating a sustainable livelihood perspective to address site-specific coastal environmental issues.
- Enabling positive community perception of decentralization policies pursued in the recent past in the administrative and political fields, and to provide an enabling environment for effective and sustainable coastal resource management through SCMA.
- Recognition of a need to formalize indigenous or traditional sustainable resources management practices within the legal and wider governance framework to minimize coastal resource depletion, overexploitation, and user conflicts.
- The SCMA process is viewed as an effective means of promoting sustainable management of coastal resources within a defined geographic setting, making it possible to deal more comprehensively and effectively with the complex management issues.
- The principles adopted in SCMA planning are the same that have been followed from ancient times in Sri Lanka as well as the internationally accepted principles relating to sustainable development.

9 SCMA COORDINATING COMMITTEE (SCMACC)

The SCMA Coordinating Committee (Fig. 2) is the main important management body for implementing the SCMA process. The divisional secretary of respective division chairs the SCMACC if the project area falls into one Divisional Secretariat Division (DSD). However,

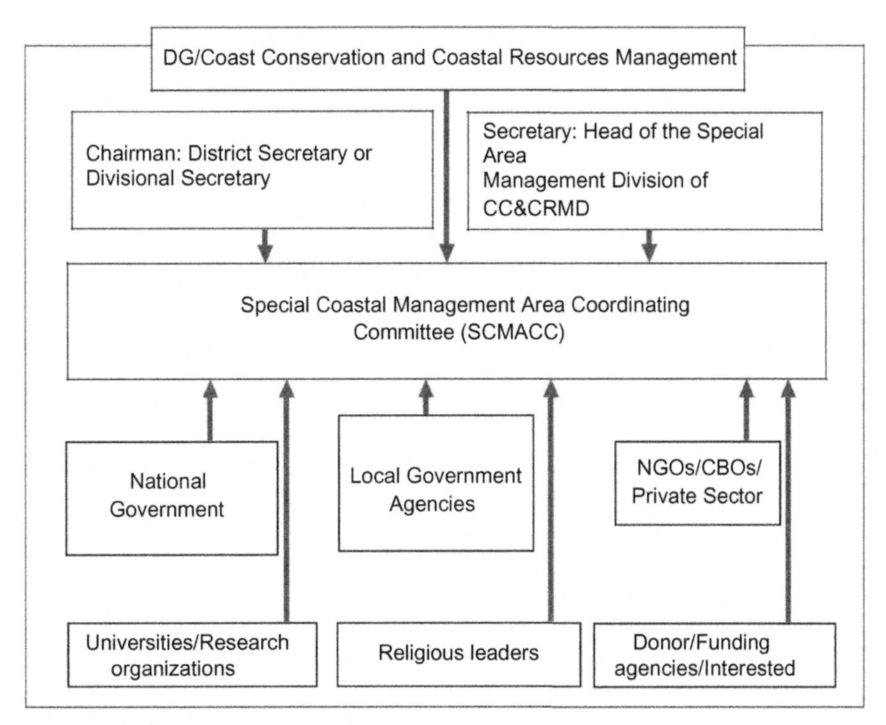

FIG. 2 Special Coastal Management Area Coordinating Committee (SCMACC).

if the SCMA belongs to many DSDs, the chairmanship of the committee is eventually given to the district secretary. In this process, the chairman has been identified and no other committee positions have been recognized. This needs to be reviewed and the CC&CRMD needs to hold the secretary position of the SMACC. Then, CC&CRMD has the authority to coordinate SCMA activities. It is the responsibility of SCMACC to continue the SCMA process once initiated with the funding sources along with community awareness (Fig. 3).

FIG. 3 Awareness for the members of the Special Coastal Management Area Coordinating Committees.

10 LESSONS LEARNT AND DRAWBACKS OF THE SCMA PROCESS IN SRI LANKA

SCMA planning and implementation initiatives related to coastal zones in Sri Lanka are driven by the government sector, which has been led by the CC&CRMD. A total of 12 SCMA plans have been formulated and implemented by CC&CRMD since 1992 under four distinct foreign-funded projects. As per the evaluation of SCMA Projects in 2014, the lessons learned and the drawbacks identified from the past experience have been highlighted in three main areas:

i. Legal and institutional.
ii. Effectiveness and impacts.
iii. Sustainability and challenges.

10.1 Legal and Institutional

Lessons and drawbacks in terms of legal and institutional aspects of the SCMA have been summarized as follows:

a. All the formulated and implemented SCMA plans clearly demonstrate that an administrative or collaborative arrangement itself is not effective without a proper legal framework to ensure sustainable institutionalization for implementation of the SCMA process. This was clearly noted in the defunct internal records Community Coordinating Committees, Internal records CC&CRMD, 2014, Evaluation of SCMA Projects (2014).
b. The Community Coordinating Committees that have oversight over the implementation of the SCMA plans are not legally recognized in the CC&CRM Act or any other statutes. The failure to either statutorily or administratively recognize the CCCs emasculates their authority to implement SCMA plans (Karunaratne, 2008).
c. The functionality of the CCC setup under the planning and implementation process of the SCMA was at a standstill in all sites in the absence of a catalytic role and the financial and technical resources of the respective projects (SCMA Projects, 2014).
d. The key roles played by the divisional secretaries in the planning and implementation process of the initial stages have disappeared during the postproject process over time, due to the absence of legal and institutional recognition of their role. In addition to that, the absorption capacity, resources, and orientation of the local level officials also not matched with the envisage roles to be played during the post project period by them (SCMA Projects, 2014).
e. The auxiliary institutional arrangements made through forming new NGOs and strengthening existing nongovernmental institutions to ensure the continuation of the SCMA process has failed in many instances due to a discontinuation of linkages with the central government institutions, especially with the CC&CRMD, and a lack of capacity (e.g., Rakawa Development Foundation, Negombo Estuary Management Authority, Maduganga Development Foundation, etc.).
f. After pulling out of the project, the CCCs have to be fully dependent on external donor support for postproject implementation, especially with respect to implementation of large interventions. The spontaneous switch between totally financially dependent

status to independent status does not provide an interim period for consolidation and evolution of institutional capacities and processes. Although a marketable plan has been produced through the SCMA process, most of the significant interventions identified for implementation in the SCMA plans could not be implemented within a reasonable time frame.

g. The lack of prospects for statutory authority and recognition as well as individual benefits for the community created negative impacts on participation in the SCMA process.

h. The lack of transparency in local organizational structure and operation negatively influenced the decision making-process and the distribution and sharing of benefits.

i. Failed to appreciate the influence of community and intergroup heterogeneity on building participatory and consensus-based resource governance institutions, due to attention primarily being placed on establishing and strengthening vertical linkage between existing/newly formed local level organizations and central government/local government institutions instead of improving the combination of horizontal and vertical linkages.

j. Past SCMA planning experience revealed that the process of negotiations with multiple stakeholder groups had a broad range of differentiating factors, including disparity in knowledge, resources, social standing, and political influence. This can constitute a substantial disruptive force at the local level and significantly influence participation and consensus building in the planning and implementation process.

k. As many critics highlighted, although the national policy documents are a guide for a locally driven collaborative management process in SCMA planning, in theory central agencies retain the decision-making functions. Similarly, strong local organizations have not emerged due to a lack of institutional and effective social mobilization.

10.2 Effectiveness and Impacts

Lessons and drawbacks in terms of effectiveness and impacts of the SCMA have been identified as follows:

a. The past experience revealed that, to ensure long-term sustainability and effectiveness, the SCMA should be part of the comprehensive national level CRM planning and management effort.

b. The status of conservation measures initiated by other agencies has been improved as a result of the SCMA process (evaluation of SCMA Projects, 2014).

c. The overall awareness of the importance of coastal resource management among stakeholders has been notably increased.

d. Largely, the livelihood development initiatives carried out under the SCMA process to ensure the social and economic well-being of the communities has created little impact due to the sustainability issues encountered.

e. While ensuring sustainability and effectiveness, greater impacts have been created through the implementation of interventions that create common/individual benefits (e.g., community water supply schemes, improvement of power supply schemes, improvement of sanitary facilities, development of landing sites, etc.)

f. In terms of objectives of the SCMA plans and the actual effectiveness, it was revealed that the majority of the investigated SCMA sites achieved about 50% of the spelled-out objectives (Rekawa 53%, Mawella/Kudawella 57%, Negombo 43%, Lunawa 43%, Puttalam Lagoon 53%, Batticaloa Lagoon 53%, RUK 37%, Kalpitiya Bar reef 57% (SCMA Projects, 2014; Fig. 4).

g. The SCMA planning and implementation process has created a sense of ownership among stakeholders and enhanced community enthusiasm.

10.3 Sustainability and Challenges

As per the evaluation carried out on the selected SCMA sites in 2014, sustainability was measured according to the following criteria;

i. Functioning of the CCC after the project period.
ii. Continuation of institutional mechanisms set up by the project.
iii. Continuation of financial support from other sources for the SCMA.
iv. Continuation of projects and programs introduced by the SCMA process.
v. Involvement of CC&CRMD after the project.

The following findings were revealed according to the evaluation on sustainability of the SCMA (Fig. 5).

- All the CCCs set up under the SCMA process have not functioned successfully after the project period, irrespective of the lead agency's involvement.
- No financial support for the SCMA process had been continued after the project period, except for a few instances.
- The majority of the projects and programs have not been continued after the project period, except in a few instances (the community water supply schemes in the Bar Reef SCMA

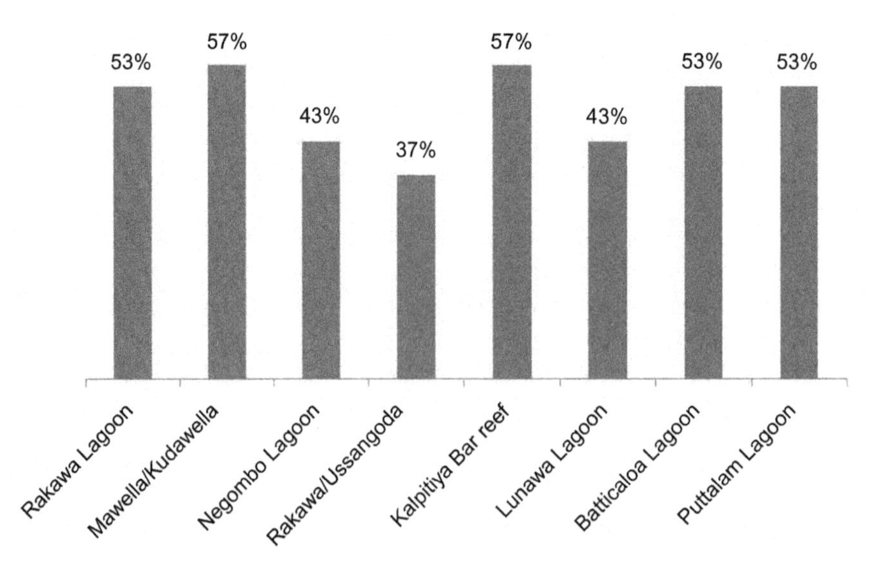

FIG. 4 Effectiveness of SCMA sites.

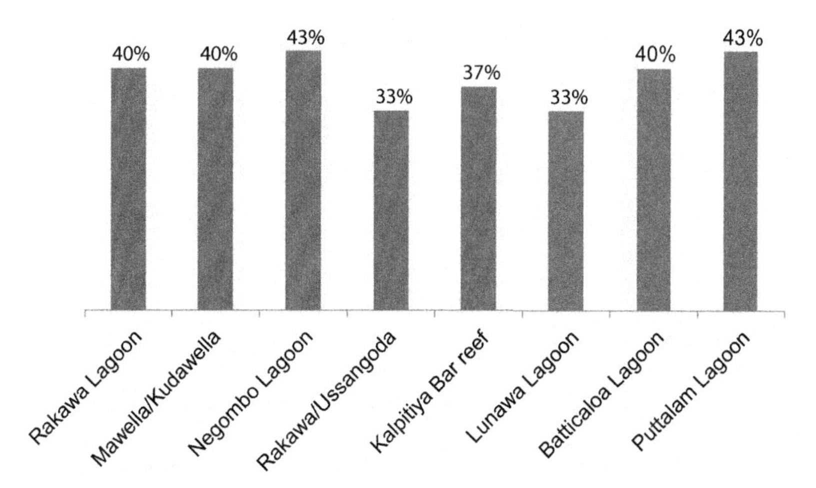

FIG. 5 Sustainability of SCMA sites.

site and the sea bass project in the Negombo SCMA site are being continued). In addition, all the visitor centers established under the SCMA projects are functioning after the project period.
- The impacts of policies related to the other sectors have negatively impacted the SCMA process, creating threats to sustainability.
- The community-based organizations (CBOs) and NGOs actively involved in the SCMA process have limits that prevent the achievement of major objectives.
- Accommodating new economic policies for SCMA planning and implementation is a major challenge due to the inadequate application of environmental valuation.
- Setting up an institutional mechanism in compliance with the existing legal provisions will be a great challenge for ensuring the sustainability of the SCMA process.

10.4 Overall Success of the SCMA Process

After many lessons learned and experiences gained, another two (02) SCMA plans were developed in the most southern districts in Sri Lanka under the Hambantota Integrated Coastal Zone Management Project (HICZMP) between 1997 and 2000, followed by the Rakawa Ussangoda and Kalamatiya Coastal Ecosystem restoration approach; this was done as one SCMA plan under GEF funding. There are eight SCMA sites implemented by the Coastal Resource Management Project under the CC&CRMD. The overall success of the SCMA sites is shown in Fig. 6 and measured by using all the values gained from appropriateness, sustainability, effectiveness, objective, and achievement of the SCMA sites.

According to the overall achievements (Fig. 6) based on the SCMA concept, the Rakawa SCMA site has shown the highest achievements (57%) while the Kalpitiya under CRMP, the Puttlam under IUCN, and the FAO projects have shown a remarkable success

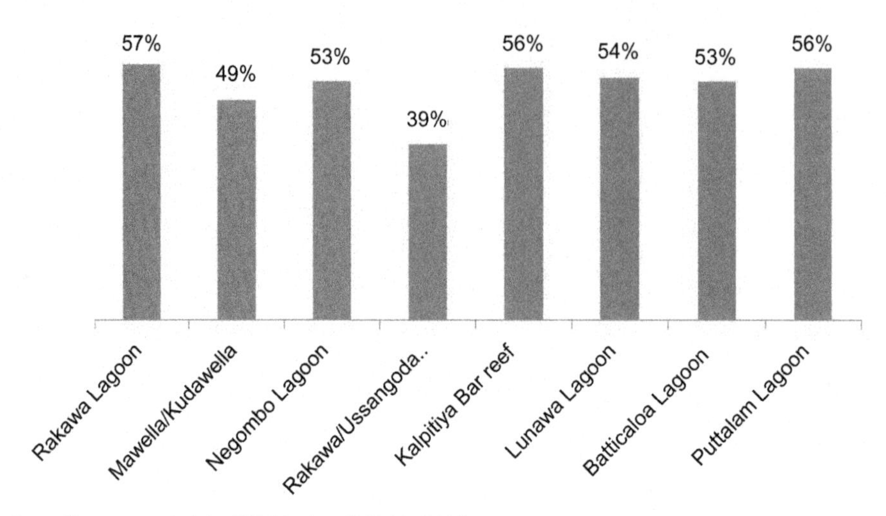

FIG. 6 Overall success of eight SCMA sites (SCMA, 2014).

rate of 56%. This is because the sustainability and objectives achievements recorded very high in these two sites. Another important finding is the fact that the Rakawa site has shown greater success as an SCMA process due to its publicity and literature. It has produced for more than two decades. The Kalpitiya site has not done such a massive campaign but it has shown some sustainability due to interventions continued by various other projects, such as being the tourist destination of the country for coral reefs and whale watching.

11 NEW LEGAL PROVISIONS FOR SCMA

Although the SCMA has been adopted by the CC&CRMD as the main policy instrument in coastal resource management in Sri Lanka, there was no proper legal provision for formulating and implementing the concept until 2011. The absence of proper legal provisions under the CC&CRMA or any other statutes created negative impacts on the sustainability of the SCMA process in the long run, as described in the preceding section. The absence of proper legal provisions created constraints on administrative and/or institutional arrangements that had been adopted by the CC&CRMD in formulating and implementing the SCMA planning process in the past with the collaboration of divisional secretaries and other relevant agencies.

However, when the required provisions were introduced through Part 3c of the Coast Conservation Amendment Act 49 of 2011, new opportunities came to be for the continuation of SCMAs as a sustainable and effective supplementary planning tool for coastal resource management in Sri Lanka. As per the new legal provisions (Section 22E (1)), SCMAs can be declared covering land within the coastal zone, adjacent to the coastal zone, or comprising both through gazette notification. The new legal provisions have also recognized the adoption of a collaborative approach for planning resources management in the defined SCMAs. According to the new legal provisions, SCMAs will be declared only if such areas are included in the coastal zone and the CRMP.

To formulate an effective institutional structure for planning and management of SCMAs, new regulations have to be prepared in compliance with the legal provisions provided through Part 111c Subsection 2 of the Amendment Act 49. To achieve the desired objectives of SCMA planning, regulations should be framed and brought before Parliament as soon as convenient. If the regulations are not approved by this way, it is deemed to have rescinded without prejudice to any act had done under the regulation. The new regulations have to be formulated prescribing the manner and mode in which and the persons by whom such SCMAs should be administered, and the persons entitled to have access to these areas and the activities that can be carried out within such areas (under the provision in section 22E-2 of Amendment Act 49).

Accordingly, the existing legal provisions have to be used with the effective participation of stakeholders comprised of the relevant community directly attached to the SCMA site; local institutions, including both government and nongovernmental; outside beneficiaries; and the central government institutions.

12 CONCLUSION AND RECOMMENDATION

The most important areas to be considered for the SCMA process are:

- Most SCMA activities had been implemented by CC&CRMD under the foreign-funded projects. The sustainability of SCMA activities is questionable after the funding is over. Therefore, it is very important to recommend adopting the program-based approach instead of a project-based approach for the SMA process to avoid this problem by focusing more realistic budget and time plan suite to the local, regional, and national setting during the planning stage of the process.
- Community mobilization is very important in the process but in most cases, their participation has not been to the level that the project had anticipated. This is due to poor or less emphasis on social mobilization. The necessary programs should be organized for the awareness and educational improvement of local communities for active participation in the SCMA process.
- Most often, participation in the local community has been done with some kind of organized selected group. Therefore, it should clearly identify the resource abusers and obtain their active participation for the SCMA process.
- SCMA is a good process that is applied in resource management with more institutional backing from the initial stage of the project and even after the project period. These leading agencies should monitor the SCMA process for better effectiveness and sustainability of the relevant programs, at least until the desired result is obtained.

Acknowledgments

We are extremely grateful to Mr. B.K. Prabath Chandrakeerthi, Director General of the CC&CRMD; Dr. Anil Pramaratne, former director general of the CC&CRMD; Indra Ranasingha, former director general of the Ministry of Fisheries and Aquatic Resources in Sri Lanka; and Kapila Gunarathna, Director of Green Planet and Ecoconsultant. We are also grateful for the assistance of the staff of the CC&CRMD. The contribution of research partners and officers of the government and nongovernment organizations are also gratefully acknowledged.

References

Atlas of Sri Lanka, 1997. Surface Water. Arjuna Consulting Co. Ltd, Colombo.

CC&CRMP, 2018. Coast Conservation and Coastal Resources Management Department, Colombo.

CZM Plan, 2004. Coast Conservation and Coastal Resources Management Department, Colombo.

Karunaratne, P., 2008. A Review of Coastal Zone Laws and Implementation Experience in Sri Lanka: Coastal Zone Law—Laws and Implementation—Sri Lanka. Pannipitiya Stamford Lake, Pannipitiya.

SCMA Projects, 2014. Special Management Area Projects. Coast Conservation and Coastal Resources Management Department, Colombo.

CHAPTER

3

Coastal Development: Construction of a Public Policy for the Shores and Seas of Mexico

Norma Patricia Muñoz Sevilla, Isaac Azuz Adeath†, Maxime Le Bail*, Alejandra Cortés Ruiz†*

*Instituto Politécnico Nacional (IPN)—Centro Interdisciplinario de Investigaciones y Estudios sobre Medio Ambiente y Desarrollo (CIIEMAD), Ciudad de México, México

†CETYS Universidad—Ensenada-Campus Internacional, Ensenada, México

1 INTRODUCTION

Mexico's privileged geographic position, territorial extension, varied physiography, and climatic diversity have given the country an unparalleled biological wealth, both marine and terrestrial, as well as many natural areas not seen anywhere else in the world. Marine and coastal areas strongly contribute to this environmental wealth and favor the existence of diverse productive activities that, in turn, benefit Mexico's economic development (i.e., ports; petroleum extraction, storage and transport areas; tourism activities; mining; and fishing and aquaculture). Unfortunately, the lack of a multisector and articulated long-term, sea-focused vision has resulted in a chaotic development that has damaged the coastal zones and plundered the marine areas. No coastal city has managed to take advantage of the natural resources and ecosystems in a sustainable way nor been able to incorporate as a social, economic, and cultural asset the exploitation of coastal and marine areas.

The coastal and marine areas are not perceived by the Mexican authorities as a third border of the country (León and Rodríguez, 2004) nor are they considered as key areas for cultural influence, potential generation of renewable energy, pharmaceutical research, safe-keeping of food and water, commercial minerals, atmospheric carbon capture, or key regions for commercial exchange in the globalization era. Recent but very notable exceptions such as the National Environmental Policy for the Sustainable Development of Oceans and Coasts

https://doi.org/10.1016/B978-0-12-810473-6.00003-0

© 2019 Elsevier Inc. All rights reserved.

of Mexico (SEMARNAT, 2006b), or the National Strategy for Ecological Ordinance of Seas and Coasts (SEMARNAT, 2006a) together with a wide but unclear legal frame, an overregulation, inconsistency and lack of coherence among the different regulatory instruments, judicial voids, disarticulated competencies from the different levels of government, limited financial resources, low levels of control, and inefficient surveillance, the lack of strategic vision and adequate management instruments, inhibit the appropriate development of these areas. Not recognizing these areas as special for the territory has led to very limited knowledge about them (i.e., how many regular or irregular inhabitants live in the federal marine/land area). Therefore, human capacity is focused on a very few coastal states, technological breakthroughs do not regard it as a primary object of study, no budget is allocated to long-term monitoring programs and scientific research on the phenomena and processes that take place in these areas has occurred in an unclear direction, without optimizing resources and without zone and time coverage as per its management requirements.

Every day, the anthropogenic impact on the coastline becomes more obvious. The seaboard is subject to multiple pressures (i.e., population and urban area growth, lack of beach access, construction of inadequate infrastructure) and impacts (i.e., pollution by local and remote sources, coastal erosion, loss of vegetation, "artificialization," or artificial rendering, of the coast line, waste dumping) that have led, in many places, to an irreversible deterioration of these fragile areas. On the other hand, the effects of climate change, such as the rise of atmospheric and oceanic temperatures, the accelerated rise of the sea level, the acidification of the oceans, and the higher frequency and intensity of extreme meteorological phenomena, are occurring in a more tangible way in these areas. If we consider the concept of risk proposed by (McFadden et al., 2007), which indicates that risk can be defined as the connection that exists between the probability of occurrence of a dangerous phenomenon and the vulnerability of the area of interest, we can expect that, along with the effects of climate change, the probability of occurrence of a dangerous phenomenon is greater. Also, the burden on these coastal areas renders them more vulnerable to weather, therefore the risk increases continuously, resulting in the loss of human lives as well as environmental and economic losses.

In addition to incipient environmental governance, these contextual elements made (Anta Fonseca et al., 2008) write that the government is not the only one that should be responsible for managing environmental issues. Instead, society should also participate and decision-making should be transferred to local governments, always giving preference to national interests above any other. Therefore, the existence of management tools to guarantee the sustainable development of the coastal and marine areas, seeking continuously better life conditions for current inhabitants and boosting expectations for generations to come, is essential.

In this context, the creation in 2008 by presidential decree of the permanent Interministerial Committee for Sustainable Management of Oceans and Coasts (CIMARES) has turned out to be very promising. This committee looks first to articulate the efforts from different entities of the federal public administration. It was created by the heads of the government State Cabinets, Foreign Affairs, Marine Corps, Social Development, Energy, Economy, Agriculture, Rural Development, Fishing and Food, Communication and Transport, Tourism, and Environment and Natural Resources. This committee has the historical responsibility to provide Mexico with a national policy on this matter and to ensure its adequate implementation.

2 COASTAL DEVELOPMENT

2.1 Definition

Coastal development is defined as the human-induced change of the landscape within sight of the coastline. This includes building structures that are on or near the coast in general for protection, commerce, communication, or recreation. These structures support economic and social activities that can contribute with positive or negative effects on the coastal environment.

Worldwide, 2.5 billion people (40% of the world's population) currently live within 100 km of the coast, adding increased pressure to coastal ecosystems. Coastal development linked to human settlements, industry, aquaculture, or infrastructure can cause severe impacts on near-shore ecosystems, particularly coral reefs. Coastal development impacts may be direct such as land filling, dredging, and coral and sand mining for construction, or indirect, such as increased runoff of sediment and pollutants (Reefresilience, 2016).

Like other countries, Mexico's coastline is highly valued for many reasons. Beyond its intrinsic ecological value, it also provides the country with tourism, recreation, sport, communications, real estate, transport, renewable energy, oil and gas, fisheries and aquaculture, and ports and harbors. The factors that influence development along the Mexican coast need to be better understood to ensure minimal harm to the environment (State of Nova Scotia, 2016).

2.2 Current State and Tendencies of the Coastal and Maritime Zones in Mexico

2.2.1 Geographical and Management Scope

Mexico has a privileged geographical location that sets it strategically in the world context, between the two largest oceans on the planet. Mexico's seas and coasts are a key piece of the national territory for its safety and sovereignty as well as for the country's sustainable development. These regions possess a great natural wealth that needs to be preserved and sustainably managed by using the best and greatest available scientific and technological information. Also, its development must be promoted toward the well-being of its inhabitants so as to offer future generations the possibility of getting the best use of them with a greater sense of responsibility and fairness. The exclusive economic zone of the country (2,715,012 km^2), including the territorial seas (231,813 km^2), is much larger than the continental area (1,959,248 km^2) and represents an area from the national territory that needs to be sustainably managed for the benefit of the nation (CIMARES-SEMARNAT, 2012, 2016) (Fig. 1).

Seventeen of the 32 states (federative entities) that constitute the Mexican Republic have access to the sea and represent 56% of the national territory. Within these states, 156 municipalities have beachfront shores and represent approximately 21% of the continental area of the country. The insular (islander) area covers 5127 km^2 (INEGI, 2009).

The country's coastline, without considering the islands, is 11,122 km long. The Pacific and Gulf of California coasts extend over 7828 km and the Gulf of Mexico and Caribbean Sea 3294 km.

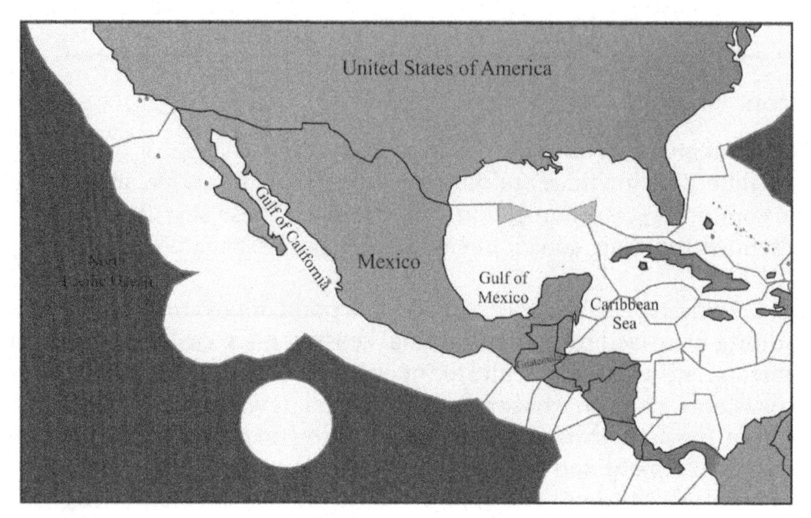

FIG. 1 Exclusive Economic Zone of Mexico and its surroundings. *Flanders Marine Institute. VLIZ (2008) Maritime Boundaries Geodatabase.*

The territorial management area of this policy is limited to the Mexican maritime zones, in accordance to what the federal law of the sea establishes and to the Mexican coastal zone, as per the definition of the country's coastal zone that this policy adopts (CIMARES-SEMARNAT, 2016).

Mexican maritime zones:

(a) The Territorial Sea.
(b) The Interior Maritime Waters.
(c) The Adjoining Zone.
(d) The Exclusive Economic Zone.
(e) The Continental Platform and Insular Platforms.
(f) Any other allowed by international law.

2.2.2 Demographic Tendencies and Characteristics

The dynamics of the population from the coastal zones of Mexico follow the world's tendencies that indicate the migration of human population toward the coastal zones. In 2010, the population of the coastal states was 51,577,111 inhabitants, almost 6 million more than in 2000 (Fig. 2), and this number is expected to reach 58 million by 2030 (Partida Bush, 2006; Azuz Adeath and Rivera Arriaga, 2007). The population of the coastal municipalities, as defined by the National Policy for the Seas and Coasts of Mexico, grew 33.4% from 1995 to 2005, whereas the population from the urban municipalities increased by 44.9% (CIMARES-SEMARNAT, 2016). For the 155 coastal counties (counties with direct access to the sea) the population increased from 14.6 million to 17.5 million during the period 2000–2010.

During the periods of 1990–2000 and 2000–2010, the average population growth rate in more than half the coastal states was higher than the national average population growth. In both periods, the three coastal states that showed the highest growth rates were Quintana Roo, Baja California Sur, and Baja California (Fig. 3).

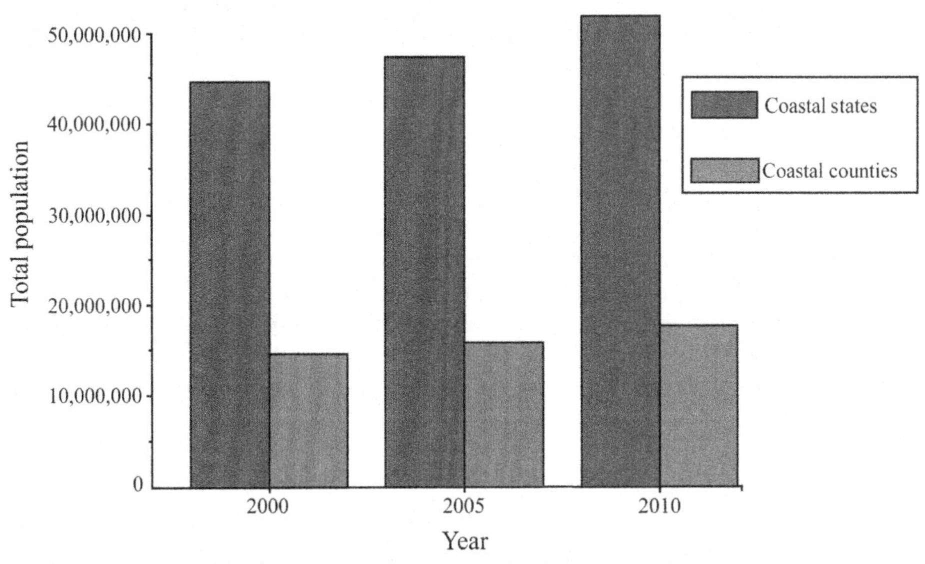

FIG. 2 Population for the coastal zone, year 2000–2010.

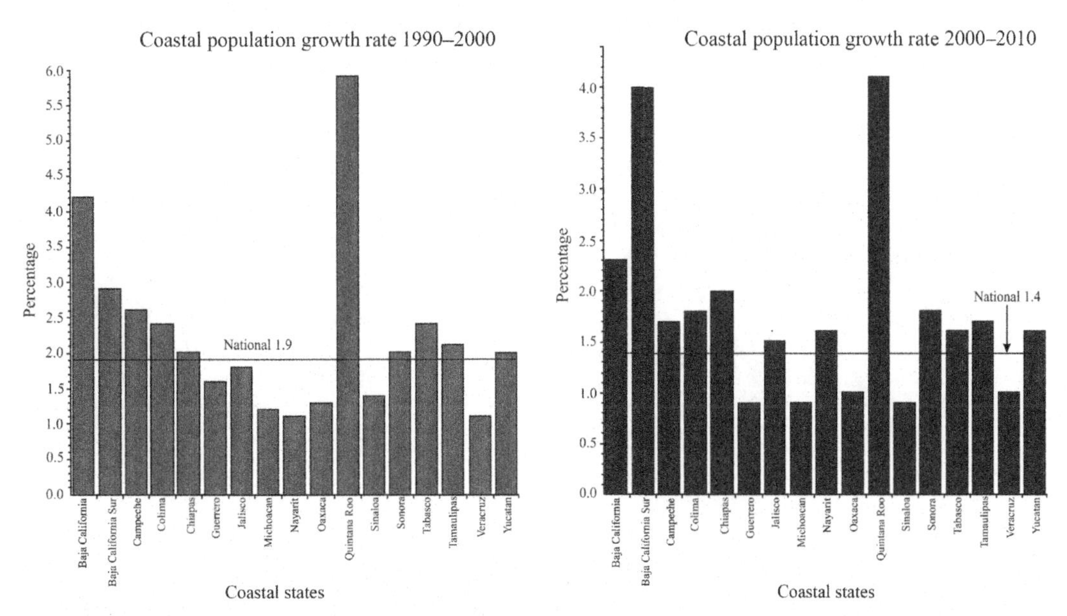

FIG. 3 Yearly average population growth rate of the coastal states during the period 1990–2000 and the period 2000–2010. *Azuz and Muñoz, with data from INEGI, 1990, 2000, 2005, and 2010.*

In general terms, the Mexican coastal zones are experiencing an especially irregular population growth that takes place in a few urban locations, which generates important economic, social, institutional, and environmental stresses over the coastal and maritime zones. An interesting example could be found in the state of Nayarit were a big tourism project was developed by the local government, called "Riviera Nayarit." Between 2000 and 2010 in the Bahía de Banderas municipality (the principal location of the new resorts), population growth was 64,397 inhabitants (108% increase) but the volume of sewage treatment plants remains at 319.5 L/s (almost the same than in 2005) and no single potable water treatment system plant exists in the municipality (CONAGUA, 2013).

In 2005, the distribution of the population in the municipalities of the coastal zones indicated that 73.5% live in the municipalities considered urban (44.8% of the total municipalities). The concentration of inhabitants increases if we identify that the 16 metropolitan areas located in the coastal zone, plus the 10 municipalities with the largest population, host 66.1% of the total population of the coastal area. In 2010, 53% of the total coastal population lived in 15 municipalities, and 20 municipalities had less than 10,000 inhabitants.

2.2.3 Economic Tendencies and Aspects

Between 1993 and 2006, the coastal states contributed 36% of the national Gross Domestic Product (GDP). This contribution to the GDP should grow if we pay attention to the historical indicators of the different economic activities of the coastal and maritime regions during past years (CIMARES-SEMARNAT, 2016) (Fig. 4).

The layout of the regional GDP by coastline shows that a small number of states contributes more than 50% of the regional GDP in each coastline. For the Pacific-Gulf of California coastline, those states are Baja California, Sonora, and Jalisco (54%). For the Gulf of Mexico-Caribbean Sea, the states are Tamaulipas and Veracruz (58%). Considering the integration of the state GPD, the manufacturing activity and the services sector are key components in the five coastal states that have the largest contribution to the regional GDPs of both

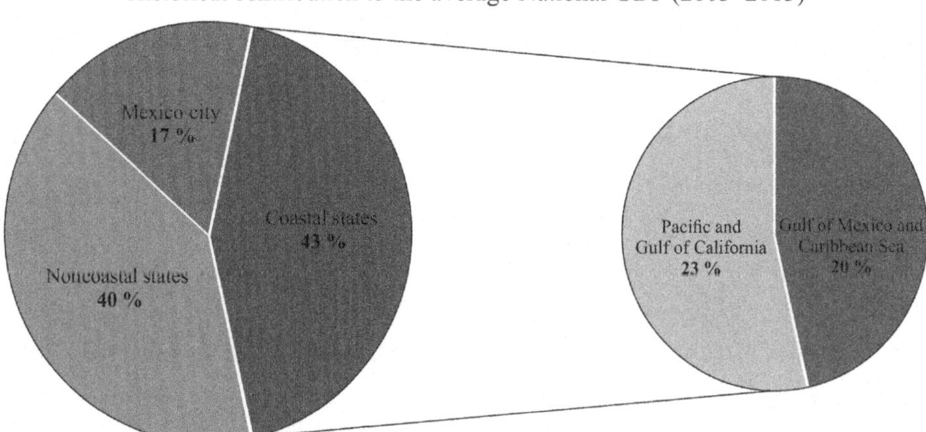

FIG. 4 Historical contribution to the average National GDP from the Federal District – Mexico City, the non-coastal States and the coastal States, with details from the Pacific and Gulf of California coastline and the Gulf of Mexico and Caribbean Sea. *Azuz and Muñoz, based on data from BIE-INEGI.*

coastlines: Jalisco, Baja California, Sonora, Veracruz, and Tamaulipas. For the Gulf of Mexico and Caribbean Sea, the mining activity stands out, associated with oil extraction in Campeche. Tourism stands out in Quintana Roo.

The average value of each activity is presented for the period between 1993 and 2006. On the left, the Pacific and Gulf of California coastline. On the right, the Gulf of Mexico and Caribbean Sea coastline. Activity 1: agriculture, forestry, and fishing; Activity 2: mining; Activity 3: manufacturing industry; Activity 4: construction; Activity 5: Power, gas, and water; Activity 6: commerce, restaurants, and hotels; Activity 7: transportation, warehousing, and communications; Activity 8: financial services, insurance, real estate, and leasing; Activity 9: municipal and social and personal services (BIE-INEGI, 2009).

In the coastal states, on average 60% of the population 14 and older belong to the economically active population (PEA) and 97% of these belong to the working population (PO). The greatest percentage of PEA is found in Quintana Roo (69.9%) whereas the lowest is found in Chiapas (55.4%) (INEGI, 2008).

Information regarding the income of the working population in the coastal states shows important structural differences (CIMARES-SEMARNAT, 2016). Even if a distinctive pattern by region hasn't been done, it is evident that in the Pacific-Gulf of California region, most workers have incomes that range from two to five minimum wages, except for Oaxaca and Chiapas; the states with the largest contribution to the GDP have the lowest percentage of workers with incomes up to 2 two minimum wages. In the Gulf of Mexico, most workers are found in the up to two minimum wage range, except for Quintana Roo and Tamaulipas (Fig. 5).

2.2.4 *Environmental Tendencies and Aspects*

Considering the environmental characteristics, such as ocean currents, geomorphology, bathymetry, climates, and productivity, among others, Mexican seas can be identified by regions. To determine these maritime regions, important pieces of work have been

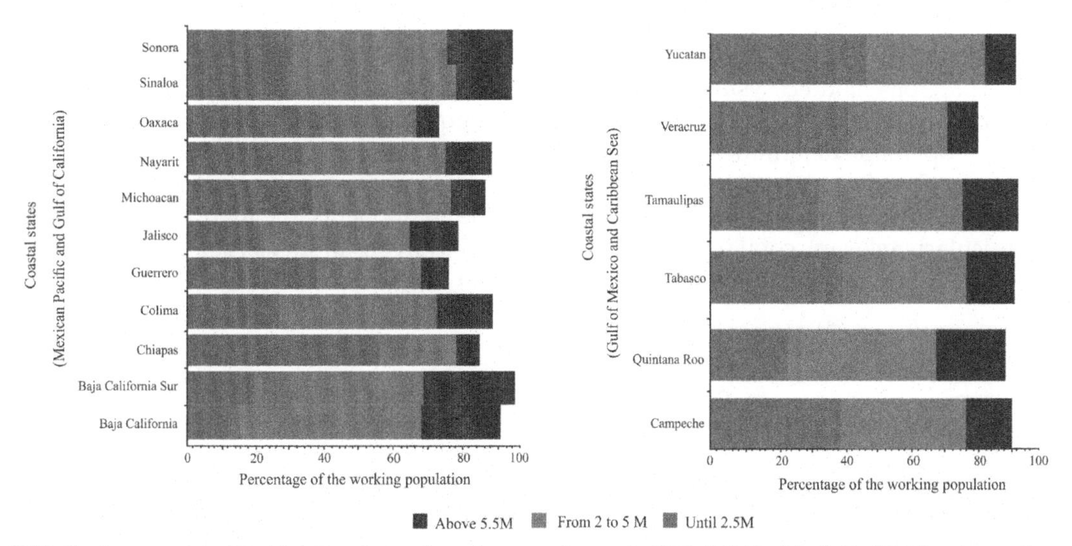

FIG. 5 Income of the Working Population from the coastal states in 2008. (*left*) Pacific-Gulf of California coastline. (*right*) Gulf of Mexico-Caribbean Sea coastline. *Azuz and Muñoz, based on data from BIE-INEGI.*

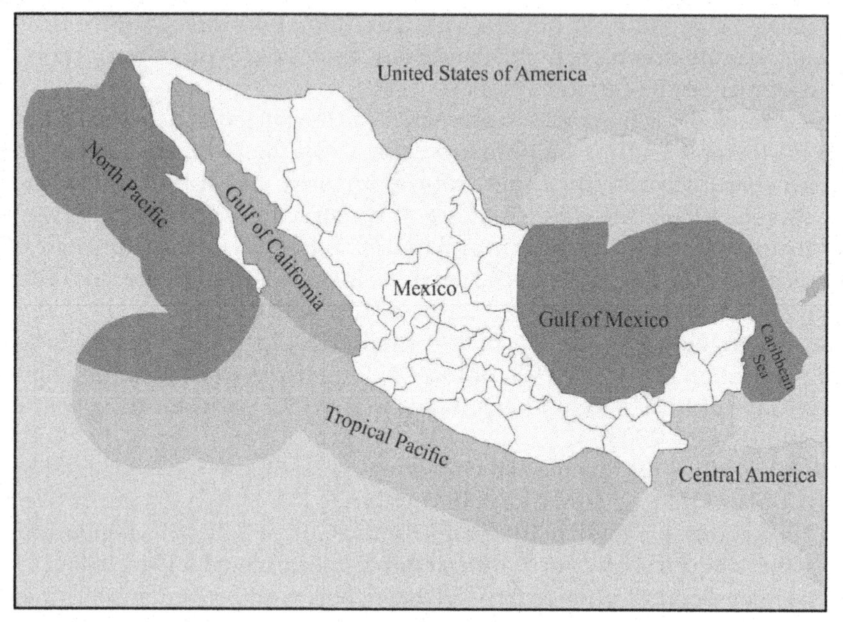

FIG. 6 Coastal-Oceanic regions of Mexico (CONABIO, 2003).

developed, such as the scientific workshops lead by the National Ecology Institute in 2004 and 2006. This exercise and others have contributed so that, while tending to the defining aspects of our maritime territory, the national policy adopts the following regionalization (Fig. 6):

Region I: North Pacific.
Region II: Gulf of California.
Region III: South Pacific.
Region IV: Gulf of Mexico and Caribbean Sea.

These regions hold ecosystems as well as marine and coastal environments and components that, regardless of quantity or size, possess a unique ecological importance. In conjunction, their operation determines the potential of economic development and social well-being of the coastal and maritime area of Mexico.

From the large number of these coastal and marine components, the following turn out to be particularly relevant: coral reefs and rock bottoms; bays, coastal shoals, and sandy beaches; coastal wetlands; and coastal dunes and the Mexican insular territory.

3 WHAT IS THE COAST

3.1 Definition

Many definitions are used for "coast." The simplest is probably "the land adjacent to the sea." But coast can mean different things in different contexts. A political definition might be the municipalities adjacent to the sea or the states adjacent to the sea. An environmental definition might be land that drains to the sea or watersheds of streams that drain directly to the sea. One scientific definition of coast is "the space in which terrestrial environs interact with marine environs and vice versa." Doubtless, many others exist. For our purposes, having one definition of "coast" or "coastal" is of no benefit. From the broadest definition to the narrowest, the definition that best fits the circumstances is used by (Volusia County Florida, 2014).

"The coast is a dynamic place and its dynamism makes it susceptible to stresses and changes in a number of ways." Because the coast is where the land interacts with the sea, it is open to the action of wind, waves, tides, and currents that not only erode the shore but also can expand it with sedimentary deposits. Storm systems gather energy from the ocean and intensify natural coastal forces with wind, waves, and rain powerful enough to severely damage property and hasten erosive processes.

The coast is made more vulnerable to these natural dynamic forces by rising sea levels. Although the sea level has been steadily rising for centuries, the process may be accelerating because of global warming. Social and economic forces also bring stresses to coastal areas. Population growth, land development, and resort development are all particularly intense along the coast.

Coastal areas are experiencing high growth rates, and the beach is a popular destination for vacations, second homes, and retirement. Property on or near the shore is always in high demand and as a consequence is usually expensive. Because this land is a valuable asset, people will go to extraordinary lengths to protect property near the shore. This behavior—the need to protect coastal property—is responsible for what is probably the greatest threat to the shoreline: the practices we use trying to stabilize it. Sea walls, jetties, groyne, riprap, and sandbags all disrupt the natural processes of the coast and exacerbate erosion and habitat destruction (Pawlukiewicz et al., 2007).

3.2 Mexican Coastal Zone

> The coastal zone is the geographical area of mutual interaction between the maritime environment, the land, and the atmosphere, comprised by: a) a continental portion defined by 264 coastal municipalities; 151 with beachfront and 113 interior municipalities adjacent to the coastal ones and that have medium and high coastal influence; b) a maritime portion defined by the continental platform limited by the 200 m Isobathic line, and c) an insular portion represented by the national islands" (SEMARNAT, 2006b; Escofet Giansone, 2008; CIMARES-SEMARNAT, 2016) (Figs. 7 and 8).

4 IMPACTS FROM COASTAL DEVELOPMENT

Ocean coastlines are breathtaking networks of plants and animals that work together to survive in an ever-changing ecosystem. The plants and animals living in this harsh environment have adapted to sandy soil, continuous salt spray, exposure to extreme weather events, and the constant ambush of ocean waves, which give way to habitat erosion.

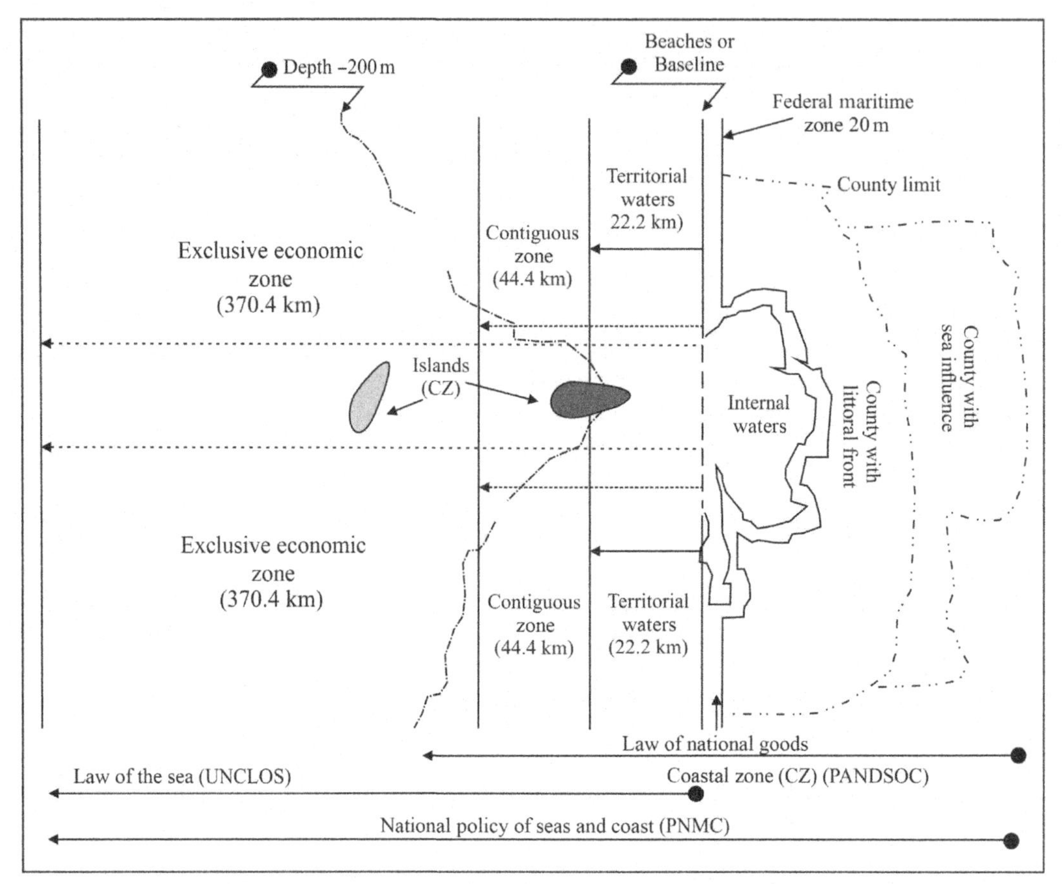

FIG. 7 Outlining from atop of the relevant regions for the National Policy of Seas and Coasts of Mexico as per current definitions (plan based on Escofet Giansone, 2008).

Coastal animals rely on these dunes for protection, shelter, and food. Each animal plays an important role in shaping the landscape. The placement of sea turtle nests near the base of dunes provides vital nutrients for the plants growing there. In return, the dune vegetation reduces erosion and retains the beach elevation needed to allow sea turtle nests to remain safe and dry from potentially devastating wave action.

These natural coastal ecosystems are also important for people living along the coast. Unlike armored beaches, natural beaches lined with sand dunes provide a strong buffer to defend against the damaging effects of strong winds and storm surge by absorbing wave energy (Volusia County Florida, op. cit).

Construction projects in coastal cities and communities may be built on land reclaimed from the sea. In many areas, wide shallow reef flats have been reclaimed and converted to airports, industrial, or urban lands, for example, Veracruz harbor. Dredging activities (e.g., deep water channels, harbors, marinas) and dumping of waste materials in the coastal and marine environment can also damage and even destroy adjacent coral reefs.

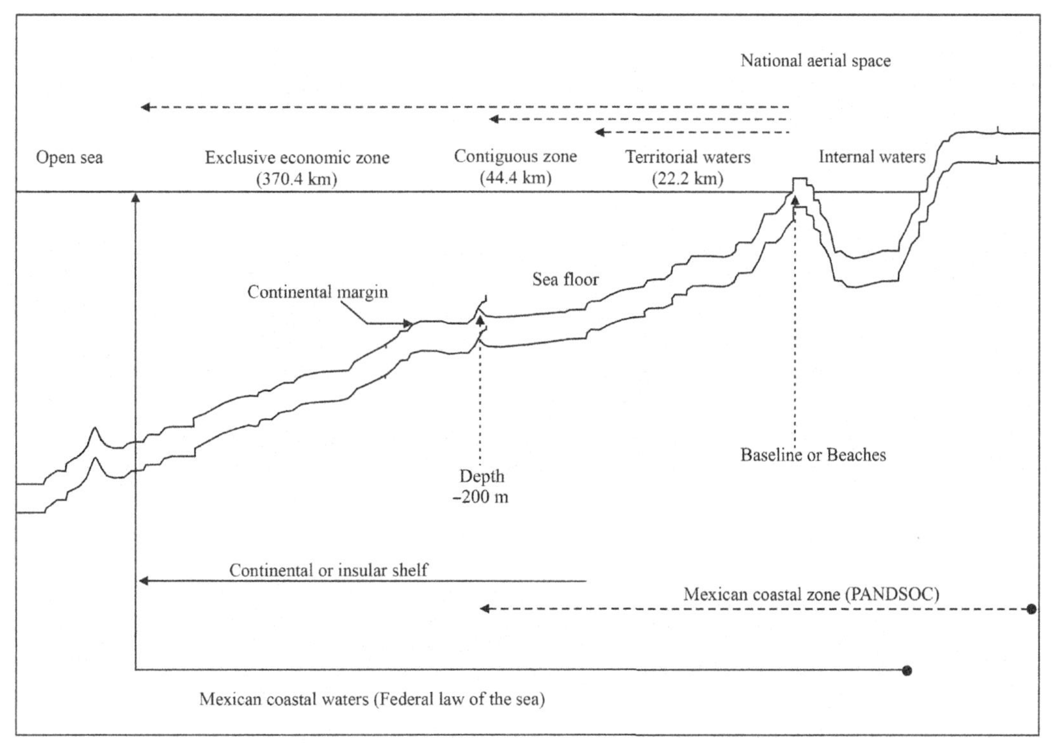

FIG. 8 Cross section Outlining of the relevant regions for the National Policy of Seas and Coasts of Mexico as per current definitions. *Azuz and Muñoz based on the Federal Law of the Sea and PANDSOC, 2006.*

4.1 Major Impacts From Coastal Development

Principal impacts that affect coastal development (Pawlukiewicz et al., 2007):

- *Construction projects (piers, channels, airstrips, dikes, land reclamation, etc.)*—can kill coral directly.
- *Degradation of coral reefs*—can result in lost tourism revenue in countries that depend on reef-based tourism and can also reduce fish populations.
- *Coastal construction*—can cause chronic sedimentation, sewage effluent, industrial discharge, and changes in water flow and runoff, which can adversely affect coral growth rates and metabolic activities as well as directly kill coral.
- *Removal of reefs*—can result in beach erosion, land retreat, and sedimentation.
- *Sedimentation*—can smother reefs or increase turbidity in coastal waters, thus reducing the light needed for coral growth and survival.
- *Pollutants*—can lead to increases in coral disease and mortality, cause changes in coral community structure, and impede coral growth, reproduction, and larval settlement; for example, nutrient runoff can lead to algal blooms that can stifle coral growth.
- *Mining of coral for construction materials*—can lead to long-term economic losses in terms of lost benefits for fisheries, coastal protection, tourism, and food security and biodiversity.

As coastal populations increase and natural coastal protection is degraded or lost, sea level rise and changes in storm patterns are likely to increase the effects of harmful coastal development activities. Local impacts of land-based sources of stress will occur in combination with global and regional stressors, such as climate change, land-use practices, and freshwater inputs, further threatening the survival of coral reef ecosystems. For example, increases in storm impacts linked to climate change could exacerbate the run-off of sediments and other pollutants.

Reducing the effects of coastal development is critically important; it threatens nearly 25% of the world's coral reefs, particularly in Southeast Asia, the Indian Ocean, and the Atlantic.

The impacts of coastal development can be drastically reduced through effective planning and land-use regulations. For example, planning and management approaches can include land-use zoning plans and regulations, protection of coastal habitats (such as mangroves), coastal setbacks that restrict development within a fixed distance from the shoreline, watershed management, improved collection and treatment of wastewater and solid wastes, and management of tourism within sustainable levels.

In Mexico, the high population growth and the accelerated development of some economic activities has led to disorder in the coastal zones, generating deterioration, pollution in the marine environment, overexploitation of natural resources in particular fisheries, and conflicts derived from land use and resources (an example of this is the settlements on dunes and wetlands) as well as several impacts on the population health, quality of life, and impoverishment of coastal communities.

4.2 Ten Principles for Coastal Development

According to the Urban Land Institute (Pawlukiewicz et al., 2007), we can consider at least 10 principles for coastal development:

- Enhance value by protecting and conserving natural systems.
- Identify natural hazards and reduce vulnerability.
- Apply comprehensive assessments to the region and site.
- Lower risk by exceeding standards for siting and construction.
- Adopt successful practices from dynamic coastal conditions.
- Use market-based incentives to encourage appropriate development.
- Address social and economic equity concerns.
- Balance the public's right of access and use with private property rights.
- Protect fragile water resources on the coast.
- Commit to stewardship that will sustain coastal areas.

5 MODELS OF COASTAL DEVELOPMENT

5.1 Theoretical Models

The sustainability of regional rural development depends on the integrated status and the coordination between rural resources/environment conditions and rural socioeconomic development (Liu et al., 2009). The authors proposed four rural development models for the eastern coastal region of China from the integrated perspective of population, resources, environment, and development: the Mentougou model, the Taicang model, the Yueqing model, and the Qionghai model (Fig. 9).

Conclusions of the study:

Model 1: This region has a very good overall condition with a high degree of coordination (0.2454) between rural resources/environment and rural socioeconomic development.

Model 2: Of all the economic development models in China, the Taicang model spurred by the collective economy is one of the most vibrant. The overall effect of rural development (0.5974), coordination degree (0.2477), and comprehensive level of coordinated development (0.3847) of this model is much superior to those of the other three models.

Model 3: The appraisal results of this model indicate that the region's rural socioeconomy is highly developed (0.7455), the highest among all four models. Owing to the very poor rural resources/environment condition (only 0.2929) and the low level of coordination between rural resources/environment and socioeconomic (0.1640), the rural comprehensive coordination has a low degree of development.

Model 4: The Qionghai model is dominated by tropical agricultural production that has a low level of development. Its rural development overall effect is about the average (0.3927), and its rural socioeconomic development lags especially behind (only 0.2601). There is a low level of coordination between rural resources/environment and socioeconomic development (0.1963). However, it has an advantageous rural resources/environment (0.5253), which is far better than the socioeconomic development level.

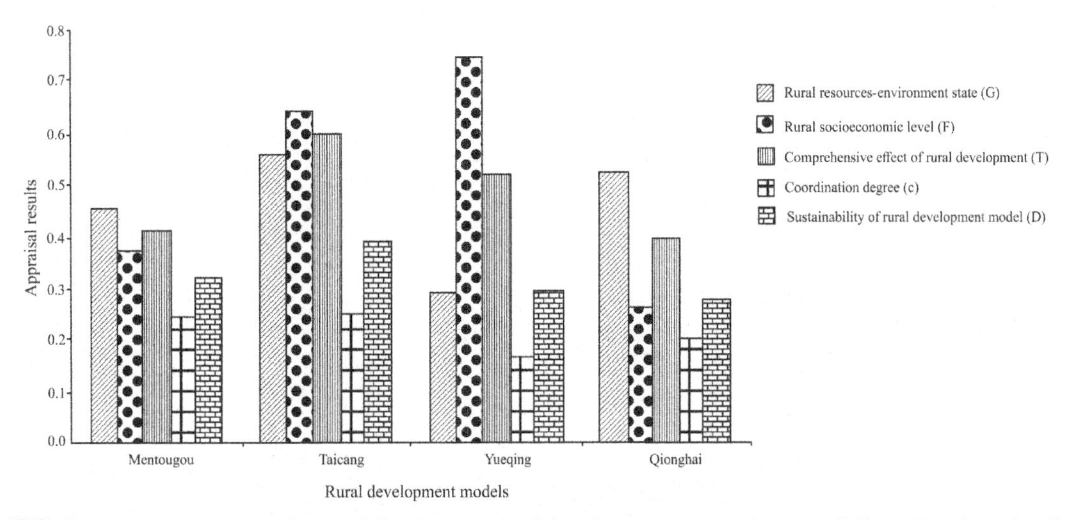

FIG. 9 Appraisal results of four rural development models in the eastern coastal region of China. *From Yansui et al. (2009).*

5.2 International Models

For New Zealand (Xu, 2015) proposes a new sustainable development model. This coastal zone development model shows that coastal land use value is better when an environmentally sustainable development plan is put in place before urban construction.

Sustainable Development Model = Low Density Resident Development Zone + Safe Zone (Conservation + Environmental sustainable tourism) + Beach Zone.

In order to understand the background of coastal region development, an investigation of international case studies was conducted, looking at the Gold Coast, Dubai, and San Ya. These coastal region development case studies demonstrated that tourism can generate coastal zone development. However, there are also numerous problems with these developments, for example, coastal zone vegetation degradation, habitat destruction, soil erosion, climate change, and population explosion.

The Pakiri case study identified a sustainable development site and, through environmentally sustainable tourism criteria, designed a holiday park to preserve the coastal zone environment and local lifestyle while encouraging development.

Finally (Xu, 2015), concludes that sustainable coastal development requires an understanding of the coastal zone and the development of an environmentally sustainable plan before the urban development stage. This diagram (Fig. 10) clearly shows that an environmentally sustainable coastal zone plan should have these criteria:

- Acceptable to the public.
- Respect for the natural environment.
- Setting a safe distance between residential areas and the natural coastal areas.
- Protecting the natural coastal areas.

For Spain, tourism is a fundamental sector of the economy. The "sun and beach" tourist product, the most important product for the economy of the Valencia Region, is based on

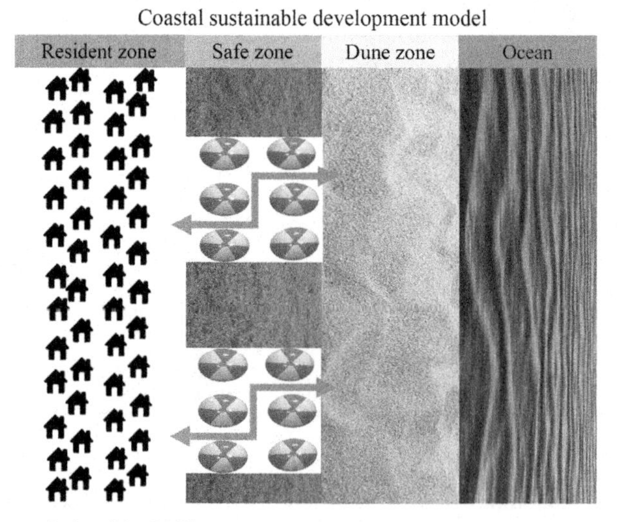

FIG. 10 Sustainable coastal plan (Xu, 2015).

a fragile natural resource: beaches (Yepes and Medina, 2005). For this region, the intensive tourism models are more efficient than extensive ones, both in economic and environmental terms, as long as the maximum capacity is not exceeded.

Damming, port breakwaters, and urban development are the principal causes of the erosion processes of 3 million m^3/year. Erosion reduces beach land by approximately 200,000 m^2/year, affecting 58% of the 178 km of sandy shoreline of the central and northern Valencia Region.

For sustainable development, urban growth must be greatly limited to a much lower rate and far stricter protection mechanisms must be established for the remaining virgin and rustic areas (Yepes and Medina, 2005).

5.2.1 Pelican Bay, Florida

Americans love the coast: almost 40% of the US population lives in coastal counties that cover less than 10% of the US land area. The 2010 census data show that Florida again has one of the highest population growth rates in the country, following a decline due to the housing crisis. Living on the coast is not always peaceful as the risk of hurricanes increases with global warming, the shoreline changes as a response of coastal dynamics, the sea level rises, and other phenomena such as the expansion of the tourism infrastructure are today a strong threat for the coastal areas around the world. Can we find a model of coastal development that minimizes storm surge risk from hurricanes, mutes the most immediate and most likely impacts of sea-level rise, and yet continues to offer high levels of amenities and preserves economic value? The answer is yes, for example, Pelican Bay, Florida, is a potential model of such "climate ready development" (Kousky, 2013).

5.2.2 Description of the Area

"Pelican Bay is located in Naples, Florida, on the Gulf Coast. The community has a quite distinct model of development compared to surrounding areas. High-rise condo buildings sit behind a mangrove forest that runs for several miles along the coast. The forest is perhaps (by my rough estimate) between 0.5 and 1 mile deep. Right in front of the development stretches "the berm," a path for walking, jogging, and bicycling. You can't go very far along the berm without seeing an alligator or one of the many species of birds, such as egrets and herons. At two spots, a boardwalk intersects the berm and proceeds out to the ocean."

"Pelican Bay was created by Westinghouse to protect much of the mangroves and concentrate development at their edge, in part due to new environmental regulations. Residents of Pelican Bay pay dues to the Pelican Bay Foundation and the Pelican Bay Services Division. Of course, as with management of all coupled human-environment systems, this one is not without challenges. Restoration efforts for the mangroves have been needed, such as flushing the mangrove forest to prevent die off. Efforts to control mosquitoes in the most environmentally sensitive way have been pursued.

Still, this appears to be a model for climate-ready coastal development and one that could potentially be replicated more broadly" (Kousky, 2013).

5.3 National Models (Mexico)

Mexico's privileged geographic location, it's territorial extension, the varied physiography, and the climatic diversity have given the country an unparalleled biological wealth, both marine and terrestrial, as well as a large number of natural areas not seen anywhere else in the world.

The fact that these coastal and marine zones are not recognized as special areas of the territory has led to very limited knowledge about them. Every day, the anthropogenic impact on the coastline is more obvious. The seaboard is subject to pressures (i.e., population and urban area growth, lack of beach access, construction of inadequate infrastructure) and impacts (i.e., pollution by local and remote sources, coastal erosion, loss of vegetation, "artificialization," or artificial rendering, of the coastline, waste dumping) that have lead, in many places, to an irreversible deterioration of these fragile areas. For many coastal areas in Mexico, the model to develop is the Spanish model known as the "Sun and beach tourist product," with the consequences of a strong damage for the coastal line itself and for the population living behind these "dreaming areas" (Fig. 11).

According to (Honey and Krantz, 2012), "the site selection for new coastal tourism construction is the development decision with the greatest environmental and social impact. This is where private sector developers, land owners, and public agencies need further education regarding the importance of site selection in ensuring sustainable development. In Mexico, the government, through the appropriate organisms, has the responsibility to choose sites that ensure their tourism projects will be not only economically but also socially and environmentally sustainable. Regrettably, several sites chosen and approved have caused environmental and social damage." They recognize the good practices in site selection and recommend to the developers to follow six factors to comply with sustainable language:

FIG. 11 Aerial view of a hotel complex in Solidaridad, Quintana Roo – Caribbean Sea. *Picture taken by: Norma Patricia Muñoz Sevilla.*

(1) commercial suitability of the site, (2) going beyond legal compliance, (3) appropriate scale of the proposed development, (4) natural heritage and environmental value of the land, (5) social context and cultural heritage of the proposed site, and (6) unintended consequences of land use change. The good practices associated with these factors have been identified as critical to ensuring that site selection contributes to long-term sustainability.

6 CONCLUSION

Mexico's oceanic and coastal areas must be understood as areas of priority attention, and their management should be addressed at a national security level that includes physical risks caused by natural disasters, mainly hurricanes and floods, as well as landslides and earthquakes. Among the main natural hazards in the Mexican coastal zones, those originating from climate change, such as hurricanes, are the most dangerous due to their effects on the population, infrastructure, and ecosystems.

Public policies around the coastal zone in Mexico have been widely sectorial and poorly integrated, generally provoked by political speeches and not by a vision for economic and social development. As long as Mexico continues to develop the coastal zone with the "sun and beach" strategy, we will risk losing this important and fragile ecosystem. More planning and local and regional ordinance is necessary, particularly with the participation of all sectors of the society.

The federal intervention with the National Policy for Seas and Coasts and the development of a coastal law seem to be essential for coastal development in our country.

References

Anta Fonseca, S., Carabias, J., et al., 2008. Consecuencias de las políticas públicas en el uso de los ecosistemas y la biodiversidad. In: CONABIO, (Ed.), Capital natural de México, vol. III: Políticas públicas y perspectivas de sustentabilidad. Comisión Nacional para el Conocimiento y Uso de la Biodiversidad (CONABIO), México, D.F., pp. 87–153.

Azuz Adeath, I., Rivera Arriaga, E., 2007. Estimación del crecimiento poblacional para los estados costeros de México. Papeles de Población 51, 187–211.

BIE-INEGI, 2009. Banco de Información Económica (BIE) - Instituto Nacional de Estadísticas y Geografía (INEGI). Retrieved from, http://www.inegi.org.mx/sistemas/bie/.

CIMARES-SEMARNAT, 2012. Política Nacional de Mares y Costas de México: Gestión Integral de las regiones más dinámicas del territorio nacional. Comisión Intersectorial para el Manejo Sustentable de Mares y Costas de México (CIMARES) y Secretaría de Medio Ambiente y Recursos Naturales (SEMARNAT), Ciudad de México.

CIMARES-SEMARNAT, 2016. Política Nacional de mares y costas de México: gestión integral de las regiones más dinámicas del territorio nacional. Comisión Intersectorial para el Manejo Sustentable de Mares y Costas de México (CIMARES) y Sceretaría de Medio Ambiente y Recursos Naturales (SEMARNAT), Ciudad de México.

CONAGUA, 2013. Estadísticas del Agua en. México, Edición 2013. Retrieved from, https://www.gob.mx/conagua.

Escofet Giansone, A.M., 2008. Las aguas marinas interiores: rescate de una figura de la Ley Federal del Mar e indagación de su valor operativo. In: Escobar, M.E. (Ed.), El Ordenamiento Territorial: Perspectivas Internacionales. Secretaría de Medio Ambiente y Recursos Naturales (SEMARNAT).

Honey, M., Krantz, D., 2012. Alternative Development Models and Good Practices for Sustainable Coastal Tourism: A Framework for Decision Makers in Mexico. Center for Responsible Travel, Washington, D.C.

INEGI, 2008. Información sobre Trabajo, Remuneración y Empleo. Retrieved April 2016, from, http://www.inegi.org.mx/inegi/default.aspx?s=est&c=125.

INEGI, 2009. Instituto Nacional de Estadísticas y Geografía. Retrieved from, http://www.inegi.org.mx/.

Kousky, C., 2013, March 18. Resources for the Future. Retrieved May 2016, from, http://www.rff.org/blog/2013/climate-ready-coastal-development-model-pelican-bay-florida.

León, C., Rodríguez, H., 2004. Ambivalencias y asimetrías en el proceso de urbanización en el Golfo de México: Presión Ambiental y concentración demográfica. In: Caso Chávez, I.P.M. (Ed.), Diagnóstico Ambiental del Golfo de México. INE, INECOL A.C., Harte Research Institute for Gulf, pp. 1043–1082.

Liu, Y., Zhang, F., Zhang, Y., 2009. Appraisal of typical rural development models during rapid urbanization in the eastern coastal region of China. J. Geogr. Sci. 19, 557–567. https://doi.org/10.1007/s11442-009-0557-3.

McFadden, L., Penning-Rowsell, E., Nicholls, R., 2007. Setting the parameters: a framework for developing cross-cutting perspectives of vulnerability for coastal zone management. In: McFadden, L., Nicholls, R., Penning-Rowsell, E. (Eds.), Managing Coastal Vulnerability. Elsevier, Oxford, pp. 1–13.

Partida Bush, V., 2006. Proyecciones de la Población de México 2005-2050. Consejo Nacional de Población (CONAPO), México, D.F.

Pawlukiewicz, M., Gupta, P., Koelbel, C., 2007. Ten Principles for Coastal Development. ULI – the Urban Land Institute, Washington, D.C.

Reefresilience, 2016. Coastal Development. Retrieved from, http://www.reefresilience.org/coral-reefs/stressors/local-stressors/coastal-development/.

SEMARNAT, 2006a. Estrategia Nacional para el Ordenamiento Ecológico del Territorio en Mares y Costas. Secretaría de Medio Ambiente y Recursos Naturales (SEMARNAT), México, D.F..

SEMARNAT, 2006b. Política Ambiental Nacional para el Desarrollo Sustentable de Océanos y Costas de México. Estrategias para su conservación y uso sustentable. Secretaría de Medio Ambiente y Recursos Naturales (SEMARNAT), México, D.F.

State of Nova Scotia, 2016. The State of Nova Scotia's Coast Report. Retrieved from, http://www.novascotia.ca/coast/.

Volusia County Florida, 2014. Beach environment. Retrieved May 2016, from, http://www.volusia.org/services/growth-and-resource-management/environmental-management/natural-resources/sea-turtles/beach-environment.stml.

Xu, G., 2015. The New Coast: How can an environmentally sustainable model of coastal development be developed? PhD ThesisUnitec Insititue of Technology, Auckland.

Yepes, V., Medina, J.R., 2005. Land use tourism models in Spanish coastal areas. a case study of the Valencia region. J. Coast. Res. 49, 83–88. Retrieved from, http://www.jstor.org/stable/25737409.

CHAPTER

4

Governance of the Nautical Sector on the Coasts of Bahias de Huatulco in Oaxaca, Mexico

*María Concepción Martínez Rodríguez**, *Alfredo Salazar López*†, *Omar Mayorga Pérez**

*Centro Interdisciplinario de Investigaciones y Estudios sobre Medio Ambiente y Desarrollo (CIIEMAD), Instituto Politécnico Nacional (IPN), Ciudad de Mexico, Mexico

†Escuela Superior de Comercio y Administración (ESCA), Instituto Politecnico Nacional (IPN), Ciudad de Mexico, Mexico

Acronyms

PROCODES	Conservation and Sustainable Development Program
CGPMM	Coordination of Ports and Marina Merchant
PMDP	Port Development Master Program
Cofece	Federal Commission of Economic Competition
PROFECO	Federal Consumer Prosecutor's Office
PROFEPA	Federal Environmental Protection Agency
FIFONAFE	Fund of the National Fund of Ejidal Development
CGPMM	General Coordination of Ports and Merchant Marine
DGFAP	General Directorate for the Promotion of Port Administrations
DGP	General Directorate of Ports
DGFAP	General Directorate of Ports Administration
PNH	Huatulco National Park
PMPNH	Huatulco National Park Management Program
API	Integral Port Administration
CIP	Integrally Planned Center
Conanp	National Council of Natural Protected Areas
Fonatur	National Fund for the Promotion of Tourism
PMDP	Port Development Master Plan
ANP	Protected Natural Areas

https://doi.org/10.1016/B978-0-12-810473-6.00005-4

© 2019 Elsevier Inc. All rights reserved.

RLGEEPAANP	Regulation of the General Law of Ecological Equilibrium and Protection of the Environment in the Area of Natural Protected Areas
CROC	Revolutionary Confederation of Workers and Peasants
Semarnap	Secretariat of Environment, Natural Resources and Fisheries
SHCP	Secretariat of Finance and Public Credit
SCT	Secretariat of Communications and Transportation
Semarnat	Secretariat of Environment and Natural Resources
SEDATU	Secretariat of Territorial and Urban Agrarian Development
SRA	Secretariat of the Agrarian Reform
SEMAR	Secretariat of the Navy
Sectur	Secretariat of Tourism
INEGI	The National Institute of Statistics and Geography

1 INTRODUCTION

The municipality of Santa Maria Huatulco is located in the south of the state of Oaxaca in the coastal region and district of San Pedro Pochutla. Its extension comprises 579.22 km^2, which represents 0.53% of the total territory of Oaxaca. It adjoins with the municipalities of San Pedro Pochutla, San Matero Piñas, Santiago Xanica, and San Miguel del Puerto as well as the Pacific Ocean. The population of Santa Maria Huatulco, at the last count of the INEGI (2015), was 45,680 inhabitants, which represents 1.15% of the total population of Oaxaca which is 3,967,889.

Santa María Huatulco received the last Integrally Planned Center (CIP-Huatulco) built by the National Fund for the Promotion of Tourism (Fonatur). The CIP-Huatulco is located 227 km from the capital of Oaxaca and 763 km from Mexico City. Its construction began in 1984 with the expropriation of 40% of the territory of the municipality, the result of an authoritarian act by the federal government in complicity with the state, municipal authorities, and simulated civil society associations.

The project covers nine bays, a national park declared a protected natural reserve, and beautiful landscapes of mountains, valleys, and hillsides irrigated by the Coyula, San Agustín, and Copalita rivers that offer the tourist, as a whole, a great paradise to discover (Talledos, 2012). The conformation of the beaches is determined by a substrate of sands of light colors and fine textures, although there are some that are stony. The slope of the bays tends to be reduced while the most exposed beaches are very steep (Ramírez-González, 2005). With regard to the submarine environment, there are areas of shallow water whose values vary from 14 to 40 m. The shallow water zone, close to the coast, has as subunits the rocky reef areas, the coral formations, the mollusk and decapod communities, the area of the Copalita river mouth, the Tangolunda pond area, and areas of sandy and rocky bottoms. These subunits are distinguished from each other by their physical characteristics and diverse compositions of marine flora and fauna (Competitiveness Agenda of the destination Bahías de Huatulco, 2013).

The tourist activities in Huatulco Bahias have been developed in accordance with all these particularities, characterized by contact with the low deciduous forest by contemplating its flora and fauna or through adventure tourism. Such activities take advantage of the physical landscape of the zone: visits to beaches and bays, boat trips, visits to coffee farms, rappelling in the mountains, and rafting along the Copalita river. Fonatur is currently in the process of extinction, which limits its articulation with other authorities of the same level for the operation of the CIP; these authorities include the Ministry of Navy, the Harbor Master's Office, Migration, Customs, and the authorities of the Protected Natural Areas (ANP) of the National Council of Natural Protected Areas (CONANP) of the Secretariat of Environment and Natural Resources (Semarnat). This condition has accelerated disorder, problems of overpopulation, and deregulation of the providers of nautical services in the port of Huatulco, exceeding the original structural conditions and impacting the Huatulco National Park.

In this paper we address the problem that has arisen in the port of Huatulco as a result of the expropriation of land where the lack of definition of obligations and rights has led to the confrontation of the free market imposed by the neoliberal model and the sustainable development model to conserve natural resources. In the first part, we touch on the issue of environmental governance as the way in which decisions are made, inclusive planning is built with transparent processes and accountability about the intentions of all stakeholders in the management of natural resources. Later, we examine the case study with the antecedents of the zone, the way the port of Huatulco operates, the regulatory framework, the protected natural areas, and the management program of the Huatulco national park, concluding, among other things, that: *"The best way to deal with environmental issues is with social participation at the appropriate level."*

2 ENVIRONMENTAL GOVERNANCE

Is there a history of social participation in the shaping of the public and environmental policies of Mexico? Actually, the words "social participation" appeared in a government speech of the 1970s, where it was required to heal the wounds caused by authoritarian governments. Participation had among its objectives the legitimacy of decision systems (through voting) and the dissemination of public policies and government programs.

Social participation is the collective action, in this case around the management and conservation of natural resources, that arises from a process of negotiating interests, defining rules, and building commitments among various social actors based on their rights and obligations (Paré and Fuentes, 2007). Developing a participation scheme that includes local actors, and the local, state, and federal governments as well as nongovernmental organizations and the business sector is important as a first step. However, this can be a failure if those actors are not representative, are not organized around their resources with clear and collectively established and respected norms, and are not legitimate institutions. Some situations faced in Mexico are the lack of legitimacy of the representatives and the opacity of their operation. Although it is possible to have communication channels between the different actors, it turns out that these are not the representatives of their community or sector or there is incompetence in the construction of agreements. Environmental governance refers to "The processes of decision-making and exercise of authority in which governments intervene at their various levels or decision-making bodies, but also other interested parties that belong to civil society and that have to see the development of regulatory frameworks and the establishment of limits and restrictions on the use of ecosystems" (Piñeiro, 2004).

Governance implies developing inclusive planning and transparency mechanisms, accountability regarding the intentions of all stakeholders, instruments for coresponsibility between the state and society for the management of natural resources, and the participation spaces for democratic planning. The center of many environmental conflicts originated by *economic* activities is located in a struggle for the definition of a development model that is never the object of a consensus because the actors in opposition share different worldviews and rational logic (Martínez-Rodríguez, 2015). The contradictions between the interests and rights of the different actors as well as the different perceptions around these rights and obligations make it difficult to exercise coresponsibility in the management of the territory and the management of the resources on which the success of policies rely (Paré and Fuentes, 2007).

The legitimacy of the policies then rests on the consensus or discussion that takes place with the actors involved (Jardón-Medina et al., 2017). It is intended that participatory planning and the search for consensus among the actors involved in management might help in the better functioning of tourist destinations. Environmental governance is a continuous process and its evolution depends on social participation and political agreement, becoming an alternative to the predominant market path that arises in a context of growing weakness of the state, which it forces civil society to act and self-organize in order to conserve natural resources that provide environmental services (De la Mora-De la Mora and Montaño Salazar, 2016).

3 BACKGROUND OF SANTA MARIA HUATULCO

The development of the tourist destination of Bahías de Huatulco began in 1960 in the process of innovation of tourist destinations such as Cancun, Ixtapa, and Puerto Escondido, with the support of the World Bank and the Inter-American Development Bank, under the command and coordination of the National Fund of Tourism (FONATUR). FONATUR is a federal institution whose main objective was to promote the growth of the tourism sector in previously selected regions of the country through various prospective studies in regions such as the Pacific coast and the Mexican Caribbean. The first step toward development was to legalize the ownership of large areas of territory to build the so-called Integrated Planned Centers (CPI). The power of the state allowed overcoming the alienation of agrarian territories in favor of foreigners as well as the expropriation of communal territories, all this under the discourse of boosting economic activities that generate high levels of income and job creation, at the same time should provide development of regions with serious backwardness and marginality (Talledos, 2012).

The previous concept became the public practice to offer security in the possession of the earth driven by tourism. Based on development plans, the spaces were planned and ordered with a positivist argument, displacing the peasant communities as a strategic space for tourism activity. The implementation process of CIP-Huatulco, according to the Huatulco National Park Management Program (PMPNH), began with the expropriation of 21,163 ha for Santa Maria Huatulco (of a total area of the municipal territory of 51,511 ha) through a presidential decree on May 29, 1984.

In 1985, work began on converting Bahías de Huatulco into the fifth CIP of FONATUR. The master plan of this development contemplated 9.9% for the tourist area, 3.3% for the urban area, 40.6% for ecological conservation, and 46.1% for other uses. The development adopted a tourism-real estate, rather than tourist-port, so there are different aspects that in the analysis for commercial ports are essential, but for the port of Bahias of Huatulco are irrelevant as those are associated with the handling of commercial cargo (PMDP, 2012-2017, p. 14).

A Tourist Zone and an Urban Zone were built, which consisted of building infrastructure in nine natural bahias: Santa Cruz, Chahue, Tangolunda, Conejos, Chacahual, Cacaluta, San

Agustín, the Organ, and Maguey, followed by the construction of the airport, the golf course, and the Marina Chahue. In Fig. 1, you can clearly see the distribution of the spaces.

For compensation for the expropriated territory, the Trust Fund of the National Fund of Ejidal Development (FIFONAFE) and the Mixed Liquidating Commission were created. They were made up of representatives of the Oaxaca state government, the Secretariat of Agrarian Reform, the Secretariat of Tourism (Sectur), and FONATUR, which were responsible for making the appraisals and paying for the goods adhered to the land in its entirety of the affected inhabitants. Those in charge of negotiating were Communal Assets and the Municipal Presidency, the latter of which had previously made agreements with the federal government to facilitate the conviction of the comuneros so that they will quickly accept the sale of their properties.

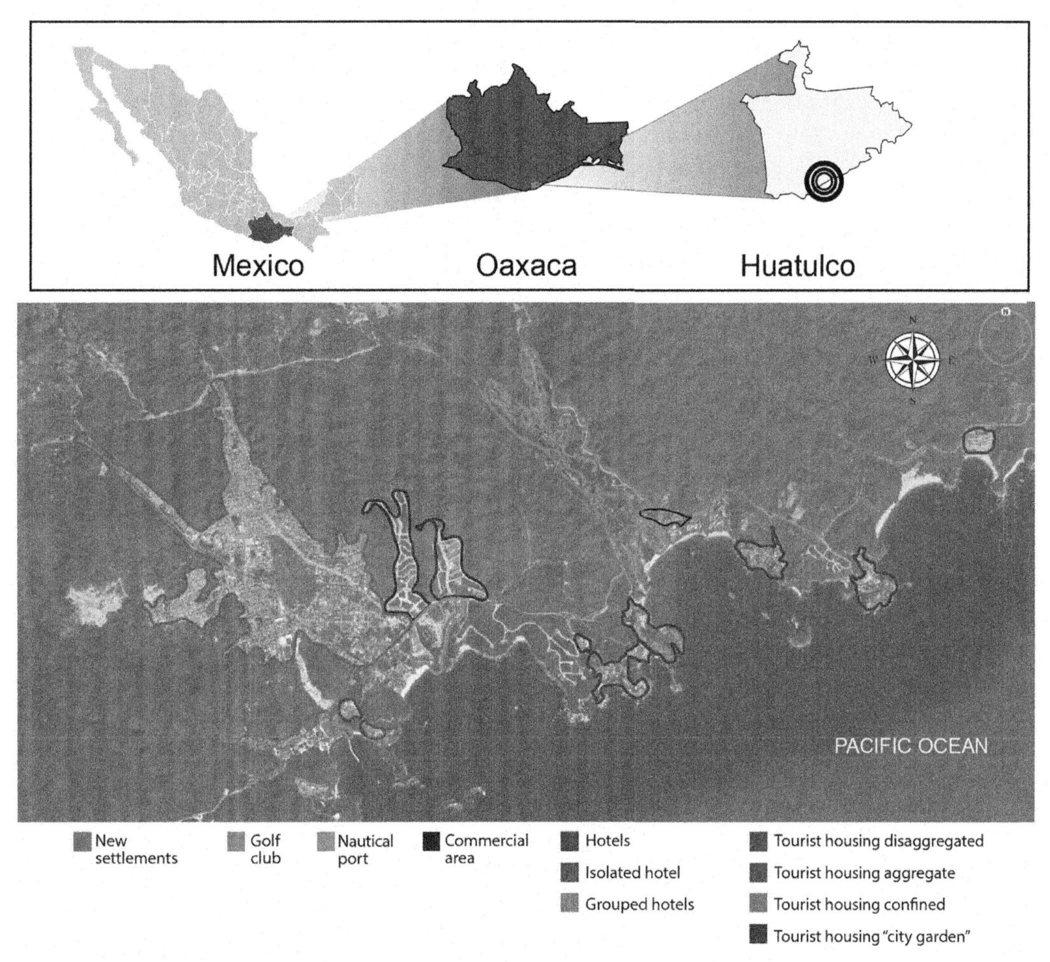

FIG. 1 Polygon from FONATUR in the study area of Huatulco, Mexico.

In this context, it was operated to reproduce a new unequal and deeply segmented space that had dispossession and violence as two of its foundations of consolidation. On Jul. 24, 1998, the Huatulco National Park (PNH) was created as a protected natural area (ANP) due to the constant pressure from the nongovernmental and academic organizations that settled in the municipality after the creation of the tourist complex (Talledos, 2012). The creation of the CIP-Huatulco that includes the nine bays and the national park also created a spatial configuration in the municipality of Santa Maria Huatulco that generated the conditions so that the capital could circulate without any legal, political, economical, or environmental obstacles. The new organization of the territory has brought success both financially and socially.

4 POLICIES IN THE PORT OF BAHIAS DE HUATULCO

The ports have standards issued by the same authorities through the operations committee or planning committee, rules that are binding to the port's endogenous conditions. Likewise, in a guiding way, the Port Development Master Program (PMDP) interprets the port operation rules. The port of Bahías de Huatulco and the port of Huatulco are located in the model of ports administered by FONATUR. It is denominated this way to the ports that are within the polygon of the CIPs created by FONATUR. They are atypical in port legislation and are integrated by an Integral Port Administration (API), which meets the guidelines of the General Directorate of Ports (DGP) and the General Directorate for the Promotion of Port Administrations (DGFAP), both dependent on the Coordination of Ports and Marina Merchant (CGPMM); examples of these types of ports are found in Los Cabos and Zihuatanejo.

The nature of the port of Huatulco is located as: Height based on the Law of Ports: ARTICLE 9 - The ports and terminals are classified:

> I. For browsing in: a) of height, when boats, people and goods in navigation between ports or national and international points attend, and... (de Puertos, 2016).

4.1 Coordination and Cooperation Between Authorities of the Port of Bahias de Huatulco

The efficiency of the operations of the port of Huatulco are represented by a scaffolding of national and international legal ordinances. This has given confidence to the shipping lines of foreign cruise ships that arrive on average between 50 and 74 times a year (PMDP, 2015). The case of the port and the problems it presents in relation to the management of the nautical tourism service is specified in the diagnosis issued by the Port Development Master Program (PMDP) in section 3.1 of the port situation:

> ... that has about the recent works that have been made for smaller fishing boats and riparian that occupy a space of 21,433 m^2 that in its perimeter of 787.58 total meters allows only 126 berths for fishing sport fishing riverine, causing with this a supersaturation of more than 43% of the vessels, permanent interference in inland navigation and the increase in the culture of non-payment ... (PMDP, 2015).

The above represents collision risks in addition to the impossibility of reacting to the threat of a natural phenomenon (tsunami, hurricane, sinking, etc.) or terrorist attack; it also exposes users, visitors, or tourists to situations that are otherwise unsafe. Another element is the culture of nonpayment of the boat owners. The update of the PMDP affirms that the cause of such conditions is the lack of cooperation and coordination of the API with the port captaincy. The lack of cooperation and coordination of the port authorities with jurisdiction to deal with any event that puts security at risk within the port is evident. These authorities have not managed to approach the permit holders of smaller vessels through the operations committee or committee of port planning in order to achieve consensus. There is no historical record where the port authority has initiated a procedure of warning, verification, infraction, or sanction against the permit holders of smaller vessels. In addition, there is also no record that an opinion has been requested from the General Directorate of Ports (DGP) to establish rates and prices for the provision of the nautical service or consultation with the Federal Commission of Economic Competition.

The participation of various authorities such as federal, regional, state, municipal, and local at different levels in the administration of the port of Huatulco, such as API, DGP, and DGFAP, fostered by the order led by FONATUR as the federal authority that has subordinated the local authority. This dilutes the responsibility and the intervention of these, a situation that has allowed an excessive and disorganized growth of the smaller ships operated by locals who were displaced from their places of origin and take this activity as a right because their land has been expropriated.

5 HUATULCO MUNICIPALITY AND COMMON GOODS OF SANTA MARIA HUATULCO

After a long agrarian conflict between communities of various populations, the territory of Santa María Huatulco was recognized in favor of a group of 1523 "*comuneros*" (comuneros/commoners; the beneficiaries of agrarian rights). They are the rightful possessors of the land. They belong to the nucleus of the community, and they benefit from the use of the land and the agrarian law by means of a presidential resolution dated May 25, 1984, which was registered and titled as owners of 51 thousand 510.9 ha of deciduous forest. The resolution was executed on May 28, 1984. The fourth article of the resolution states:

> ...It is hereby declared that the communal lands that are recognized and titled are inalienable, imprescriptible, and unteachable and that only to guarantee the enjoyment and enjoyment thereof by the community to which it belongs, will be subject to the limitations and modalities that the Agrarian Law in vigor establishes for ejidal land ... (DOF, 1984).

According to Héau Lambert (2014), there are four types of village nuclei of the inhabitants of Santa María Huatulco:

(1) The original nucleus of the town of Santa María (2000 inhabitants before the foundation of the tourist center in 1984).
(2) The surrounding communities that were founded between 1930 and 1950 from migrations of agricultural workers from Miahuatlán (Oaxaca) and were integrated into the municipality; they are currently part of the 1882 commoners de Santa María.

(3) The creation of the town of Crucecita from the land acquisition in 1984 by the Ministry of Tourism of the coast belonging to Santa María Huatulco, to deliver it to the government agency FONATUR, which administers it. Crucecita now has 15,131 inhabitants from various parts of the Mexican Republic who are mostly engaged in various areas of tourism services.

(4) New colonies far from both Santa Maria and Crucecita. The creation of this tourist center, out of nothing, has given rise to forms of different and specific territorial appropriations.

Within these four groups, those in the CIP polygon are the expropriated that were divided into three communities: Santa Cruz, El Arenal, and Coyula, the children of the expropriated who claim some benefits of tourism. Currently, there is a large group of migrants who go mainly to the destination for employment, whether tourism or not but which is basically created by the existence of the CIP (Fig. 2).

With respect to the economic activities of the municipality, it is necessary for the primary sector (agriculture, livestock, forestry, fishing) 5.73%, secondary sector activities 19.29%, activities in the tertiary sector or service sector have a 79.78% (Competitiveness Agenda of the destination Bahías de Huatulco, 2013).

Consequently, as is normal in tourist places, most of the population is engaged in service sector activities as merchants, sales employees, personal services, and monitoring, among others. However, despite the amount of investment with national and international capital and the increasing arrival of tourists, the living conditions of the local population have not improved significantly. From 2000 to 2015, Huatulco practically remained more or less at the same national ranking in the social delay index. Out of the 2458 municipalities (INEGI, 2017) in the country, Huatulco ranked 1597th in 2010 in the national social lag index. In 2005, it was 1453, and in 2000 it was 1531 (CONEVAL, 2010).

Thus, the bonanza of tourism does not benefit the entire population. There is a population group that is outside tourism that is constituted of farmers and indigenous people. Among those who obtain benefits, there are the investors, hoteliers, and owners of large businesses and the villagers who work as waiters, cooks, and boatmen, among others.

5.1 Police and Good Governance of Santa Maria Huatulco

When we study the different actors that participate in the space of Santa María Huatulco, we find FONATUR that manages the CIP, which represents more than 40% of the real estate areas of Tourist Chahue, Tangolunda Tourism, and Urban Tourism. In this space, it is responsible for water management, garbage collection, maintenance of common areas, and sewage, among others, all part of the powers that Article 115 Constitutional recognizes in favor of the municipality (CPEUM, 2017).

Within the territory of Santa María Huatulco, the municipal government and the Commissioner of Communal Property administer the 60% that FONATUR and PNH do not manage. Both institutions have innovative regulatory frameworks. Even though they are weakened in their legitimacy, the reality shows a great social contrast where they have not allowed the involvement of the original groups. As a result of weakening the powers of the local authority, there is no one who really strengthens the cooperation and coordination in social cohesion and defends the interests of native people. There is no safeguard for local and common development.

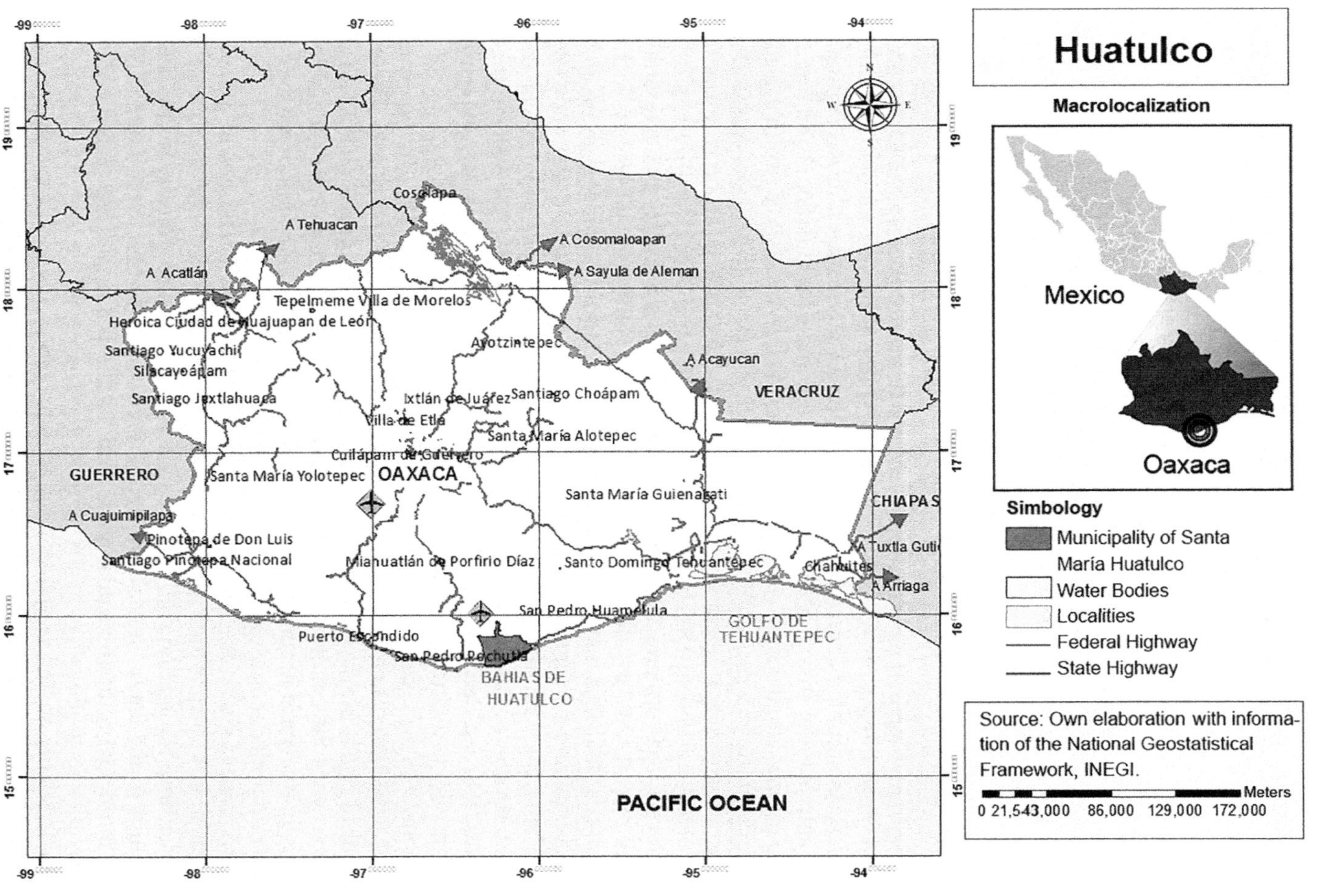

FIG. 2 Macro localization of Huatulco.

With the reforms of the constitution at the municipal level in 2013, the new Police and Good Government of Santa María Huatulco came into force, making a special emphasis on the regulation of tourist activities. It is the seventh title, chapter one, where it states in general the authorizations or permits (definitive or provisional) to carry out industrial, commercial, or tourist activities (Periódico Oficial Oaxaca, 2013a,b).

The municipality has the conditions to influence the ordering of tourism providers. However, its requirement has not been approved to comply with the rest of the federal authorities. This condition obliges us to meet in a space where various authorities concur. However, institutional barriers limit the cooperation and coordination between them, with the native population the most affected of the municipality.

5.2 Municipal Tourism Regulations of Santa Maria Huatulco, Pochutla, Oaxaca

The most important thing in the history of Huatulco is its effort to become an icon at an international level, meet international standards, and position itself as a tourist destination. Under this context, the municipal government has issued an innovative tourism regulation for the municipality in order to coincide with various commitments of national and international certifications: municipal ecological territorial order, a clean beaches committee, a community system of protected areas, and Blue Flag and Earth Check, international organizations of tourism and environment that acknowledge when the site is an attractive tourist destination as well as being ecological and sustainable (Rodríguez, 2013) (EarthCheck Certification, 2016).

There are various legal systems that are responsible for coordinating and cooperating authorities at all three levels, in addition to the high participation of civil society such as WWF, Costa Salvaje, Rotarians, etc. On the other hand, Da Jandra comments in his documents: Huatulqueños and Samahua, mainly (Da Jandra, 1991, 1997) the devastation of an ecologically rich area, but with internal complications among indigenous peoples who compete for natural resources against the tourism industry, ecologists and a government that is incapable of protecting and defending the lands and their people; The losers are the natural resources that are slowly dwindling in the area.

The reality about the management of the authority at the front from 1957 to the current administration is considered as a weakened act against the FONATUR parastatal. The municipal president of Santa María Huatulco has been considered employee of FONATUR. In the same sense, so is the commissariat of communal property that has not been able to be a counterweight in favor of its members in order to capitalize the momentum for the consolidation of a community ecological territorial order.

For the specific case of study, the Municipal Tourism Regulation establishes in Article 9, section XXII, obligations for maritime service providers, in an expanded manner to the Regulation of the Maritime Navigation and Trade Law (LNCM, 2016), specifically for the attention of cruise ships in the Santa Cruz wharf or the registration of municipal tourism. The validity of the municipal tourism regulation as well as the police and good government started the day after its publication in the official newspaper of the state of Oaxaca on Mar. 2, 2013. To date, it has not been applied in the activity of the providers of nautical service; contrary to this, there are demonstrations and sometimes violent acts threatening to take over the facilities of the Fonatur offices, the Port Administration, the Harbor Master's Office, the

Huatulco National Park, port facilities, access to Huatulco, etc. It is urgent to restore order in the port of Santa Cruz Huatulco so that boats and service providers are regularized and this is reflected in the care and safety of national and foreign tourists (Pacheco, 2017).

5.2.1 Nautical Service Providers

The entrance of FONATUR for the construction of the CIP-Huatulco implied changes in the activities of the original inhabitants, the natural wealth of the zone allowed to take advantage of the abundance of the resources that the space offered them. With the land acquisition, a confrontation between the natives and the government was formed. The former could no longer have their home, their land, or work fishing and hunting while the latter were championing a pole of "development" that they imposed but never agreed.

The declaration of a Protected Natural Area (ANP) of the Huatulco National Park (PNH) further limited the activity of fishing and hunting in the expropriated area. However, before the CIP-Huatulco, the municipal authorities promoted the hunting and processing of shark as one of the main economic activities as well as installing refrigerators and encouraging the formation of cooperatives of local fishermen and processors for sale. The imposition of the tourism project triggered processes of conflict and social segregation (Lorena-Rodiles and López-Guevara, 2015).

Displaced by the hotel infrastructure and the new dynamics of the tourism industry, the fishermen and farmers joined forces, having to negotiate indemnities imposed on them until they joined the new precarious labor market in hotels, restaurants, and nautical tourism services. This causing the eventual loss, for fishermen, of spaces on several beaches that they once used to carry out their activities.

While the fishermen have been displaced from the beaches, the federal agencies and the local government took on the task of assigning the collective a dock space at the Santa Cruz dock, which, according to the fishermen, is expensive and highly insufficient to tie all the boats. The comments of the fishermen also highlight that since 2009 (the year in which the space was granted at the local dock), there has been no formal delivery agreement, which means that the group does not have legal possession of the mooring positions. The fishermen assume that, in the absence of such an act, there is a risk that at any moment the port authorities will withdraw the berthing permit and, with that, the availability of a space to carry out the operations associated with their activity.

Interviews that are repeated in different works (Lorena-Rodiles and López-Guevara, 2015; López, 2010; Mendoza et al., 2011) and that are added to the present work collect the negotiations that FONATUR made with fishermen. The fishermen of the bays have relocated their boats in the port of Santa Cruz along with the other cooperatives, which has meant assuming an expense for payment for dock use. There is no record with FONATUR that any agreement has been signed in which the space of the port of Santa Cruz (Port of Huatulco) is provided as a form of compensation for the construction of hotels. Within the agreements for the expropriation of their land, the fishermen were discounted the price of their boats that they had acquired with previous credits under the assumption of the cooperatives that were going to fish and industrialize the product. Then the fishermen were left without their land, without their source of work, with little money, and with boats but no place to use them.

The testimonies collected shows that the other cooperatives have experienced processes of fragmentation due to the confrontation of interests among their members. Few fishermen are dedicated full time to this activity. Of the fishermen interviewed, 40% are engaged in fishing, 20% combine fishing with other activities related to urban transportation (taxis) and the provision of services in restaurants while the remaining 40% are retired fishermen without any pension or benefits (Lorena-Rodiles and López-Guevara, 2015).

In this picture, it is appreciated how the life of the original inhabitants went, until incorporating itself to the tourism related activities, with the protection of a collective contract of the Revolutionary Confederation of Workers and Peasants (CROC), that acts as a white or *charro* union (political corporatism that serves as a control system to sustain and reproduce authoritarian and corrupt political regimes), in the case of workers who work in hotels.

On the other hand, for informal workers who participate in services such as taxis, tourism services, restaurants, and others, the reality is what the superior auditor of the federation has communicated, derived from the audits of the budget of public expenditures that have been invested. Likewise, for informal workers who participated in such services, the reality is what the superior auditor of the federation has communicated derived from the audits of the budget of public expenditure that have been invested in works without recording the impact they have had on society. This is precisely the question for which there are no indicators of how the local society has benefited in a sustainable manner while INEGI indicators confirm that poverty in Santa María Huatulco and neighboring communities remains in a marginalized condition.

5.2.2 Port Captaincy and Integral Port Administration of Bahias de Huatulco

Currently, the port of Bahías of Huatulco has two authorities, one port and one maritime. The first falls to the Ministry of Communications and Transportation (SCT), through the CGPMM, which is deposited in the General Directorate of Ports (DGP) based on article 16 of the Law of Ports (de Puertos, 2016), and the General Directorate of Ports Administration (DGFAP), which is only for corporate aspects that are exercised through the API.

The second authority in the port is the maritime one, which with the recent amendment to the Maritime Navigation and Commerce Law (LNCM) of Dec. 19, 2016, is the Secretariat of the Navy (SEMAR), which at the same time exercises through the Port Captaincy, as provided in Article 7.8, 8Bis, and 9 of LNCM (2016). The delimitation of the port area is proposed by the DGP to the Secretariat of Territorial and Urban Agrarian Development (SEDATU), based on article 7 of the Law of Ports (de Puertos, 2016) and 49 of the General Property Law Nationals (LGBN, 2016).

In the case of Bahías de Huatulco, it is in the case of the federal port of height and is administered by the API (the figure of the Mercantile Society adopted by the Ministry of Communications and Transportation) for the constitution of a *Land Lord* within the ports in order to efficiently manage the space that was constituted through the API that the 16 federal ports, five state and one private, which recognizes the Federal Law of Parastatal Entities in articles 32, 33 and 34, who is responsible for reporting to a Board of Directors to the DGP.

The API of Bahías de Huatulco formally began operations on Dec. 24, 1999, once the SCT, through the CPMM and with the attention of the DGP and DGFAP, provided in favor of FONATUR-BMO, S.A de C.V. the concession title. The API of Bahías de Huatulco designed its Master Port Development Plan (PMDP) based on articles 41 of the Ports Act (2016) and 39 of its regulations, as well as in the ninth condition of the concession title. The last modification was made with the approval of the Operation Committee of the Bahias de Huatulco port facility, adopting its recommendations at the XXIII Regular Meeting held on Jan. 31, 2006, in accordance with article 58, paragraph II of the Ports Act (2016).

In order to promote the PMDP, the coordination and cooperation of the port for greater efficiency in maneuvers and operations, Article 11 of the Law of Ports (2016) establishes that the Operations Committee will issue the port operation rules that should observe the permit holders and concessionaires as well as any other person. The rules of operation are an endogenous, binding order that formulates the API and authorizes the DGP, whose purpose is to govern the maneuvers and operations that are carried out within the port area without prejudice to the specific cases that the PMDP states, the concessions, respective permits, or contracts, in the official Mexican standards.

Based on Article 58 of the Ports Act, the Operations Committee becomes a space to resolve conflicts arising from the operation of the port, within the framework of local regulations issued by the same committee. If the disputes cannot be settled through the port committees, the port authority, deposited in the DGP in turn in the Harbor Master's Office, has the power provided by articles 63, 64, and 65 of the Ports Law (2016). This consists of verifying, apperceiving, infringing, and sanctioning until the cancellation and nullity of the permit or partial assignment of rights.

6 DÁRSENA OF SANTA CRUZ

By means of the decree published in the Official Gazette of the Federation on Jul. 21, 1997, the port of Bahías de Huatulco was activated for commercial maritime traffic and height on the Pacific coast, in the state of Oaxaca. With a joint agreement between the Ministry of Communications and Transportation and the then Ministry of Environment, Natural Resources, and Fisheries, SEMARNAT (Secretary of environment and natural resources) published in the Official Gazette of the Federation on Dec. 23, 1999, the corresponding port area was defined and determined to be the port of Bahías de Huatulco, in what is called Santa Cruz Huatulco, the municipality of Santa María, Oaxaca.

The port area was originally conceived to develop as a public marina, for which a dock was dredged 2.14 ha and 3 m deep, with vertical parameters based on concrete sheet piles. The dock is bordered by a boardwalk that is 11 m wide. The idea of FONATUR was to cover the most elementary needs of the 80 smaller boats and occasionally the reception of cruises, so there was no adequate planning for the exploitation and use of the little infrastructure described above. This is because there was no regulation traffic in the dock, areas properly conditioned and defined for the embarkation and disembarkation of people, offering the provision of port services in a disorderly manner. The total number of boats does not coincide with those stated in the rules of operation of the Port of Bahias de Huatulco, precisely in Rule 37 that states: The dock of Santa Cruz has 139 mooring places for boats, distributed from the following shape: Zone A, 38 spaces for boats up to 24 ft, Zone B, eight spaces for boats up to 24 ft, Zone C, 17 spaces for boats up to 24 ft, Zone D, eight spaces for barges, Zone E, 11 spaces for tourist yachts, Zone F, 17 spaces for boats up to 24 ft, Zone G, 18 spaces for boats up to 24 ft, Zone H, 22 spaces for fishermen (there are four different data from official sources that are: API Bahias de Huatulco, Harbor Master's Office and State Tourism Secretariat respectively report 236 vessels, 345 vessels the others).

Passengers who have entered the port area to take nautical tourism trips to the bays, which includes smaller, collective, private, or sport fishing boats, have had an important growth.

Since Dec. 2001, 29,367 trips were made; for 2002, 174,677 were registered; in 2003, 177,447 trips were recorded; in 2004 they reached the highest figure by registering 198,197 trips, since for 2005 the figure decreased to 188,884 trips. The most saturated months for tours are March, April, July, August, and December (PMDP, 2015).

The information provided by the Bahias de Huatulco Comprehensive Port Administration registered a total of 236 vessels, 70% more than what is stated in rule 37, as shown in Fig. 3 according to its use, and four by its type (corroborated by collating the list of vessels provided by the Harbor Master's Office, the API of Bahias de Huatulco and the Tourism Directorate of the Municipality of Santa María Huatulco) (Figs. 4 and 5).

As you can see in Figs. 1 and 2, it is possible to locate the class of vessel and the volume that goes beyond those provided by the port's rules of operation. This also explains the absence of authorities that are responsible for monitoring the provision of the maritime service. The overpopulation of boats in the Santa Cruz dock is a singular case because the rules have been imposed by the providers of nautical services, most of them local "*comuneros*." This is despite

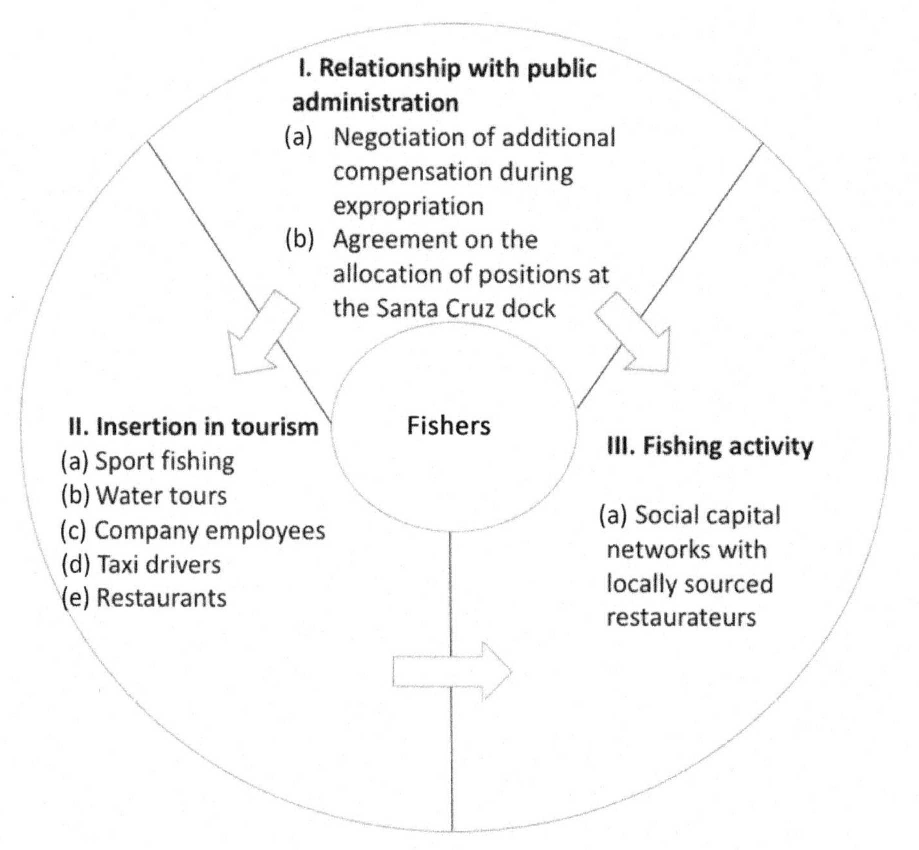

FIG. 3 Adaptive strategies undertaken by the fishermen of Huatulco. *Adapted from Lorena-Rodiles, S., López-Guevara, V., 2015. Pesca tradicional y desarrollo turístico en Bahías de Huatulco. Una lectura desde la historia oral de los pescadores locales. Invest. Turísticas, 10, 150–169.*

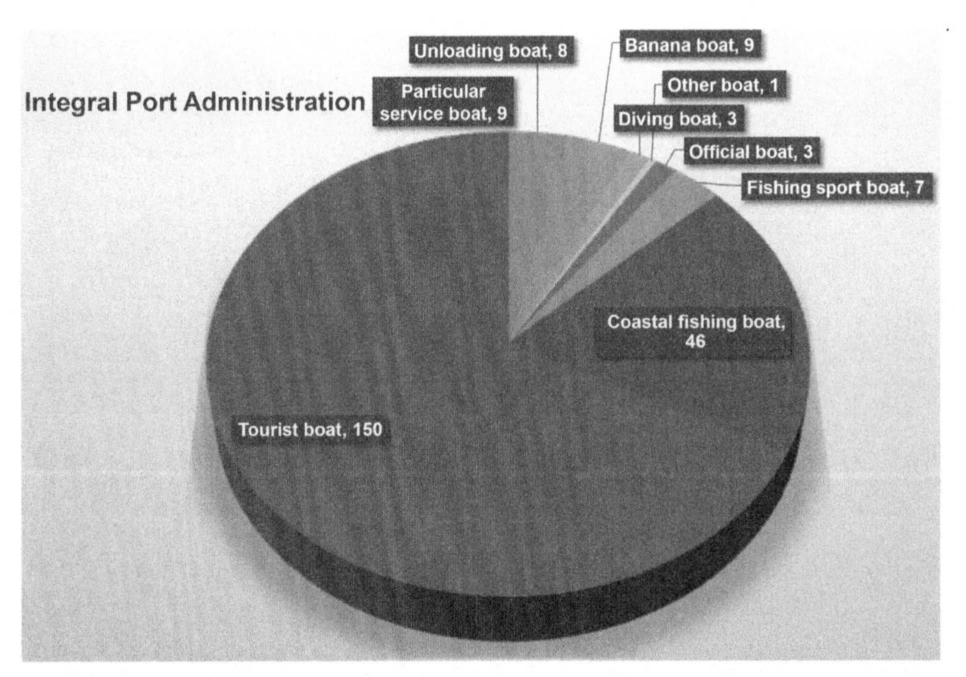

FIG. 4 Vessels by service inside the port. *Source: Own elaboration with data from Harbor master, API, and Ministry of economy and tourism from Oaxaca, 2014.*

the fact that, in light of the laws and the nature of the space, it is a polygon of a federal port. It is demonstrated in a systematic way that security provisions are violated in navigation and threats to civil protection that would make it impossible to respond to a contingency (Fig. 6).

In Fig. 3, we can verify that there is a difference in the registry of the vessels. The Harbor Master's office and the Secretary of Economy and Tourism of the government of the state of Oaxaca add up to 345 while the API lists 236. As you can see, this is a difference of 109 vessels, approximately 150% more than what is set by Rule 37.

With the above, it can be affirmed that there is no coordination between authorities to verify the number of vessels entering the dock, even though the same authorities participate in the Operations Committee (section VII of article 40,57,58 of the Law of existing Ports has extensive powers to the Operations Committee to formulate its own rules of operation as well as to issue recommendations in matters of conflict resolution, coordination for the efficient operation of the dock, operation of port schedules, prices, and tariffs and assignment of docking positions) and the Planning Committee (Article 58 bis of the current Ports Act creates a Planning Committee that together with the SEMARNAT, Harbor Master's Office, the API, assignees, providers and services that met at least three times a year to verify that the master port development program is fulfilled. Committees in the operations involved all authorities, licensees and permit holders and in planning only the Captaincy, SEMARNAT, API, assignees and permit holders), to agree on the most relevant issues of the port.

The scenario described as incapacity and institutional indifference transcends systematically contribute toward the detriment of tourists, conservation, and protection of the

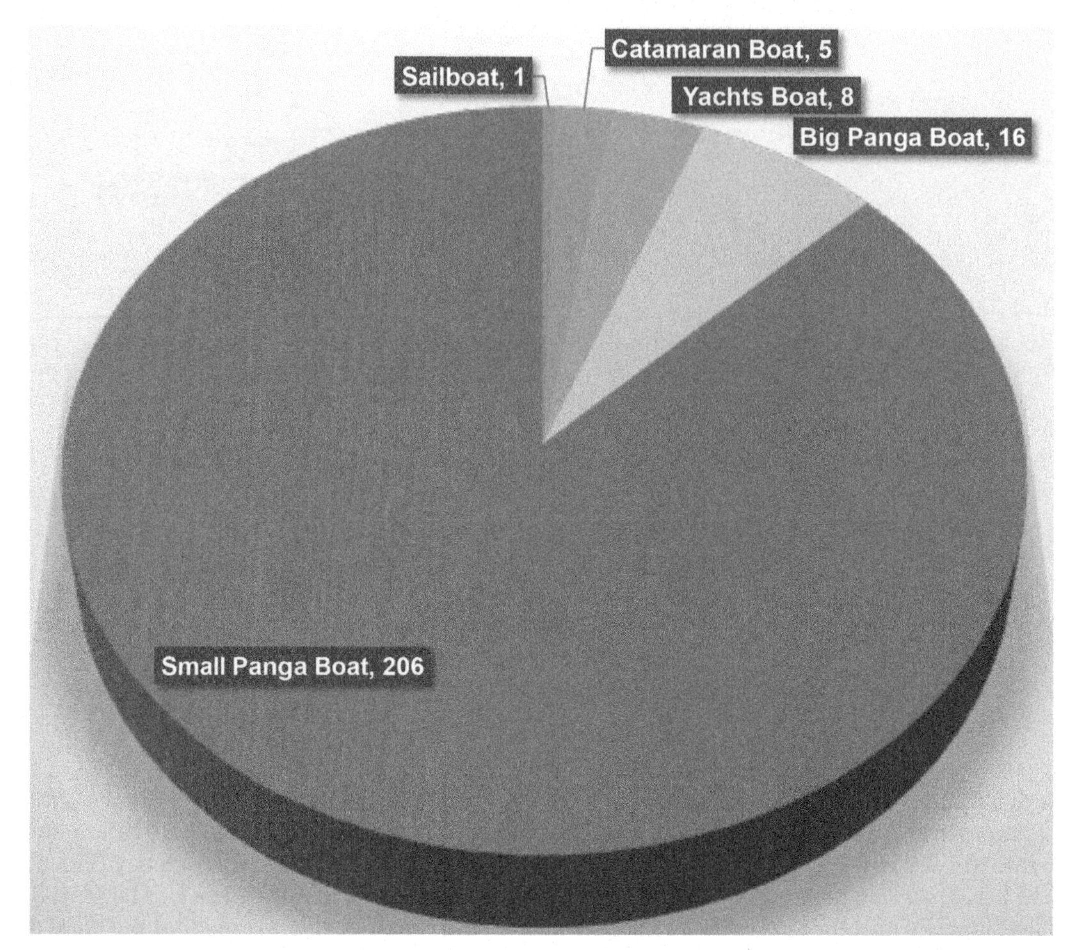

FIG. 5 Types of vessels inside the port. *Source: Own elaboration with data from Harbor master, API, and Secretary of economy and tourism from Oaxaca, 2014.*

environment. Another of the social conflicts that threaten the authorities in the particular case of the Darsena arises from the resentment of expropriation of FONATUR and, of course, in the face of the development lag that the southeast presents. The degree of social disagreement is also found in the fact that service providers do not make any contribution to the use of the port.

7 HUATULCO NATIONAL PARK (PNH)

The PNH was decreed on Jul. 24, 1998, as part of the polygon expropriated by FONATUR to create the CIP Bahías de Huatulco. It spreads over almost 12,000 ha; 6375 ha are terrestrial and 5516 belong to the marine zone, which includes five of the nine bays of Huatulco: San Agustín, Chachachual, Cacaluta, Maguey, and Órgano. The priority ecosystems for

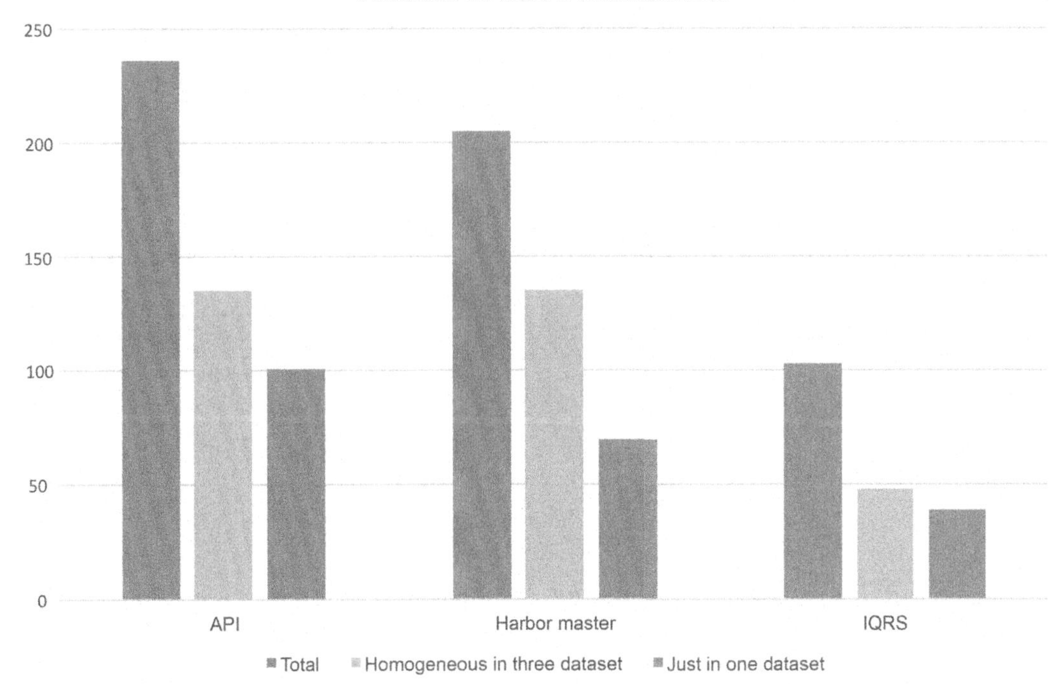

FIG. 6 Vessels (ship or boat) of three institutions. *Source: Own elaboration with data from Harbor master, API, and Secretary of economy and tourism from Oaxaca, 2014.*

conservation are low deciduous forests, reefs, mangroves, and floodplains. The HNP presents complex ecosystems characterized by the presence of endangered, endemic species with high ecological value for their conservation. In the marine portion of the park, there are natural habitats that house a large number of mammals, fish, coral reefs, algae, and various invertebrates subject to constant anthropogenic pressure, as in the case of the purple snail. Due to this biotic relevance, in 2003 the PNH was designated as Site RAMSAR 1321 (Wetlands Convention), and in 2006 it was recognized as part of the World Network of Biosphere Reserves of the United Nations Educational, Scientific, and Cultural Organization (UNESCO).

The Huatulco National Park Management Plan (PMPNH) was published on Dec. 2, 2002 (DOF, 2002). This document includes a summary of the ecological characterization of the PNH. Finally, the PNH Management Plan was published in Nov. 2003 (Comision Nacional de Areas Protegidas (CONANP), 2003). This document illustrates in detail the activities that can be carried out by tourist service providers, tourists, visitors, and scientists.

In Fig. 7, we can locate the aquatic and terrestrial space that includes the polygon of the PNH.

The territory that includes the Huatulco National Park (PNH), in which there are areas of land and sea, is located between the population of Santa María Huatulco and the expropriated polygon of FONATUR. It has an area of sea where the comuneros (they are natural holders of agrarian rights that they possess in community with other individuals or owners, they belong to the nucleus under the communal regime, and they benefit from the use of the lands

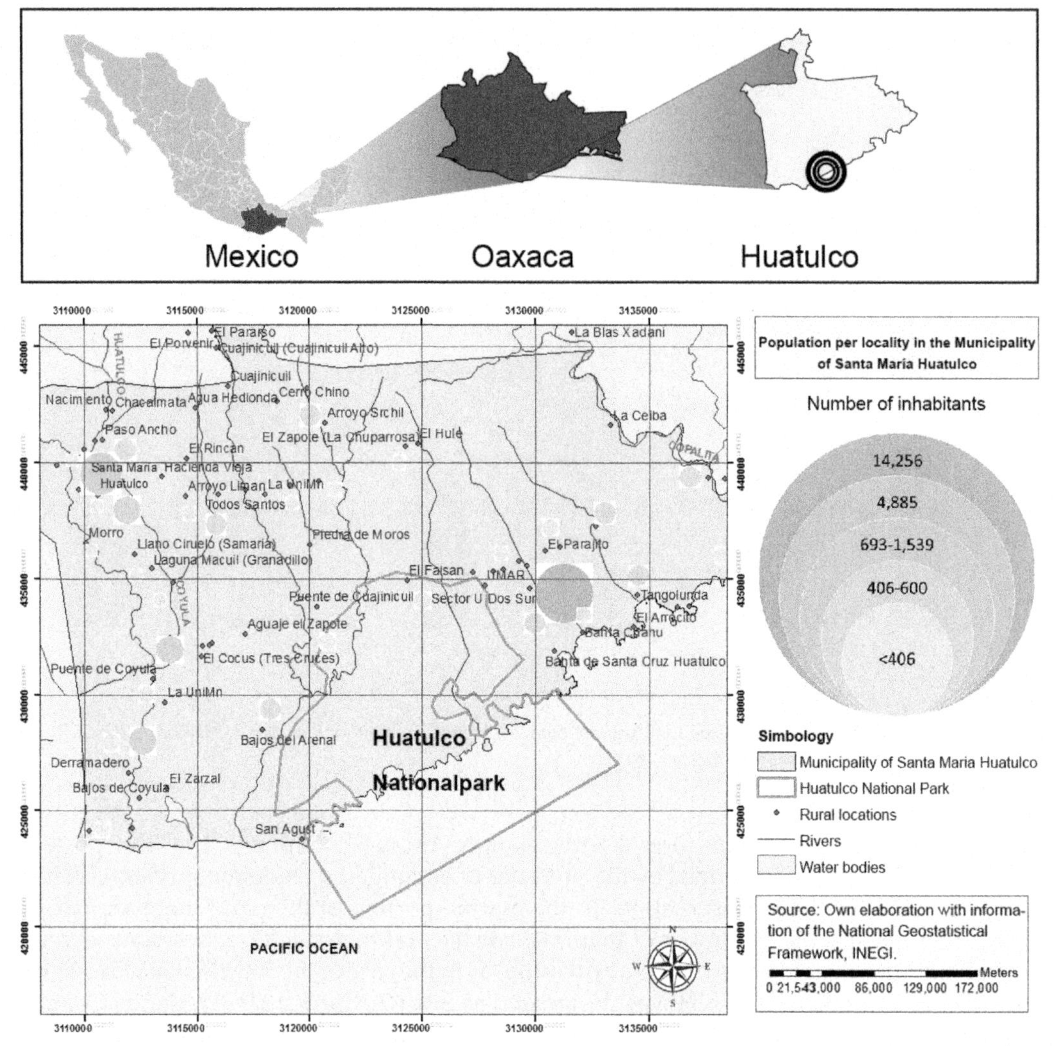

FIG. 7 Polygon of Huatulco National Park. *Source: Own elaboration with information of the National Geostatistical Framework, INEGI, and National Commission of Natural Protected Areas.*

included in the agrarian law) were authorized to provide the service of nautical transportation to tourists by boat.

7.1 Regulation of Protected Natural Areas and Management Program of the Huatulco National Park (NPH)

The maritime service providers have a close relationship with the HNP, at least within the framework of the law and in the benefits provided by resources of the Conservation and Sustainable Development Program (PROCODES, which is a subsidy program that constitutes a public policy instrument that promotes the conservation of the ecosystems and their biodiversity in the Protected Natural Areas. Priority Regions for the Conservation, through the sustainable use of biodiversity, with emphasis in the indigenous population of the localities), which they receive year after year. Based on articles 80–104 of the Regulation of the General Law of Ecological Equilibrium and Protection of the Environment in the Area of Natural Protected Areas (2014), during the month of April, natural and moral persons must have authorization to enter the PNH each year to pay rights fixed by the Congress through the Federal Law of Rights. Likewise, visitors to the PNH tourism service product that is provided either by land or water should cover rights to enter the PNH. Legislation of the General Law of Ecological Equilibrium and Protection of the Environment 1988 (LGEEPA, 1988) mentions management programs, and in its amendment of 1996, in article 65, it is stated as a management program. In the ANP Regulation of 2000 in article 3, Fraction XI defines the management program as the guiding instrument of planning and regulation that establishes the activities, actions, and basic guidelines for the management and administration of the protected natural area respective.

It is referred to in the PNH Management Program in Rule 3, Paragraph 3, and Rule 7, Section I (PMPNH, 2003); when a vessel requests a permit for the nautical tourism service in the Harbor Master's Office of Bahías de Huatulco, as established in the Regulation of the Maritime Navigation and Commerce Law (LNCM, 2016) in the Title of Maritime Tourism, demands and conditions present the travel routes as well as the authorization of the PNH. This is a case of cooperation and coordination between the two authorities. However, it is only presented in the permit procedure for nautical tourism, so it does not exist in any another area in an efficient way.

In Mar. 2014, the PNH Directorate, in coordination and cooperation with the SEMAR, the Federal Environmental Protection Agency (PROFEPA), and municipal and state authorities, in an isolated and unique manner, evidencing an act of isolated authority, verified the payment of visitors' rights by waterway, boarding, and sailing in the dock of Santa Cruz. During the operation, it was confirmed that a *catamaran* with more than 200 tourists did not request the payment of rights, a condition that demanded immediate abandonment of the PNH as well as being subject to a sanction and the irreparable condition of visitors and tourists. It should be noted that from 2009 to 2015, only the operative described was known. This confirmed that the water service providers do not pay the authorization fees each year and do not require payment of fees to tourists who enter the PNH onboard their boats. It reiterates the lack of cooperation and coordination between authorities, providers of nautical service, and users to comply with the provisions of the law of the matter of ANP orders.

Finally, the Federal Consumer Prosecutor's Office (PROFECO), API, port, and municipality captaincy contain the complaints of users for affecting their rights as consumers of the nautical tourism service that systematically violates carriers, unfair price competition, simulation of services, and of low quality that jeopardizes navigation safety, the physical integrity of operators and users, and the environment.

Therefore, the Federal Commission of Economic Competition (Cofece), the municipality, the DGP, the API, and PROFECO established the clearest conditions of the service according to the season of the year and modality of the nautical tourism service. This gives us a dimension that the rules observed by the actors (permit holders, concessionaires, users) within the dock have implications, depending on the norm that the API is not observed with the support of the Port Authority. The DGP must execute compliance with the legal provisions that are stated.

The fulfillment can be in general or individual through agreement of wills that fixes the parts in the contracts of permits or authorization for the provision of nautical or port services. Also the partial assignments of rights, for the exploitation of spaces within the port with the foregoing, we can affirm that the providers of nautical services in the Santa Cruz Bahia must have permits that request the API to make observations and finally authorize the DGP, based on the concession title held by the API, as well as such as the Port Development Master Plan (PMDP), port operation rules, ports law, ports law regulation, federal economic competition law, as well as the tariffs and prices that are set. In the such occasions when the conditions established in the permit are not observed, the API may recommend through a meeting of the Operations Committee or Planning Committee that the DGP may request verification. The operations of the port, apperceive, infringe and punished through the Harbor Master. The DGP may also cancel the permit or authorization that had been provided for their service in the port as well as inform the Ministry of Finance and Public Credit (SHCP) of the nonpayment of fees generated for the consideration of the permit that was issued.

8 CONCLUSION

In relation to the study, we were able to locate the original space, the ways in which it was modified, after the declaration of common property in 1984, in the same year 40% was expropriated by the state to decree it as an Integrally Planned Center that would be justified the impulse of development poles, which in the 60's was the bet that had been imposed for development. This is a clear example of contradictions in public policies that are issued at the federal level. On the one hand, indigenous people are granted their land rights for communal goods and on the other hand, the lands are expropriated to implement the neoliberal model, expelling the people that had been given property rights days before.

The compensations that were established did not contemplate the new sources of work, more than those of the tourist center itself, but what it did contemplate was the deduction for debit of the nautical transports acquired under the model of communal goods, granting a quantity of money very different from what was officially promulgated.

In 1998 part of the polygon that the government removed from the settlers was decreed as the Huatulco National Park, which maintains a special regime for nautical tourism operations. This has been severely impacted by the absence of authorities at the three levels of government, adding those dependencies that are concentrated in the port of Bay of Huatulco.

The PNH was formed by the demand of another group of actors ignorant of local customs. With the banner of conservationism, ecologists again imposed a model of neoliberalism called protected natural areas, once again removing sources of work for the natives and establishing new ways of operating resources. At the end of 1999, FONATUR was granted the title of concession of the port area by the DGP, based on the amendment to the Law of Ports of 1993. This began the comings and goings of authorities where the responsibility of the administration of the port was diluted. Once the land expropriation is "controlled," it is passed to the control of the maritime part, the port regime and the protected natural area are originally designed to respond to endogenous conditions, in both areas, currently the authorities in charge of compliance at the Three levels of government, say, federal, state and municipal and the authorities that are concentrated in the port of Bay of Huatulco have not achieved an atmosphere of trust that allows the cooperation and coordination of the water service providers.

The operation of the port of Bahías de Huatulco is an example of how the concurrence of a myriad of authorities with duplication of functions is unable to coordinate the operation of a port with the highest federal and high-altitude categories, causing present cases such as not knowing exactly the number of vessels that interact in the port, the number of existing concessionaires, the payments made for services, the number of cruise ships that arrive at the port, etc. The constant modifications to the laws, to the attributions, and the scant vigilance to its fulfillment of rights and obligations maintain a peculiar operation in the port of Bahías de Huatulco. Authorities at the three levels of government and those that are concentrated in the port of Bahías de Huatulco show weakness and inability to reach agreements with the providers of nautical service and users.

The solution of the conflict is not found in the regulations that govern the port area and the protected natural area. As the space is spoilt due to distrust between the parties involved in the activity of nautical service, negotiations are not carried out because the representatives are illegitimate, and the meetings are not inclusive as they are full of private interests on the collectives. The wounds that still persist in this area, given their history of expropriation, continue to be open and the authorities at all levels have not been sensitive to this situation.

References

Competitiveness Agenda of the destination Bahías de Huatulco, 2013. Universidad del Mar, Sectur, Fonatur. http://www.umar.mx/boletin/pages/diciembre2013/UMARINO_nov-dic2013.pdf. (Accessed 20 February 2017).

CONANP, 2003. National Commission of Protected Areas. https://www.gob.mx/conanp (accessed 20 February 2017).

Consejo Nacional de Evaluación de la Política de Desarrollo Social (CONEVAL), 2010. Social Lagging Index 2010. Municipal Level and by Locality. http://www.coneval.org.mx/Medicion/IRS/Paginas/%C3%8Dndice-de-Rezago-social-2010.aspx. (Accessed 21 March 2017).

Da Jandra, L., 1991. Huatulqueños. Joaquín Mortiz, México.

Da Jandra, L., 1997. Samahua. Joaquin Mortiz, México.

De la Mora-De la Mora, G., Montaño Salazar, R., 2016. ¿Hacia la construcción de una gobernanza ambiental participativa? Estudio de caso en el Área Metropolitana de Guadalajara. Intersticios Sociales n. 11.

de Puertos, L, 2016. http://www.diputados.gob.mx/LeyesBiblio/pdf/65_191216.pdf. (Accessed 16 February 2017).

Diario Oficial de la Federación (DOF), 1984. Resolución sobre reconocimiento y titulación de bienes comunales del poblado denominado Santa María Huatulco, ubicado en el municipio del mismo nombre, Oax. http://dof.gob.mx/nota_detalle.php?codigo=4670457&fecha=28/05/1984&print=true. (Accessed 15 February 2017).

Diario Oficial de la Federación (DOF), 2002. Secretaría de Medio Ambiente y Recursos Naturales. In: Plan de Manejo del Área Natural Protegida con el carácter de Parque Nacional Huatulco.www.dof.gob.mx/nota_to_doc.php?codnota=715276. (Accessed 17 February 2017).

EarthCheck Certification, 2016. EarthCheck Certification. https://earthcheck.org/news/2016/august/huatulco-2do-destino-tur%C3%ADstico-a-nivel-mundial-con-la-certificaci%C3%B3n-earthcheck-platinum/. (Accessed 9 March 2017).

Héau Lambert, C., 2014. Reconstrucción de un territorio turístico mediante un Bricolage cultural. Cult. Represent. Soc. 8(16).

Instituto Nacional de Estadística y Geografía (INEGI), 2015. Encuesta Intercensal. http://www.beta.inegi.org.mx/proyectos/enchogares/especiales/intercensal/. (Accessed 15 February 2017).

Instituto Nacional de Estadística y Geografía (INEGI), 2017. Catálogo de entidades federativas, municipios y localidades. http://www.inegi.org.mx/est/contenidos/proyectos/aspectosmetodologicos/clasificadoresycatalogos/catalogo_entidades.aspx. (Accessed 16 February 2017).

Jardón-Medina, A.G., Martínez-Rodríguez, M.C., Alvarado-Cardona, M., 2017. De las políticas públicas a la gobernanza ambiental. In: Martínez-Rodríguez, M.C. (Ed.), Políticas Públicas ambientales. Ed. Colofón, Mexico City.

Ley de Navegación Comercio Maritimos (LNCM), 2016. www.diputados.gob.mx/LeyesBiblio/pdf/LNCM_191216.pdf (accessed 16 February 2017).

Ley General de Bienes Nacionales (LGBN), 2016. http://www.diputados.gob.mx/LeyesBiblio/pdf/267_010616.pdf. (Accessed 16 February 2017).

Ley General del Equilibrio Ecológico y la Protección al Ambiente (LGEEPA), 1988. http://www.salud.gob.mx/unidades/cdi/nom/compi/l280188.html. (Accessed 17 February 2017).

López, V., 2010. La reorientación en los destinos litorales planificados. In: Caso de estudio: Bahías de Huatulco. Universidad de Alicante, Oaxaca Master thesis http://rua.ua.es/dspace/handle/10045/14977. (Accessed 19 February 2017).

Lorena-Rodiles, S., López-Guevara, V., 2015. Pesca tradicional y desarrollo turístico en Bahías de Huatulco. Una lectura desde la historia oral de los pescadores locales. Invest. Turísticas 10, 150–169.

Martínez-Rodríguez, M.C., 2015. Gobernanza Ambiental. Orígenes y Estudios de Caso, first ed. Plaza y Valdés, Mexico City.

Mendoza, M., Monterrubio, J., Fernández, M., 2011. Impactos sociales del turismo en el Centro Integralmente Planeado (CIP) Bahías de Huatulco, México. Gestión Turística (15), 47–73.

Pacheco, P., 2017. Vigilará Semar las Bahías de Huatulco. http://www.nvinoticias.com/nota/50171/vigila-semar-las-bahias-de-huatulco-por-repunte-de-crimenes. (Accessed 30 March 2017).

Paré, L., Fuentes, T., 2007. Gobernanza ambiental y políticas públicas en Áreas Naturales Protegidas: lecciones desde Los Tuxlas. UNAM, Instituto de Investigaciones Sociales, Mexico.

Periódico Oficial Oaxaca, 2013a. Bando de Policia y Buen Gobierno del Municipio de Santa María Huatulco, Pochutla Oaxaca. Tomo XCV. Oaxaca de Juárez, Oaxaca. www.periodicooficial.oaxaca.gob.mx/files/2013/03/SEC09-04TA-2013-03-02.pdf. (Accessed 30 March 2017).

Periódico Oficial Oaxaca, 2013b. Reglamento de Turismo para el Municipio de Santa María Huatulco, Pochutla Oaxaca. Tomo XCV, Oaxaca de Juárez, Oaxaca. www.periodicooficial.oaxaca.gob.mx/files/2013/03/SEC09-03RA-2013-03-02.pdf. (Accessed 30 March 2017).

Piñeiro, D.E., 2004. Movimientos sociales, gobernanza ambienal y desarrollo territorial rural. Departamento de Sociología. Facultad de Ciencias Sociales. Universidad de La República, Uruguay RIMISP.

Political Constitution of the Mexican United States (CPEUM), 2017. http://www.diputados.gob.mx/LeyesBiblio/pdf/1_150917.pdf. (Accessed 22 March 2017).

Programa de Manejo Parque Nacional Huatulco (PMPNH), 2003. Comisión Nacional de Áreas Naturales Protegidas (CONANP). http://www.conanp.gob.mx/que_hacemos/pdf/programas_manejo/huatulco.pdf. (Accessed 29 March 2017).

Programa Maestro de Desarrollo Portuario (PMDP), 2015. Programa Maestro de Desarrollo Portuario 2006–2015. Administración Portuaria Integral Bahías de Huatulco, Fonatur.http://www.sct.gob.mx/fileadmin/CGPMM/PNDP2008/doc/pms/pmdp/hua.pdf. (Accessed 29 March 2017).

Programa Maestro de Desarrollo Portuario (PMDP), 2017. Programa Maestro de Desarrollo Portuario de Bahías de Huatulco y de la ZFM de la Marina Turística Chahué, Oaxaca. 2012–2017. Fonatur.http://www.fonaturoperadoraportuaria.gob.mx/APIS/Documentos/hua/Portada_30ene13.pdf. (Accessed 30 March 2017).

Ramírez-González, A., 2005. Las bahías de Huatulco, Oaxaca, México: ensayo geográfico-ecológico. Cienc. Mar. 25, 3–20.

Rodríguez, Ó., 2013. Recibe certificación internacional playa en Huatulco. http://www.milenio.com/estados/Recibe-certificado-internacional-playa-Huatulco_0_105589515.html. (Accessed 14 January 2017).

Talledos, E., 2012. La imposición de un espacio: De la Crucecita a Bahías de Huatulco. Rev. Mex. Cienc. Polít. Soc. 216, 119–142.

CHAPTER

5

Sustainable Coastal Management for Social-Ecological Systems—A Typology Approach in Indonesia

Bernhard Glaeser

German Society for Human Ecology (DGH) and Berlin Free University (FUB), Berlin, Germany

1 INTRODUCTION AND SCOPE

Coasts and oceans have gained ever more importance during recent decades. Roughly two-thirds of the human population live, work, and produce on coasts. Oceans and coasts are major sources of food, minerals, and other resources and services. Oceans are the "unknown planet" where a census of marine life was launched. Coasts and oceans represent a maximum of biodiversity. They also represent political and economic vested interests, which produce conflicts (Knowlton, 2010).

Economic costs and social hardships induced by global change, such as climate change, appear at the local level. The terms "level" and "scale" are often used synonymously. For clarity reasons, it was suggested to distinguish between scale and level (Gibson et al., 2000), with *scales* being "the spatial, temporal, quantitative, or analytical dimensions used to measure and study any phenomenon" and *levels* "the units of analysis that are located at different positions on a scale." Scaling problems as an umbrella term relate to issues surrounding both scale and/or level. Perry and Ommer (2003: 513) discussed "characteristic spatial, temporal, and organizational scales in marine ecosystems and human interactions," noticing "the difficulties inherent in their cross-disciplinary application." They suggested that the "essential task is to discover how to combine social and natural science scale analyses to understand the impact of natural systems on people and the impact of people on natural systems." Research should focus on case studies that clarified "the need to manage marine resources in such a way as to encompass global to local scales." This contribution attempts to pursue such an approach, that is, to include the "local" and the "global" in order to arrive at a coastal typology that is based on regional case studies.

https://doi.org/10.1016/B978-0-12-810473-6.00006-6
© 2019 Elsevier Inc. All rights reserved.

Important political decisions, activities, and measures usually occur at higher levels, mostly national. Internationally, an uneven distribution of interests, benefits, and costs relating to climate change is evident. Climate impacts, poverty, and social justice are interlinked across spatially nested, hierarchical levels. The uneven distributions of wealth can be understood as nested hierarchies that are reproduced at the various levels of the socially and ecologically organized global system, beginning at the local up to the global level (Glaeser, 2016).

2 A REGIONAL EXAMPLE FROM SOUTHEAST ASIA: THE SPERMONDE ARCHIPELAGO IN INDONESIA

An Indonesian case study is presented to show the coincidence of and the link between social and ecological systems. Indonesia is located in Southeast Asia, the world's largest archipelagic country (Fig. 1). Its unsustainable use of marine resources poses threats to its coastal and marine ecosystems and its social systems. A severe decline in marine fisheries has been predicted for the decades to come, with disastrous impacts on the livelihoods of coastal populations (Glaeser, 2016; Glaeser et al., 2018).

The Spermonde Archipelago is part of the province of South Sulawesi, off the coast of the old trade and merchant city of Makassar, with a population of about 1.8 million. The archipelago covers an area of approximately 2500 km^2, consisting of about 67 low-lying coral atoll islands, 54 of which are inhabited. The islands' inhabitants are almost exclusively dependent on marine resources for their livelihood (Fig. 2).

While some islands have been inhabited for several centuries, others were settled more recently, for example, during the Japanese occupation in World War II. All islands have seen considerable population growth during the past 50–60 years. The predominant activity in the area is fishing, which occurs in a wide variety of different types using multiple fishing gear and targeting numerous fish species. However, a wide range of species is harvested in an

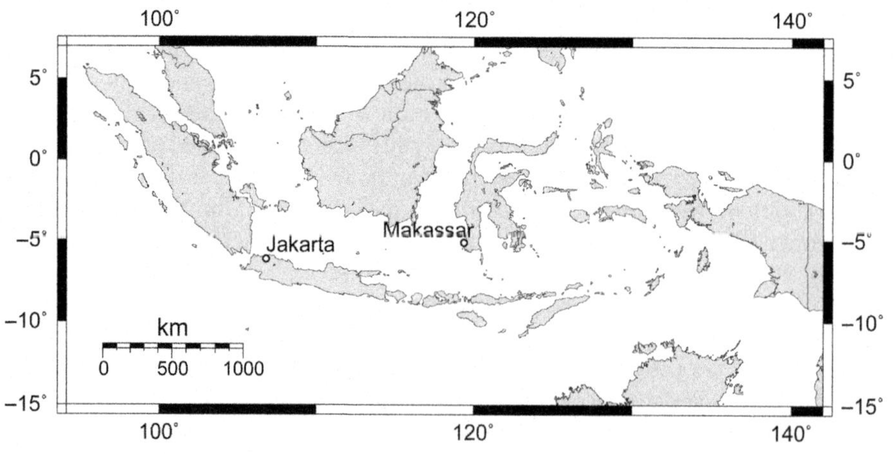

FIG. 1 The study area, the Spermonde Archipelago, is located west of Makassar, South Sulawesi, at the center. *Map: Sebastian Ferse.*

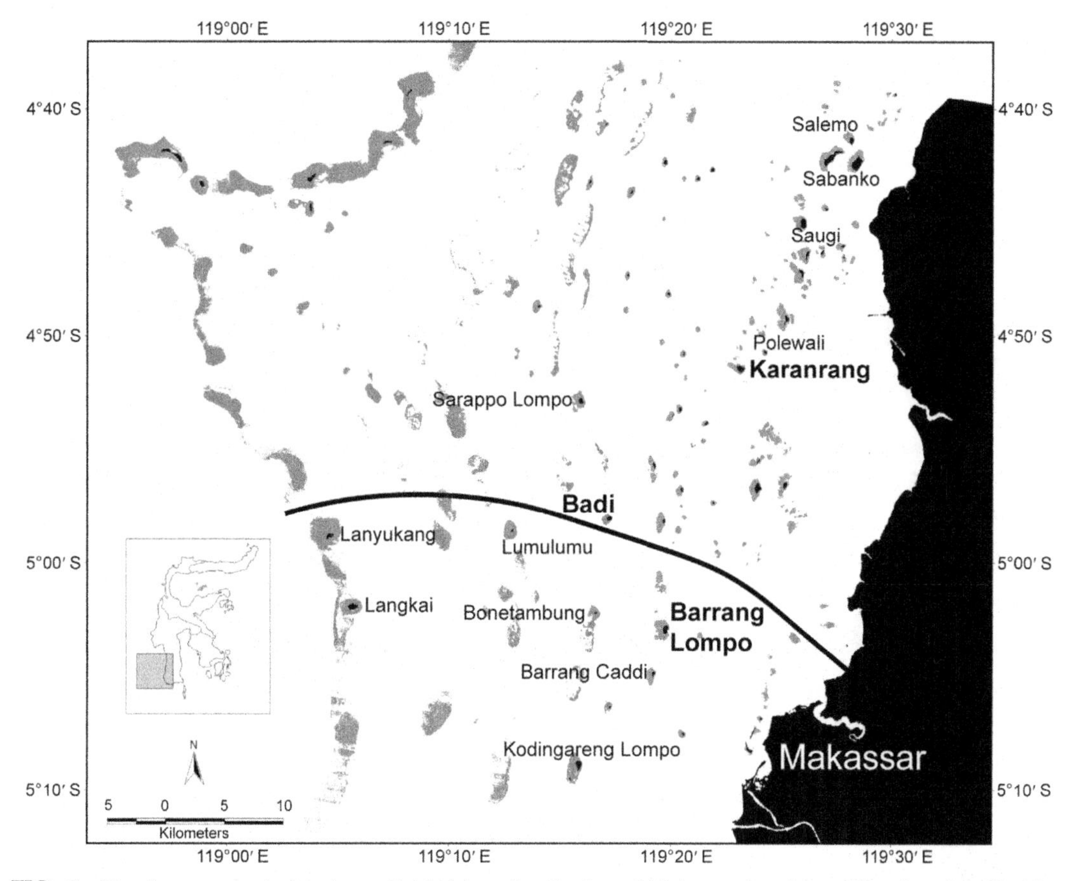

FIG. 2 The Spermonde Archipelago, divided into the districts of Makassar (south) and Pangkep (north). *Map: Sebastian Ferse.*

unsustainable and even destructive manner, leading to the degradation of marine ecosystems in the area (Figs. 3 and 4; Glaeser et al., 2018).

The Spermonde case shows in an exemplary fashion how perceived or felt global climate and environmental changes produce economic, social, and cultural dimensions of change at the regional and local levels. Fig. 5 illustrates nested social and ecological hierarchies from the local to the global in a systemic view. The spatial scale adds a third dimension (Glaeser and Glaser 2010, 2011).

According to local respondents, climate change is felt in the islands. Increased ocean temperature, rising sea levels, acidification, and coral bleaching attributed to climate change are affecting not only the region's fragile ecosystems and species, but also the human communities that rely on fishing for their livelihoods. Ecological problems and resource shortages are first noticed locally. Fish stocks and their diversity diminish. Fishing becomes riskier and more time intensive. Fishers leave the coast and move toward the open sea. Coasts and beaches face erosion, streets are destroyed, and near-shore houses are washed away. Wave breakers of varying quality have been installed to protect beaches and coasts (Figs. 6–8; Glaeser 2016; Glaeser et al., 2018).

FIG. 3 Most Spermonde islands need an extended pier to be accessed: Lumulumu Island. *Photo: Bernhard Glaeser.*

FIG. 4 In lack of building materials, islanders mine their coral reef for island protection: Lumulumu Island. *Photo: Bernhard Glaeser.*

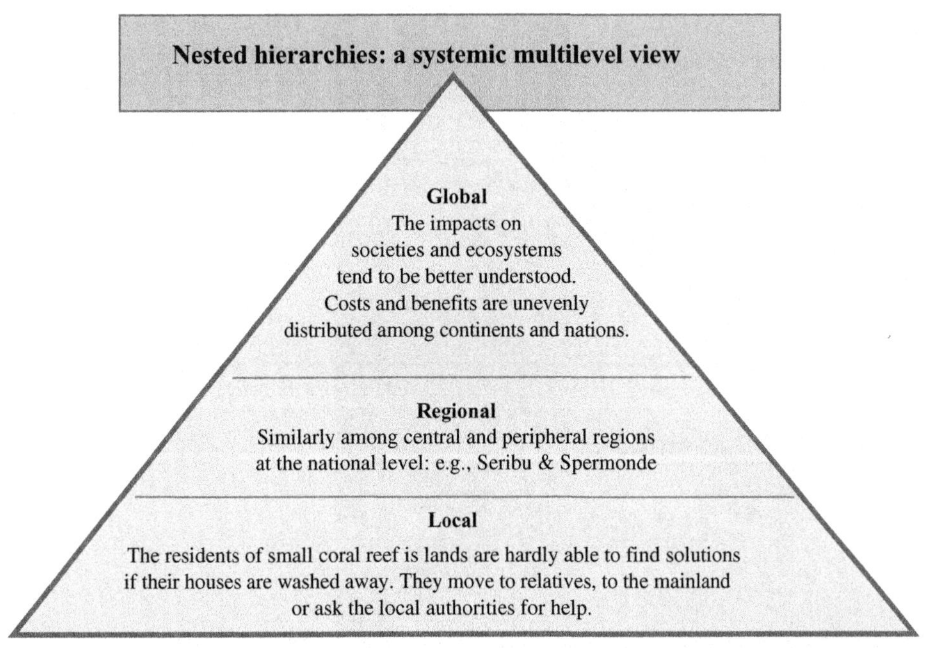

FIG. 5 Nested hierarchies—a systemic multilevel view. *Modified from Glaeser, B., 2015. Klimawandel und Küsten – Humanökologisch-systemisch betrachtet am Beispiel Indonesien. In: K.H. Simon, K.H., Tretter, F. (Eds.), Systemtheorie und Humanökologie, Positionsbestimmungen in Theorie und Praxis. Edition Humanökologie. Vol. 9. oekom, Munich, pp. 316–336.*

3 COMBINING SOCIAL AND ECOLOGICAL FACTORS

In the following, I integrate problem-focused social-ecological systems and regional analysis. The shelf area, the basis of the ecological system, is confined to the east by the Sulawesi mainland and to the west by the deep waters of the Makassar Strait. Habitats consist of coral reefs, seagrass beds, and a few mangroves along the shore. Most mangroves have been destroyed for shrimp and fish ponds, and reefs have been declining in coral cover as well as associated fauna over the past four decades.

It is most difficult to define the extension of the social system. Regarding economic aspects and depending on what activity is considered, the "system" extends over a very large geographic area. Fishing, fish processing, and fish trading are the major occupational activities. Around 80% of the population depends directly on fishing. Approximately 70% of the fishery operations are small-scale (one to four people per boat), with the remainder of the fishers engaged in medium-scale operations (purse seining and mobile lift-net fishery with 10–20 people per boat) (Glaeser et al., 2018).

The focus is on a regionally based social-ecological systems analysis that links to global drivers and global change in order to build the foundations of a coastal and marine typology by comparing an undefined number of local and regional case studies. The case studies vary

FIG. 6 Beaches are eroding, and trees are being derooted: Lanyukan Island. *Photo: Bernhard Glaeser.*

according to problem focus, social agents, and geographical location, including different climate zones or ecozones. Comparative case studies appear to provide a reasonable approach to reconcile a concept-driven agenda with an empirical baseline, a combined top down and bottom up approach.

The goal is to use a "global sustainability research matrix" (Glaser and Glaeser, 2014) to arrive at a coastal and marine typology that is based on problem types and ecozones and combines descriptions, responses, and appraisals for the natural, social, and governing systems. As a reality check, I use an exemplary tropical case study from Indonesia. Its specific generic problems are climate change, storm surges, overfishing, and coastal poverty. Global problems are directly linked to the stakeholder regional and local agendas (Glaeser, 2016).

FIG. 7 Public toilets were washed away: Bone Tambung Island. *Photo: Bernhard Glaeser.*

FIG. 8 Large wave breakers were installed as a protecting harbor to host fishing boats: Polewali Island. *Photo: Bernhard Glaeser.*

4 CONSTRUCTING A COASTAL TYPOLOGY

4.1 Two-Dimensional Research-Based Hierarchical Typology

By employing a comparative analysis, I create links across hierarchical levels on the one hand (*vertical analysis*), and across geographical areas or regions on the other hand (*horizontal analysis*) to construct a two-dimensional matrix as the conceptual frame for a research-based hierarchical typology. A problem focus is needed as suggested above, as there can be no general (unspecific) definition for a social-ecological system. A social-ecological system definition for comparative purposes is feasible if and only if limiting boundaries are set and the and the problem focus is defined, the territorial and climate conditions plus scaling are included, and an interdisciplinary and natural and social science approach is used (Glaeser, 2016: 375).

At the national level, a political and administrative response to the felt pressure may link the global and local levels. It is, finally, at the regional and local levels where the ecological and social impacts of global change are encountered. This is the place for political and administrative measures or for responses by the affected population itself. The responses to hazardous or threatening situations reach from passive adaptive strategies, such as migrating, to proactive responses, such as coastal protection, change of professions, or educational efforts (Table 1).

4.2 Two-Dimensional Research-Based Policy Typology

In a second step, I search for interfaces and feedback loops between global change and local dynamics in an interdisciplinary mode, in interplay with levels of action. Eventually, all global change is induced and produced locally. For the double feedback (global to local

TABLE 1 Two-Dimensional Research-Based Hierarchical Typology: The Example of Indonesia

Social-Ecological Multilevel Typology for Coastal and Marine Systems: **With Respect to-Specific Features Under the Condition of Climate Change Indonesia (SPICE Case Study) as a Tropical Zone Example**			
	Ecozones/Climate Zones		
Spatial Scale /Level	**Temperate Social Ecological**	**Tropical/Subtropical Social Ecological**	**Polar Social Ecological**
Global		*Global* climate and environmental change, linked to the economic, social, and cultural dimension of change	
National		*National* (e.g., Indonesian) coastal, ocean, and environmental policies, linking ocean developments and coastal threats to climate change	
Regional and Local		*Local* (e.g., Sulawesi) coastal developments and livelihoods *Social:* Earning opportunities for women, including aquaculture (sea cucumbers, fish cages, algae cultivation) *Ecological:* Beach protection, mangrove plantation, and reef management and rehabilitation to reduce beach erosion	

Based on Glaeser, B., 2015. Klimawandel und Küsten – Humanökologisch-systemisch betrachtet am Beispiel Indonesien. In: K.H. Simon, K.H., Tretter, F. (Eds.), Systemtheorie und Humanökologie, Positionsbestimmungen in Theorie und Praxis. Edition Humanökologie. Vol. 9. oekom, Munich, pp. 316–336.

and vice versa), once again, a comparative analysis is needed. This time, on the one hand, it's *across the policy cycle of actions and reactions taken*, including their different levels, scales, and units (vertical approach), and on the other hand, it's across the various types of social-ecological systems, typified study areas, and regions, such as urban metropolitan areas, river mouths/catchments, small islands, and open seas (horizontal approach) (Glaeser, 2016: 375).

As a comparative analysis using both a vertical and a horizontal approach, I develop a second roadmap that leads to different social-ecological coastal and marine typology, including different focal areas. Once again, typology includes ecological and social factors and is oriented toward governance features. This second variant, however, does not proceed in a hierarchical multilevel order. Instead, it follows a systems-type sequence of actions and reactions whereby either society or nature can be the actors. The social-ecological zones are refined and custom-made to fit the regional-local level. The emerging cells represent social-ecological systems and behave dynamically (Table 2).

5 A GOVERNANCE BASELINE AND INDICATORS

The comparative approach at the regional and local case study level needs further specification by means of the involved subsystems. The typological matrices provide or will provide comparative data for the three social-ecological subsystems: natural, socioeconomic, and governance. They also incorporate different approaches and tools: various quantitative and qualitative methods, including multivariate analysis or fuzzy logic (Cf. the interrelated methodologies in Bailey, 1994: 66–76). Values and norms play a role when it comes to identifying goals and objectives. System-specific issues need to be identified. Systems boundaries vary from subsystem to subsystem and may overlap. Levels may range from local to global (Table 3).

Changes involve a time horizon whereas the mere descriptive social and ecological features do not. Changes call for, or have triggered, responses by the governance system. The governance system consists of three elements: (1) government/administration at different levels, (2) markets at different levels, and (3) civil society at the national or subnational level. The responses have happened or happen over time. This means that the nature of the changes and the response cells is different from that of the descriptive social and ecological cells; they are dynamic and not static. Timelines need to be developed. The example is taken from the above Indonesian case study (Table 3).

The term *governance* is used in contemporary contexts within several social sciences, particularly in economics and political science. The concept is quite versatile but usually refers to the threefold exercise of power: to actors, persons, or departments that constitute a body for administering purposes; to the act of governing or exercising authority; and finally to the means used to govern, such as rules or regulations. *Political governance* is often connected to notions of good governance, which includes multilevel participation, legitimacy, accountability, and transparency. *Coastal governance* is about the creation and change of the legal and institutional framework for coastal management, that is, to influence objectives, policies, laws, or institutions. *Coastal management* implements objectives, policies, and laws provided by coastal governance at a specific site or in a specific region or country. It deals with competing issues and attempts to resolve conflicts among stakeholders, including the local

TABLE 2 Two-Dimensional Research-Based Policy Typology: The Example of Indonesia

Social-Ecological Policy Cycle for Coastal and Marine Systems: **At the Regional Level With Respect to Climate Change Spermonde Archipelago, Indonesia (SPICE Case Study) as a Small Island Example**				
	Social-Ecological System (SES) Types			
Policy Cycle	***Urban Areas*** **(Settlements, Ports)**	***River-Mouth Systems*** **(Estuaries, Deltas, Lagoons)**	***Small Islands: Fisheries*** **(Life, Ornamentals),** ***Aquaculture*** **(Finfish, Shells, Sea Cucumber),** ***Resource Use*** **(e.g., Corals),** ***Conservation*** **(e.g., No-Take Zones),** ***Tourism*** **(National, International)**	***Ocean*** **(Open Sea)**
Governing system (as related to the natural system)			*National*: Indonesia, the world's largest archipelagic country, divided into decentralized provinces *Regional-local*: Spermonde Archipelago with two major administrative units, Makassar Municipality and Pangkep Regency, both of which extend beyond the geographical boundaries of the natural system	
Drivers, Pressures (Anthropogenic and natural)			*Global* climate and environmental change, linked to the economic, social, and cultural dimension of change: *Regional-local*: Storm surges, overfishing, destructive fishing methods	
Changes, Impacts			*Local* (e.g., Sulawesi small islands) coastal and marine developments and livelihoods: – Reef destruction – Fish depletion – Coastal erosion – Collapsing houses	
Responses			*National* (e.g., Indonesian) coastal, ocean, and environmental policies, linking ocean developments and coastal threats to climate change *Local* (e.g., Sulawesi islands) *Social:* Income opportunities for women, including aquaculture (sea cucumbers, fish cages, algae cultivation) *Ecological:* Beach protection, mangrove plantation, and reef management and rehabilitation to reduce beach erosion	
Appraisal (Outputs, Outcomes)			*Social:* Number of income opportunities achieved, positive and negative side effects *Ecological:* Amount of reef rehabilitation achieved, effects for fish stocks and biodiversity	

Modified from Glaeser, B., 2015. Klimawandel und Küsten – Humanökologisch-systemisch betrachtet am Beispiel Indonesien. In: K.H. Simon, K.H., Tretter, F. (Eds.), Systemtheorie und Humanökologie, Positionsbestimmungen in Theorie und Praxis. Edition Humanökologie. Vol. 9. oekom, Munich, pp. 316–336.

TABLE 3 Comparative Subsystems Typology: The Example Indonesia

Systems Features	Natural System	Social System	Governance System
Objectives	Stock preservation	Decent livelihood	Equitable regulations
System specific issues	Climate change	Overfishing	Regulations not implemented
Scale: system boundary	Ecological boundaries	Kinship, trade, market boundaries	Administrative boundary
Scale overlaps	Fishing	Markets	Rules/regulations
Level	Local to international	Local to global	Regional to national
Change: major changes/impacts	Storm surges and beach erosion	A patron-client relation evolved among traders and fishermen	Change of administration, markets, civil society
Responses: governance subsystem (government, market, civil society)	Government regional to national, Markets no, Fisher community partly	Government little effort, Markets local to global, Fisher community and patron-client relation	Government national, Markets no, NGOs national and international
Appraisal: outputs, outcomes, ensuing goals and visions	Enabling conditions, changed behavior, achievements, sustain. development	Enabling conditions, changed behavior, achievements, sustain. development	Enabling conditions, changed behavior, achievements, sustain. development

Based on Glaeser, B., 2016. From global sustainability research matrix to typology: a tool to analyze coastal and marine social-ecological systems. Reg. Environ. Change. 16 (2), 367–383. doi:10.1007/s10113-015-0817-y.

population, fishermen, involved scientists, NGOs, or local/regional and national governments. Multiple competing issues are tourism, fisheries, aquaculture, harbor development, nature protection, offshore mining, and windpower. The most urgent issues need to be identified (Glaeser, 2006).

It is timely to separate synchronic/static and diachronic (temporal)/dynamic elements and isolate descriptive elements from change, response, and appraisal. Governance responds to social and/or ecological changes in an attempt to attain certain societal goals and objectives, such as sustainable development. In other words, desired outcomes (visions) determine responses, which include a timeline perspective. Any governance baseline (R1...n) complements descriptive biophysical and socioeconomic characteristics in an interdisciplinary and possibly transdisciplinary approach. A policy cycle—or management cycle at the local level—moves from descriptive to prescriptive elements, evaluates their outputs and outcomes, and assesses a new SES state, which calls for new responses according to a prescribed or preset normative societal goal (Glaeser, 2016: 376f).

The appraisal typology (Table 4) evaluates responses and achievements of measures taken by the governance system, first by local and regional level civil society measures at the village, town, or district level. Government actions at the provincial and national level would then also have to be included. Markets for fish and other resources at different levels will certainly play a role and influence outputs and outcomes. In all three governance segments (government,

TABLE 4 Subsystems Appraisal Typology: The Example of Indonesia

Climate Zone, Ecoregion, SES			Outputs and Outcomes at the Case Study Level: Across the Natural, Social and Governance System				
Climate Zone	**Eco-System Type**	**SES Type**	**Outputs (Immediate Results of Responses)**	**Outcome: First Order (Enabling Conditions)**	**Outcome: Second Order (Behavior Change)**	**Outcome: Third Order (Visions Achieved)**	**Outcome: Fourth Order (Sustainable Development)**
Polar	Shelf	Estuary					
Tropical	Coastal	Island	Local people construct wave breakers, use coral stone as building material, which led to a stop of government support. Governmental measures: Dams have been constructed to reduce waves, to prevent coastal erosion, and to establish harbors for fishing boats.	Fishers develop intricate patron-client links and informal rules on resource use. The patron-client system provides fishers with an alternative type of social security, allowing them to cope with short-term stressors (storms, loss of gear, localized overfishing). However, capacity to cope with more extensive stressors (complete loss of ecosystem services: fisheries resources, protection from erosion) appears limited. Few alternative livelihood options. low level of education (Glaser et al., 2010)	Locally enforced rules for a specified but small sea territory surrounding the respective islands in the whole archipelago. These regulations provide rules-in-use for marine resource use, creating a polycentric structure for area-based marine governance. However, their formation and enforcement depend on local leaders, which—in conjunction with the missing horizontal connectivity—creates vulnerability.	Informal governance system: NGOs are advocating for no-take zones, introduce (1) new technologies (coral management, mariculture cultivation of sea weeds), (2) credits to empower self-employment in mobile restaurants (soups, snacks, drinks) or shops (drinks, snacks, tools). Introduction of COREMAP (Coral Reef Rehabilitation and Management) program) to improve environmental awareness, alternative livelihood options, establish community-based MPAs (Ferse et al., 2014).	Not yet achieved
Temperate	Open Sea	Urban					

SES, social-ecological systems.

Design: Bernhard Glaeser.

From Glaeser, B., Ferse, S., Gorris, P., 2013. Case Study Indonesia. Spermonde Archipelago: Island development and livelihoods. Template for the description of ADApT_A case studies. IMBER HDWG: Version 2013-05-17: F. Capacity-Response-Appraisal; Ferse, S.C.A., Glaser, M., Neil, M., Schwerdtner Mánez, K., 2014. To cope or to sustain? Eroding long-term sustrainability in an Indonesian coral reef fishery. In: Glaser, M., Glaeser, B. (Eds.), Linking Regional Dynamics in Coastal and Marine Social-Ecological Systems to Global Sustainability. Reg. Environ. Change 14 (6), 2127–2138. (special issue); Glaser, M., Baitoningsih, W., Ferse, S.C.A., Neil, M., Deswandi, R., 2010. Whose sustainability? Top-down participation and emergent rules in marine protected area management in Indonesia. Mar. Policy 34, 2053–2066.

market, civil society), appropriateness, effectiveness, and acceptability of outputs and outcomes are appraised and valued (Glaeser, 2016: 377).

The effects of the responses (*appraisal*) that differentiate between *outputs* and *outcomes* are monitored and evaluated. They are also time-dependent unless a specific point in time is set and defines the appraisal at a certain moment along the time scale: $T_1 = T_{x \ldots y} + Tn$. Responses refer to the three governance subsystems: government, market, and civil society. Whereas an output is ascribed to the immediate result of a response, outcomes refer to its social and ecological consequences (Table 4).

Olsen differentiates between four orders of outcomes: enabling conditions for sustained implementation, implementation through changed behavior, the "harvest" (environmental and societal outcomes achieved), and finally sustainable development, that is, "optimal equilibrium between environment and society" (Olsen et al., 2009: 34). The case studies, first described by their respective social-ecological setting within different climate zones/ecoregions and secondly compared according to their issue and problem focus (Tables 1–3), need to be monitored and appraised according to outputs and different level outcomes across the natural, social, and governance system (Table 4).

The achievements and their acceptability may be judged differently by different groups of stakeholders: by local affected people, varying according to profession, age, or gender, by local to national administrators, politicians, or scientists. Different stakeholders will come up with varying recommendations, which may include measures for adaptation or mitigation. A new social-ecological "state" can be assessed and described in the natural, social, and governance subsystems. The new description is followed by new responses. The whole process, which is a policy and/or management cycle, starts all over again and turns to a second round, like an upward winding spiral (Cf. Olsen et al., 2009: 31–32; GESAMP, 1996).

In order to compare the selected case studies throughout the entire policy cycle on an empirically grounded basis, a limited number of indicators needs to be selected. The indicators need to be sufficient in number as to fill the typology cells and to correspond to the magnitude of type differentiation. At the same time, it may even be more important to limit the indicators selected to the absolute minimum to arrive at a limited number of types or classes, which can still be understood and compared. The "garbage can approach" (Buddemeier) to select variables or variable indicators may yield an empirically profound picture of a multitude of social-ecological systems but will fail to classify and compare variegated case studies. Glaser et al. emphasized that key status and process indicators need to be regularly monitored to assess the sustainability of coastal and marine social-ecological systems (Glaser et al., 2012: 302–303, Tables 1 and 2). These are the minimum yet not to be proliferated indicator requirements.

Social-ecological *indicators* to construct case study typologies consist of (1) *focal areas*, namely climate zone (tropical/subtropical, temperate, polar), ecosystem type (urban, estuaries, coasts/small islands, shelf/open sea), and the main issue (overfishing, climate change, mass mortality); (2) *stressors (drivers, pressures):* natural pressures, social drivers, governance drivers; (3) *Natural system (state, impact):* system boundaries, scale issues, changes in habitat (impact on biotope and biocoenosis); (4) *Social system (state, impact):* system boundaries, scale issues, changes in demography and livelihood (impact); (5) *Governance system (state, impact):* system boundaries, scale issues, changes in governance and power relations (impact); (6) *Adaptive capacity, response:* natural system, social system, governance system; and (7) *Appraisal (outputs, outcomes):* natural system, social system, governance system (Glaeser, 2016: 378f).

The DPSIR framework (Driver-Pressure-State-Impact-Response) is a widely used tool to structure information by simplifying the relations when analyzing social-ecological systems in an integrated way. It is not a modeling tool, but conceptually links causes (drivers and pressures) to environmental effects (state and impacts) and social activities (response, policies, and decisions) (among others, see Bowen and Riley, 2003; MA, 2003; Burkhard and Müller, 2008). It may be noted that the classic DPSIR framework was amended by an *appraisal* category. "Appraisal" measures or evaluates the outputs and outcomes of a response that eventually leads to a new state.

6 CREATING A POLICY SCENARIO AND FUTURE STEPS

Coastal and marine sustainability is the imperative goal that drives our research and determines the mehodology selected. The worldwide near-collapse of many marine fisheries impoverishes local communities all over the globe at the local level as well as at a global scale. These unfortunate developments may be taken as paradigms for local environmental problems that increasingly reach global social dimensions. The impulses that fuel this development as well as the associated perceptions and knowledge generation processes develop from the global to the local level in a top-down manner as well as from the local to the global in a bottom-up manner (Glaeser, 2016: 381).

Let us assume that we aim at "responses" to create a policy scenario. In this case, we need to identify the issues affecting pressures and perceived problems. The issues vary, and they are natural-environmental, sociocultural, and governance related, including climate change, fish depletion, illegal fishing, and a multitude of environmental hazards. The responses refer to administrations, the markets, and civil societies, all divided into regional, national, and international levels. At the international level, multinational and crossnational development aid groupings and initiatives, and political foundations or church-related organizations represent aspects of an international civil society. They have been playing an ever-increasing role, yet were controlled more recently and constrained by national governments (Table 5).

The next step will be to collect additional regional case studies and to evaluate them in a comparative way. The comparative analysis will be done by means of the tested typology concept that is applied to the social-ecological subsystems–natural, social and governance–and to each step of the complete social-ecological policy cycle. The elements of the policy cycle are drivers/pressures, state/impact, adaptation/response, and evaluation/appraisal of outputs and outcomes. They represent a process, that may induce a new secondary policy cycle initiated by the appraisal result to become the new driver. The whole process will serve as a governance tool (Glaeser, 2016: 381; Bundy et al., 2015).

7 SUMMARY AND CONCLUSION

Local hazards are tied to global change—socially, economically, and ecologically. A coastal typology at different levels and with differentiated variables can be useful as an analytical

TABLE 5 Comparative Response Typology to Create a Policy Scenario

Issues	Government Regional-National-International	Market Regional-National-International	Civil Society Regional-National-International
Climate change – Ocean acidification – sea level rise – sea temperature	International agreements		
Environmental hazards – Ocean debris – Marine contaminants – Mass mortality – subsea level noise – Storm surges – Reef destruction – Coastal erosion – Mangrove decrease	National policies and regional management		International organizations interfere (e.g., Greenpeace)
Overfishing – Near-shore – Offshore		Market regulations through environmental marketing	Regional informal fisheries agreements
Regulations not implemented – Harvest quotas – Illegal fishing – Ban on destructive fishing	National and international enforcement		Social pressure on wrongdoers
Cultural identity – Poverty – Social justice – Employment – Women's involvement – Settlement density		Investments, small enterprises, and shops, Fair Trade	Informal rules and subsidies by regional communities

Design: Bernhard Glaeser.

instrument, but also as a planning tool. It is to be integrated into a common frame and applied to case studies by means of a multilevel and a cross-scale analysis.

This contribution proposed to study coastal and marine social-ecological environments at different scales on different levels. This was exemplified by a case study from Indonesia. The Indonesian case showed paradigmatically how perceived or felt global climate and environmental changes produce economic, social, and cultural dimensions of change. Scales are the space, time, or governance-related dimensions. Levels are located at different positions on a scale and include the global level as well as local to regional level comparative case studies. Social-ecological sustainability problems are caused by drivers from multiple levels of the Earth's system. A coastal and marine typology was presented to understand and compare the ecological, sociocultural, and economic aspects of coastal and marine systems.

Reviewing the typologies presented, it becomes apparent that nested hierarchies organize the number of variables in a hierarchical mode. They form different scenarios depending on selected objectives. In that sense, typology is a tool that takes different shapes according to different issues or responses. It does not—by any means—represent a fixed structure. The typologies presented started from concepts, but were empirically backed up. Conceptual versus empirical typologies may be dichotomous but not contradictory. One could also start empirically and cluster up to arrive at an equally valid result.

Distributional imbalances at different levels can be identified as climate change impacts: the "climate divide." On that basis, science should be linked to policy. Such a process involves different levels again, on a governance scale: from local management decisions to national law-making and international to global governance production and producers, such as the European Union or the United Nations.

Acknowledgments

This paper follows up on discussions and joint publications with Marion Glaser (LOICZ and ZMT Bremen) as well as on a submitted publication in cooperation with Sebastian Ferse (ZMT Bremen) and Philipp Gorris (Osnabrück University). I am grateful for discussions and joint work on the typology theme in our IMBER working group on human dimensions (IMBER-HDWG), specifically to Alida Bundy, Ian Perry, Patrice Guillotreau, and Ratana Chuenpagdee.

References

Bailey, K.D., 1994. Typologies and Taxonomies: An Introduction to Classification Techniques (Sage Universities paper series on quantitative application in the social sciences. Series no. 07/102). Sage, Thousand Oaks/London/New Delhi.

Bowen, R.E., Riley, C., 2003. Socio-economic indicators and integrated coastal management. Ocean Coast. Manage. 46, 299–312.

Bundy, A., Chuenpagdee, R., Cooley, S.R., Defeo, O., Glaeser, B., Guillotreau, P., Isaacs, M., Mitsutaku, M., Perry, R.I., 2015. A decision support tool for response to global change in marine systems: the IMBER-ADApT Framework. Fish Fish. 17 (4), 1183–1193. https://dx.doi.org/10.1111/faf.12110.

Burkhard, B., Müller, F., 2008. Drivers-pressure-state-impact-response. In: Jørgensen, S.E., Fath, B.D. (Eds.), Ecological Indicators. In: Encyclopedia of EcologyVol. 2. Elsevier, Oxford, pp. 967–970.

Ferse, S.C.A., Glaser, M., Neil, M., Schwerdtner Mánez, K., 2014. To cope or to sustain? Eroding long-term sustrainability in an Indonesian coral reef fishery. In: Glaser, M., Glaeser, B. (Eds.), Linking Regional Dynamics in Coastal and Marine Social-Ecological Systems to Global Sustainability. Reg. Environ. Change 14 (6), 2127–2138 (special issue).

GESAMP, 1996. The contributions of science to integrated coastal management. Reports and studies No. 61, Food and Agriculture Organization (FAO) of the United Nations, Rome.

Gibson, C.C., Ostrom, E., Ahn, T.K., 2000. The concept of scale and the human dimensions of global change: a survey. Ecol. Econ. 32 (2), 217–239.

Glaeser, B., 2006. Coastal management and coastal governance: what is the difference?.Presentation in Makassar/Indonesia on March 27, 2006, Summer School on Coastal and Marine Management at Hasanuddin University, Center for Coral Reef Research.

Glaeser, B., 2016. From global sustainability research matrix to typology: a tool to analyze coastal and marine social-ecological systems. Reg. Environ. Change 16 (2), 367–383. https://dx.doi.org/10.1007/s10113-015-0817-y.

Glaeser, B., Glaser, M., 2010. Global change and coastal threats: the Indonesian case. Human Ecol. Rev. 2, 135–147.

Glaeser, B., Glaser, M., 2011. People, fish and coral reefs in Indonesia. A contribution to social-ecological research. GAIA 2, 139–141.

Glaeser, B., Ferse, S., Gorris, P., 2018. Fisheries in Indonesia between livelihoods and environmental degradation: coping strategies in the Spermonde Archipelago, Sulawesi. In: Guillotreau, P., Bundy, A., Perry, R.I. (Eds.), Global change in marine systems. Integrating natural, social and governing responses. Routledge: Studies in Environment, Culture, and Society, London/New York.

Glaser, M., Baitoningsih, W., Ferse, S.C.A., Neil, M., Deswandi, R., 2010. Whose sustainability? Top-down participation and emergent rules in marine protected area management in Indonesia. Mar. Policy 34, 2053–2066.

Glaser, M., Christie, P., Diele, K., Dsikowotzky, L., Ferse, S., Nordhaus, I., Schlüter, A., Schwerdtner Mánez, K., Wild, C., 2012. Measuring and understanding sustainability-enhancing processes in tropical coastal and marine social-ecological systems. Curr. Opin. Environ. Sustain. 4, 300–308.

Glaser, M., Glaeser, B., 2014. Towards a framework for cross-scale and multilevel analysis of coastal and marine social-ecological systems dynamics. In: M. Glaser, M., Glaeser, B. (Eds.), Linking Regional Dynamics in Coastal and Marine Social-Ecological Systems to Global Sustainability. Reg. Environ. Change 14 (6), 2039–2052.

Knowlton, N., 2010. Citizens of the Sea. Wondrous Creatures from the Census of Marine Life. National Geographic, Washington.

MA (Millennium Ecosystem Assessment), 2003. Ecosystems and Human Wellbeing. A Framework for Assessment. Island, Washington DC.

Olsen, S.B., Page, G.G., Ochoa, E., 2009. The Analysis of Governance Responses to Ecosystem Change: A Handbook for Assembling a Baseline. vol. 34 LOICZ Report and Studies, GKSS Research Center, Geesthacht, Germany.

Perry, R.I., Ommer, R.E., 2003. Scale issues in marine ecosystems and human interactions. Fish Oceanogr. 12 (4/5), 513–522.

Further Reading

Glaeser, B., 2015. Klimawandel und Küsten – Humanökologisch-systemisch betrachtet am Beispiel Indonesien. In: Simon, K.H., Tretter, F. (Eds.), Systemtheorie und Humanökologie, Positionsbestimmungen in Theorie und Praxis. In: Edition Humanökologie, vol. 9. oekom, Munich, pp. 316–336.

Glaeser, B., Ferse, S., Gorris, P., 2013. Case Study Indonesia. Spermonde Archipelago: Island development and livelihoods. Template for the description of ADApT_A case studies. IMBER HDWG: Version 2013-05-17.

CHAPTER

6

Evaluation and Management Strategies of Tourist Beaches in the Pacific Coast: A Case Study From Acapulco and Huatulco, Mexico

I. Retama, S.B. Sujitha, D.M. Rivera Rivera, V.C. Shruti, P.F. Rodríguez-Espinosa, M.P. Jonathan

Centro Interdisciplinario de Investigaciones y Estudios sobre Medio Ambiente y Desarrollo (CIIEMAD), Instituto Politécnico Nacional (IPN), Ciudad de Mexico (CDMX), Mexico

1 INTRODUCTION

Coastal management demands a sound integrated understanding of the countless biotic and abiotic factors that govern the shores (Vallejo, 1993). Extending for approximately 440,000 km, the world's coastline comprises less than 0.05% of the total combined landmasses. With nearly half the global population within less than 100 km of the coastline, coastal zones are disputably the most critical and analyzed part of the planet in terms of global economy, conservational strategies, and sustainable management.

A coastal system is an orthogonal system signifying shore face (Swift et al., 1985), extending from the landward limit of the swash to the depth at which the wave action ceases to be competent to transport noncohesive seabed sediments. These systems are experiencing growing population and exploitation pressures where nearly 40% of the global populations thrive in the coastal environments, resulting in the compromise of various ecological services that are crucial to the well-being of coastal economics and people (Agardy et al., 2005). Coastal areas are also vulnerable to different physical factors such as increased flooding, accelerated erosion, and seawater intrusion as a result of the climatic variations in the present century. Various heterogeneous ecosystems embodied in the coastal systems are dynamic in nature

https://doi.org/10.1016/B978-0-12-810473-6.00007-8
© 2019 Elsevier Inc. All rights reserved.

and have been threatened in recent times by the huge impact of urbanization and industrialization. Theoretically, inputs (marine, geological, atmospheric, and people), processes (deposition, erosion, and weathering), and outputs work together to create a coastal equilibrium. In general, management plans and strategies for coastal resources and human impacts are ineffective, leading to numerous conflicts, a decline in services, and less resistance from natural resources to altering environmental conditions.

In coastal geomorphology, beaches are depositional landforms that are constantly changing. A beach is defined as a narrow, gently sloping strip of land that lies along the edge of an ocean, lake, or river. Beaches are valued and treasured for their intrinsic ecology that binds the adjacent terrestrial and marine ecologies. Moreover, beaches are a tourism hub for a country's economic development. Each year, a large percentage of holidaymakers heads to the coastlines around the world, where they have an enormous impact on marine resources. Research also suggests that people living in the coastal areas experience a higher well-being and are considered to be the wealthiest populations, with an income that is four times per capita greater than the inland areas. With distinct ecosystem values (Fig. 1), beaches call for a better strategic management plan to maximize the supply of services that has been inadequate due to various human-induced influences. Recent trends on using science to incorporate management plans in beach conservation will be optimal for society by considering the ecological values and imbalances.

Mexico is a country that offers growth in its tourism industry in spite of the international financial crisis and the social image that the country experiences. With its vast advantageous geographic expanse that includes mountain snows, deserts, forests, rain forests, colonial cities, metropolitans and warm beaches all year long, Mexico has an exceptionally diverse range of tourism products. Mexico was ranked number 10 in international tourist arrivals in 2014. It is the first international destination for US citizens and the second destination for Canadians. The country witnessed revenues of 19,571 million dollars from international visitors in 2016 with an increased growth of 10.4% from 2015 (SECTUR, 2017). Thus, tourism has proved to be

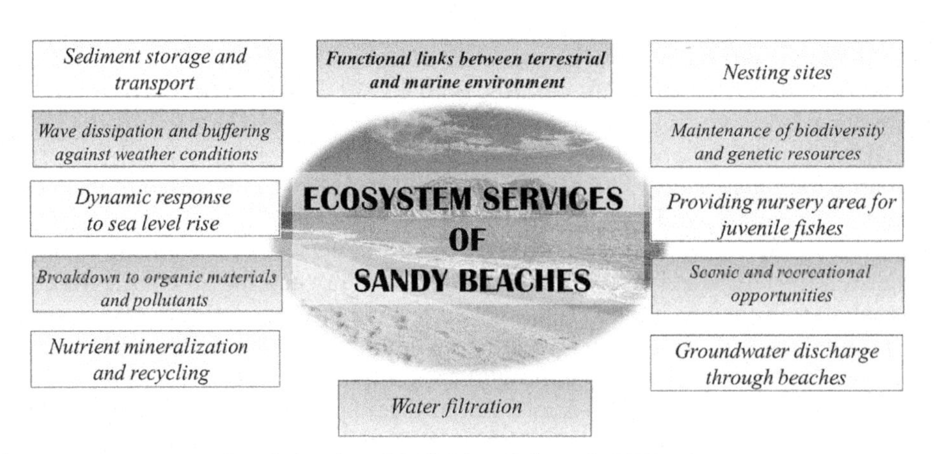

FIG. 1 Ecosystem services of sandy beaches. *(Modifed from Defeo et al., 2009).*

one of the most spirited industries in Mexico. Additionally, more than 450 beaches make up the Mexican coast overlooking the Caribbean and Gulf of Mexico on the east as well as the Pacific Ocean and the Gulf of California on the west. These beaches are interweaved with desert or lush jungle, colonies of ancient art and culture with unmeasurable silence that take the holiday makers for an experience of warmth and paradise. Beach tourism in Mexico is well known among foreign travelers as "sun and relaxation," which attracts between 40% and 50% of tourists annually. With numerous discrepancies and shortages in development for a better market value in the tourism industry, Mexico still finds a major tourist destination among travelers.

In this chapter, the causal factors of environment and human-induced sources that affect the tourism development of two important beach destinations in Mexico will be discussed, along with a better job of incorporating science and technology for the conservational and management plans.

2 STUDY AREA

This chapter encompasses the influence of environmental characteristics for the sustainability of the tourist beaches in Acapulco and Huatulco, Mexico, and to perceive the most relevant measures for their conservation and management. Mexico well known for its beach centers are highly significant in terms of the frequency of visitors (40%–50%) and economy of the country. The states of Oaxaca and Guerrero host the beaches of Huatulco and Acapulco, respectively, and are positioned on the Pacific coast of Mexico, which spreads across about 7828 km of the country (Fig. 2).

2.1 Acapulco

Acapulco is regarded as the "Queen of Mexican beach resorts" for its glamorous, spontaneous, and splashy nature. The region is more urbanized than resort-like and is considered to

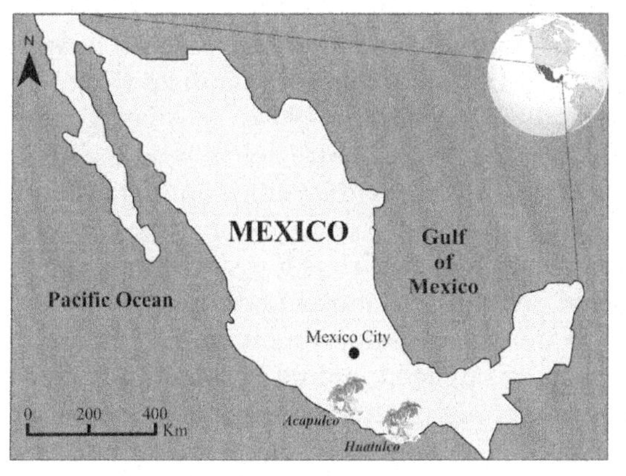

FIG. 2 Mexico map representing the locations of study areas of Acapulco and Huatulco.

be the country's 16th largest metropolitan area. Since the 1950s the beaches have been the most preferred destination of domestic and foreign travelers. Acapulco is located on the Pacific coast in the state of Guerrero in Southwestern Mexico. It is a semicircular bay (49 km^2) encompassed by the mountainous highlands of the Sierra Madre del Sur. The city of Acapulco is located south of Chilpancingo (the capital city of Guerrero), between 16° 41′05″N: 99°30′W, 17°11′ 37″N: 99°59′49″W. Geologically, the region embraces formations of metamorphic (47.28%) rocks, igneous rocks (29.35%), and sedimentary limestone (0.07%). In Acapulco, a warm humid climate prevails with an average annual maximum temperature of 28°C and a minimum of 22°C, whereas rains are concentrated in the summer with a variation of 2–15 mm.

The region includes diversified species of flora and fauna with a predominant vegetation of deciduous forests composing various species. Acapulco lies within the hydrologic regions of Rio Verde, Costa Chica and Costa Grande. The soils are mainly regosol (70.28%), leptosol (5.36%), phaeozem (5.11%), luvisol (3.94%), arenosol (1.27%), solonchak (0.52%), and fluvisol (0.59%). The population approaches 1 million with 810,669 habitants in 289 localities (INEGI, 2015). The economic activities of the majority of the population are in the service sector, including hotels, transport and communications, financial services, insurance, real estate, banking, and community, social, and personal services.

2.2 Huatulco

Huatulco is regarded as one of "Mexico's most relaxed destinations" with the focus on the city's series of nine bays encompassing 36 beaches with a unique setting and ecology. The tourism development is mainly concentrated in the bays located in the Santa Maria municipality of Oaxaca. The rugged terrain of the region divides the coastline into a number of bays, including Conejos, Tangolunda, Chahué, Santa Cruz, Órgano, Maguey, Cacaluta, Chachacual, San Agustín, and the beaches and open seas at Bajos de Coyula and El Arenal.

Huatulco is located on the coast of Oaxaca at the verge of the Sierra Madera del Sur mountains and covers nearly 35 km of Mexican Pacific coastline between the Coyula and Copalita rivers, extending from 18°39′N; 93°52′W to 15°39′N; 98°32′W. Locally, one can spot the outcrops of igneous and metamorphic rocks that are predominantly gneisses, granites, schists, quartzites, and phyllites belonging to the basal complex metamorphic Paleozoic age (FONATUR, 1984). Huatulco experiences a tropically hot and subhumid climate (Köppen classification) with a high percentage of summer rainfall greater than 90%. The average annual temperature is 28°C. Its location in the foothills of the Sierra Madre del Sur and the high altitudinal gradient make the rainfall torrential and short-lived, with average annual rainfall reported to be between 1000 and 1500 mm, of which almost 97% occurs during the summer (June–October) (CONANP, 2003). The major ecosystem in Huatulco is low deciduous vegetation comprising coastal dunes, riparian vegetation, mangroves, and sand dunes. The demographics of the region include 45,680 habitants, as per the census estimate in 2015 (INEGI, 2015). In addition, the inhabitants of Huatulco initially engaged mainly in agriculture, forestry, hunting, and fishing, Now, however, they are occupied in trade, tourism, and services (National Institute for Federalism and Municipal Development, State Government of Oaxaca, 2009).

3 TOURISM AND ITS DEVELOPMENT IN MEXICO

With a slogan of "*Live it to believe it*," Mexico ranked number 10 based on the international tourist arrivals in 2014. The country endowed with many natural and culturally significant places ranks eighth and sixth in terms of natural and cultural UNESCO World Heritage sites, respectively. Tourism is considered a pillar of the Mexican economy with a contribution of 14.8% to the country's GDP growth (WTTC, 2015). The history of Mexican tourism dates back to the postrevolution government of the 1920s, when the government chose tourism as a primary strategy for achieving modernization and economic development (Berger, 2006). Literature cites Mexico as a nation of contrasts with pre-Columbian architecture, ancient traditions, indigenous cultures, and yet a modern nation with comfortable and modern amenities. From 13,892 visitors in 1929, the country proudly claimed 6,855,624 visitors from January–October 2015, an increase of 16.4% from 2014 (Fig. 3).

4 HISTORY OF TOURISM

The name Acapulco is derived from ancient Náhuatl words meaning "*Place of giant reeds.*" Some 3000 years ago, the region was an unspoilt paradise where only a few animals lived with the sun, sea, and vegetation making up a green and pleasant landscape. Acapulco was discovered by the Spanish conquerors on Dec. 13, 1521 (Santa Lucia Day) by Francisco Chico on the order of Hernan Cortes. They built port and shipbuilding facilities, taking advantage of the naturally protected and deep harbor, with an eagerness to find a commercial route to Asia for trade development. The region remained isolated from the rest of the country until a paved road linking Mexico City was built in 1927. By the 1950s, Hollywood came calling and Acapulco became more extravagant. However, during the late 1970s, overdevelopment and overpopulation resulted in high pollution. Later, with tremendous revitalization programs that poured millions into cleaning up the bay, the region is returning to its pinnacle. In spite of being the paradise of the Pacific Coast, the region wears an image of drug-fueled violence and street crime that has given the area a reputation for high homicide rates moreso

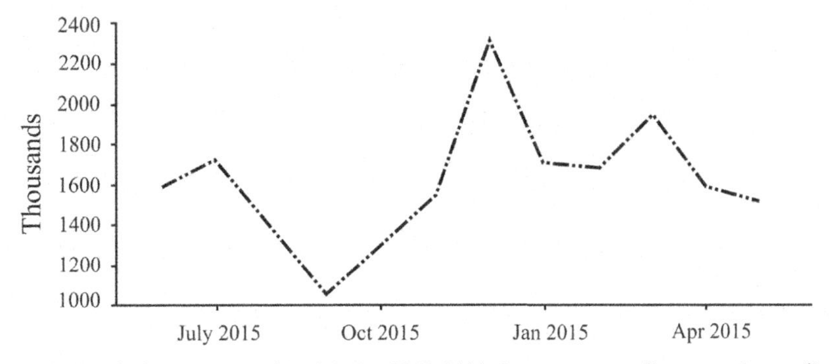

FIG. 3 Statistical record of tourist arrivals in Mexico 2015–2016. *Source: www.tradingeconomics.com/Banco de Mexico.*

than high-end glamour. However, the golden sands of the Acapulco beaches guarantee a warm, soothing holiday.

The bays of Huatulco were said to be robbed by the Spanish pirates, but that they didn't succeed in stealing its beauty. The region comprised a successful port under the control of Cortes, serving as a vantage point for Spanish conquerors and as a distribution center for all the good supplies on the Pacific Coast. However, the development of tourism in Huatulco began in 1967 as a response to the US demand for beach vacations. Mexico's central bank identified the five best new places for tourism. Among Cancun, Loreto, Ixtapa, and Los Cabos, Huatulco also patented its place for its unspoiled pristine beauty. Thus Huatulco is one of Mexico's newest tourist destinations.

5 PROMOTING AND DEPROMOTING FACTORS OF TOURISM

Since 1929, tourism has been considered to be one of the main economic activities of Mexico, after oil production and exports. The country contains a diversity of characteristics that naturally make it a potential important tourism destination. The country embraces palm-fringed beaches, chili-spiced cuisine, dusky jungles, bustling cities, and parties, making it a nation of vivid dreams.

For a nation to remain 10th in world tourism (Mexico tourism board, 2012a, 2012b), it must include varied promoting and depromoting factors. Mexico proves to be a country that utilizes tourism as an important strategy for its economic development. Mexico is the world's fourth country in term of biodiversity, with jungles, smoking snow-capped volcanoes, cactus-filled deserts, 10,000 km of coasts with sandy beaches, and wildlife rich lagoons, making the nation an endless adventure. The welcoming warm hearts of Mexicans are more often positively charming. The most advantageous factor of tourism in Mexico is that it includes different aspects of social tourism, child tourism, natural tourism, business tourism, youth tourism, colonial tourism, and beach tourism. Moreover, the country caters to all types of visitors, from toddlers to senior citizens. Also for the tourists, getting from one place to other is tranquil with comfortable buses and an extensive flight network (86 airports). The country's pride also sits on the different pre-Hispanic civilizations and the wonderful handicrafts of the indigenous population.

Acapulco, known as the "Pearl of the Pacific," is one of the most visited vacation spots in Mexico. The beach resorts are full of energy, vibrancy, and color that never stops day and night. The promoting factors of Acapulco include an attractive climate all through the year, golden beach sands, and famed attractions.

The region also hosts a wealth of activities including:

- Diving and snorkeling
- Jet skiing
- Power boating
- Fishing
- River rafting
- Golf

The fame of Acapulco is almost universal and the area has one of the most stunning bay views offered by any coastal resort. Huatulco, with unspoiled bays and beaches, is the new resort area and is on target to become one of Mexico's key beach attractions for people in search of well-preserved natural surroundings and ecocenters. The region is less expensive with all commercial resorts but not a high rise like that of Cancun and Los Cabos. However, the region efficiently manages to cater the needs of the people without any compromises.

The beaches of Huatulco mostly revolve around relaxation, ecotourism, and water sports such as snorkeling and sailing. Unlike the other resorts, Huatulco enjoys a less crowded vibe and is popular for its peaceful, scenic, and restful beauty. Huatulco is the only resort in Mexico with international green globe certification for sustainable tourism with more than 70% of the region being considered for preservation as green zones. In terms of distinction and additional recognition obtained, it is important to note that, in 2011, Huatulco became the first beach in Latin America (and the third in the world) to receive the environmental certification Earth Check Gold, which is specifically designed for the tourism industry.

For several years, Mexico's tourism industry has battered the storm of violence in the country with negative impacts on tourism activities. Despite of rising crime and an international reputation for drug-fueled violence, there is an augmentation of foreign visitors to the country. Acapulco, which has been a fundamental factor in opening Mexico as a natural destination for international tourists, has also been attacked by violence these days. However, the sunny beach destinations serve to be popular tourist places in Mexico. With all the differences in security, the country promises for a well-managed tourism with the encouragement of local communities and residents to preserve the environment, historical and cultural sites. Other than the violence prevalent in the nation, Huatulco can be very hot or very muddy and like much of the country, it is seated in a seismically active zone as small and moderate earthquakes frequently occur in this region.

6 GOVERNMENTAL PLANS AND POLICIES

The tourism activities on the beaches of Mexico flourished well before the country implemented regulations for sustainable expansion, resulting in an unplanned growth of tourism. The nation formulated its first law on ecological conservation in 1988 (Ley General del Equilibrio Ecológico y Protección al Ambiente), which deals with ecological balances and environment protection. Now, various governmental agencies such as SEMARNAT (Secretary of environment and natural resources), SEDESOL (Secretary of social development), SECTUR (Secretary of Tourism), and SEMARNAT-CONAGUA (National commission of water) are the authoritative organizations for sustainable tourism and conservation. In order to keep up tourism development, the government of Mexico has employed many laws that satisfy the explicit growth of population and tourist activities (Fig. 4).

Acapulco claimed the highest number of tourist arrivals (4,590,910 people) in 2014 (Source: Ministry of Tourism Promotion) by broadening marketing efforts. Thus, the Mexican Tourism Promotion Council (CPTM) has succeeded in attracting more visitors from other countries. The government is striving to implement ecological urban planning and tourism for the Acapulco bay to improve its developmental conditions.

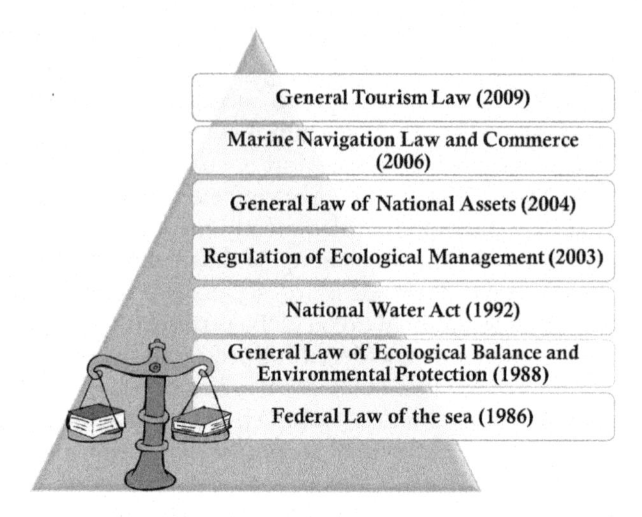

FIG. 4 Governmental laws pertaining to the conservation of resources and tourism development (SEMARNAT, 2015).

Huatulco witnessed 90,375 tourists with a total income of 541 million pesos in 2014 and 110,000 tourists in 2015, leaving behind 659 million pesos. A rise in the arrival of tourists to Huatulco is attributed to the renewal of its international airport linking various foreign destinations including Canada, the United States, and Europe. Considering the high profitability that Huatulco offers, FONATUR (National Fund for Tourism Development) has proposed to generate a new image for this destination and revive its economic activity and tourism in order to make it one of the most successful tourism developments of the country. The goal is to double the number of hotels through sustainable projects of high quality and low density to maintain the respect of the environment.

7 BEACH QUALITY—*BLUE FLAG BEACHES*

The quality of beaches is evaluated by the notable organization of Blue Flag, operated under the Foundation for Environmental Education and headquartered in Copenhagen, Denmark. Blue Flag is a world-renowned ecolabel trusted by millions around the globe. Out of the 4154 certified Blue Flag beaches and marinas in 49 countries (2016), Mexico proudly hosts 35 certified Blue Flag beaches and two marina (CONAGUA, 2018). This certification exemplifies a serious and profound commitment to the people and the environment. Having reached 32 criteria for the prestigious Blue Flag status, these beaches pose a sense of pride by promoting environmental issues and awareness. Beaches in Chahue, Huatulco, and El Chileno, Acapulco, are also certified as Blue Flag beaches, benefitting the campaign of environmental health, safety, services, and the improvement of its image and competiveness, providing an ideal tourist destination.

8 CLIMATIC IMPACTS

8.1 Hurricanes

Positioned between two warm oceans, Mexico has been battered throughout history by storms and hurricanes. Although Acapulco claims to be a tropical city and a prime escape for snowbound Americans and Canadians, it is blessed with warm temperatures all through the year. The official storm season begins in Acapulco on May 15 and runs through Nov. 30, overlapping with the Eastern Pacific hurricane season. The bay experiences high intense tropical depressions, tropical storms, and hurricanes, especially during the *El Niño* years. Among the various hurricanes that have struck the Acapulco coast, Pauline in 1997 and Manuel in 2013 are considered to be the deadliest, causing at least $750 million in damages. The impact of disaster and poor urban planning in the Acapulco bays is one of the reasons for the setback in tourism development. In the case of Huatulco, it's very rare that the region is influenced by a hurricane due to its positioning in the area where most of the East Pacific hurricanes originate. By the time they evolve as a tropical depression, they are a few hundred kilometers away from the coast. The most intense hurricanes that affected the region were Pauline in 1997 (category 3) and Carlotta in 2012 (category 2) with minimal destruction.

Hurricanes and storms potentially discourage tourism, as many of the resorts offer outdoor recreational activities resulting in huge amounts of waste in the beach sands. The vulnerability of coastal regions has increased in recent years as more people tend to move to the coasts in search of a better economic lifestyle as part of beachfront tourism development. Mexico also witnesses high population growth in the coasts frequented by tropical cyclones, causing risk to life and property. Countries such as Mexico with an economy that depends on tourism are particularly at risk during these climatic events. Changes in the climate have major implications on the tourism industry as they alter the charm of various destinations. With the present global warming conditions, oceans are heating up and more hurricanes are expected in the future, signaling more appropriate assessments to understand the intensity and frequency of hurricanes for better management policies.

8.2 Coastal Inundations

Inundations take place when the most seaward dune and the shoreline of a beach are completely submerged under the rising water due to storm surges, tides, storms, hurricanes, and tsunamis (Ciro Aucelli et al., 2016). During the 19th century, the process of inundation was mainly due to the impact of anthropogenic emissions of greenhouse gases (Titus, 2008).

However, several factors such as undersea earthquakes, landslides, volcanic eruptions, atmospheric and oceanic processes, tropical cyclones, and storms also affect the coasts. These inundations also threaten the viability of many coastal regions affecting tourism, signifying a two-way relationship between tourism and climate change.

From modeling and historical documents, it is stated that during the Great Mexican Tsunami of 1787 (Nuñez-Cornú and Ponce, 1989; Nuñez-Cornú et al., 2008), at several locations in Acapulco an inundation of 4–13m was observed. The coast of Acapulco is highly prone to earthquakes for its proximity to the subduction zone where two plates (cocos and pacific) collide, also the region is vulnerable to hurricanes. The hurricanes of Cosme (1989) and Paulina

(1997) brought vast devastation to Acapulco, resulting in numerous deaths and millions of dollars of damage to properties. In the case of Huatulco, the region does not experience serious inundations for its geographical setting, which only serves as a site for the formation of storms rather than their direct impact.

9 ANTHROPOGENIC IMPACTS

9.1 Metal Pollution

The birth of the Industrial Revolution, rapid urbanization, and multifold demographic growth have resulted in the ubiquitous environmental contamination of metals,threatening more than 50% of the global population. Throughout history, it is well known that people have inhabited the coasts to take advantage of the amenities that the coasts offer. The discovery of high mercury levels in coastal environments has spurred widened research of the possible presence of other toxic elements. Distribution of metals in aquatic sediments depends on a number of geological, mineralogical, hydrological, biological, and anthropological processes (Węsławski et al., 2000; Bigus et al., 2016) as well as sources controlled by both internal and external factors.

Baseline studies were carried out in the popular beaches of Acapulco (Jonathan et al., 2011) and Huatulco to evaluate the concentration levels of metals. Acapulco beach sediments (Fig. 5) presented the following order of metal levels (all values in μg/g except Fe): Fe (6549 mg/kg) > Mn (90.47) > Zn (19.04) > Cr (17.86) > Cu (6.33) > Pb (3.73) > Ni (3.41) > Cd (1.61) > As (0.84). The higher concentrations of metals in the sands of Acapulco are due to the impact of small tourist boats, antifouling paints, a ship-repairing region, influence from the leaching of used electronic goods in hotels/restaurants, and the remnants of various construction work taking place on the coast.

Unlike Acapulco, the beaches of Huatulco were studied in two seasons (April and December) to evaluate the influence of tourist activities in the concentration of metals. In general, an average high concentration of metals was observed during the Easter vacations of April, which was also augmented by the activity in the zone favoring high precipitation of metals from the beach waters due to the changes in water pH. The presence of Huatulco bays (Fig. 6) close to the zone of subduction also impacts the concentration of high levels of Cr, Zn, Ni, and Zn. The beach sediments of Huatulco bay presented an order of (avg. for April and December, all values in mg/kg) Fe (6501.21) > Cr (69.89) > Mn (61.37) > Zn (14.62) > Ni (7.37) > Pb (5.97) > Co (4.91) > Cu (3.04) > As (0.73) > Cd (0.60) > Hg (0.00). Apart from the impacts of anthropogenic activities such as beach tourism, industrial and domestic discharge, commerce, agricultural effluents, and intense marine traffic (Jonathan et al., 2011; Nagarajan et al., 2013; Retama et al., 2016a,b), the region on a highly diversified geological, tectonic, and subduction zone is substantially enriched with metals.

9.2 Microplastics

The occurrence of small plastic particles on beaches and in coastal waters was first reported in the 1970s, although the term "microplastics" was formulated very recently in the 2000s to

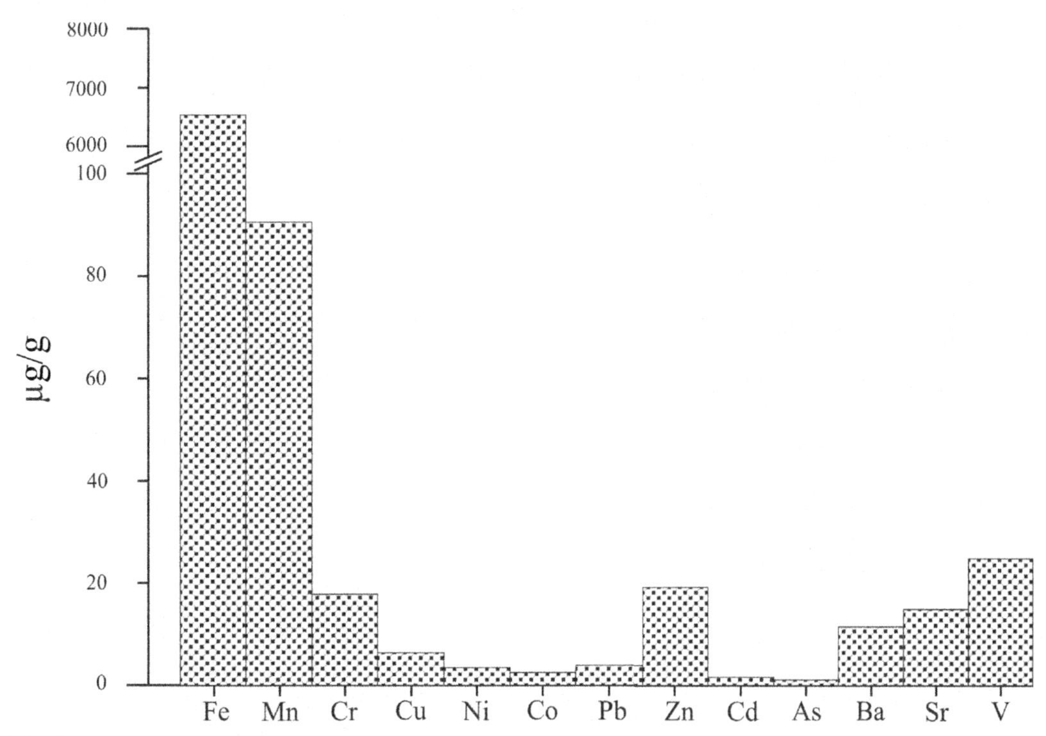

FIG. 5 Average metal concentrations in surface sediments of Acapulco beaches.

define tiny plastic particles up to 5mm in diameter (UNEP, 2013). For the high potential impact of microplastics in the marine ecosystems, research has gained momentum during the past few years. This interesting study was also carried out on the beaches of Huatulco as a pioneer study on the Mexican Pacific coast (Fig. 7). The abundance of these fibrous materials was presented as numbers per 30 g of dry sediment sample. On average, the beach sands sampled during the Christmas vacations yielded a higher amount of microplastics (14 fibers/30 g of the sample) than April, which is due to the intense activities of tourism and the accumulation of plastics for for more than one year from April 2013 to December 2014 (Retama et al., 2016a, 2016b). The appearance of microplastics can be due to the deterioration of larger plastic fragments, cordage, consumer products, and films, with or without the assistance of UV radiation as well as mechanical forces such as wave action, high energy shorelines, or though biological activity. The fibers, also classified based on their color and abundance, were in the order of white > black > blue > red > light brown, representing risks to the organisms as they may be ingested by marine organisms due to the resemblance to prey. Studies on these tiny fibers are also important for their high influential impact on the marine food chain from oysters to whales, where they swallow these particles while filtering large amounts of water. Despite a rapid growing knowledge on the effect of microplastics on the marine environment, the studies still provide valuable information on the pathway of harmful chemicals in the food web.

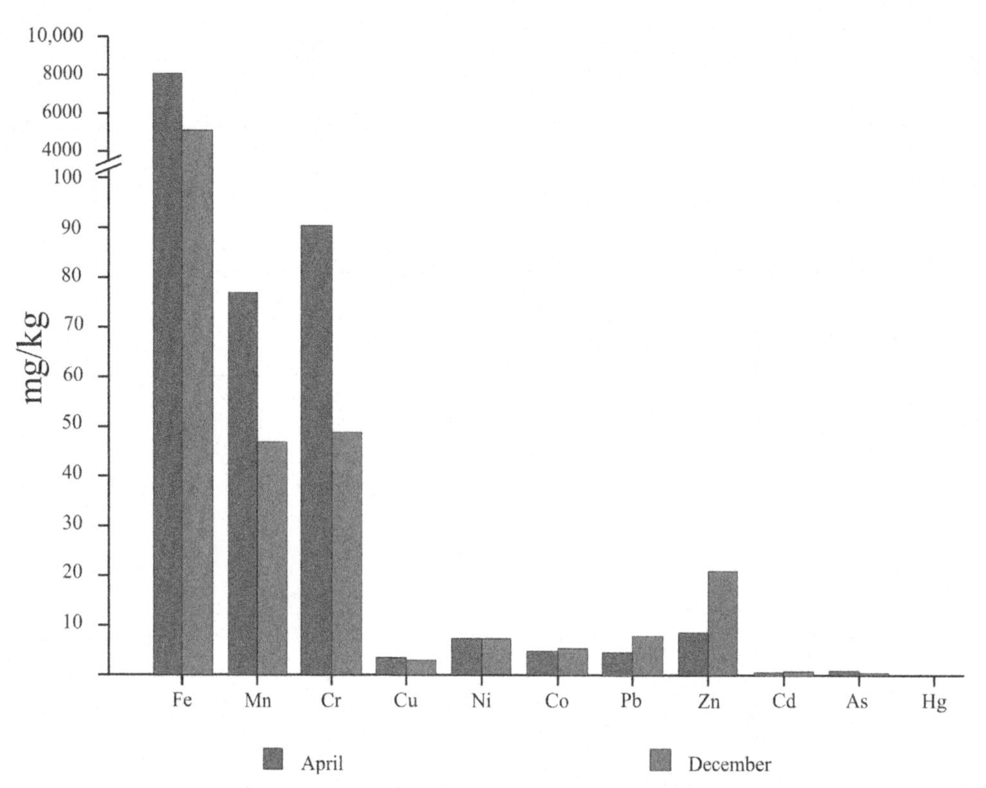

FIG. 6 Average metal concentrations in surface sediments of Huatulco beaches collected during the months of April and December.

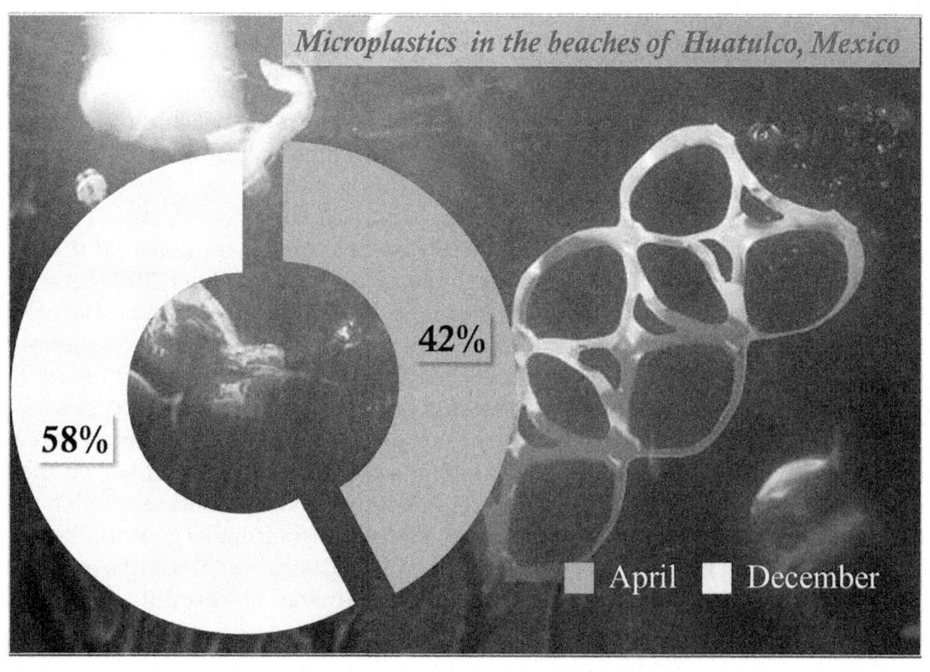

FIG. 7 Percentages of microplastics present on the beaches of Huatulco, Mexico.

9.3 Solid Waste

Solid waste such as picnic plates, plastic bags, bottles, and cigarette butts that is left behind by tourists also poses a menace to the beaches. Solid wastes such as plastics, rubber, foam materials, and metals take hundreds of years to degrade, posing a serious threat to marine organisms. However, solid wastes are the easiest to prevent by simply maintaining the discipline of throwing them into the trash. According to official sources, Acapulco records 722 tons of solid waste, which increases during the vacations and weekends, whereas in Huatulco, 45,000 kg of solid waste is collected. If stringent measures are undertaken by the hotel communities, government officials, and general public, the high amounts of solid waste generated could be controlled, adding advantages to the quality of beach tourism.

10 SUSTAINABLE TOURISM AND BEACH MANAGEMENT

Beach management aims to maintain or improve the conditions of a beach as a recreational resource and a means of coast protection while providing facilities that meet the demands of those who use the beach (Bird, 1996). Tourism growth has been intensely vivid over the past 50 years, so much so that it has been claimed as the largest industry worldwide for its economic and cultural benefits. The concept of sustainable forms of tourism has gained awareness and reputation since the 1992 Earth Summit in Rio de Janeiro. To sustain the charm and productivity of coastal ecosystems, management policies are required to be comprehensive and holistic in nature. With massive influxes of tourists and overdevelopment of tourism, coasts are under enormous pressure, resulting in an ecological imbalance. The benefits of management strategies are well amplified from the economic to social sectors. They contribute to government revenues, foreign exchange, employment opportunities, direct financial contribution for its protection, and further environmental management and planning. Moreover, they also prove to be a scope for strengthening communities, revitalizing culture and tradition, and social involvement.

11 CONCLUSION

In recent years, beaches have been considered to be the major driving force behind the economic welfare of a nation, in spite of the fact that they are a productive and protective ecosystem. Modern research defines that the success of beach management is dependent on understanding its ecological processes and structures. In Mexico, coastal beaches, territorial seas, and resources are under federal jurisdiction. Each Mexican beach has various conservational efforts undertaken to fulfill the criteria put forth by different agencies for certification as a clean beach. Article 27 of the Mexican constitution establishes that the federal government has complete dominion over all the water resources of the country.

The present chapter about the quality of beaches in Huatulco and Acapulco suggests that more remedial measures and nourishment processes have to be carried out for accomplishing a better sustainable environment. Studies also note that the presence of high levels of metals in the beaches is due to the influence of anthropogenic and natural enrichments. The novel studies on the concentration of microplastics in the Huatulco beaches suggest immediate actions to protect the ecological life of the ecosystem. With several important conservation plans undertaken by the Mexican government, it is validated that they are still bringing in innovative ideas and management plans to keep up the tourist flavor globally, for example by the slogan, "Planning for a sustainable future in tourism: Let's put the environment first."

References

Agardy, T., Alder, J., Dayton, P., Curran, S., Kitchingman, A., Wilson, M., et al., 2005. Coastal systems. In: Reid, W.V. (Ed.), Millenium Ecosystem Assessment Current State and Trends. In: vol. 1. Island Press, Washington, DC; Covelo; London, pp. 513–519.

Berger, D., 2006. The Development of Mexico's Tourism Industry: Pyramids by Day, Martinis by Night. Palgrave Macmillan, New York.

Bigus, K., Astel, A., Niedzielski, P., 2016. Seasonal distribution of metals in vertical and horizontal profiles of sheltered and exposed beaches on Polish coast. Mar. Pollut. Bull. 106, 347–359.

Bird, E.C.F., 1996. Beach Management. John Wiley & Sons, Chichester.

Ciro Aucelli, P.P., Di Paola, G., Incontri, P., Rizzo, A., Vilardo, G., Benassai, G., Buonocore, B., Pappone, G., 2016. Coastal inundation risk assessment due to subsidence and sea level rise in a Mediterranean alluvial plain (Volturo coastal plain – Southern Italy). Estuar. Coast. Shelf Sci., 1–13.

CONAGUA, 2018. https://www.gob.mx/conagua/articulos/mexico-tiene-37-sitios-blueflag?idiom=es.

CONANP (Comisión Nacional de Áreas Naturales Protegidas), 2003. Programa de Manejo de la Reserva de la Biosfera Barranca de Metztitlán. Comisión Nacional de Áreas Naturales Protegidas, SEMARNAT, México.

Defeo, O., McLachlan, A., Schorman, D.S., Schlacher, T.A., Dugan, J., Jones, A., Lastra, M., Scapini, F., 2009. Threats to sandy beach ecosystems: a review. Estuar. Coast. Shelf Sci. 81, 1–12.

FONATUR, 1984. Análisis de la operación de las cadenas hoteleras en México.

INEGI, 2015. Instituto Nacional de Estadística, Geografía e Informática. Retrieved from, http://www.inegi.gob.mx.

Jonathan, M.P., Roy, P.D., Thangadurai, N., Srinivasalu, S., Rodriguez-Espinosa, P.F., Sarkar, S.K., Lakshumanan, C., Navarrete-Lopez, M., Munoz-Sevilla, N.P., 2011. Metal concentrations in water and sediments from tourist beaches of Acapulco, Mexico. Mar. Pollut. Bull. 62, 845–850.

Mexico tourism board, 2012a. Mexico poised to break tourism records in 2012. (Press release) retrieved from, http://www.prnewswire.com/news-releases/mexico-poised-to-break-tourismrecords-in-2012-138200009.html.

Mexico tourism board, 2012b. The role of travel and tourism lauded by G20 leaders. (Press release) retrieved from, http://www.prsnewswire.com/news-releases/the-role-of-travel-tourism-lauded-by-G20-leaders-160000175.html.

Nagarajan, R., Jonathan, M.P., Roy, P.D., Wai-Hwa, L., Prasanna, M.V., Sarkar, S.K., Navarrete-Lopez, M., 2013. Metal concentrations in sediments from tourist beaches of Miri City, Sarawak, Malaysia (Borneo Island). Mar. Pollut. Bull. 73, 369–373.

National Institute for Federalism and Municipal Development, State Government of Oaxaca, 2009. Enciclopedia de los Municipios de México, Estado de Oaxaca, Santa María Huatulco, H. Ayuntamiento de Santa María Huatulco.

Nuñez-Cornú, F., Ponce, L., 1989. Zonas sísmicas de Oaxaca, México: Seísmos máximos y tiempos de recurrencia para el periodo 1542–1988. Geofis. Int. 28, 587–641.

Nuñez-Cornú, F., Ortiz, M., Sanchez, J.J., 2008. The great 1787 Mexican tsunami. Nat. Hazards 47, 569–576.

Retama, I., Jonathan, M.P., Roy, P.D., Rodríguez-Espinosa, P.F., Nagarajan, R., Sarkar, S.K., Morales-García, S.S., Muñoz-Sevilla, N.P., 2016a. Metal concentrations in sediments from tourist beaches of Huatulco, Oaxaca, Mexico: an evaluation of post-Easter week vacation. Environ. Earth Sci. 75, 375.

Retama, I., Jonathan, M.P., Shruti, V.C., Velumani, S., Sarkar, S.K., Roy, P.D., Rodríguez-Espinosa, P.F., 2016b. Microplastics in tourist beaches of the Huatulco Bay, Pacific Coast of Southern Mexico. Mar. Pollut. Bull. 113 (1–2), 530–535.

SECTUR, 2017. https://www.gob.mx/sectur/prensa/registra-turismo-cifras-historicas-en-2016-35-millones-de-visitantes-y-19-571-mdd-en-divisas.

SEMARNAT, 2015. Secretaría de Medio Ambiente y Recursos Naturales. Dirección General de Política Ambiental e Integración Regional y Sectorial, Mayo.

Swift, D.J.P., Niedoroda, A.W., Vincent, C.E., Hopkins, T.S., 1985. Barrier island evolution, middle Atlantic shelf, USA. Part 1: Shoreface dynamics. In: Oertel, G.F., Leatherman, S.P. (Eds.), Barrier Islands. Mar. Geol.In: vol. 63, pp. 331–361.

Titus, J.G., 2008. Greenhouse effect, sea level rise, and coastal zone management. Coast. Zone Manage. J. 14, 147–171.

UNEP, 2013. UNEP Year Book 2013: emerging issues in our global environment. In: Microplastics. Retrieved from, www.unep.org/yearbook/2013.

Vallejo, S.M., 1993. The integration of coastal zone management into national development planning. Ocean Coast. Manag. 21, 163–182.

Węsławski, J.M., Urban-Malinga, B., Kotwicki, L., Opaliński, K., Szymelfing, M., Dutkowski, M., 2000. Sandy coastlines—are there conflicts between recreation and natural values? Oceanol. Stud. 29, 5–18.

WTTC (World Travel and Tourism Council), 2015. Travel and Tourism Economic Impact 2015: World. WTTC, London.

CHAPTER

7

Evaluation of Decadal Shoreline Changes in the Coastal Region of Miri, Sarawak, Malaysia

A. Anandkumar, H. Vijith*, R. Nagarajan*, M.P. Jonathan†*

*Department of Applied Geology, Faculty of Engineering and Science, Curtin University Malaysia, Miri, Malaysia

†Centro Interdisciplinario de Investigaciones y Estudios sobre Medio Ambiente y Desarrollo (CIIEMAD), Instituto Politécnico Nacional (IPN), Ciudad de Mexico (CDMX), Mexico

1 INTRODUCTION

The Earth's surface features are undergoing rapid change due to natural processes and anthropogenic activities. The surface of the Earth includes a variety of natural and artificial geographical features such as ecosystems, landforms, human settlements, and engineered constructions. Land use and land cover (LULC) analysis is a general term used to depict Earth surface cover, whether it is natural or manmade. According to the United Nations' Food and Agricultural Organization (FAO, 1999), land cover is the observed physical and biological cover of the Earth's land, as vegetation or man-made features. Land use is the total of arrangements, activities, and inputs that people undertake in a certain land-cover type. The assessment of changes in LULC over different time periods attained more importance in the last few decades. A number of studies have been conducted to assess the spatial patterns of land use/land cover (Xiubin, 1996; Carlson and Arthur, 2000; Lambin et al., 2003; Tan et al., 2010; Mendoza-González et al., 2012; Avalar and Tokarczyk, 2014; Khan et al., 2015; Rawat and Kumar, 2015; Parsa and Salehi, 2016; Yang et al., 2017). The causes of LULC changes can be either natural or human activities.

The LULC of an area is influenced by human development activities or other changes taking place in the region. The temporal changes in LULC in an area can be assessed by using historical maps and, more recently, aerial photos or satellite images (Mas, 1999; Shalaby and Tateishi, 2007; Yang et al., 2010). The advancement of satellite imaging technology and geographical

© 2019 Elsevier Inc. All rights reserved.
https://doi.org/10.1016/B978-0-12-810473-6.00008-X

information systems (GIS) made the process of change detection quicker and more accurate than in the past. Temporal images available from a number of satellites with different spatial and spectral resolutions have complemented the effective assessments of LULC made by human activities and natural phenomenon as well as helped to predict future LULC changes in the region (Veldkamp and Lambin, 2001). LULC studies can be adopted to monitor major environmental problems such as deforestation, sea-level rise, watershed management, acidification, forest fires, greenhouse effects, eutrophication, and desertification by using different resolutions of satellite images (Fonji and Taff, 2014). High-resolution satellite images are normally used to monitor the biodiversity loss of an area to distinguish small contrasting areas. LULC analysis is an important technique used by many professionals for proper planning and sustainable management of land resources. LULC assessment can also be applied to monitor the changes of the Earth's terrestrial surface due to climate change, loss in biodiversity, and the impact of pollution on terrestrial and aquatic environments (Berlanga-Robles and Ruiz-Luna, 2002).

Coasts are the narrow stretches of land that serve as the contact zone between the ocean and land forms. They are highly dynamic natural systems where ocean processes dominate over land processes. The human impact on coastal environments has increased over the last few decades, resulting in negative impacts on coastal ecosystem such as coastal erosion, accretion, land reclamation, and coastal pollution (Bird and Teh, 1990). Studies related to coastal geomorphology include coastal change analysis and LULC changes in defined coastal areas over a period of time. Describing coastal regions through coastal geomorphological analysis and mapping has rapidly gained acceptance recently (Blodget et al., 1991; Boak and Turner, 2005; French and Burningham, 2009; Hapke et al., 2010; Payo et al., 2016; Maanen et al., 2016). Changes in shorelines can be caused by natural causes as well as by human-induced development activities; these changes increase the risk factor for coastal communities (Malini and Rao, 2004). Shoreline changes are a topic of environmental concern when considering development projects such as harbors, jetties, ports, embankment facilities, etc. Monitoring shoreline changes is necessary in the short term as well as the long term, either seasonally or annually to provide efficient data on erosion and deposition in regards to coastal zone management and to improve sustainable shoreline management plans (Kankara et al., 2014).

Malaysia has a 4809 km coast line (excluding smaller islands) in which 1035 km are in the Sarawak state (EPU, 1985). The Malaysian coastal zone is coming under great pressure as Malaysia strives to achieve developed status by 2020 (Bird and Teh, 1990). The coastal region of Miri has experienced various problems such as coastal erosion, accretion, and pollution in recent years due to urbanization-associated industrial development and LULC changes in the upstream of the major rivers. The stress and changes in the climatic conditions in the region have induced changes in the coastal shoreline over the last few decades. The NE part coastal zone experiences dynamic environment by the influences of the Baram River (second largest river in Sarawak state) and Miri River. The Sarawak Department of Irrigation and Development has indicated that the Miri coastal zone is facing a major coastal erosion problem between the two river mouths. However, the coastal erosion is natural, the combined effect of different land uses and economic activities. The Department of Irrigation and Development (www.did.sarawak.gov.my) has carried out some pilot studies; however, more studies are needed for both short-term and long-term solutions for the coastal erosion. In addition, the changes in the Miri coastal region are related to sediment input from the major rivers (the Baram, Miri, and Sibuti Rivers) and smaller streams, which carry and drain enormous amount of sediment and suspended matter into the Miri coastal region. Thus, the main objective of the present study is to assess the shoreline and coastal morphological changes.

2 STUDY AREA

The study area extends from the Baram River mouth (Kualabaram) to Bungai Beach, which covers a coastline of approximately 74km. Traveling from northeast to southwest, the study area consists of 11 important tourist and commercial beaches such as Kuala Baram Beach, Fish Landing Center, Lutong Beach, Piasau Boat Club, Miri River Estuary, Park Everly Beach, Tanjung Lobang Beach, Esplanade Beach, Hawaii Beach, Kampong Baraya, Tusan Beach, and Bungai Beach (Fig. 1). This environment encompasses sandy beaches, rocky shores, sand dunes, sea cliffs, arches, sea caves, and developed waterfront areas. All these important beaches are connected with tourist and economic developments in this region and have a strong interaction between local communities and the developmental activities taking place in the coastal region. In the study area, two major rivers (the Baram and Miri) flow directly into the South China Sea near Kualabaram and Miri, respectively. A high amount of sediment and suspended solids is carried from the upstream regions of the Baram River and discharged at the Miri coast. The Miri River contains less sediment input compared to the Baram River. These two rivers flow through many palm oil plantation sites and transport a significant amount of organic and agricultural wastes, which is then discharged into the Miri coast.

Miri is located in the northern part of the study area and its influences are common until the Hawaii beach. The other two beaches (Kampong Baraya to Tusan Beach and Bungai Beach) are located further away from Miri. However, these two beaches attain tourist importance due to having sea cliffs, arches, and small sea caves. Human development in the coastal zone has been rapidly increasing over the last several years due to the discovery of oil fields. Enormous developmental activities such as the construction of Marina Bay, luxurious residential complexes, a recreational waterfront area, ocean-view apartments, squatter houses along the Miri River, and breakwater structures are some of the major man-made activities evident in the coastal area of Miri over the past few decades. Primary forests are removed by logging companies for timber production and then the land is cleared to introduce rubber and palm oil plantations as a major commercial endeavor (Lambin and Geist, 2008; Gaveau et al., 2014). Such forests are adjacent to the Baram and Miri Rivers. Siltation is mainly a result of the mining of sand and land development in conjunction with deforestation, particularly in the upper Baram region, while harbor dredging (UNEP, 2005), which is clearly evident by the high turbidity in the Baram River, is visible on satellite images. The study area has a tropical climate characterized by southwest (May–August) and northeast monsoons (November–April), with semidiurnal tides and a temperature range of 23–32°C. The annual mean rainfall in the study area varied from 2247 to 3499mm between 1981 and 1990 (avg. 2715mm), 2228–3265mm for 1991–2000 (avg. 2682mm), 2516–3267mm in 2001–10 (avg. 2916), and 3246–3125mm through 2011–13 (avg. 3022), respectively (Source: Jabatan Meterologi, Malaysia, Miri). The tidal range in Sarawak is generally <2m and the sea is relatively calm during SW monsoons. The nearshore sediments are sand, interrupted by silty clay (near the river mouth) with the influence of local geology (sedimentary terrane).

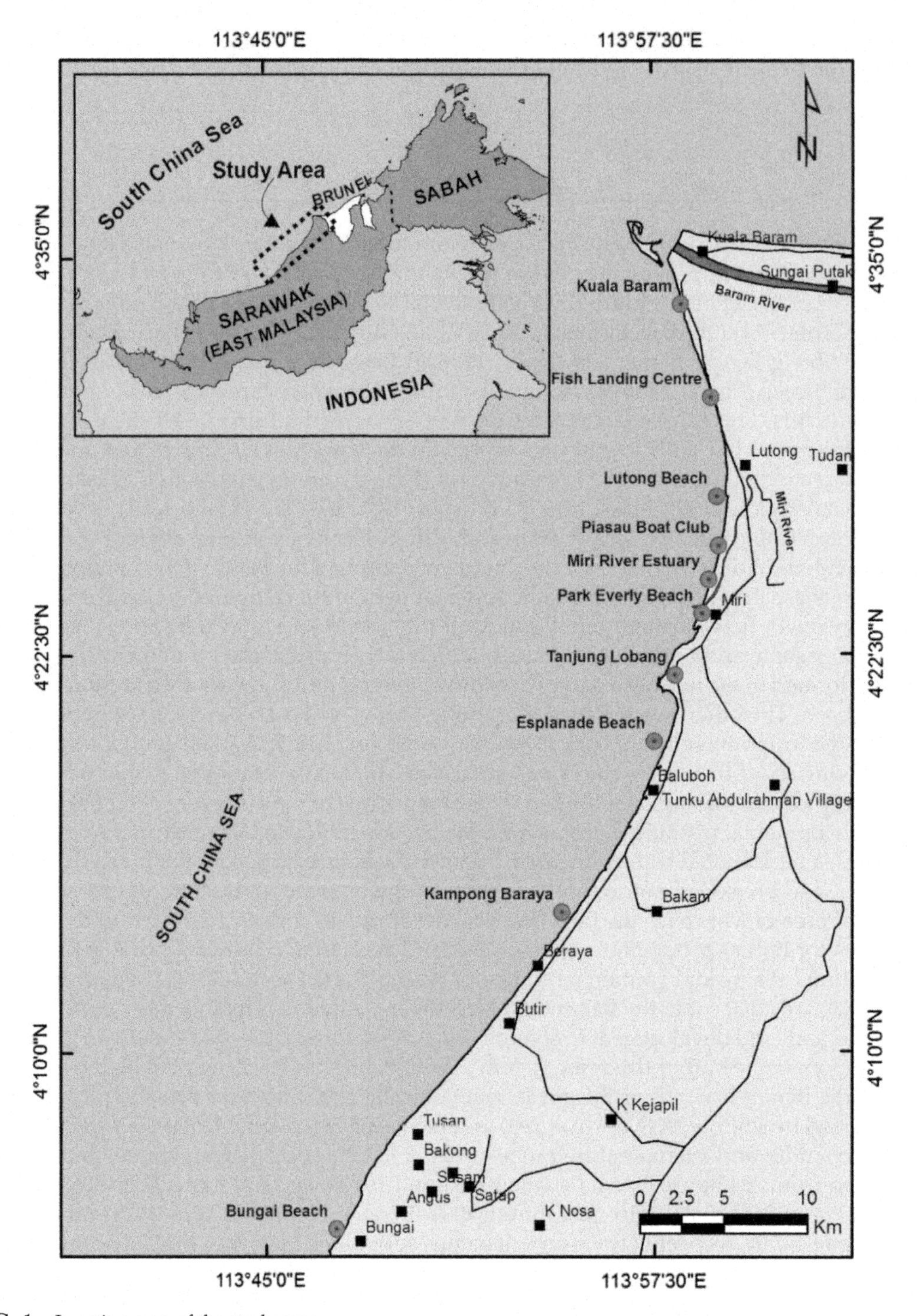

FIG. 1 Location map of the study area.

2.1 Geology and Geomorphology of the Study Area

The study area falls under the northern Sarawak province, based on regional geological classification. The North Sarawak region is located north and east of the Rajang-Baram watershed. The adjacent offshore area is named the Luconia Block (Hutchison, 1989). This North Sarawak region is underlain by Neogene sediments of the NW Borneo Basin and the sediments are younging toward North and NNE with Oligocene base rocks. The North Sarawak region is underlain by thick sequences of shallow and deep marine sediments, such as sand and shale formations, respectively (Liechti et al., 1960). The formations are Nyalau, Setap, Tangap, Sibuti, Belait, Lambir, Miri, and Tukau and range in age from Oligocene to Pliocene. Among these, the Nyalau, Belait, Lambir, Tukau, and Miri formations have formed under shallow marine to deltaic environments, whereas the other formations are considered to have been deposited in open marine environments. This is based, among other things, on the relative absence of planktic foraminifera in shallow marine to deltaic sediments and their presence in the open marine environments. In the Baram Delta region of NE Sarawak, there was a very dramatic change in sedimentation from Setap shale to the very sandy Lambir formation at the SE margin of the Lambir hills. The sediments of NW Borneo are mainly derived from the Rajang-Croker accretionary prism as a recycled source with some input from mafic-ultramafic rocks (Van Hattum et al., 2013; Nagarajan et al., 2014, 2015b, 2017a,b).

In the study area, the (1) Sibuti, (2) Lambir, (3) Tukau, and (4) Miri formations are covered with numerous sea cliffs belonging to these formations. The Sibuti formation mainly consists of shale with sandstone and minor limestone/marl limestone lenses. This formation is underlain by the Suai formation (fault contact) and overlain by the Lambir formation along a conformable, abrupt boundary (Banda, 1998; Hutchison, 2005). The Sibuti formation is folded gently to moderate and form shallow anticlinal and synclinal folds (Wannier et al., 2011). Common fossils reported and observed from the Sibuti formation are shell fragments, bivalves, gastropods, fossil crabs, corals, foraminifera, and trace fossil burrows as well as minor fossils such as fish and shark's teeth, sand dollars, bryozoans, and sea urchin spikes (Wannier et al., 2011; Nagarajan et al., 2015b). The age of the Sibuti formation is Early Miocene to Late Middle Miocene (Hutchison, 2005; Simon et al., 2014). The Sibuti formation is time-stratigraphically equivalent to the Setap shale formation and is termed the Sibuti member of the Setap shale formation (Hutchison, 2005). However, lithogically, the Sibuti formation is different from the Setap shale formation (Hutchison, 2005; Nagarajan et al., 2015b) and also can be distinguished from that formation based on high fossil content, predominant marl lenses, and thin limestone beds (80 m thickness; abandoned quarry near Kampong Opak). Concretions are common in the Sibuti formation and these concretions can be grouped into Mn-rich and phosphate-rich concretions (Peng et al., 2004; Nagarajan et al., 2015b).

The Lambir formation was deposited during the Early Mid Miocene in deltaic and fluvial environments. It consists of sandstone, shale, and some limestone. The Lambir formation is underlain by the Sibuti or Setap Shale formations and is overlain by the Tukau formation. The Lambir formation is dominated by sandstones with a number of cycles with hummocky cross-bedding at the basal part while the top of the sandstone consists of low angle planar cross-bedding, indicating a beach depositional environment (Hutchison, 2005). The sequence shows a distinctive shallowing upward transition from marine to coastal conditions, associated with the change in orientation of the coastline and the uplift of interior Borneo (Change from

Cycle III to IV) (Hutchison, 2005). The change in coastline orientation and uplift may have had an impact on the provenance of sediments deposited in the basin. Foraminifera, Ophiomorpha, and coal laminations are observed within these sediments (Nagarajan et al., 2017a).

The Tukau formation is exposed in the western part of Miri and along the beaches, which extend inland. The basal part of the Tukau formation conformably overlies the Lambir formation near Sungai Liku in the eastern Lambir Hill area, whereas it is conformably overlain by the Liang formation near Miri. The age of the Tukau formation is assumed to be Upper Miocene to Lower Pliocene. The absence of foraminifers (except some brackish water forms), the presence of lignite layers, and amber balls imply a coastal plain depositional environment for the Tukau formation (Hutchison, 2005). The Tukau formation mainly consists of sandstone, shale and alternating interbedded shale and sandstone with minor conglomerates. Pyrite concretions and amber balls are common in these sediments (Nagarajan et al., 2017b).

The Miri formation is geologically a very complex NE-SW elongated anticline, bound to the east by a steep NW dipping fault known as the Shell Hill Fault. The downthrown block of the Shell Hill Fault is divided into a series of steep NW dipping additional faults and a series of flat SE dipping antithetic faults. The rocks exposed in and around Miri belong to the Miri formation (Middle Miocene), and are the uplifted part of the subsurface, oil-bearing sedimentary strata of the Miri oilfield (Jia and Rahman, 2009). The Miri formation is divided into upper and lower formations that consist of marine sandstone and shale alternations. The Upper Miri formation is more arenaceous than the lower formation and consists of repeated and irregular sandstone-shale alternations, with the sandstone beds passing gradually into clayey sandstone and sandy or silty shale (Hutchison, 2005). The Lower Miri formation consists of well-defined beds of sandstone and shale, with the shale slightly dominant. The sandstones vary from very fine grained, laminated, 1 cm thick tidal deposits to medium-grained, massive cross-bedded, or bioturbated shoreface and bar deposits; their lateral continuity is generally at the scale of the field. Quaternary terrace deposits are observed along the northern Sarawak coastline, deposited in fluvial to marginal marine depositional environments; these deposits were uplifted above the present-day sea level (Kessler and Jong, 2014). These deposits mark an angular unconformity on the top of the Sibuti, Lambir, Tukau and Miri formations that may have originated as an intratidal abrasion surface (Kessler, 2005).

2.1.1 The Coastal Plain

The alluvial coastal plain can be observed along the shoreline of the Miri coast and the alluvium formation can be seen on the banks of the rivers in the study area. The shoreline areas near the Baram and Miri River are characterized by peat soil, mangrove, nipah, swamp forest, and tidal inundation. As a result of its composition and the tropical monsoon climate in central Borneo, erosion rates in this range have been among the highest in the world since the Eocene (Sandal, 1996; Straub et al., 2012). The coast of Miri consists of well-developed sandy beaches resulting from a strong SW longshore drift and a relatively high offshore wave action. The beaches in this region are dominated by sandy texture with open stands of *Casuarina equisetifolia*, coarse grasses, and shallow swamps running parallel to the coast in most places. The Miri-Kualabaram highway, part of the Lutong beaches area, the Miri River estuary area, Marina bay, and the Marriot Resort and Spa are protected by seawalls and breakwater structures to prevent wave erosion (DID, 2009). Parts of the shoreline, such as the crude oil station in Lutong and the Tanjung Lobang Beach area, are protected with groins. The purpose of these groins is to interrupt the water current flows and limit the movement of sediments. In some places (the Kualabaram River mouth and Tusan beach), wave erosion is commonly noticed during the monsoon (NE monsoon) season due to strong waves from the South China Sea while sediment input from the Baram River is likely to be deposited all along the coast due to longshore currents (Nagarajan et al., 2015a). This is indicated by a low salinity regime prevailing across the coastal region (though there are seasonal variations). Rock shore faces are observed along the coastline such as in Tanjung Lobang, Beraya, Tusan, and Bungai beaches.

3 MATERIALS AND METHODS

In the present study, the LULC changes of a 713 km^2 area, which included a 74 km coastal belt adjacent to Miri, were analyzed by GIS for a period covering 51 years using two different datasets. Information relating to the LULC of the study area was extracted from two sources: a topography map for the year 1963 (scale 1:50,000), produced by the Department of Survey and Mapping, Malaysia, and, for more recent LULC information, a Landsat 8 operational land imager (OLI) sensor image acquired on May 23, 2014. Besides the topography map and 2014 satellite image, satellite images from three different period corresponding to 1988, 1991, and 2001 were also used to detect the temporal changes in the shoreline. Any LULC changes were analyzed by comparing the area statistics of different LULC classes and shoreline changes by the overlay analysis in the GIS, using different temporal shoreline maps derived from various sources. For change detection studies, a 1963 topography map from the Malaysian Department of Survey and Mapping was used to delineate the base LULC and shoreline of the study area. Landsat satellite images of path 119 and row 057 covering an area between 113°46′59″E and 114° 03′19″E longitude and 4°00′03″ N and 4°36′56″N latitude were also used for this study. The Landsat datasets corresponding to the study area were downloaded from the Earth Explorer website (http://earthexplorer.usgs.gov/), which hosts a multitude of digital datasets. Because the area is characterized as a tropical rainfall region, there was some difficulty in obtaining a cloud-free image of the study area. The image chosen for the present study is comparatively cloud-free and was used for further analysis. The details of the datasets collected and used in the analysis are given in Table 1 and also shown in Fig. 2.

TABLE 1 Sources of Datasets Used in the Analysis of Its Characteristics

Data Source	Year	Scale	Bands	Date of Acquisition	Resolution (m)
Toposheet	1963	1:50,000	–	–	–
LandSat-2	1988	1:50,000	2,3,4	1/17/1988	60
LandSat-5	1991	1:50,000	2,3,4	4/18/1991	30
LandSat-5	2001	1:50,000	2,3,4	7/10/2001	30
LandSat-8	2014	1:50,000	3,4,5	5/23/2014	30

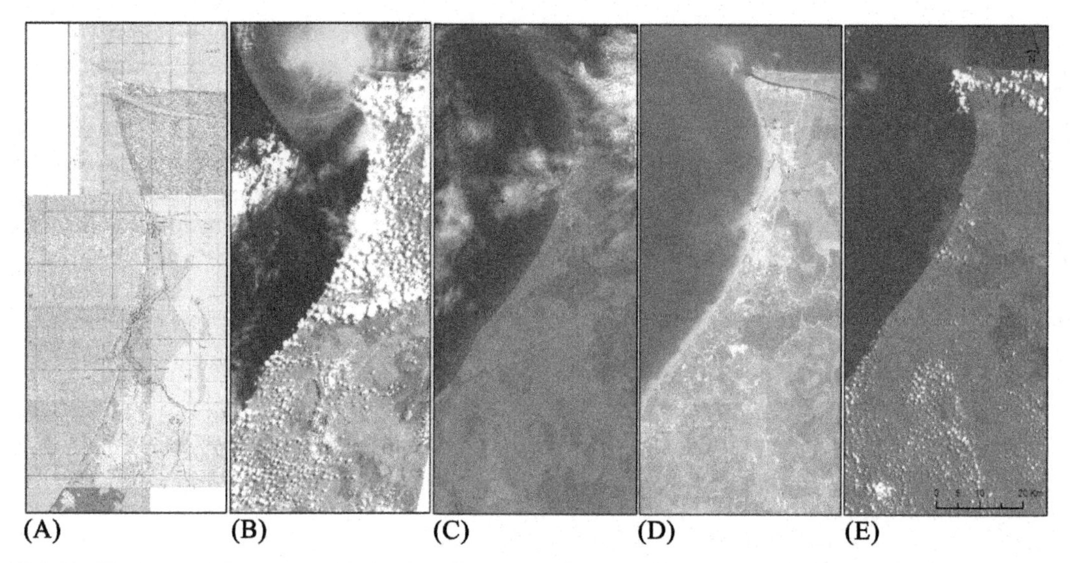

FIG. 2 The temporal datasets used in the shoreline change detection. (A) Topographical maps—1963, (B) Landsat 2 image—1988, (C) Landsat 5 image—1988, (D) Landsat 5 image—2001, and (E) Landsat 8 image—2014.

3.1 Data Processing

The historical dataset available for the study region (topography map) was georeferenced to create a seamless image of the study area. Initially, the known coordinates of the topo-map were entered as decimal degrees and the images were then projected in rectified skew orthomorphic (RSO) with datum as Timbale 1948. The extent of the study area was fixed by designating a 10km buffer from the shoreline, which existed in 1963 with an area of 713 km^2. The downloaded Landsat images were coregistered to the topography map in order for use in the LULC analysis. Basic image correction techniques were applied to the satellite images to minimize the errors associated with the reflectance values. The interpretation of historical LULC from the topography map and more recent LULC changes from the additional images was carried out as an on-screen visual interpretation and digitization process using ArcGIS 9.3 software.

For the shoreline analysis, the interval was fixed at every 10 years, but due to the absence of satisfactory and continuous Landsat images covering 1972–80, the first change year was kept as 1988. After 1988, satellite images were collected for 1991, 2001, and finally 2014. The baseline data from the topography map was fixed as the master data layer and all the images were coregistered to the topography map by cross-matching using identifiable and detectable common points. Once the coregistration process was over, the individual datasets were projected in RSO with datum as Timbale 1948 and used to interpret the shoreline area for the corresponding years. The generation of the shoreline area for the corresponding years started with the digitization of the topography map-based shoreline for 1963 in ArcGIS. The same technique was used to delineate the shoreline layers corresponding to the satellite images collected; this resulted in five different shoreline vector files based on the year (shape file format supported by the ArcGIS was used for this purpose).

In order to identify, determine, and calculate the shoreline changes (area changes) between the different time periods, the merge tool available in the Arc Toolbox was used. The merging of different time shorelines has resulted in a new output of overlapping lines in some areas. Due to the variation or changes in shorelines in different time periods, the change calculation in the area was facilitated by converting merged output to polygons by using the feature-to-polygon conversion tool in the Arc Toolbox. This resulted in a polygon layer with different numbers of polygons in the overlapping regions and the automatic geometry calculation tool was used to attribute areas to each polygon created. Each polygon was assigned with erosion and accretion labels for better understanding. Ground truth field trips were conducted during the sediment sample collection for the geochemical study (2014–15) to ensure the classified land uses and land cover in the study area.

4 RESULTS AND DISCUSSION

4.1 LULC—Base Map Versus 2014

The various LULC types identified from the topographical maps were artificial surfaces, settlements (urban/semiurban area), plantation (rubber), forest (primary and secondary), grass lands, swamps, river/water bodies, tidal/barrier islands, beaches or tidal flats, and rocky cliff beaches. The distribution of identified LULC classes in 1963 is presented in Fig. 3 and tabulated in Table 2. The distribution of LULC areas in the particular time period provides information about the spatial distribution of various kinds of LULCs. Among the delineated LULC types, primary forest covers >66%, followed by secondary forest (~27%), and swampland at (~3%) of the total area. All other LULC classes together cover <5% of the total area. Furthermore, during 1963, plantation activities such as rubber and coconut were taking place and rubber plantations in particular covered >1% of the total study area. The extracted information from the 1963 topography map was considered as the baseline for further LULC change analysis in the study area.

The assessment of the temporal pattern of LULC in the study area was facilitated with the use of recently available satellite imagery (Landsat image 2014). The major types of LULCs identified from the satellite images were artificial surfaces, settlements (urban/semiurban area), plantations (palm oil and chili), forests (primary and secondary), grass lands, swamps, river/water body (lakes), tidal/barrier islands, beaches or tidal flats, rocky cliff beaches, household agriculture, cleared areas, afforested areas, secondary mixed forests with oil palms, and golf courses. Few new types of LULC were introduced during 2014 compared to baseline data from 1963 and other existing LULC areas were modified to other types. Table 2 shows the LULC types identified during 2014 with the area covered by each type and their spatial distribution is shown in Fig. 4. Among the LULC patterns identified for 2014, the majority of the area is covered by secondary mixed forest with oil palm (45%) followed by settlements (urban + semiurban 19%), palm oil plantations (14%), and afforested areas (11%). It was also noted that the extent of the remaining primary forest was reduced to 3% while the secondary forest was 2% of the total area.

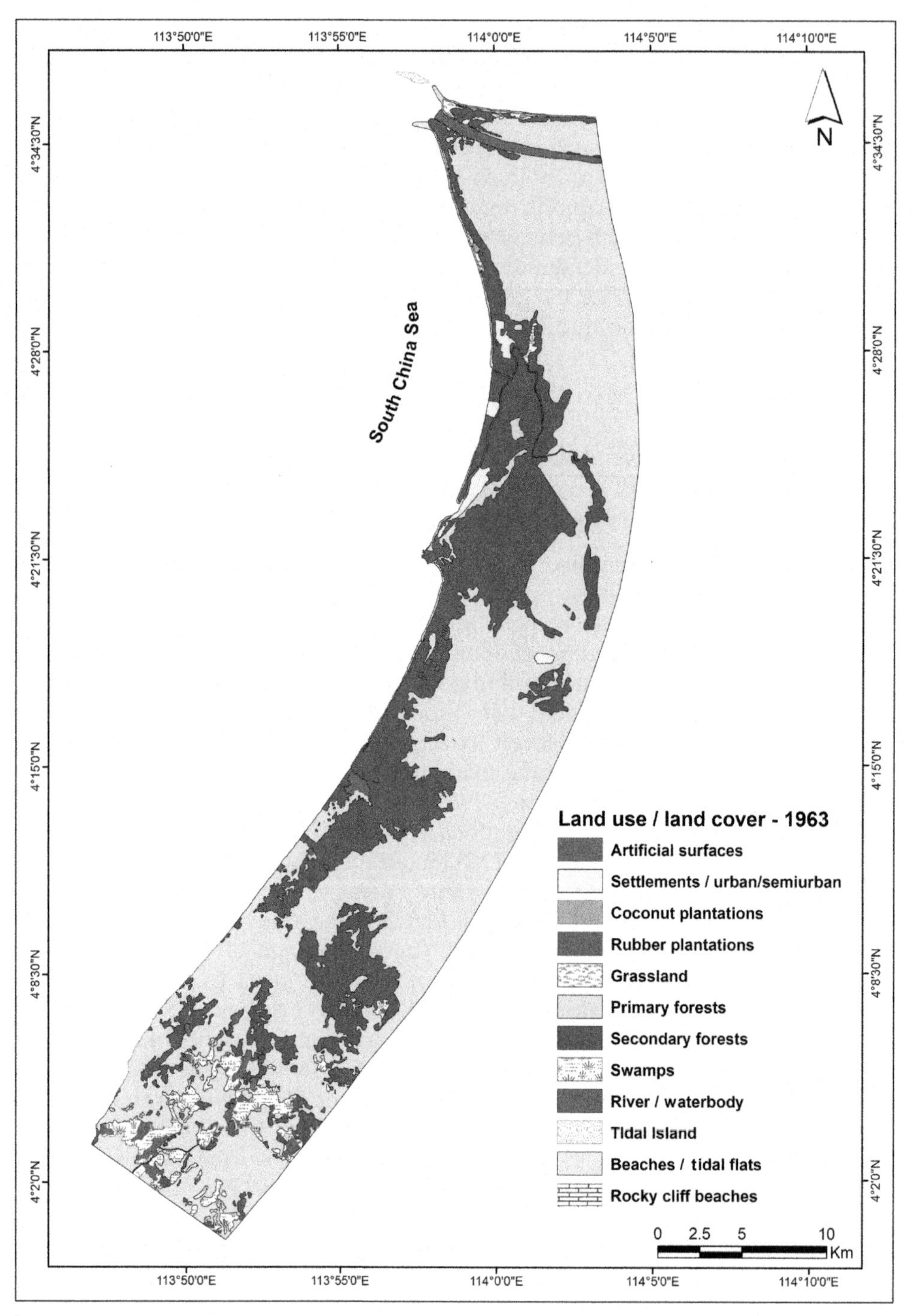

FIG. 3 LULC map of 1963. *Source: Topographical map by Dept. of Survey and Mapping, Malaysia.*

TABLE 2 LULC Types During 1963 Versus 2014

	1963		2014		
Land Use/Land Cover	**Area (km^2)**	**Area (%)**	**Area (km^2)**	**Area (%)**	**Change % 1963–2014**
Artificial surfaces	0.089	0.01	2.29	0.32	+0.31
Settlements/urban/semiurban	**5.05**	**0.71**	**136.68**	**19.16**	**+18.46**
Coconut plantations	0.28	0.04	–	–	−0.04
Rubber plantations	9.53	1.34	–	–	−1.34
Palm oil plantations	**–**	**–**	**104.98**	**14.72**	**+14.72**
Chilly plantation	–	–	0.08	0.01	+0.01
Grassland	1.89	0.26	–	–	−0.26
Primary forests	**471.77**	**66.11**	**26.56**	**3.72**	**−62.39**
Secondary forests	**191.41**	**26.82**	**14.33**	**2.01**	**−24.82**
Secondary mixed forest with oil palm	**–**	**–**	**327.38**	**45.90**	**+45.90**
Swamps	21.9	3.07	–	–	−3.07
River/water body	6.79	0.95	8.31	1.17	+0.21
Tidal/barrier island	0.6	0.08	0.95	0.13	+0.05
Beaches/tidal flats	4.04	0.57	1.82	0.26	−0.31
Rocky cliff beaches	0.22	0.03	1.21	0.17	+0.14
Household agriculture	–	–	1.06	0.15	+0.15
Golf courses	–	–	1.46	0.20	+0.20
Cleared area	–	–	1.48	0.21	+0.21
Afforested area			**84.61**	**11.86**	**+11.86**
Total	**713.56**	**100**	**713.56**	**100**	

The bold letters indicate significant changes in land use / land cover (increase/decrease during the period of analysis).

4.1.1 LULC Analysis—1963–2014

The changes in the LULC in an area can be analyzed by comparing LULC maps between different time periods. In the present study, in order to identify the intensity and significance of LULC changes for the last 51 years (1963–2014), LULC maps of both years were overlaid in the ArcGIS software using the analysis support tools. This process facilitated the assessment of spatial changes for different types of LULCs from one time period (1963) to more recent times (2014) and also provided the area statistics of each class. The area calculation derived from the overlay analysis is given in Table 2. In 2014, few new classes of LULC types are identified and they occupy 91% of the total area. While comparing the LULC maps between 1963 and 2014, it become quickly apparent that the most obvious changes occurred to the primary

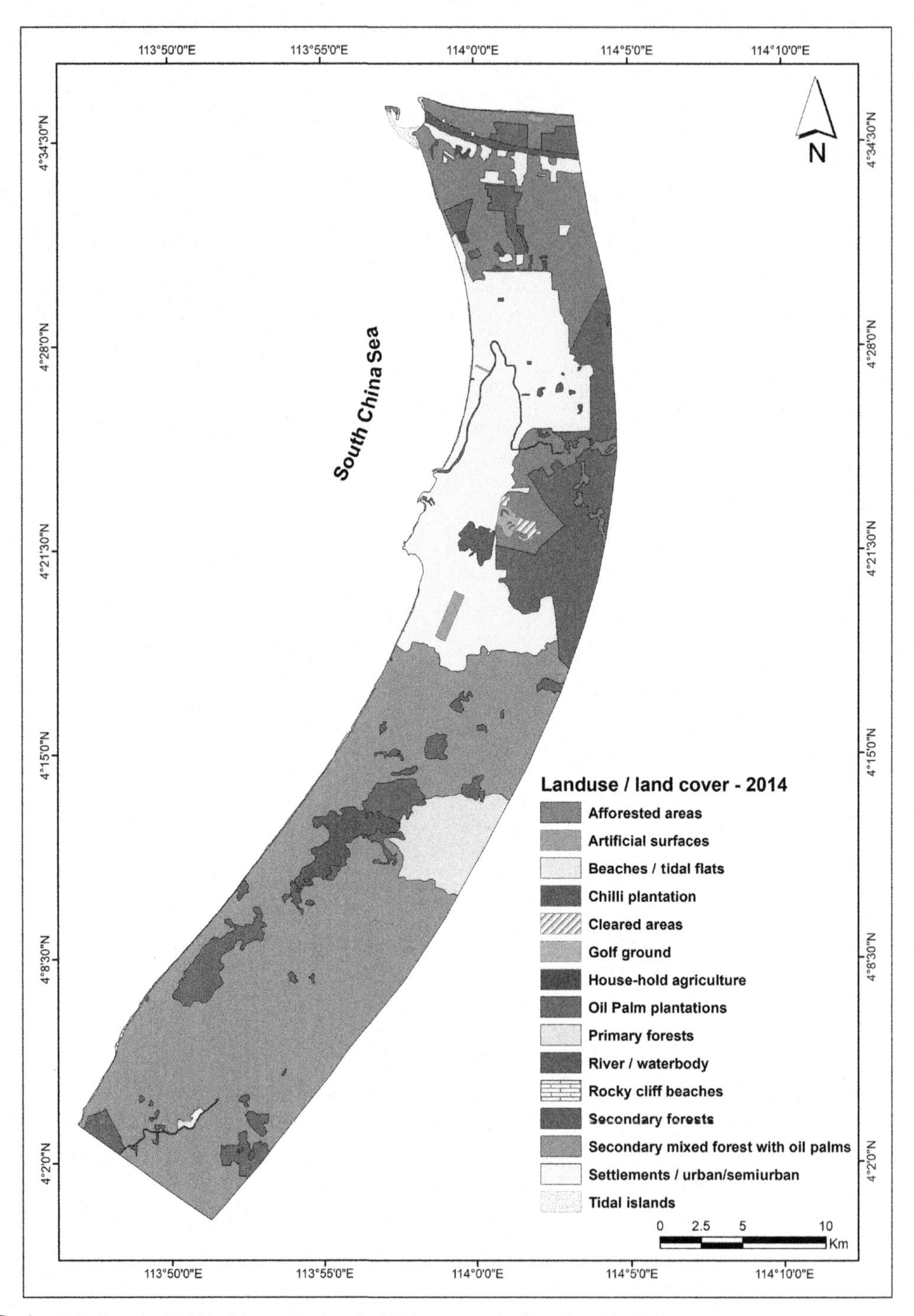

FIG. 4 LULC map of 2014. *Source: Landsat 8, OLI image acquired on June 23, 2014.*

and secondary forest cover types. Primary forests, which covered nearly 90% of the total area in 1963, were reduced to 5% of the total area by 2014.

Southeast Asian forests experience a high rate of degradation, which has resulted in less than half the original forest cover remaining (i.e., 49% in 2000; Achard et al., 2002; Corlett and Primack, 2008). The rate of deforestation is about 0.64% per annum, which is related to the expansion of palm oil plantations, increasing at the rate of 10.2% annually (Hon and Shibata, 2013). Primary and secondary forests have given way to palm oil plantations, settlement (urban/semiurban) developments, afforestation, and secondary mixed forests with oil palms. This forest conversion process continued as primary forest areas were cleared by logging companies for timber and replanted as palm oil plantations; this has emerged as a major commercial crop. The distribution of changes in the LULC during the time period covered was calculated in percentages and any variations were indicated as plus (increase) or minus (decrease) in Table 2; the major changes during the time period are shown in Fig. 5. The results of this comparison point toward the destruction of tropical primary forests for conversion to palm oil plantations in the area, which ultimately contributes to environmental changes in the area. According to the recent study by Gaveau et al. (2014), Sarawak has 36.4% logged forest and 17.4% intact forest as of 2010. Also, Achard et al. (2002) has indicated that Borneo has lost the forest nearly twice as fast as the world's humid tropical forest, in part to create and expand palm oil plantations (Gaveau et al., 2014). The present study resembles the similar thread in

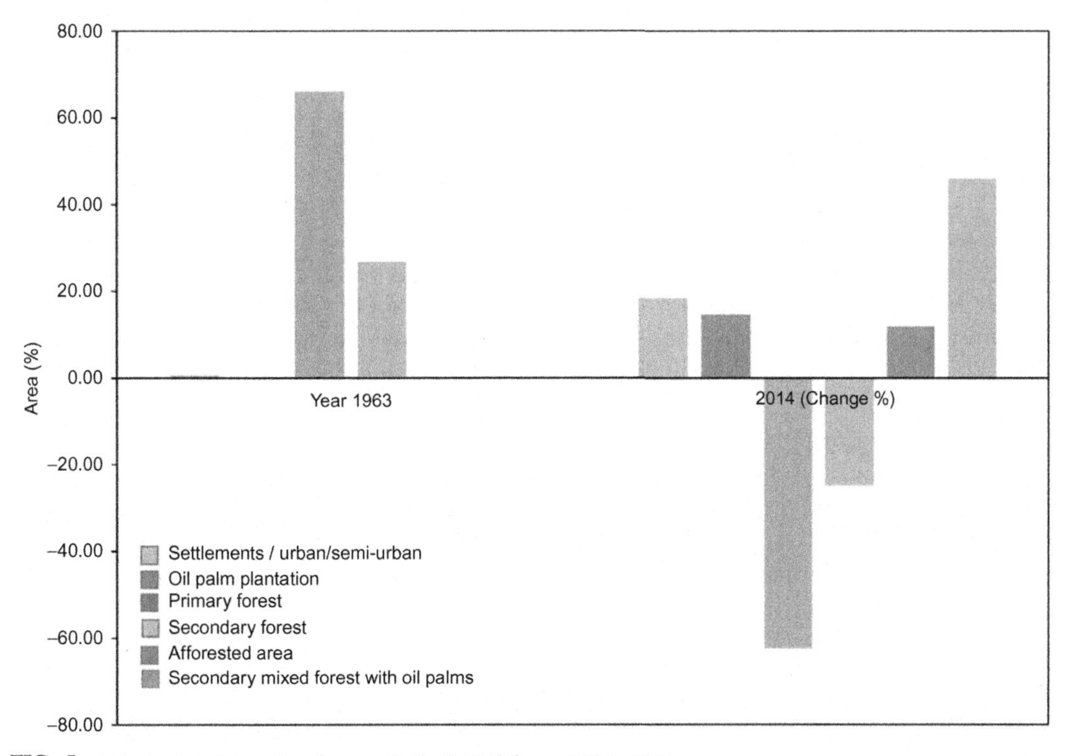

FIG. 5 Graph showing major changes in the LULC from 1963 to 2014.

the coastal region of Miri. Upon discovery of oil resources in the early 19th century in Miri, there was an increase in migration from the rural areas, which led to urban infrastructure developments all along the Miri coast. The construction of offshore oil and gas extraction plants and other industrial developments in the Miri region caused further degradation of the natural environment and resources along the coast. Most of the initial settlements by these new immigrants in this region were observed near the river banks and coastal areas. Miri is rapidly becoming one of Sarawak's most popular tourist destinations and is the gateway to many national parks, such as Gunung Mulu National Park, Lambir National Park, Niah National Park, and Loagan Bunut National Park. Beautiful beaches, exotic tropical rainforests, colorful coral reefs, and underwater gardens are some of the attractions for divers and tourists from all over the world. Miri has 12 diving spots that attract tourists to explore the beauty of underwater wonders. Many areas of live coral are visible along the Miri coast (Jacqueline, 2013). Thus, urban development due to the development of the oil and gas industries as well as the tourist sector has influenced LULC changes in the study area, in particular in Miri.

4.2 Shoreline Changes

Based on the analysis techniques used in the present study, the changes in the shoreline are assessed in two different approaches. In the first analysis, shoreline changes between the years was carried out and in the second approach, total changes from the base year 1963 through the final year of analysis in 2014 were performed. The results show spatially significant changes occurred along the Miri coast, including both erosion and accretion, which can be attributed to natural and man-made activities. In order to make the assessment easy and more understandable, the shoreline of the study area is divided into six regions/beaches (A. Baram River mouth, B. Miri River estuary, C. Tanjong Lobang beach, D. Esplanade beach, E. Hawaii beach, and F. Bungai beach), as shown in Fig. 6.

The simple overlay analysis of shorelines at different years (1963, 1988, 1991, 2001, and 2014) shows variations from year to year and also from the base year (1963) to respective temporal years. This facilitated the identification of erosion and accretion zones along the shoreline and their quantification through estimation of the area under each class. Some areas of the shoreline (in relation to the base shoreline in 1963) have expanded outward, indicating accretion, and in other areas the shoreline has contracted inland, indicating erosion.

4.2.1 Assessment of Periodic Shoreline Changes

The characteristics of the shoreline during the periods studied (1963–2014) were assessed by comparing the fluctuations between shorelines in respective years. This provided cumulative shoreline variable maps of compared timeframes with prominent erosion and accretion zones along the total stretch of the study area. The area was calculated for each timeframe to understand the variability in shoreline changes (Table 3).

The periodic changes of the study area due to erosion and accretion were clearly defined and are shown in Fig. 7. During the period from 1963 to 1988, erosion and accretion are recorded as 634 and 215 acres, respectively. Similarly, the erosion and accretion of the shoreline area between 1988 and 1991 show a reversed pattern with less erosion (147 acres) and high accretion (356 acres). The accretion is related to the high sediment input from the Baram

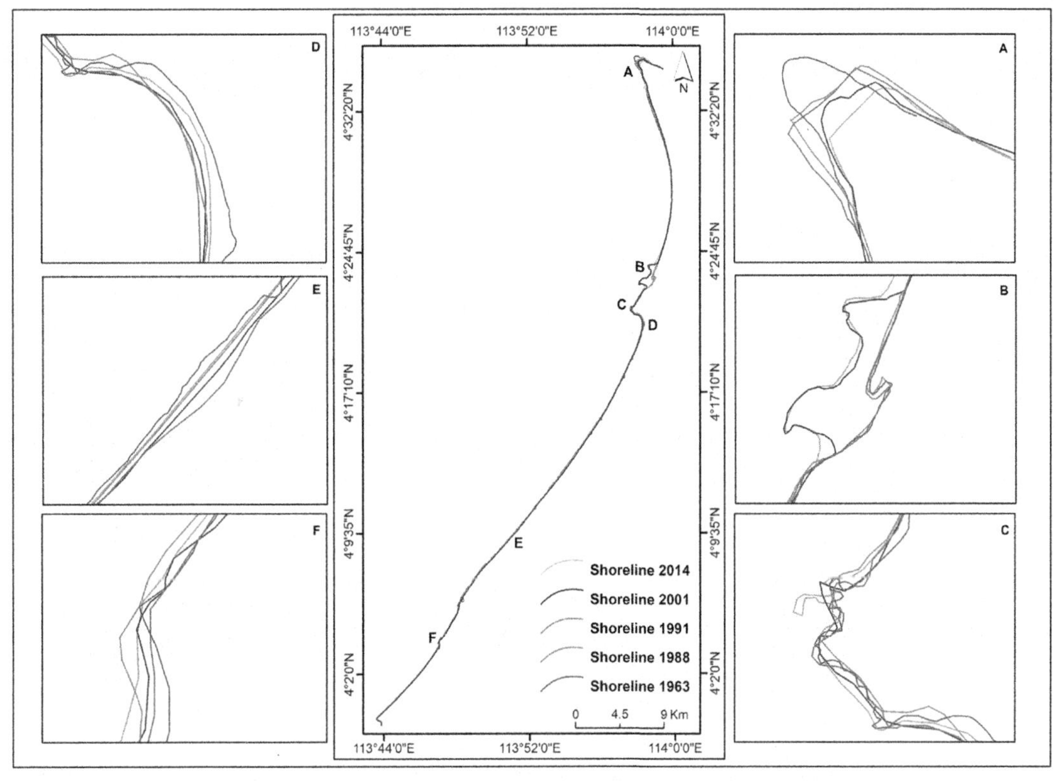

FIG. 6 Shoreline for the four time periods with selected segments (A. Baram River estuary, B. Miri River estuary, C. Tanjong Lobang beach, D. Esplanade beach, E. Hawaii beach, and F. Bungai beach).

TABLE 3 Quantification of Erosion and Accretion Between Different Years

Year	Erosion[a]	Accretion[a]
1963–1988	634.04	215.87
1988–1991	147.19	356.18
1991–2001	660.87	464.12
2001–2014	291.97	709.73

[a] *All units are in acres.*

River, where timber extraction was high during this period or earlier. High erosion in the Baram River catchment area after the timber extraction and before the land use changed to either secondary forest or palm oil plantations has supplied surplus sediment and suspended matter, which resulted in accretion in the coastal area of Miri where the Baram confluences to South China Sea. However, during the period from 1991 to 2001, the eroded area was larger than the accretion area. During this period, several rock revetment programs (along the

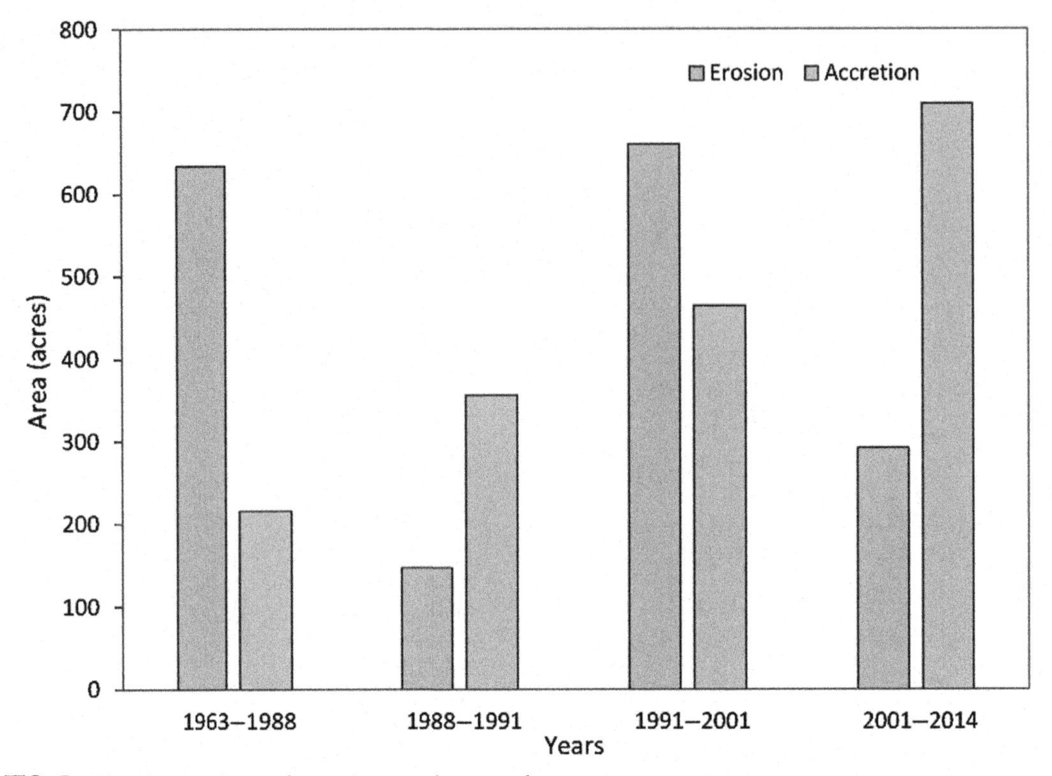

FIG. 7 Shoreline erosion and accretion rates between the years.

Lutong-Kualabaram Road) and construction of a parallel breakwater structure (at Miri River mouth) were carried out along the shoreline, which escalated the process of erosion and accretion simultaneously due to the effect of the modified shoreline. The effects of these activities were reflected in the analysis as accretion and erosion totalled a maximum of 464 acres and 660 acres, respectively. From analyzing the entire stretch of shoreline in the study area, it was determined that the shoreline protection measures covered approximately 25 km from the Baram River mouth to Marina Bay. These measures have augmented erosion in unprotected areas and depositions between the parallel breakwater structures. This supports the concept of decades of erosion and accretion along the shoreline. This concept is again reinforced by assessing the areas eroded and accreted during 2001–14. During this period, erosion was down to 291 acres with increased accretion of 709 acres, and this can be attributed to large-scale land reclamation that took place at the Marina Bay, located near the Miri River estuary. During the analyzed time frame, maximum erosion was recorded between 1991 and 2001 (660 acres) and the maximum accretion was from 2001 to 2014 (709 acres). The spatial analysis of erosion and accretion classified the Baram River mouth area as the area of maximum erosion and the Miri River estuary area as the point of maximum accretion. Kampong Baraya beach shows a trend of erosion whereas the Esplanade, Tanjong, and Bungai beaches show evidence of shoreline accretion to varying degrees.

4.2.2 Shoreline Changes from the Base Year (1963)

The second approach analyzed the changes of the entire shoreline from the base year, which provides a historical perspective of the erosional and accretion processes that have taken place along the total stretch of the study area with regards to natural and man-made changes. In order to generate the statistics and also assess the spatial changes, individual shoreline areas from respective years are cross-compared with the base shoreline map; the analysis results are shown in Table 4 and Fig. 8.

The results show similar characteristics of shoreline changes between different times. The comparison of the shoreline in 1988 with the base year shows more erosion than accretion at 634 and 215 acres, respectively. The analysis reveals a major change in the Kualabaram region (the shape of the Baram River mouth itself changed), which reflects greater erosion taking

TABLE 4 Quantification of Variation in Erosion and Accretion From the Base Year (1963)

Base Year	Change Years	Erosion[a]	Accretion[a]
1963	1988	634.04	215.87
	1991	447.32	234.61
	2001	816.25	545.23
	2014	546.72	723.35

[a] *All units are in acres.*

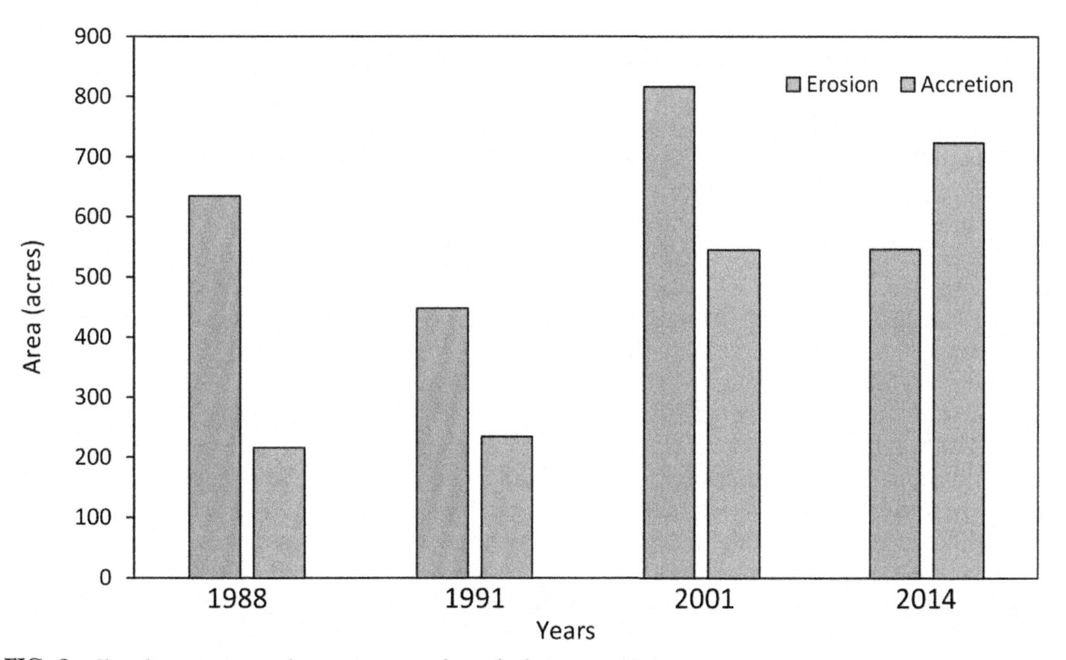

FIG. 8 Shoreline erosion and accretion rates from the base year 1963.

place. But a shoreline characteristic from 1991 was entirely different compared to 1988 and the base year 1963, showing decreased erosion at 447 acres and slightly increased accretion at 234 acres. At the same time, the comparison of the base year shoreline with 2001 data showed that shoreline erosion with an area of 816 acres and accretion of 545 acres. This correlates with the shoreline protection measures implemented in selected stretches such as rock revetments and breakwaters, which exaggerate both erosion and accretion by intervening in the natural process of ocean currents. The highest rate of accretion (723 acres) in 2014 also had a comparatively higher rate of erosion (547 acres). The rate of accretion shows an increasing trend during 2001–14 because more coastal areas have been reclaimed in the Miri River estuary area. Landfill reclamation has changed the coastal morphology and has been reported from many coastal cities of Malaysia. Large-scale reclamation for unban expansion or as a land bank has taken place in Malaysia over the recent past (Bird and Teh, 1990). These reclamation projects have caused erosion in other areas along the shoreline due to the differential wave action. A closer examination and cross-comparison of the analyzed results of both approaches show a natural cycle of decades of reversal in erosion and accretion along the shoreline studied. Earlier, during 1963–2001, ocean currents determined the erosion and accretion process and followed a more natural general pattern of coastal processes. However, after 2001, human-induced activities (such as rock revetments, construction of groins, and breakwater structures) along the shoreline have changed the total scenario by changing the natural erosion and accretion processes.

4.2.3 *Spatial Shoreline Changes During the Period 1963–2014*

An attempt has been made to detect the total spatial changes in the shoreline for the past 51 years (1963–2014) with the help of GIS. The major changes were identified and the total area of erosion and accretion was also calculated. During those 51 years, 547 acres of shoreline were eroded and 723 acres of shoreline were accreted. This facilitated the identification and differentiation of natural (strong waves) and man-made changes (seawall construction, groins, and breakwaters) along the total length of the shoreline studied. Fig. 8 shows the major areas of erosion and accretion in the study region during the cumulative period of assessment. The data, show that the maximum variation of either erosion or accretion of the shoreline is concentrated at four major beach areas: the Baram River mouth, the Miri River estuary, the Tanjong Lobang beach, and the Esplanade beach. The maximum change in shoreline characteristics and shape was observed in two areas, the Baram River outlet and the Miri River mouth. The first is the site of maximum erosion and the latter is the site of maximum accretion.

Close examination of changes in the Baram River estuary region indicates the changes of shoreline to natural to man-made by the shape itself (Fig. 9). Due to the heavy rainfall in the upper Baram region, sediments and organic debris are washed from the deforested agricultural lands, mixed with the Baram River and finally drained into the Miri coast. An enormous amount of sedimentation has led to the creation of a barrier island in front of the Baram River. During the monsoon seasons, a heavy input of finer sediments and suspended particles builds up this barrier island and sand spit, extending it southward (Nagarajan et al., 2015a). On the mouth of the Baram River, a barrier island has been formed that is connected to the beach by the elongation of spits, which is evidence of accretion. Though the migration of spit and the thinning of the barrier island is also common, this depends on the sediment

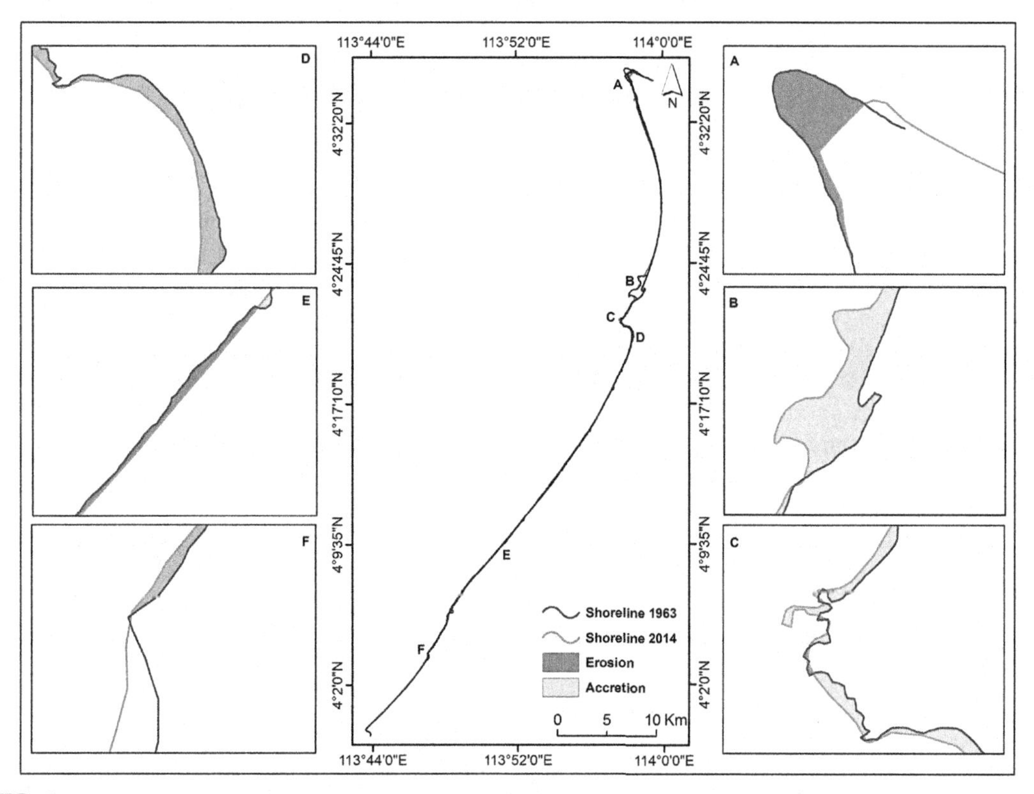

FIG. 9 Major changes in shoreline during 1963–2014 (A. Baram River estuary, B. Miri River estuary, C. Tanjong Lobang beach, D. Esplanade beach, E. Hawaii beach and F. Bungai beach).

supply from the upstream side of the Baram River. The location of the barrier island has moved over four decades and is currently connected with the beach by the sand spit extending southwest (Nagarajan et al., 2015a). The accretion and erosion in the Baram River mouth can be linked with the development of the spit (accretion) and the migration/morphological changes of the barrier island (erosion). According to Bird and Teh (1990), the elongation of spits and their successive development as a beach ridge adjacent to the river mouths are the most prominent geomorphic evidence of accretion. Though the erosion and accretion processes are natural, man-made activities (e.g., logging and land-use changes) induced the erosion in the upstream. The catchment area of the Baram River covers approximately 22,800 km^2 and the average discharges is estimated as 1590 m^3/s of water and 2.4×10^{10} kg/yr of sediment load (Sandal, 1996; Straub and Mohrig, 2009). The input of sediments from the Baram River into the coastal zone mainly consists of fine sand, where as silt dominates in the SW side of the Miri River (Anandkumar, 2016). The Baram River drains into the South China Sea at Kualabaram, and serves as the main means of transportation for timber, sand, and rock from regions upstream. The increased sediment delivered by the Baram River into the confluence point may disrupt the efficient navigation of big ships by reducing the depth of the river delta through sediment deposit. In order to make navigation more efficient and safer, frequent dredging activities and shoreline protection measures are being implemented in the area by the port authorities (DID, 2009). This extended the shoreline into the South China Sea to transform the original shape into a comparatively straight shoreline with protective measures against erosion (Fig. 10).

The DID Manual (2009) classified the coastal erosion in the Marina Bay resort area as category 1, based on the National Coastal Erosion Study program (1986): "Shoreline is currently in a state of erosion and shore-based facilities or infrastructure are in immediate danger of collapse or damage"). Similar to the Baram River mouth region, the Tusan Cliffs nearby (Kampong Baraya beach) showed continued erosion caused by strong ocean wave action. The Tusan beach consists of dominant sandstone, moreso than the Peliau to Bungai beaches. Coastal erosion features such as sea caves, arches, and failure of arches are evidence of natural coastal erosion. Many sandstone cliffs have eroded, and, in particular, shale beds have eroded more rapidly than sandstone cliffs in a short span of time due to wave erosion and also by cliff retreat owing to rainfall-induced gully erosion from the land surface (Fig. 11). This is classified as shoreline erosion caused by natural processes. However, a detailed study is required to investigate the coastal erosion processes in the rocky shore beaches.

The Miri River estuary region exhibits the combined effect of natural and man-made factors that contribute to shoreline accretion. The construction of parallel breakwater structures and the reclamation of land for the Marina Bay projects (2001–14) were the main man-made activities observed along the Miri coast in the last decade. The river mouth was also dredged at least once to allow bigger boats to enter easily. In addition, in the early 2000s, the land

FIG. 10 Rock revetment projects in the shoreline of Miri (A). Rock revetment Miri—Kualabaram Road (B). Rock revetment near Lutong beach, (C) Rock revetment and groins in Tanjung Lobang beach (D). Rock revetment in Marina Bay.

FIG. 11 Effect of erosion on the shoreline of Miri; (A) Erosion due to strong waves between Baraya and Tusan beach (Lambir formation); (B) Erosion and accumulation of organic debris near Tanjung Lobang beach; (C) Sea arch structure with the most eroded mudstone beds than sandstone at Tusan beach (Lambir formation); and (D) Cliff retreat and rainfall induced gully erosion at Tusan Beach (Lambir formation).

reclamation project remodeled the Miri river mouth completely. This includes seawalls being built, the river mouth being relocated, and the addition of a marina bay. Previous topographical features of Miri River mouth indicates that it was deflected towards the south by the sand spit due to the sand supplied by the Baram River. This spit is considered the most intensively developed spit in Malaysia. However, land reclamation has changed the shape of the Miri spit and a new channel has been dredged on the neck of the spit; meanwhile, the old mouth of the river has been filled (Teh and Boon, 2003). This construction activity at the Miri River estuary region has changed the natural balance of erosion and accretion through the extended introduction of wave protectors in the sea as well as increased the accretion process along one side of the extended river mouth. The characteristic shape of the shoreline studied indicates an open concavity toward the South China Sea. This indicates the presence of spatially separated open bays in the region. The open bay shape acts as zones of higher erosion through the littoral drift. The mechanism behind the erosion is the powerful waves that interact with the shoreline at an inclined angle (varying with a range of 35–450), which scoop the material in the wash and backwash zones (Tang and Lee, 2010). This eroded material is deposited near the breakwaters or extended structures, which reduces the sediment-carrying capacity of waves by obstructing the path. This major man-made project can be linked with the major accretion of sediments at the mouth of the Miri River. In addition, the accumulation of peat soil on the beach near the Miri River is also common following the monsoon season.

Observations on the other regions south of Kampong Baraya have shown that the shoreline seems unbroken with little concavity toward the South China Sea. Shorelines in these regions undergo receive frequent strong wave events, which results in the retreat of rocky shorelines (south Baraya beach, Tusan beach, and Peliau-Bungai beach) and the associated landward extension of beaches. The eroded sediments are carried away by the seasonal longshore drifts into different directions.

5 SUMMARY

An attempt has been made to characterize the shoreline in the study area through the assessment of LULC patterns and shoreline changes over a period of 51 years (1963–2014). A significant variations in the LULC classes are related to conversion of primary and secondary forests into new LULC classes such as a rapid increase in the number of settlements in the urban and semiurban areas, introduction of palm oil plantations, afforestation, and secondary forests with palm oils etc.as recorded in 2014. Similar trends in LULC changes were also noticed along the shoreline of the region that has either eroded or accreted. Shoreline change detection over the different time periods shows a decade-long cyclic reversal of erosion and accretion along the shoreline of Miri. Man-made changes along the shoreline, particularly near the Miri River mouth (Marina Bay area), rapidly expanded after 2001 as construction of seawalls, groins, breakwater structures, and reclamation activities occurred, whereas the coupled effect of natural and man-made factors caused the erosion and accretion process along the shoreline of Tanjong Lobang and Esplanade beaches.

The assessment of shoreline erosion and accretion within the different periods showed a maximum erosion of 660 acres, which was observed during 1991–2001, and the maximum accretion was 709 acres of shoreline observed from 2001 to 2014. Compared to the baseline map, the maximum erosion of 816 acres and accretion of 723 acres are recorded during 2001–14. This can be attributed to the implementation of rock revetments, groins, and breakwater structures along the eroding beaches, and land reclamation near the Miri river mouth. Simultaneously, a large quantity of sediment input from the Baram River, carried by the longshore currents, was deposited along the beaches wherever the breakwater structures were present. The total shoreline changes during the last 51 years show more accretion than erosion with variable effects by natural, man-made, or combined factors. During the study period, a total of 546 acres of shoreline was eroded and 723 acres of shoreline were constructed or accreted. The spatial pattern of shoreline changes during the period facilitated the identification of four major zones of erosion and accretion along the total stretch of shoreline studied: the Baram River mouth (erosion), the Miri River estuary, and the Tanjong Esplanade beaches (accretion). The long-shore currents and strong waves (natural) were amplified by man-made intervention in the form of construction of the resort city and sports complex at Marina bay, development of recreational areas (Bungai and Tanjong beach), and dredging activities at the Kualabaram River mouth (human activities); these are the major causes of the erosion/deposition process along the Miri coast.

The present findings of the study serve as background information for the planning of future developmental activities by the coastal management board, environmental protection department, and sustainable development authorities. Based on this regional study, it is recommended that more studies should be carried out at different beaches to understand the multiple factors involved with coastal erosion in this region.

Acknowledgments

First author (AAK) would like to thank Curtin Sarawak Research Institute, Curtin University for the Scholarship through CSIR research project (CSRI2011/Grants/01 Dr.Ramamsay Nagarajan) to carry out PhD degree under Department of Applied Geology, Curtin University, Malaysia. Our sincere thanks to Ms, Mary Jiew, Jabatan Meterologi Malaysia for providing the rainfall and wind data.

References

Achard, F., Eva, H.D., Stibig, H.J., Mayaux, P., Gallego, J., Richards, T., Malingreau, J.P., 2002. Determination of deforestation rates of the world's humid tropical forests. Science 297, 999–1002. https://dx.doi.org/10.1126/science.1070656.

Anandkumar, A., 2016. Ecological Risk Assessment of the Miri Coast, Sarawak, Borneo—A Biogeochemical Approach. (Ph.D. thesis). Curtin University, Malaysia. 353 pp.

Avalar, S., Tokarczyk, P., 2014. Analysis of land use and land cover change in a coastal area of Rio de Janeiro using high-resolution remotely sensed data. J. Appl. Remote. Sens. 8 (1), 1–15.

Banda, R.M., 1998. The Geology and Planktic Foraminiferal Biostratigraphy of the Northwestern Borneo Basin Sarawak, Malaysia (Ph.D. thesis). University of Tsukuba, Japan.

Berlanga-Robles, C.A., Ruiz-Luna, A., 2002. Land use mapping and change detection in the coastal zone of Northwest Mexico using remote sensing techniques. J. Coast. Res. 18 (3), 514–522.

Bird, E.C.F., Teh, T.S., 1990. Current state of the coastal zone in Malaysia. Malays. J. Trop. Geogr. 21 (1), 9–24.

Blodget, H.W., Taylor, P.T., Roark, J.H., 1991. Shoreline changes along the Rosetta-Nile promontory: monitoring with satellite observations. Mar. Geol. 99 (1), 67–77.

Boak, E.H., Turner, I.L., 2005. Shoreline definition and detection: a review. J. Coast. Res. 21 (4), 688–703. https://dx.doi.org/10.2112/03-0071.1.

Carlson, T.N., Arthur, S.T., 2000. The impact of land use—land cover changes due to urbanization on surface microclimate and hydrology: a satellite perspective. Glob. Planet. Chang. 25 (1), 49–65.

Corlett, R., Primack, R., 2008. Tropical rainforest conservation: a global perspective. In: Carson, W., Schnitzer, S. (Eds.), Tropical Forest Community Ecology. Blackwell Publishing, West Sussex, UK, pp. 442–457.

Department of Irrigation and Drainage (DID), 2009. Coastal Management. vol. 3. Jabatan Pengairan dan Saliran, Malaysia, Kuala Lumpur. 519 pp. Retrived from, http://forum.mygeoportal.gov.my/smanre/aduan/Volume%203_Coastal%20Management.pdf.

Economic Planning Unit (EPU), 1985. National Coastal Erosion. Government of Malaysia, Kuala Lumpur.

FAO, 1999. State of the World's Forests. Rome.

Fonji, S.F., Taff, G.N., 2014. Using satellite data to monitor land-use land-cover change in North-Eastern Latvia. Springer Plus 3, 61. https://dx.doi.org/10.1186/2193-1801-3-61.

French, J.R., Burningham, H., 2009. Coastal geomorphology: trends and challenges. Prog. Phys. Geogr. 33 (1), 117–129. https://dx.doi.org/10.1177/0309133309105036.

Gaveau, D.L.A., Sloan, S., Molidena, E., Yaen, H., Sheil, D., Abram, N.K., Ancrenaz, M., Nasi, R., Quinones, M., Wielaard, N., Meijaard, E., 2014. Four decades of forest persistence, clearance and logging on Borneo. PLoS One. 9(7):e101654https://dx.doi.org/10.1371/journal.pone.0101654.

Hapke, C.J., Himmelstoss, E.A., Kratzmann, M., List, J.H., Thieler, E.R., 2010. National assessment of shoreline change; historical shoreline change along the New England and Mid-Atlantic coasts. In: U.S. Geological Survey Open-File Report 2010-1118. 57 pp. Available at http://pubs.usgs.gov/of/2010/1118/.

Hon, J., Shibata, S., 2013. A review on land use in the Malaysian state of Sarawak, Borneo and recommendations for wildlife conservation inside production forest environment. Borneo J. Res. Sci. Technol. 3 (2), 22–35.

Hutchison, C.S., 1989. The Palaeo-Tethyan realm and Indosinian orogenic system of Southeast Asia. In: Şengör, A.M.C. (Ed.), Tectonic Evolution of the Tethyan Region. NATO ASI Series (Series C: Mathematical and Physical Sciences), vol. 259. Springer, Dordrecht. https://dx.doi.org/10.1007/978-94-009-2253-2_25.

Hutchison, C.S., 2005. Geology of North-West Borneo, Sarawak, Brunei and Sabah. Elsevier, Amsterdam. 421 pp. https://doi.org/10.1016/B978-044451998-6/50001-6.

Jacqueline, R., 2013. Miri Offers Great Diving Sites. Borneo Post Online. 28.07.2013, http://www.theborneopost.com/2013/07/28/miri-offers-great-diving-sites/.

Jia, T.Y., Rahman, A.H.A., 2009. Comparative analysis of facies and reservoir characteristics of Miri Formation (Miri) and Nyalau Formation (Bintulu), Sarawak. Bull. Geol. Soc. Malaysia 55, 39–45.

Kankara, R.S., Selvan, S.C., Rajan, B., Arockiaraj, S., 2014. An adaptive approach to monitor the shoreline changes in ICZM framework: a case study of Chennai coast. Indian J. Mar. Sci. 43 (7), 1266–1271.

Kessler, F.L., 2005. In: Comments on the evolution of Bukit Lambir area.PGCE Bulletin 2005, Kuala Lumpur.

Kessler, F.L., Jong, J., 2014. Habitat and C-14 ages of lignitic terrace deposits along the northern Sarawak coastline. Bull. Geol. Soc. Malaysia 60, 27–34.

Khan, M.M.H., Bryceson, I., Kolivras, K.N., Faruque, F., Rahman, M.M., Haque, U., 2015. Natural disasters and land-use/land-cover change in the southwest coastal areas of Bangladesh. Reg. Environ. Chang. 15 (2), 241–250.

Lambin, E.F., Geist, H.J., 2008. Land-Use and Land-Cover Change: Local Processes and Global Impacts. Springer-Verlag, Berlin, Heidelberg, p. 222.

Lambin, E.F., Geist, H.J., Lepers, E., 2003. Dynamics of land-use and land-cover change in tropical regions. Annu. Rev. Environ. Resour. 28 (1), 205–241.

Liechti, P.R., Roe, F.W., Haile, N.S., 1960. The geology of Sarawak, Brunei and the western part of North Borneo. Br. Borneo Geol. Surv. Bull. 3, 1–360.

Maanen, B., Nicholls, R.J., French, J.R., Barkwith, A., Bonaldo, D., Burningham, H., Murray, A.B., Payo, A., Townend, I.H., Walkden, M.J.A., 2016. Simulating mesoscale coastal evolution for decadal coastal management: a new framework integrating multiple, complementary modelling approaches. Geomorphology 256, 68–80.

Malini, B.H., Rao, K.N., 2004. Coastal erosion and habitat loss along the Godavari delta front—a fallout of dam construction. Curr. Sci. 87 (9), 1232–1236.

Mas, J.F., 1999. Monitoring land-cover changes: a comparison of change detection techniques. Int. J. Remote Sens. 20 (1), 139–152.

Mendoza-González, G., Martínez, M.L., Lithgow, D., Pérez-Maqueo, O., Simonin, P., 2012. Land use change and its effects on the value of ecosystem services along the coast of the Gulf of Mexico. Ecol. Econ. 82, 23–32.

Nagarajan, R., Roy, P.D., Jonathan, M.P., Lozano, R., Kessler, F.L., Prasanna, M.V., 2014. Geochemistry of Neogene sedimentary rocks from Borneo Basin, East Malaysia: paleo-weathering, provenance and tectonic setting. Chem. Erde-Geochem. 74 (1), 139–146.

Nagarajan, R., Jonathan, M.P., Roy, P.D., Muthusankar, G., Lakshumanan, C., 2015a. Decadal evaluation of a spit in the Baram River mouth in eastern Malaysia. Cont. Shelf Res. 105 (15), 18–25.

Nagarajan, R., Armstrong-Altrin, J.S., Kessler, F.L., Hidalgo-Moral, E.L., Dodge-Wan, D., Taib, N.I., 2015b. Provenance and tectonic setting of miocene siliciclastic sediments, Sibuti Formation, Northwestern Borneo. Arab. J. Geosci. 8 (10), 8549–8565. https://dx.doi.org/10.1007/s12517-015-1833-4.

Nagarajan, R., Armstrong-Altrin, J.S., Kessler, F.L., Jong, J., 2017a. Petrological and geochemical constraints on provenance, paleo-weathering and tectonic setting of clastic sediments from the Neogene Lambir and Sibuti Formations, NW Borneo. In: Mazumder, R. (Ed.), Sediment Provenance: Influences on Compositional Change From Source to Sink. Elsevier, The Netherlands, Amsterdam, pp. 123–154. https://dx.doi.org/10.1016/B978-0-12-803386-9.00007-1.

Nagarajan, R., Roy, P.D., Kessler, F.L., Jong, J., Dayong, V., Jonathan, M.P., 2017b. An integrated study of geochemistry and mineralogy of the upper Tukau Formation, Borneo Island (East Malaysia): sediment provenance, depositional setting and tectonic implications. J. Asian Earth Sci. 143, 77–97. https://dx.doi.org/10.1016/j.jseaes.2017.04.002.

Parsa, V.A., Salehi, E., 2016. Spatio-temporal analysis and simulation pattern of landuse/land cover changes, case study: Naghadeh, Iran. J. Urban Manag. 5 (2), 43–51. https://dx.doi.org/10.1016/j.jum.2016.11.001.

Payo, A., Hall, J.W., French, J., Sutherland, J., van Maanen, B., Nicholls, R.J., Reeve, D.E., 2016. Causal loop analysis of coastal geomorphological systems. Geomorphology 256, 36–48. https://dx.doi.org/10.1016/j.geomorph.2015.07.048.

Peng, L.C., Leman, M.S., Hassan, K., Nasib, M.B., Karim, R., 2004. Stratigraphic Lexicon of Malaysia. Geological Society of Malaysia, Kuala Lumpur. 162p.

Rawat, J.S., Kumar, M., 2015. Monitoring land use/cover change using remote sensing and GIS techniques: a case study of Hawalbagh block, district Almora, Uttarakhand, India. Egypt. J. Remote Sens. Space Sci. 18 (1), 77–84. https://dx.doi.org/10.1016/j.ejrs.2015.02.002.

Sandal, S.T., 1996. The Geology and Hydrocarbon Resources of Negara Brunei Darussalam, second ed. Syabas Publisher, Brunei.

Shalaby, A., Tateishi, R., 2007. Remote sensing and GIS for mapping and monitoring land cover and land-use changes in the northwestern coastal zone of Egypt. Appl. Geogr. 27 (1), 28–41.

Simon, K., Hakif, M., Hassan, A., Barbeito, M.P.J., 2014. Sedimentology and stratigraphy of the Miocene Kampung Opak limestone (Sibuti Formation), Bekenu, Sarawak. Bull. Geol. Soc. Malaysia 60, 45–53.

Straub, K.M., Mohrig, D., 2009. Constructional canyons built by sheet-like turbidity currents: observations from offshore Brunei Darussalam. J. Sediment. Res. 79, 24–39.

Straub, K.M., Mohrig, D., Pirmez, C., 2012. In: Architecture of an aggradational tributary submarine channel network on the continental slope offshore Brunei Darussalam.Application of the Principles of Seismic Geomorphology to Continental-Slope and Base-of-Slope Systems: Case Studies from Seafloor and Near-Seafloor Analogues, SEPM, Special Publication 99, pp. 13–30.

Tan, K.C., San Lim, H., MatJafri, M.Z., Abdullah, K., 2010. Landsat data to evaluate urban expansion and determine land use/land cover changes in Penang Island, Malaysia. Environ. Earth Sci. 60 (7), 1509–1521.

Tang, F.E., Lee, V.Z., 2010. In: A study of coastal areas in Miri, Sarawak. World Engineering Congress 2010, Conference on Natural and Green Technology. International Association of Engineers (IAENG), Kuching, Sarawak, Malaysia, pp. 56–65.

Teh, T.S., Boon, Y.H., 2003. South East Asia: Malaysia introduction. In: The World's Coasts: Online. Geostudy Pty Ltd., Australia. 47 pp.

UNEP, Wilkinson, C., DeVantier, L., Talaue-McManus, L., Lawrence, D., Souter, D., 2005. South China Sea, GIWA Regional Assessment 54. University of Kalmar, Kalmar, Sweden 104 pp.

Van Hattum, M.W.A., Hall, R., Pickard, A.L., Nichols, G.J., 2013. Provenance and geochronology of cenozoic sandstones of Northern Borneo. J. Asian Earth Sci. 76, 266–282.

Veldkamp, A., Lambin, E.F., 2001. Predicting land-use change. Agric. Ecosyst. Environ. 85, 1), 1–6.

Wannier, M., Lesslar, P., Lee, C., Raven, H., Sorkhabi, R., Ibrahim, A., 2011. Geological Excursions Around Miri, Sarawak. Ecomedia Software, Miri, Sarawak, Malaysia. 279 pp.

Xiubin, L., 1996. A review of the international researches on land use/land cover change. Acta Geograph. Sin. 51 (6), 553–558.

Yang, J., Seo, D., Lim, H., Choi, C., 2010. An analysis of coastal topography and land cover changes at Haeundae Beach, South Korea. Acta Astronaut. 67 (9), 1280–1288.

Yang, Y., Zhang, S., Liu, Y., Xing, X., Sherbinin, A.d., 2017. Analyzing historical land use changes using a historical land use reconstruction model: a case study in Zhenlai County, northeastern China. Sci. Rep. 741275. https://dx.doi.org/10.1038/srep41275.

Further Reading

Brookfield, H., Byron, Y., 1990. Deforestation and timber extraction in Borneo and the Malay Peninsula: the record because 1965. Glob. Environ. Chang. 1 (1), 42–56.

Website, http://www.did.sarawak.gov.my/modules/web/pages.php?mod=webpage&sub=page&id=477.

CHAPTER

8

A View on South Africa's KwaZulu-Natal Coast: Stressors and Coastal Management

E. Vetrimurugan, M.P. Jonathan†, V.C. Shruti†, B.K. Rawlins**

*Department of Hydrology, University of Zululand, Kwa Dlangezwa, South Africa
†Centro Interdisciplinario de Investigaciones y Estudios sobre Medio Ambiente y Desarrollo (CIIEMAD), Instituto Politécnico Nacional (IPN), Ciudad de Mexico (CDMX), Mexico

1 INTRODUCTION

South Africa has some of the finest beaches in the world, encompassing a pristine coastal stretch of Cape Vidal in KwaZulu-Natal to the Eastern Cape's famous wild coast, from the penguin colony of Boulders beach to sun-drenched Camps Bay in the Western Cape.

The KwaZulu-Natal (KZN) province of South Africa, also known as the "garden province," is an extravagant and enticing destination flanked by the warm Indian Ocean on the east and soaring peaks to the west. The province stretches from Port Edward in the south to the borders of Swaziland and Mozambique to the north.

2 REGIONAL SETTING

2.1 Physiography

The KZN province located in the southeastern part of South Africa is bordered by Mpumalanga, Swaziland, and Mozambique in the north, Eastern Cape in the south, Free State and Lesotho in the west, and the Indian Ocean in the east with an areal extent of 94,361 km^2 (Fig. 1).

Coastal Management
https://doi.org/10.1016/B978-0-12-810473-6.00009-1

© 2019 Elsevier Inc. All rights reserved.

FIG. 1 Map showing the KwaZulu-Natal province located in the south eastern part of South Africa.

It is South Africa's third smallest yet one of its most populous provinces, with a population of 10,919,100 that is 19.9% of the total population of South Africa (Stats SA, 2016).

The KZN coast that stretches 800 km along the east of South Africa is divided into 11 districts with Pietermaritzburg as its capital. It includes major urban centers such as Durban, which is its largest city, and is vibrant and buzzing with activity year round. Richards Bay/Empangeni form the prime industrial hubs. The coastline is split at Durban into the North Coast and South Coast, each of which is punctuated with a multitude of small towns and settlements offering a wide variety of tourism facilities and activities. The provincial economy thrives on agriculture, forestry, mining, and tourism.

The KZN province can be subdivided into three distinct geographical areas: (a) the lowland Indian Ocean coastal region, which is very narrow in the south but gets wider in the northern part of the province, (b) the central Natal Midlands, which encompass undulating hilly plateaus that rise toward the west, and (c) the mountainous areas comprising Drakensberg in the west and Lebombo in the north (Kruger, 1983). Drakensberg forms one of the UNESCO World Heritage sites for its magnificent beauty and San Bushman rock art found in the caves, which is the richest concentration of such art on the entire African continent.

2.2 Rivers, Lakes, and Dams

The major rivers draining in the province include the Tugela, Mfolozi, Mgeni, Msunduzi, Mkomaas, and Mzimkulu. The Tugela forms the largest river sourced from a small stream on top of Mont-aux-Sources, flowing some 502 km from west to east across the center of the province through the midlands and out into the Indian Ocean. Lake Sibaya in Zululand is the deepest natural freshwater lake in South Africa, forming part of the Greater St. Lucia Wetland Park. Numerous dams have been built on several rivers to provide water for drinking, agriculture, and industries. These include (a) the Midmar dam near Pietermaritzburg, (b) the Inanda dam near Durban, (c) the Jozini dam in Zululand, (d) the Woodstock dam near Bergville, and (e) the Spioenkop dam near the Drakensberg.

2.3 Climate

The KZN province situated amidst the Indian Ocean in the east and the high Drakensberg escarpment in the west experiences varying weather patterns. It displays a relatively balanced climate and rainfall. The KZN province is classified as a summer rainfall region and it is also the wettest part of South Africa. The rainfall is strongly seasonal, with an average of 720 mm in the west and 890 mm in the east with more than 80% falling between October and March, although the maximum is observed during January. The average air temperatures are 15 and 18°C in the west and east regions, respectively (Jury, 1998; Mathieu and Yves, 2003; De Villiers, 2005).

The undulating terrain of the KZN province results in localized climatic variations. Generally, subtropical conditions prevail along the coastline with inland regions becoming progressively colder. As the altitude rises from the east to the great escarpment (over 3000 m in the Drakensberg Mountains above sea level), there is a drop in temperature. The Zululand north coast is characterized with the warmest climate and highest humidity. The temperature begins to drop toward the midland areas with Pietermartizburg being much cooler than the coastal areas in winter. The Ladysmith and Drakensberg escarpments experience very dry, cold, and very cold conditions in winter.

2.4 Geology

The geological history of KZN dates back approximately 3.5 billion years (Fig. 2). The stratigraphy represents an evolution of geological events from the Archaezoic to the present. The geological foundation of KZN is represented by the Kaapvaal Craton, which was formed when the Earth's crust was intruded by granite and is preserved as greenstone fragments within the granite. The Pongola Supergroup overlying the basement is a succession of basalt, sandstone, and minor limestone. During the Cambrian to Ordovician periods, the first sedimentary sequence of the Natal group was deposited. The Dwyka group overlying the Natal group is a thick unit of tillite that was deposited about 300 million years ago in a glacial environment by retreating ice sheets (Visser, 1990). The Ecca group is represented by the Pietermaritzburg (carbonaceous shales), Vryheid (fluviodeltaic deposits), and Volkrust formations (argillaceous unit). The overlying Beaufort group is made up of argillaceous and arenaceous rocks of the Permian-Triassic age. The breakup of the Gondwana supercontinent approximately 180 Ma led to outpourings of massive lava, forming the Drakensberg and Lobombo groups. The regional volcanism, uplift, and faulting resulted in the separation of Africa and Antarctica. In the newly opened Indian Ocean, the marine sediments of the

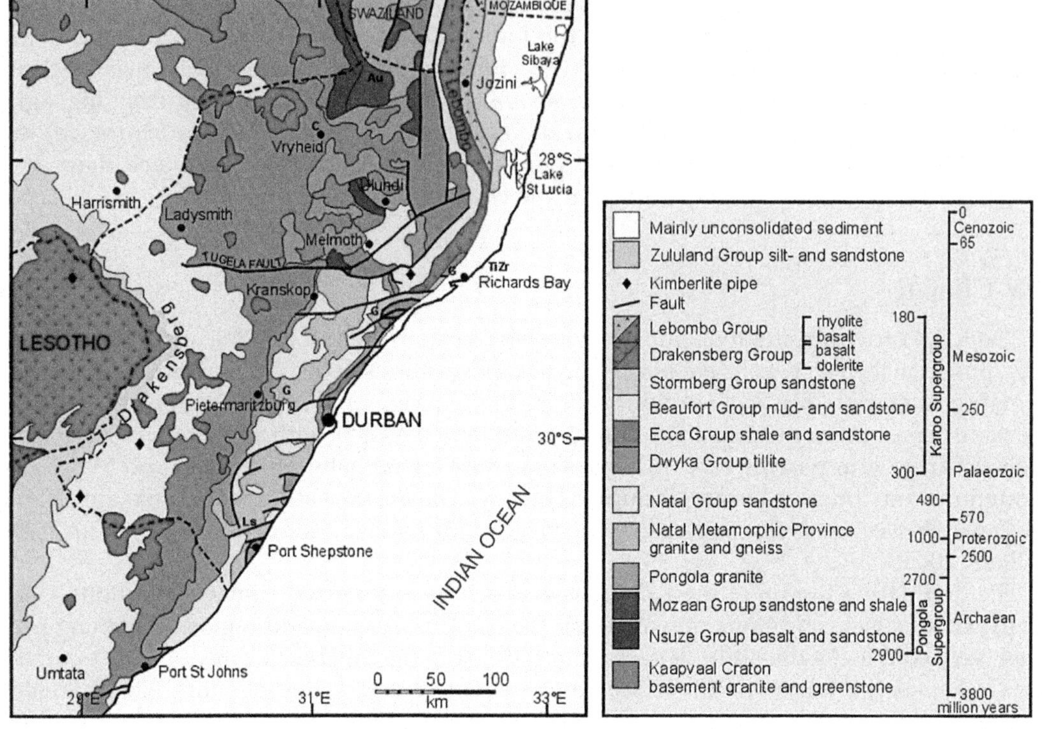

FIG. 2 Geological map of KwaZulu-Natal province. *Geological Survey, 1984, Pretoria, Government Printer, NE & SE sheets.*

Cretaceous Zululand group were subsequently deposited. Erosion played a major role in forming the present topography of KZN and this process continues to be active.

2.5 Ecosystems

The KZN province embraces a unique mosaic of ecosystems, including swamps, lakes, beaches, coral reefs, wetlands, woodlands, coastal forests, and grasslands, all supporting an astounding diversity of animal, bird, and marine life. This coastline forms one of the most valuable natural assets of South Africa and is utilized for residential, infrastructural, commercial, and nature conservation purposes (Cooper, 1995).

2.5.1 Sandy Beaches and Rocky Shores

Nearly 80% of the North Coast of KZN province is dominated by sandy beaches. These are dynamic and mobile ecosystems comprising sand particles as well as coral and shell fragments that are always on the move. They are continuously transformed by ocean currents moving sand on and offshore, by waves crashing the beach face, and by wind transporting sand to and from dunes. The unique biodiversity of sandy beaches, composed of the hidden life that is specially adapted to this harsh beach environment, consists of representatives from all the major food web components, including decomposers (bacteria and fungi), primary producers (chlorophyll containing diatoms), filter feeders (crabs and clams), carnivores (worms called polychaetes), scavengers (snails), and predators (birds and fish).

The rocky shores are dominated mainly on the South Coast of KZN province, either in the form of headlands, wide wave-cut platforms, or simply as rocky outcrops separated by sandy shores. As this region is traversed with the rise and fall of tides, the animals are exposed to different periods of inundation and exposure, resulting in vertical zonation with the hardiest and most tolerant organisms highest up on the rocks. The organisms attached to these rocky shores generally include barnacles, mussels, limpets, and mobile animals such as crabs, small fish, sea urchins, and sea stars.

2.5.2 Estuaries and Subtidal Reefs

The coastline of KZN is flecked with 73 estuaries of different types and includes river mouths such as the uThukela, estuarine bays such as Durban Bay, and lake systems such as the Kosi lakes (Whitfield, 2000).

The reef ecosystems are encountered just below the surface of the KZN coast. Mostly, these are comprised of rock heavily encrusted with marine life with a distinct difference between northern reefs and those in the south and central parts of the province. In the north there is higher percentage cover of fauna (mainly coral) at shallower depths and greater coverage of algae on deeper reefs. In the southern reefs, algae dominates at shallower depths whereas filter-feeding epifauna are more prevalent at deeper depths (Lawrence, 2005). These reef ecosystems embrace a wide diversity of species such as fish, rock lobsters, and other invertebrate animals with the highest percentage at 10m in the northern coral-dominated region while in the central/south region it peaks in the intermediate depth zone of 15–25m.

3 DEMOGRAPHICS

The 2011 census demonstrates the demographics of South Africa, which encompasses about 53 million people of diverse origins, cultures, languages, and religions. The demographic growth in South Africa from 1960 to 2015 is represented in Fig. 3.

Midyear population estimates of South Africa for each province are represented in Table 1. The estimates reveal that Gauteng has the highest share of population followed by KZN, the Western and Eastern Capes. In KZN province, the midyear population estimates project that the population has grown by approximately 10% from 2001 to 2010 (Table 2; Stats SA, 2010).

Though KZN has the second largest share of population, they have limited employment opportunities and substantial infrastructure backlogs. The main sectors accounting for the provisional economy are agriculture, construction, manufacturing, and mining. The average annual growth in these sectors during the period of 2004–11 and 2011–14 is presented in Fig. 4A and B, respectively. The fastest-growing sector in KZN province was construction and the province benefited from major rail, road, and port projects under the National Infrastructure Plan.

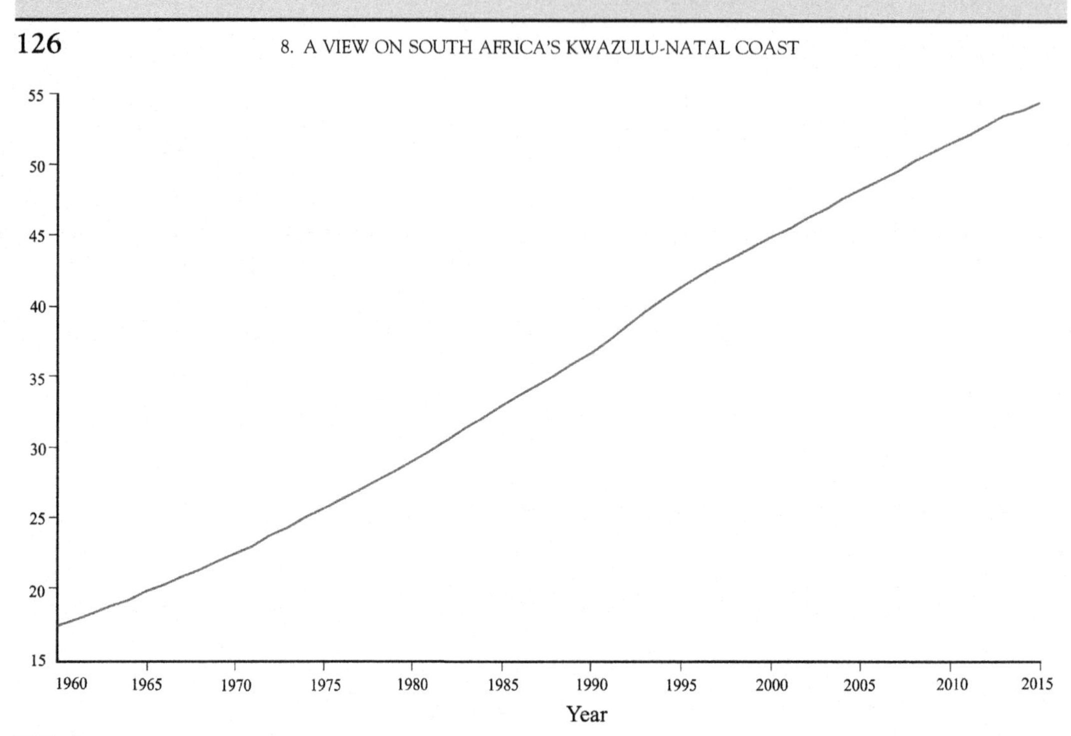

FIG. 3 Demographic growth in South Africa from 1960 to 2015 (Stats SA, 2016).

TABLE 1 Midyear Population Estimates of Individual Provinces, South Africa

	Population Estimates	% of Total Population
Eastern Cape	6,916,200	12.6
Free State	2,817,900	5.1
Gauteng	13,200,300	24.0
KwaZulu-Natal	10,919,100	19.9
Limpopo	5,726,800	10.4
Mpumalanga	4,283,900	7.8
Northern Cape	1,185,600	2.2
North West	3,707,000	6.7
Western Cape	6,200,100	11.3
Total	54,956,900	100

TABLE 2 KZN Medium Midyear Population Estimates 2001–10

Year	Medium Midyear Population Estimates by Year	Year-on-Year Growth (%)
2001	9,557,165	
2002	9,659,485	1.07
2003	9,752,211	0.96
2004	9,835,710	0.86
2005	9,910,636	0.76
2006	9,974,344	0.64
2007	10,045,594	0.71
2008	10,105,436	0.60
2009	10,449,300	3.40
2010	10,645,400	1.88
Average	9,993,528	1.21
Median	9,942,490	0.86
St Dev	339,980	0.91
Range	1,088,235	2.81

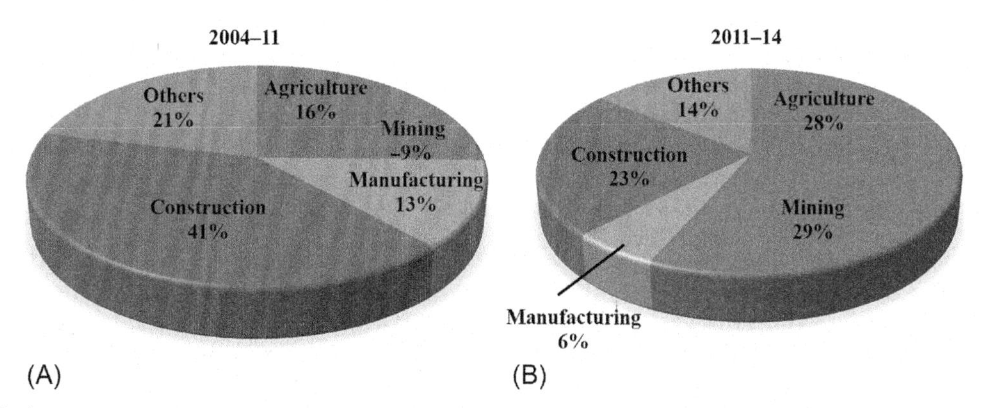

FIG. 4 Average annual growth in major sectors during 2004–11 and 2011–14 (Stats SA, GDP Annual and Regional Tables 2016).

4 TOURISM

In South Africa, the SATOUR (South African Tourism) national body was set up in 1994, initially to deal with the tourism sector. A provisional body created IPTSC for setting up the KwaZulu-Natal Tourism Authority in KZN province. In 1998, the first-ever brochure regarding tourism in South Africa and more specifically for KZN was produced by this provisional body.

4.1 Tourism Perspective of KZN Province

Tourism forms an extremely fragile economic sector that can be easily destroyed or rapidly changed and is subjected to the perceptions of the people who wish to travel for specific purposes. The KZN province is considered by the traditional foreign markets as being a long-haul destination. The major economic role in this province emerges from tourism, primarily related to traditional "sea, sun, and sand" forms of tourism. Historically, the tourism economy of KwaZulu-Natal surfaces around beach products (Grant and Butler-Adam, 1992).

The core experiences in KZN province include:

Beach experience: The sun, sand, and warm ocean waters of KZN offer people an antidote to overwork, stress, and the cold of the northern hemisphere winters. Therefore, the beach as well as coastal and resort tourism are extremely valuable tourism sectors in this province. KZN offers 600 km of coastline for tourism, out of which three tourist destinations are proposed for coastal experiences: the South coast, the North coast, and the Elephant coast.

Natural beauty and wildlife experience: The two important resources that stimulated tourism flow in the KZN province are the Drakensberg mountain range and the iSimangaliso Wetland Park. The wildlife experience is unique as it combines land and marine wildlife. Safaris are undertaken in and around the iSimangaliso wetland park and several other activities in Drakensberg Park such as avitourism, dive tourism, and additional recreational pursuits (walking, hiking) attract plenty of tourists to this province.

Cultural and heritage attractions: Covering a wide spectrum of natural and cultural landscapes, the Midlands provide a wealth of tourism through crafts, culture, and history whereas Battlefields hosts tourist adventures such as battlefield routes based on the Anglo-Boer and Anglo-Zulu wars in this area. The exclusive blend of Zulu and San cultures marks the cultural richness of this province.

Business tourism: The opening of the Durban International Convention Center in 1997 led to an expansion of business tourism (Nel and Rogerson, 2005; Preston-Whyte and Scott, 2007).

In KZN province, the most visited destination is Durban, followed by Zululand, the Elephant Coast, the Midlands, the Drakensberg mountains, the South Coast, and the North Coast. KwaZulu-Natal ranks as the third-most visited province by domestic tourists. The leading reasons drawing domestic tourists are visiting friends and relatives (VFR) and leisure purposes (Kohler, 2009). Visitors from the United Kingdom, the United States, Germany, and France account for the international tourists attracted to the coastal areas of the province looking for sun and warm ocean waters to relax in and escape the Northern Hemisphere winters (Kohler, 2010).

4.2 Tourism Trend in KZN

The two principal markets governing the tourism sector in KZN province are the domestic and foreign markets, of which the domestic market is larger compared to the foreign. This is mainly because the KZN province is essentially a long-haul destination where foreign numbers would be comparatively small. The total foreign market at present is around 1.2 million visitors while the domestic market is around 11 million. The peak tourism season for South Africa and KZN is during October–November and February, whereas for domestic visitors it is during December and April–July (Fig. 5).

The purpose of a visit to KZN is very strong in terms of domestic VFR visitors while the foreigner visit is typically for holiday and leisure purposes. The second-most important sector for foreign visitors is business tourism while it is holiday purposes for domestic visitors. The trends shown in Fig.6 indicate that percentage of foreign visitors for beaches, nature reserves and art and craft centers are higher than compared to domestic visitors. After 2004, the reasons people visited showed a different trend, with a shift to shopping, nightlife, and beach visits. The nightlife and shopping activities were highly popular among the foreign visitors more than the domestic. Overall, beach tourism was more prominent in KZN due to warm temperatures and inviting oceans that led several foreign and domestic visitors to this destination.

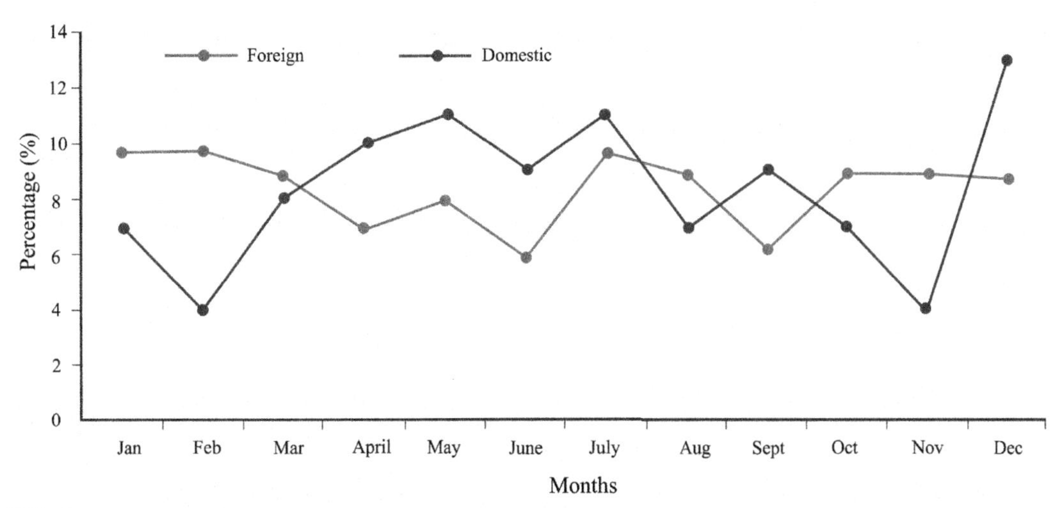

FIG. 5 Foreign and domestic seasonal flow in KwaZulu-Natal province, South Africa.

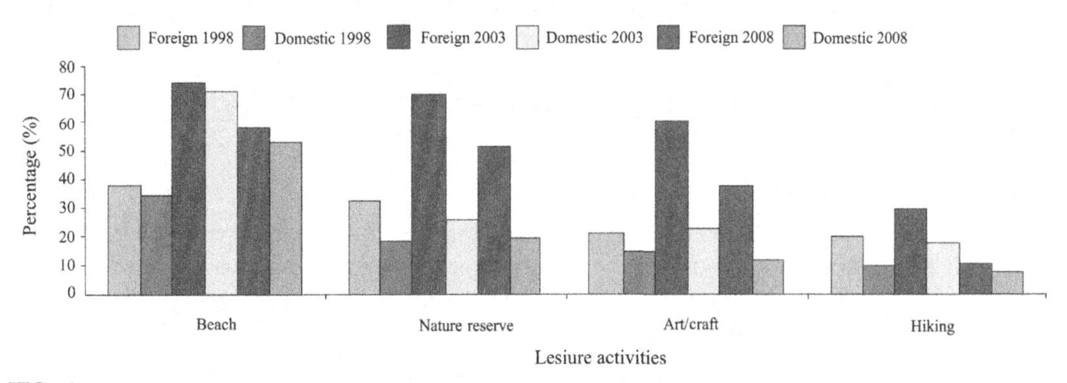

FIG. 6 Selected activities of domestic and foreign visitors during 1998, 2003, and 2008.

5 MAJOR THREATS/STRESSORS TO THE COAST

The rising allure of the coast as a destination for residence as well as recreation joins with South Africa's popularity for international tourism to act as driving factors for further development along the KZN coast. In addition to natural stressors such as climate change, the role of humans play a substantial role in pressurizing the coastal strip through high rates of urbanization, over exploitation of marine species threatening marine biodiversity and induce coastal pollution.

5.1 Climatic Extremes

5.1.1 Coastal Lows

One of the most important synoptic-scale weather systems formed through the interaction between large-scale atmospheric flow and the escarpment is known as cut-off lows or coastal lows. These systems are often associated with strong southwesterly gusts moving northward in an anticlockwise direction. The poleward flowing Agulhas current combined with intense frontal systems causes high energy swells along the coastline. The Agulhas current is one of the largest and swiftest currents on Earth, reaching a depth of 1000 m. This affects the weather patterns, storm events, and the diversity of marine life along the KZN coast. In the KZN province, these coastal lows are responsible for most of the large-scale extreme precipitation events and have caused more flood damage than tropical cyclones (Smith et al., 2010).

5.1.2 ENSO Effect

The El Niño Southern Oscillation (ENSO) is a naturally occurring phenomenon involving warmer (El Niño) or cooler temperatures (La Niña) than that are normally found in the ocean; this significantly affects weather patterns around the world. South Africa's climate is strongly effected by El Niño events, which occur irregularly and may last from 12 to 18 months. This causes a shift in rainfall patterns and is often associated with extreme weather such as floods and droughts (Jury, 2002; South African Weather Service, 2008). It generally produces warm, dry conditions through westerly winds that increase the evaporative losses over the eastern parts. It is more profound during the summer season in the months of December–February. El Niño years usually bring below-normal rainfall when compared to La Niña. ENSO only explains 30% of the rainfall variability, which means other factors should also be taken into account when predicting seasonal rainfall. For example, in 1997–98 El Niño was the strongest on record but not all South Africa received below-normal rainfall. Some regions had an abundance of rainfall because of moist air imported from the Indian Ocean.

5.1.3 Tropical Cyclones

This province is characterized with occasional tropical cyclones, mainly from November to April with a peak in January and February. Only the cyclones moving into the Mozambique Channel influence South Africa's weather. The tropical cyclones are accompanied by torrential rain wherein deep pressure gradients create strong winds. The interior regions usually experience dry weather due to the subsiding air that surrounds a tropical cyclone. The severe impacts of these cyclones not only result in damage due to strong winds and torrential rainfall but also due to storm surges and waves generated by these cyclones. Significant tropical cyclones that had a substantial impact were Domoina in January 1984, Imboa in February 1984, and Eline in February 2000.

5.1.4 Sea-Level Rise

The threat to sandy beaches through climate change, particularly a rise in sea levels, has considerable implications. The Intergovernmental Panel on Climate Change (IPCC) estimates global sea level rise at 3 mm per year while the sea level off the Durban coast of KZN province is rising by 2.7 mm every year (Mather, 2007). The northern region of KZN has the highest predicted retreat in response to the sea level rise compared to central and southern regions, as this region is undeveloped and thus responds to the sea level naturally without compromising the sandy beach ecosystem.

5.2 Disturbing Ecosystems

A range of ecosystems in the coastal stretch is greatly impacted by human interference, with a few of these activities being sand winning, breaching the mouths of estuaries, and changes to natural vegetation.

Sand winning: There is a considerable extraction of sand from estuaries, riverbeds, and coastal areas in KZN, mainly for building purposes. Sand poses an important coastal ecosystem function by constantly replenishing the beaches and sustaining the sandy beach ecosystem. The excessive extraction of sand from the beaches results in the depletion and vulnerability of these beaches to storms.

Artificial breaching: In KZN province, several estuaries close periodically as part of their natural cycle and open again normally after rain. Under closed conditions, upstream flooding may occur, urging in the need for artificially breaching the estuary mouth. Artificial breaching may also make the system less productive and less able to deliver goods and services.

Changes to natural vegetation: The coastal vegetation acts as a buffer to the coast, protecting it from several natural hazards. It contributes to the biodiversity of the region and is provision for goods and services. This highly productive zone is being exploited for urban development, overgrazing, and cultivation of riverbanks and floodplains, which is all of great concern. One of the impacts of these activities is higher silt loads in the waterways, causing estuaries to split up and leading to the degradation of these ecosystems.

5.3 Coastal Development

Based on the development, the KZN coast is classified into ~56% undeveloped in the northern region, mainly comprising the St. Lucia and Maputaland Marine Reserves; ~33% developed on the primary dune; and ~10% developed beyond the primary dune (Harris, 2008). A majority of development is observed in the southern region, which is mainly built around the railway line that runs midway up the primary foredune for most of the length of the coastline. High to intensively developed areas constitute the central part of the province where the major urban nodes are set up, for example, Durban, Umhlanga rocks, Umdolti, and Ballito.

Being largely undeveloped, the northern region consists of near-pristine beaches as they are naturally protected by the marine reserves. The sand dunes of these beaches are undisturbed, making them highly adaptive and inherently resilient to a rise in sea levels. In contrast, the southern and central regions are potentially far more ecologically compromised in the face of climate change. This is due to the higher prevalence of coastal infrastructure closer to the beach and nodes of intensively developed urban areas, conferring greater threats of coastal squeeze in response to sea level rise. In the last 12 years, KZN has experienced 22% of land-use change from natural vegetation to urban development in the 100 m strip inland of the high water mark (Harris, 2008). With respect to the conservation status of vegetation, this province (except the northern region) has transformed from endangered to critically endangered in the last five years (Kohler, 2005). Thus appropriate coastal management in KZN is mandatory to protect its beaches from losing in the coastal squeeze, along with its biodiversity and ecosystem goods and services.

5.4 Pollution/Contamination

5.4.1 *Marine Litter*

The marine debris found in coastal areas and oceans across the world consists of slowly degradable or nondegradable substances that inevitably accumulate in the environment. It originates from sea-based (oil spills, trash from vessels and fishing nets) or land-based sources (municipal and industrial waste, tourism and leisure activities), causing a wide spectrum of environmental, economic, safety, health, and cultural impacts. In South Africa, land based sources contribute nearly 80% of the all the marine debris. KZN province has the most important port and industrial hub in Durban, which poses several environmental threats due to oil spills and disposal of waste materials in the surrounding regions. The very slow rate of degradation of most marine litter, especially plastics, is believed to be a source of persistent toxic substances. One of the serious effects of plastics is entanglement or ingestion by marine animals. In KZN, during 1978–2000, of almost 29,000 sharks from 14 species, eight species became entangled and 10 species were found to have ingested litter. The amount of entanglement remains constant at 60–70 per year. The majority of land-based pollution finds its way into the oceans through rivers. The KZN coastline is highly susceptible as it is a densely populated region with a high number of rivers.

5.4.2 *Trace Metals*

The KZN province holds major ports and industrial hubs that are exerting tremendous pressure over the coastal environments. The human activities concentrated in the urban zones and the industrial activities greatly influence the distribution of toxic metals in marine environments (Frignani and Belluci, 2004; Lim et al., 2013). Trace metal pollution is one of the grave concerns under the scientific spotlight due to its deleterious effects over marine ecosystems. These are considered to pose potential threats to humans and the marine ecosystem due to their persistent nonbiodegradable nature, which consequently gets bioaccumulated and biomagnified along the food chain (Rainbow, 2007; Wang and Rainbow, 2008; Christophoridis et al., 2009; Bednarova et al., 2013; Lai et al., 2013; Hwang et al., 2016).

A preliminary study on metal concentrations in several beaches along the entire coast of KZN was conducted by Vetrimurugan et al. (2016, 2017, and 2018) in order to identify the influence of tourism, industrial, and mining activities as well as other possible sources

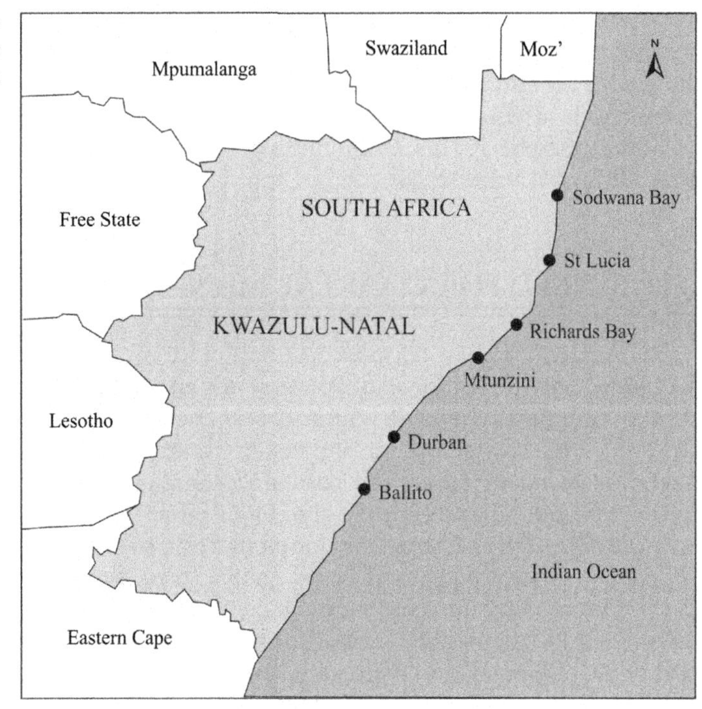

FIG. 7 Sampling locations of sediments from various beaches along the KwaZulu-Natal coastline of South Africa.

in this province (Fig. 7). Seven beach regions were selected for the present study: Sodwana Bay (32 samples), St Lucia (34 samples), Richards Bay (53 samples), Mtunzini (0.32 samples), Durban North (45 samples), Durban South (43 samples), Ballito North (51 samples), and Ballito South (53 samples). The order of metal concentrations in these seven beach regions is represented in Table 3. The results indicated that, overall, the beach sediments are enriched with Fe, Mg, Cr, Ni, Mn, Co, and Zn. The beach sands of KZN province are extensively mined

TABLE 3 Order of Metal Concentrations in KZN Beaches, South Africa

S. No	Beach Locations	Metal Concentrations
1	Sodwana Bay	Fe > Mg > Cr > Ni > Mn > Co > Zn > Pb > Mo > Cu > Hg > Cd
2	St. Lucia	Fe > Mg > Cr > Ni > Co > Mn > Zn > Pb > Mo > Cu > Hg > Cd
3	Richards Bay	Fe > Zn > Mn > Pb > Cr > Ni > Co > Cu > As > Cd
4	Mtunzini	Fe > Mg > Cr > Ni > Mn > Co > Zn > Pb > Cu > Mo > Cd
5	Durban North	Fe > Mg > Zn > Mn > Ni > Pb > Co > Cr > As > Mo > Cd
6	Durban South	Fe > Mg > Cr > Ni > Mn > Co > Zn > Pb > Mo > Cu > Hg > Cd
7	Ballito North	Fe > Mg > Cr > Ni > Co > Mn > Zn > Pb > Mo > Cu > Cd > Hg
8	Ballito South	Fe > Mg > Zn > Mn > Ni > Pb > Co > Cr > As > Cd

for heavy minerals and there are several important ports mainly exporting coal, ferroalloys, chrome ore, steel, and the heavy metal groups of titanium, rutile, and zircon throughout the year (Fair and Jones, 1991; Greenfield et al., 2011). The higher concentrations of metals in the beach sands of KZN is mainly attributed to the natural source rock weathering as well as anthropogenic inputs from mining, tourism, and industrial activities (Vetrimurugan et al., 2016, 2017, and 2018).

6 NEED FOR COASTAL MANAGEMENT IN SOUTH AFRICA

South Africa forms a popular destination for residential, industrial, transport, nature conservation, and recreational purposes with more than one-third of eastern and southern Africa's population residing within 100 km of the coast (UNEP/Nairobi Convention Secretariat, 2009). The coastal and marine resources are under considerable threat and are severely degraded in many areas. The coastal population has developed a strong reliance on the coastal resources that contribute about R 57 billion (US $5.7 billion) to the South African economy (UNOPS, 2011). Urban development along the coastal zone creates environmental pressures by deteriorating coastal water quality, threatening human and ecosystem health. Any development along the coast should be within the terms of NEMA Environmental Impact Assessment (EIA) regulations. The prime resources such as the marine fishing industry, port and harbor development, and recreational and tourism opportunities account for 34% of the total gross domestic product (GDP). As the development pressure increases, the natural functioning of the coast decreases (O'Connor et al., 2009). Large commercial ports have triggered extensive industrial and urban development, which forms the source points for sewage and storm water outfalls harming coastal resources. Approximately 63 ocean sewage outfalls are located along the South Africa coast. With fishing being the primary economic activity in South Africa, this is under risk as some fish stock has been overharvested and several species face local extinction. Shipping forms another major activity in this country, which can lead to hazardous activities such as oil spills and waste discharges from vessels, thereby making the coast less suitable for life and human use.

The degradation of coastal and marine resources in many areas of South Africa is mainly due to unsustainable development. This drives the need for adopting sustainable coastal development methods for reconstruction and development in South Africa through maintaining diverse, healthy, and productive coastal ecosystems.

6.1 Evolution of Integrated Coastal Management Act

Coastal management in South Africa has undergone various shifts since the 1970s.

1970s: This period presented a sectoral approach to coastal management related to resource exploitation and its effective management.

1980s: Aspects such as the ecological character of the coast and the threats posed by coastal resorts, townships, and related infrastructural development were focused in the coastal management plans during this period.

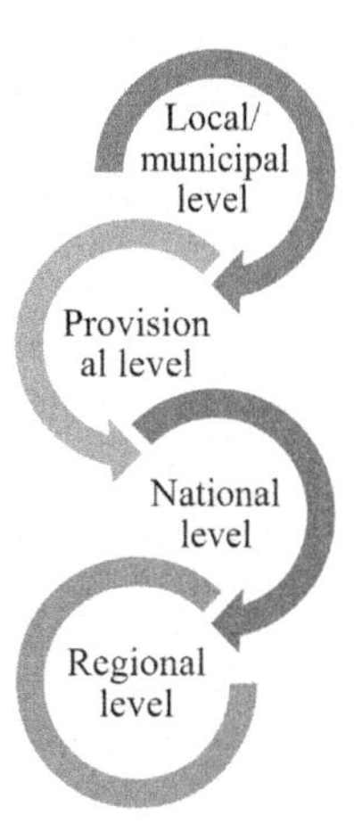

FIG. 8 Illustration of coastal management units at various tiers within which respective coastal management programs may apply.

1990s: The policies were closely in relation with the political priorities of poverty eradication and job creation. The government was promoted to invest in coastal management wherein the coastal issues moved from the political periphery toward the government.

1992–97: In 1995, the Consulative National Environmental Policy Process (CONNEPP) was launched as the primary legislation governing the environment in South Africa, resulting in the development of the coastal management program Green Paper (1998) and White Paper for Sustainable Coastal Development (2000). Currently, the integrated coastal management (ICM) of South Africa is guided by a national policy, "White Paper for Sustainable Coastal Development in South Africa." An integrated coastal management act was launched in 2008 that includes norms, standards, and policies in order to promote an integrated coastal and estuarine management in South Africa by all spheres of government (Fig. 8).

6.2 KZN Perspective in ICM

The KZN coast is recognized for its beaches that provide several recreational activities, attracting a greater number of tourists worldwide. The coast comprises the Hibiscus coast, the Durban metro, the Zululand coast, and the Maputaland coastal regions. The short coastline of ~562 km has witnessed significant economic growth in recent years. Major industrial and manufacturing activities are observed in Durban and Richards Bay. The Durban, Hibiscus, and Dolphin coasts are popular for tourism and recreational purposes. Oddly, although Zululand and Maputaland are the poorest regions in KZN province, they have spectacular coastal environments as well as the potential for community-based tourism, agriculture, and fishing.

6.2.1 Marine Reserves

The Marine Protected Areas (MPA) are developed as a management tool to safeguard marine biodiversity, protect critical hotspots, and look after marine and coastal resources.

The MPAs refer to those areas under Section 43 of the Marine Living Resources Act 18 of 1988 (MLRA). The KZN coast hosts four MPAs: St. Lucia, Maputaland, Aliwal Shoal, and Trafalgar, which together cover more than 25% of KZN's coastline. The St. Lucia and Maputaland marine reserves form a continuous protected area stretching 150 km from the Mozambique border southward to Cape Vidal. They are recognized as a globally important conservation site. These reserves form a part of the iSimangaliso Wetland Park designated as a World Heritage site. The Trafalgar marine reserve is a tiny area extending only 6 km along the KZN coast and 500 m offshore. It primarily serves to protect the coastal features, including cretaceous fossils and the marine-estuarine interface (Mann et al., 1998). The MPAs are primarily important for fisheries management by efficiently building up spawning stocks of commercially important species within their boundaries (Halpern, 2003; Russ et al., 2008; García-Charton et al., 2008; Harmelin-Vivien et al., 2008). The MPAs also provide substantial benefits in the form of ecosystem services such as coastal protection, waste assimilation, and flood management. They also play a vital role in protecting marine habitats and biodiversity.

6.2.2 Boat Launching

The off-road vehicles that launch small vessels into the sea are utilized to access the offshore marine resources of KZN (Penney et al., 1999; Dunlop and Mann, 2013), which can have a potentially negative impact on the coastal environment (Celliers et al., 2004). The use of off-road vehicles in the coastal zone of South Africa was limited during 2002 under the National Environmental Management Act (Act No. 107 of 1998). The South African Maritime Safety Authority (SAMSA) controls the safety of boat launching by setting operational limits on boating. Initially, the management of launching boat sites in KZN was based on safety and logistic considerations with little or no emphasis on environmental factors (Town and Regional Planning Commission, 1988; Goble et al., 2014). In 2003, the Department of Agriculture and Environmental Affairs (DAEA) had new regulations placed under the Coastal and Biodiversity Management Unit, which had a significant focus over the environmental aspects.

6.2.3 Blue Flag Beaches

The Blue Flag program was first started in France in 1985 for beaches and marinas by a nongovernmental and nonprofit organization, the Foundation for Environmental Education. In 2001, South Africa was the first country outside Europe to join this program. The Blue Flag program is an international standard for measuring the quality of beaches in terms of environmental management, safety, amenities, water quality, and environmental education. In South Africa, the Blue Flag program has been run by WESSA (the Wildlife and Environment Society of South Africa) since 2001. It integrates the tourism and environmental sectors at the local, regional, and national levels. The South African Blue Flag season for 2015–16 announced 39 beaches, nine boats, and six marinas with full status and a further 30 beaches were awarded pilot status. The maximum number of 49 Blue Flag sites is in the Western Cape with 19 sites in KwaZulu-Natal, and 16 sites in the Eastern Cape.

7 CONCLUSION

The KZN coast of South Africa has experienced rapid development over a relatively short period of time. Moreover, the development is uneven and is largely concentrated in the southern section of the coast. In the future, the remaining natural or undeveloped northern coast would be under the developmental limelight. The presence of numerous ports and industrial hubs in this province favors tremendous demographic growth along the coastal cities. This consequently leads to several anthropogenic pressures over the coastal environments. Marine pollution is presently less of a problem; however, the effects seems to be relatively localized. The recent studies over metal concentrations along the coastal beaches of KZN suggest that the concentration of metals is mainly attributed to the natural source rock composition and these regions are recently affected by the external inputs. The KZN coast has significant impacts of extreme climatic events. Even though coastal development is inevitable and good from the economic point of view, it has to be effectively regulated and managed. The KZN province has the opportunity to manage its coastal development in a sound manner, keeping in mind the environmental aspects in operation with the ICM Act.

References

Bednarova, Z., Kuta, J., Kohut, L., Machat, J., Klanova, J., Holoubek, I., Jarkovsky, J., Dusek, L., Hilscherova, K., 2013. Spatial patterns and temporal changes of heavy metal distributions in river sediments in a region with multiple pollution sources. J. Soils Sediments 13, 1257–1269.

Celliers, L., Moffatt, T., James, N.C., Mann, B.Q., 2004. A strategic assessment of recreational use areas for off-road vehicles in the coastal zone of KwaZulu-Natal, South Africa. Ocean Coast. Manag. 47, 123–140.

Christophoridis, C., Dedepsidis, D., Fytianos, K., 2009. Occurrence and distribution of selected heavy metals in the surface sediments of thermaikos gulf, N. Greece. Assessment using pollution indicators. J. Hazard. Mater. 168, 1082–1091.

Cooper, J.A.G., 1995. Sea-level Rise and its Potential Physical Impacts on the Shoreline of KwaZulu-Natal: Tugela Mouth to Mtamvuna River Mouth. Natal Town and Regional Planning Commission Report, 80, The Natal Town and Regional Planning Commission, Pietermaritzburg, South Africa, p. 138.

De Villiers, S., 2005. The hydrochemistry of rivers in KwaZulu-Natal. Water SA 31 (2), 193–198.

Dunlop, S.W., Mann, B.Q., 2013. An assessment of the offshore boat-based linefishery in KwaZulu-Natal South Africa. J. Mar. Sci. 35 (1), 79–97.

Fair, D., Jones, T., 1991. The Ports of Sub-Saharan Africa and Their Hinterlands: An Overview. KR Litho, Pretoria, pp. 32–35.

Frignani, M., Belluci, L.G., 2004. Heavy metals in marine coastal sediment: assessing sources, fluxes, history and trends. Anal. Chem. 94, 479–486.

García-Charton, J.A., Pérez-Ruzafa, A., Marcos, C., Claudet, J., Badalamenti, F., Benedetti-Cecchi, L., Falcón, J.M., 2008. Effectiveness of European Atlanto-Mediterranean MPAs: do they accomplish the expected effects on populations, communities and ecosystems? J. Nat. Conserv. 16 (4), 193–221.

Goble, B.J., van der Elst, R.P., Oellermann, L.K. (Eds.), 2014. Ugu Lwethu e Our Coast. A Profile of Coastal KwaZulu-Natal. KwaZulu-Natal Department of Agriculture and Environmental Affairs and the Oceanographic Research Institute, Cedara, p. 202.

Grant, L., Butler-Adam, J., 1992. Tourism and development needs in the Durban region. In: Smith, D.M. (Ed.), The Apartheid City and Beyond. Routledge, London.

Greenfield, R., Wepener, V., Degger, N., Brink, K., 2011. Richards Bay harbour: metal exposure monitoring over the last 34 years. Mar. Pollut. Bull. 62, 1926–1931.

Halpern, B.S., 2003. The impact of marine reserves: do reserves work and does reserve size matter? Ecol. Appl. 13 (1), S117–S137.

Harmelin-Vivien, M., Ledireach, L., Baylesempere, J., Charbonnel, E., Garciacharton, J., Ody, D., Perezruzafa, A., 2008. Gradients of abundance and biomass across reserve boundaries in six Mediterranean marine protected areas: evidence of fish spillover? Biol. Conserv. 141 (7), 1829–1839.

Harris, L.R., 2008. The Ecological Implications of Sea Level Rise and Storms for Sandy Beaches in Kwazulu-Natal. M.Sc. Thesis, School of Biological and Conservation Sciences, University of KwaZulu-Natal, Westville.

Hwang, D.-W., Kim, S.-G., Choi, M., Lee, I.-S., Kim, S.-S., Choi, H.-G., 2016. Monitoring of trace metals in coastal sediments around Korean Peninsula. Mar. Pollut. Bull. 102, 230–239.

Jury, M.R., 1998. Statistical analysis and prediction of KwaZulu/Natal climate. Theor. Appl. Climatol. 60, 1–10.

Jury, M.R., 2002. Economic impacts of climate variability in South Africa and development of resource prediction models. J. Appl. Meteorol. 41 (1), 46–55.

Kohler, K., 2005. B & B Development in KwaZulu-Natal. Tourism KwaZulu-Natal Occasional Paper No. 32, Durban.

Kohler, K., 2009. A Short History of Tourism Statistics in KwaZulu-Natal. Tourism KwaZulu-Natal Occasional Paper No. 71, Durban.

Kohler, K., 2010. Kwa-Zulu Natal's Hotel Occupancy Rates—A Brief Comparative Study. Tourism KwaZulu-Natal Occasional Paper No. 76, Durban.

Kruger, G.P., 1983. Terrain Morphological Map of Southern Africa 1:2 500 000. Department of Agriculture, Pretoria.

Lai, T.M., Lee, W., Hur, J., Kim, Y., Huh, I.A., Shin, H.S., Kim, C.K., Lee, J.H., 2013. Influence of sediment grain size and land use on the distributions of heavy metals in sediments of the Han river basin in Korea and the assessment of anthropogenic pollution. Water Air Soil Pollut. 224, 1609–1621.

Lawrence, C., 2005. Biodiversity Survey Towards Conservation of Subtidal Reef Habitats in Kwazulu-Natal: Biogeography and Depths Patterns. M.Sc. Thesis, University of Cape Town, South Africa.

Lim, D., Jung, S.W., Choi, M.S., Kang, S.M., Jung, H.S., Choi, J.Y., 2013. Historical record of metal accumulation and lead source in the southeastern coastal region of Korea. Mar. Pollut. Bull. 74, 441–445.

Mann, B.Q., Taylor, R.H., Densham, D., 1998. A synthesis of the current status of marine and estuarine protected areas along the Kwazulu-Natal coast. Lammergeyer 45, 48–64.

Mather, A.A., 2007. Linear and nonlinear sea-level changes at Durban, South Africa. S. Afr. J. Sci. 103, 509–512.

Mathieu, R., Yves, R., 2003. Intensity and spatial extension of drought in South Africa at different time scales. Water SA 29 (4), 489–500.

Nel, E., Rogerson, C.M. (Eds.), 2005. Local Economic Development in the Developing World: The Experience of Southern Africa. Transaction Press, New Brunswick, NJ.

O'Connor, M.C., Lymbery, G., Cooper, J.A.G., Gault, J., Mckenna, J., 2009. Practice versus policy-led coastal defence management. Mar. Policy 33, 923–929.

Penney, A.J., Mann-Lang, J.B., van der Elst, R.P., Wilke, C.G., 1999. Long-term trends in catch and effort in the KwaZulu-Natal nearshore linefisheries. S. Afr. J. Mar. Sci. 21, 51–76.

Preston-Whyte, R., Scott, D., 2007. Urban tourism in Durban. In: Rogerson, C.M., Visser, G. (Eds.), Urban Tourism in the Developing World: The South African Experience. New Brunswick, NJ, Transaction Publishers, pp. 254–264.

Rainbow, P.S., 2007. Trace metal bioaccumulation: models, metabolic availability and toxicity. Environ. Int. 33, 576–582.

Russ, G.R., Cheal, A.J., Dolman, A.M., Emslie, M.J., Evans, R.D., Miller, I., Sweatman, H., Williamson, D.H., 2008. Rapid increase in fish numbers follows creation of world's largest marine reserve network. Curr. Biol. 18, R514–R515.

Smith, A.M., Mather, A.A., Bundy, S.C., Cooper, J.A.G., Guastella, L.A., Ramsay, P.J., Theron, A., 2010. Contrasting styles of swell-driven coastal erosion: examples from KwaZulu-Natal, South Africa. Geol. Mag. 147 (6), 940–953.

South African Weather Service, 2008. Annual report. http://www.weathersa.co.za/images/documents/Annual_Reports/SAWS-Annual-Report-08-09.pdf.

Stats SA, 2010. Statistics South Africa Annual Report 2010. https://www.statssa.gov.za/publications/P0302/P03022010.pdf.

Stats SA, 2016. Statistics South Africa Annual Report 2016–2017. https://nationalgovernment.co.za/department_annual/203/2017-statistics-south-africa-(stats-sa)-annual-report.pdf.

Town and Regional Planning Commission, 1988. Survey of the ski-boat launching sites along the natal coast. In: Interim Policy Statement. Natal Town and Regional Planning Commission, KwaZulu-Natal, South Africa, p. 116. Revision of Report 6.5.1.

UNEP/Nairobi Convention Secretariat, 2009. Transboundary Diagnostic Analysis of Land-Based Sources and Activities Affecting the Western Indian Ocean Coastal and Marine Environment. UNEP, Nairobi, Kenya, p. 378.

UNOPS (United Nations Office for Project Services), 2011. Turpie, J., Wilson, G. (Eds.), Draft Cost/Benefit Assessment of Marine and Coastal Resources in the Western Indian Ocean: Mozambique and South Africa. Anchor Environmental Consultants, South Africa. Produced for the Agulhas and Somali Current Large Marine Ecosystems Project. June 2011.

Vetrimurugan, E., Jonathan, M.P., Roy, P.D., Shruti, V.C., Ndwandwe, O.M., 2016. Bioavailable metals in tourist beaches of Richards Bay, Kwazulu-Natal, South Africa. Mar. Pollut. Bull. 105, 430–436.

Vetrimurugan, E., Shruti, V.C., Jonathan, M.P., Roy, P.D., Kunene, N.W., Villegas, L.E.C., 2017. Metal concentration in the tourist beaches of South Durban: An industrial hub of South Africa. Mar. Pollut. Bull. 117, 538–546.

Vetrimurugan, E., Shruti, V.C., Jonathan, M.P., Roy, P.D., Rawlins, B.K., Rivera-Rivera, D.M., 2018. Metals and their ecological impact on beach sediments near the marine protected sites of Sodwana Bay and St. Lucia, South Africa. Mar. Pollut. Bull. 127, 568–575.

Visser, J.N.J., 1990. The age of the late Palaeozoic glacigene deposits in Southern Africa. S. Afr. J. Geol. 93, 366–375.

Wang, W.X., Rainbow, P.S., 2008. Comparative approaches to understand metal bioaccumulation in aquatic animals. Comp. Biochem. Physiol. C Toxicol. Pharmacol. 148, 315–323.

Whitfield, A.K., 2000. Available Scientific Information on Individual South African Estuarine Systems. WRC Report 577/3/00, Water Research Commission, Pretoria.

CHAPTER

9

Integral Management of the Coastal Zone to Solve the Problems of Erosion in Las Glorias Beach, Guasave, Sinaloa, Mexico

*Jiménez-Illescas Ángel R**, *Zayas-Esquer Ma Magdalena*[†], *Espinosa-Carreón T. Leticia*[‡]

*Centro de Interdisciplinario de Ciencias Marinas (CICIMAR), Instituto Politénico Nacional (IPN), La Paz, Mexico

[†]Instituto de Investigación en Ambiente y Salud de la Universidad de Occidente, Los Mochis, Mexico

[‡]Instituto Politécnico Nacional, Centro Interdisciplinario de Investigación para el Desarrollo Integral Regional (CIIDIR) Unidad Sinaloa, Guasave, Mexico

1 INTRODUCTION

The coastal zone, due to its environmental complexity and productive potential, is a place where this multiplicity of options is naturally given; an adequate planning of concerted uses presents several opportunities for development (Moreno-Casasola et al., 2005; Klemas, 2013). In some coastal areas, the presence of rivers and streams, the erosion of neighboring areas, landslides, and rock weathering, as well as the height and period of waves and wind erosion all play a very significant role. These factors change throughout the year, predominating in some areas (Splinter et al., 2012). However, another important process is the transport of sediment originated by coastal currents (González-Ruelas et al., 2005; Anderson et al., 2014). Under equilibrium conditions, the beach is practically stable, but this stability is often altered by short-term phenomena such as storms, periods of calm, and the year-to-year variation in sediment input rates. The construction of civil works on the coast or the construction of dams

https://doi.org/10.1016/B978-0-12-810473-6.00010-8
© 2019 Elsevier Inc. All rights reserved.

produces changes that affect sedimentation or loss rates by coastal transport and generate changes in the coastline, with those changes being very fast (Zayas-Esquer, 2010).

The climatic change and anthropogenic activities also affect the natural processes of sustaining beaches and coasts. The removal of the sand from the sand-sharing system by a tsunami or storm surge is not lost property, because it takes months or years to quantify the impact in the coastal area (Prasad and Kumar, 2014). The monitoring of the changing position of the coast is elemental to scientists, engineers, and managers. Also, it is important to know the answer of coastal dynamics to the hurricane presence (Stockdon et al., 2009). The change of the coastal dynamic by the erosion/accretion processes is global trouble (Barnard et al., 2012). For example, in Southern California diminished river flow during droughts resulted in beach erosion, but the beaches were restored during intervening wet years when the fluvial sediment supply revived (Orme, 1985). Reduction of the fluvial sediment supply commonly results from the construction of dams to impound water upstream. These intercept fluvial sediment discharge and so cut off the supply of sand and gravel to beaches at and near the river mouth (Hansen et al., 2013). This leads to the onset of erosion on beaches that were formerly maintained or prograded by the arrival of this fluvial sediment. Erosion develops more quickly—and becomes more severe—where there is strong longshore drift of sediment away from the river mouth. In 2004, the European Union suffered damage to its coastal areas due to erosion caused by the transport of sediments (Massenlink et al., 2016). Many places have erosion around 15,100 km, and the surface lost by erosion is 15 km^2 $year^{-1}$ (Eurosión, 2005). In Louisiana, coastal erosion is a silent killer that effectively has taken away the life of the Gulf Coast region; a loss rate of 0.326 km^2 $year^{-1}$ and a total of 158 beach nourishment episodes in 60 locations have been recorded (Anderson et al., 2014). Harley et al. (2017) said that extreme coastal erosion during an anomalous extratropical storm in southeast Australia meant that a total of 11.5 million m^3 of sand was eroded from a subaerial beach. During the El Niño event of 2015–2016, Barnard et al. (2017) reported important erosion on the California coast, which was more severe than other events (Barnard et al., 2015). In the upper zone of beaches, the erosion affect in the biology, for example, in Southern California, during the erosion the invertebrates are restricted distribution and dispersal (Hubbard et al., 2014), making habitats vulnerable in the zones affected by erosion. A regional understanding of littoral cell boundaries and sand budges can be an important tool in coastal engineering and land-use management decisions (Patsch and Griggs, 2008).

In Mexico, we have few studies of civil works, but the problem is serious because the construction of large breakwaters has caused erosion processes in more coastal areas, for example in Bahía Banderas, Jalisco (González-Ruelas et al., 2005). In Sinaloa, Mexico, there are some coastal areas with erosion, and we focus on Las Glorias Beach because its coastal dynamics have been broken at least four times since 1972, and its erosion/accretion processes are present today.

The review is aimed at know the change in the patterns of coastal transport and its affectation with the construction of the breakwater in the Las Glorias Beach in Sinaloa State, Mexico, and to propose alternative solutions.

1.1 Study Area

Las Glorias Beach is the most important recreation center of the municipality of Guasave, Sinaloa, Mexico. Over two decades, the beach line has traveled, that is to say the sea to cattle terrain, losing more than 300 m of beach by erosion. Las Glorias Beach is located 42 km from Guasave (Fig. 1). This beach, which also gives sustenance to the communities of Boca del Rio and La Pitahaya (Fig. 2), is increasingly deteriorated by the clashes of tidal waves. Each year from June to October, these waves are present with force and are constantly modifying the coastline. Playa las Glorias is located in a zone of climatic transition between semidry, very warm, and very dry climates (García, 2004). The average annual temperature in the records is 24.2°C, with a maximum annual average of 32°C. The warmest months are July and August and the coldest months are January and February. As for precipitation, the annual average is 459 mm, with the maximum rainfall in the months of August and September and the minimum rainfall in the months of April and May. The dominant winds coming from the southeast are generally cyclonic and reach speeds up to $24\,m\,s^{-1}$ (Fluviomarítima (Consultores Ingeniería) S.A., 1992).

The nine most intense hurricanes that have occurred in Mexico during 1988–2011 are shown in Table 1. "Isidoro," "Kenna," "Emily," and "Vilma" occurred the same year. The hurricanes that affected Sinaloa directly are "Paúl" in 1982 and "Ismael" in 1995. Moshinsky et al. (2014) emphasizes the greater density of these phenomena in the Pacific Ocean; in areas of interest for Mexico, the maximum density of the Pacific becomes six times the maximum density of the Atlantic. The Mexican land areas with more than 10 tropical cyclones in 52 years are mainly the coasts of Guerrero, Michoacán, Colima, Jalisco, and Baja California Sur as well as Quintana Roo. However, land areas with more than five tropical cyclones in 52 years already

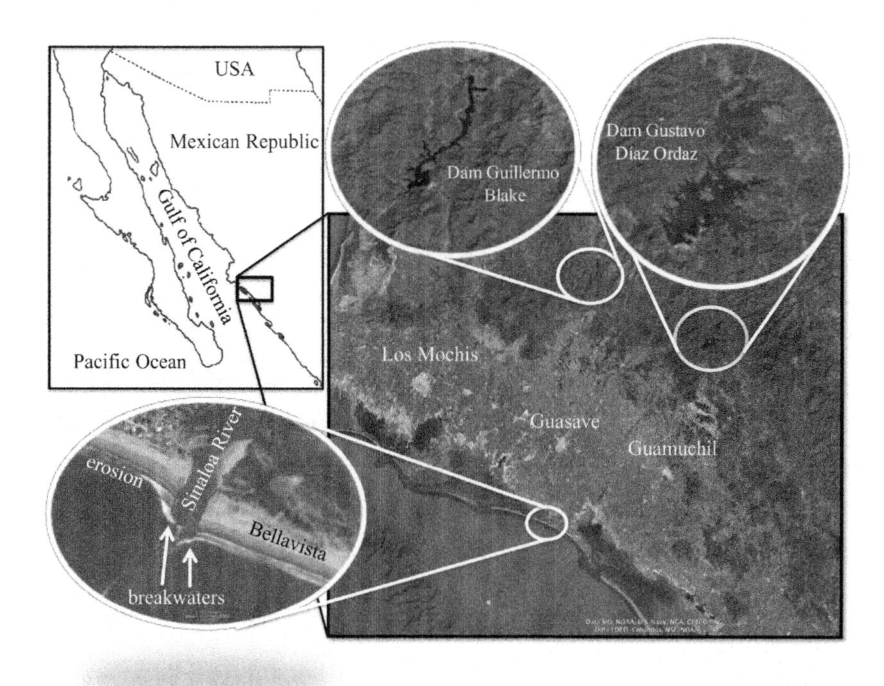

FIG. 1 Study area in Sinaloa, Gulf of California, Mexico. Dams Guillermo Blake and Diaz Ordaz. In the left is observed the breakerwaters in the discharge of the Sinaloa River. *From Google Earth 2017.*

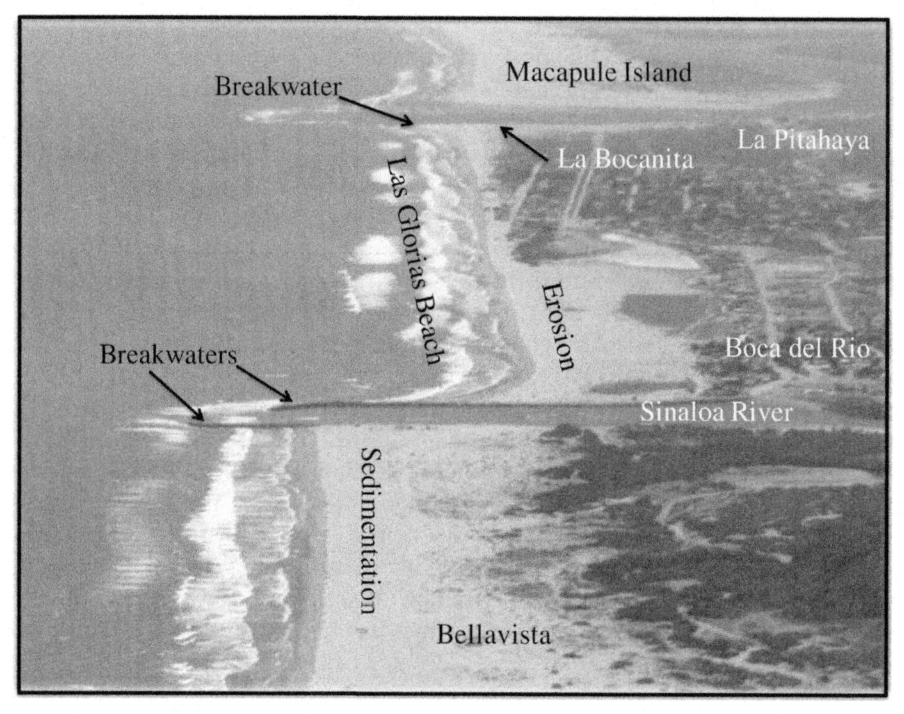

FIG. 2 Las Glorias Beach, Sinaloa. Note the breakwaters at Sinaloa River and in La Bocanita. *Modified from Espinosa-Carreón, T.L., 2013. Estudio Ambiental-Estudio de monitoreo de Procesos de Erosión en la Zona Costera de la Isla Macapule. SEMARNAT-CONANP. Informe técnico, 42 p.*

TABLE 1 Strong Hurricanes Occurred in Mexico During 1988–2011

Hurricane	Time	Affectation
Gilberto	September 1988	Category 5 (EHSS), affected Jamaica, Mexico, damages for 5.5 millions of dollars
Paulina	October 1997	Category 4 (EHSS), affected Mexico
Isidoro	September 2002	Category 3 (EHSS), affected Jamaica, Cuba, United States, Mexico
Kenna	October 2002	Category 5 (EHSS), affected Mexico, damages for 101 millions of dollars
Emily	July 2005	Category 5 (EHSS), affected Mexico, damages for 988 millions of dollars
Wilma	October 2005	Category 5 (EHSS), affected Cuba, Haití, Jamaica, México, United States, Bahamas
Lane	September 2006	Category 3 (EHSS), affected Mexico
Dean	August 2007	Category 5 (EHSS), affected Jamaica, Haití, Belice, México
Jova	October 2011	Category 1 (EHSS), affected Mexico

include the entirety of the Atlantic coast of Mexico and the whole of the Pacific coast of Mexico, except Sonora and Baja California. During the study time, the national territory had the presense of a least one tropical cyclone.

2 PROBLEM STATEMENT

In the municipality of Guasave, Sinaloa, in the locality of Las Glorias Beach, different types of actions have been carried out to promote the tourist development of the area because the site stands out for its beauty, magnitude, and impact in the region. Each year, the demand for facilities has increased for recreation and rest. Las Glorias Beach is located between the mouth of the Sinaloa River and La Bocanita, south of Macapule Island. Various tourist and housing establishments have been installed at Las Glorias Beach that support economic activity. In the 1990s, the local government of Guasave requested the participation of the National Fund for Tourism Promotion (FONATUR) for the elaboration of a "Master Plan of the Tourist Development Project of Las Glorias, Sinaloa," with the objective of increasing the tourist activity of the region at the national and international levels. Consultants in rivers, coastal, and maritime engineering carried out a study that presented a complex tourist development in Las Glorias Beach, proposing the construction of hotels, condominiums, residences, a golf course, a trailer park, and jetties. However, so far these plans have not been developed and implemented in the region (Zayas-Esquer, 2010).

The coastal dynamics of Las Glorias Beach have suffered several imbalances. The first one was due to the construction of the Guillermo Blake and Gustavo Díaz Ordaz dams in 1972 and 1981, respectively, because the contribution of sediment to the beaches was practically eliminated. Since then, an imbalance in the dynamics of this littoral cell was created, causing erosion of the beach. The second, from the construction of the breakwaters in 1992, further accelerated the process of beach recession (Fig. 1), generating a loss of land of 104 m (perpendicular to the beach) over 10 years, with an average erosion of the beach up to 37 m $year^{-1}$ (Zayas-Esquer, 2010).

Engineer Daniel Cervantes Castro elaborated the first project of the breakwaters, where it was proposed to stabilize the beachfront in Las Glorias and give certainty to potential investors to develop tourism businesses. The study indicated that, in 1992, the mouth of the Sinaloa River presented problems of horizontal stability. This is why the Marine Army began the works for the construction of the rocks and the dredging of an access channel to keep the river mouth open. Due to the joint action of the storm surge and the orientation of the discharges of the river, by July 1993 the western breakwater was isolated and formed an estuary along the west beach. In August 1993, both breakwaters were broken, forming two islands; therefore the reconstruction of the jetties was planned. By measuring wave data and applying a littoral transport model, the net direction of coastal transport from east-southeast to west-northwest was obtained. The construction of the breakwaters created a physical barrier for the passage of material from the east, shown by the accumulation of sediment from east of the mouth of the river, known as Bellavista. This caused the Sinaloa River to be contained, so the sand transport was reduced between the mouth of the river and La Ensenadita, causing the loss of the beach in Las Glorias (Fig. 2).

In 2000, the sea began to gain ground and eroded the beach more quickly, until in 2003 it reached the first line of houses and restaurants, affecting them in an important way. The owners of the affected restaurants said that the breakwaters constructed between 1992 and 1993 (Fig. 1, left side) to keep open the inlet channel and the navigability of the river Sinaloa are the cause of the erosion at Las Glorias Beach. Starting in 2004, a group of service providers from Las Glorias and Boca del Río asked the three levels of government to stop erosion and rescue the beach (Espinosa-Carreón, 2013).

With the objective of restoring the processes of beach dynamics and restoring the land to the sea, in 2006 the construction of a breakwater in La Bocanita (northern limit of Las Glorias beach) began. Of the 311 m of the original project, only 147 were built, which increased in speed and force the phenomenon of erosion, making this the third time that the coastal dynamics broke. The start zone of the breakwater began to erode, threatening to separate the breakwater from the coast. In 2008, one of the strongest effects of the erosion process was presented (Espinosa-Carreón, 2013). This led the Federation of Fishing Cooperatives of Guasave and the inhabitants to express great concern about the serious erosion at Las Glorias Beach, which has sometimes reached the limit of restaurants while the intense waves have destroyed the foundations of some buildings.

In 2012, construction of a continuation of the breakwater was planned, aiming to close the area of La Bocanita. However, the Mexican Transportation Institute and the local government of Guasave indicated that the budget was insufficient to finish the civil work. Nowadays, the high waves invade the restaurant area and this has to be added to the erosive effect of the hurricanes, which have increased in frequency and intensity. It was necessary to carry out an evaluation of the effects of the construction of the breakwater in the coastal dynamics of the region in order to propose remedies and avoid further deterioration of the coastal zone that affects not only Las Glorias beach but also Macapule Island, which is part of the natural protected area of the Macapule Lagoon—San Ignacio—Vinorama System (CONANP, 2010).

The flow chart of Table 2 describes the procedure carried out and the equipment used.

TABLE 2 Flow Chart Used in This Work

Process	Date	Equipment Used	Photographs
Topography profiles	April and November 2008. April and November 2009	Station total GPS Trimble	
		Mojoneras	
Tides	April 2009	Haller tide mark Lobo Instruments	

TABLE 2 Flow Chart Used inThis Work—Cont'd

Process	Date	Equipment Used	Photographs
Currents	2008	Sontek Argonauta MD4	
	2009	Interocean S4	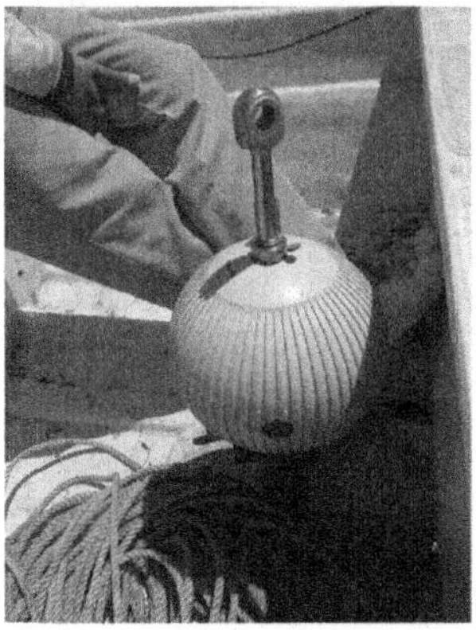
	2008, 2009	Drift bodies	
Coastlines	1999, 2003, 2007, 2008, 2010, 2012	Photographs, satellite images	

3 BEACH PROFILES

In 2008, 30 topographic monuments (PVC pipes filled with concrete) were installed along Las Glorias, La Bocanita, and Macapule Island (Fig. 3). their positioning was accomplished with the GPS TRIMBLE 5800 topographic equipment, with the WGS 84 datum of the UTM coordinates, 12 region (Universal Transverse Mercator). The georeferencing of the mojoneras was performed with a topographic RTK GPS with an accuracy of 1 mm. In each sampling season, the 30 monuments were identified (Table 2). The leveling was verified using a total electronic topographic station and a prism (Montes de Oca, 1985). To perform the beach profiles, on each level bank sections were carried out perpendicular to the coastline by recording a point of each change of slope in the terrain. In each profile, the dimensions of the natural terrain with variable distances were measured. This was carried out until the break zone, usually during low tide, in order to cover a greater possible length of beach, reaching usually below the zero of the Low Lower Bajamar Level (NBMI). The data obtained were processed with the programs Autocad and Civilcad 2008 and profiles were obtained. Erosion processes were established to estimate soil loss/gain gradients in the area in April and November 2008 and 2009.

Fig. 4 shows representative profiles at Las Glorias Beach. When comparing the seasonally obtained profiles in April and November 2008 (Fig. 4A), in general, they show an accretion in the berm area and beach face up to 1 m on the vertical scale. The highest sedimentation was in the zone of the beach face with approximately 15,409 m^3 of sedimentation and a growth in the length of the profile of up to 60 m. Erosion constituted only 707.64 m^3, corresponding to a minimal part of the seasonal variability. The accretion represents the greater weight in the seasonality from spring to autumn in 2008. However, the profiles obtained in April 2008 and

FIG. 3 Beach profiles in La Bocanita, Las Glorias Beach, and Macapule Island. *From Google Earth 2017.*

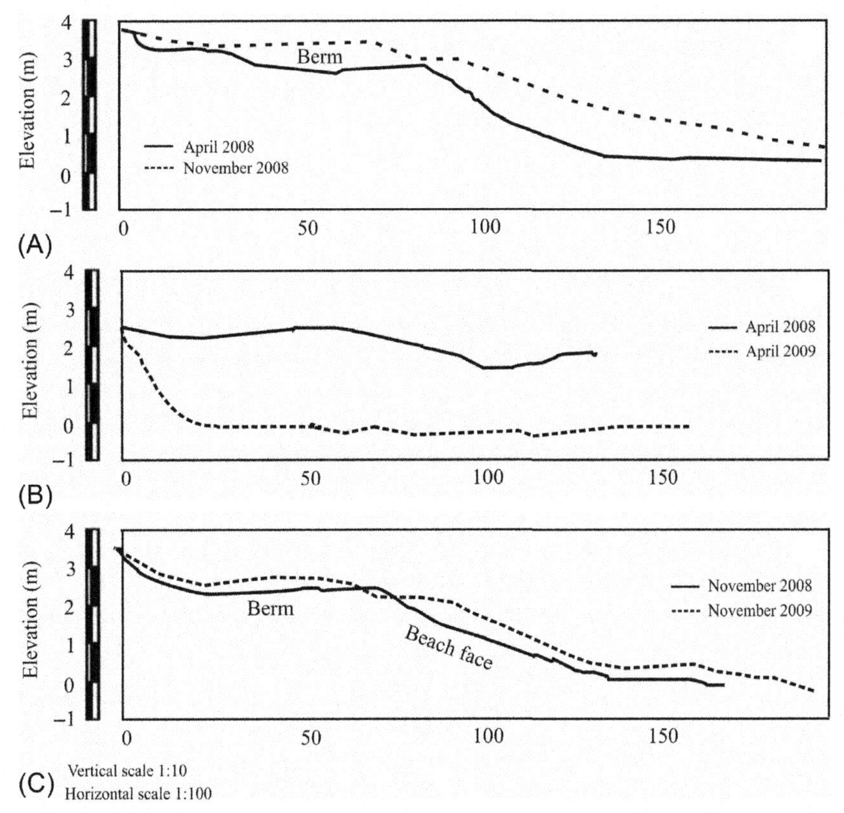

FIG. 4 Beach profiles in Playa Las Glorias, (A) April through November 2008, (B) April 2008 through April 2009 and, (C) November 2008 through November 2009.

April 2009 show an erosion of almost 2.5 m between of the 15–150 m along the section (Fig. 4B). The length of the profiles decreased approximately 40 m from April 2008 to April 2009, indicating that in the area adjacent to the breakwater, there was a large erosion of approximately 4703.48 m^3 while the accretion in the remote profiles of the breakwater represented approximately 1635 m^3. Most of the profiles made in November 2008 and November 2009 (Fig. 4C) show a similarity in their behavior from the berm zone to the beach face; however, it is notable that the length of the profile decreases to 25 m in November 2008 compared to November 2009. The contribution of sediments by the northwest winds is probably maintained, allowing the dune areas to erode; that sediment is deposited both on the berm and the beach face.

The beach profiles of April and November 2008 and 2009 showed a seasonal variation of the coastline. In April, erosion was observed as well as a decrease of approximately 50–60 m in the beach length followed by an accretion up to 1 m in November, which is considered uncommon and an inverse behavior. On the contrary, in other coastal areas in winter, particularly in the month of November, there is generally more erosion and in April (summer) more deposition. This behavior has also been reported by Alcántar-Elizondo (2007).

The interannual variability shows a possible impact from 2008 to 2009 while, in the months of April 2008 and April 2009, the beach profile showed erosion. In November 2008 and November 2009, although there was no significant change in the level of the different profiles, there was a decrease in the length of the profile on average of 40 m that suggests a gain of the sea on the beach.

4 TIDES

In April 2009, in La Bocanita has installed a tide gauge (Haller Lobo Instruments) at 2 m depth. These took wave records every half second with a pressure sensor and averaged the data every minute to record them in an internal memory. The amplitude of the tide, the average level of the sea, the low average low tide, and the average high tide were estimated. The records were compared to the tide calendar of the Topolobampo forecast provided by the Center for Scientific Research and Higher Studies of Ensenada (CICESE) on the website http://redmar.cicese.mx/.

When comparing the tides obtained in La Bocanita and Topolobampo (Fig. 5), a lag of 0.5–1.5 h was observed, that is, the high tides occur in La Bocanita and delay propagating from 0.5 to 1.5 h before being recorded by the tide gauge of Topolobampo.

5 CURRENTS

During the 2008 seasonal sampling, a Sontek Argonauta MD4 Doppler current meter with a coppled SeaBird CTD sensor was installed in the mouth of La Bocanita to measure the tide currents (Eulerian Method). In 2009, an Interocean S4 electronic current meter was installed in the deepest part of the channel, and drifter bodies with GPS were launched to trace lagrangean trajectories (Table 2) to calculate the tide velocities in La Bocanita. An ADP (Doppler acoustic profiler) was installed in the bottom, taking the readings upward, with layers programming one each meter until the surface.

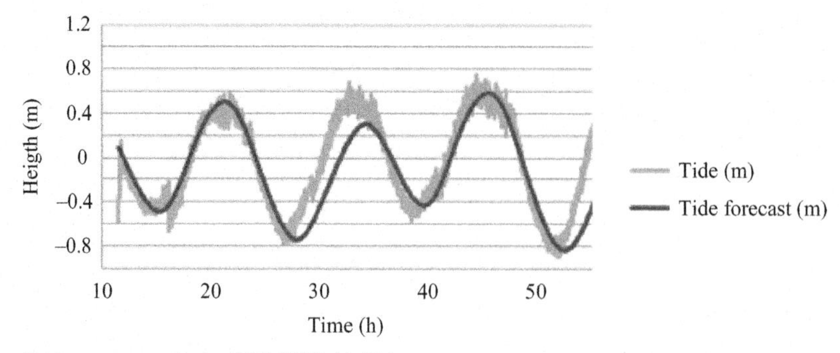

FIG. 5 Tides in La Bocanita on April 22, 2009, and tide forecast from Topolobampo, Sinaloa.

The maximum measured speed with drifter bodies in January 2009 was $1.2\,\mathrm{m\,s^{-1}}$ in La Bocanita. As it enters the navigation channel (flow), the current velocity increases and as it approaches the breakwater, it produces eddies next to the tip. The inlet flow pattern consists of a south-north current with speeds slightly less than $1\,\mathrm{m\,s^{-1}}$, and the ebb or reflux current flows over the deeper part of the navigation channel between Macapule Island and La Bocanita; its velocity increased to $1.2\,\mathrm{m\,s^{-1}}$ as it approached the mouth (Fig. 6).

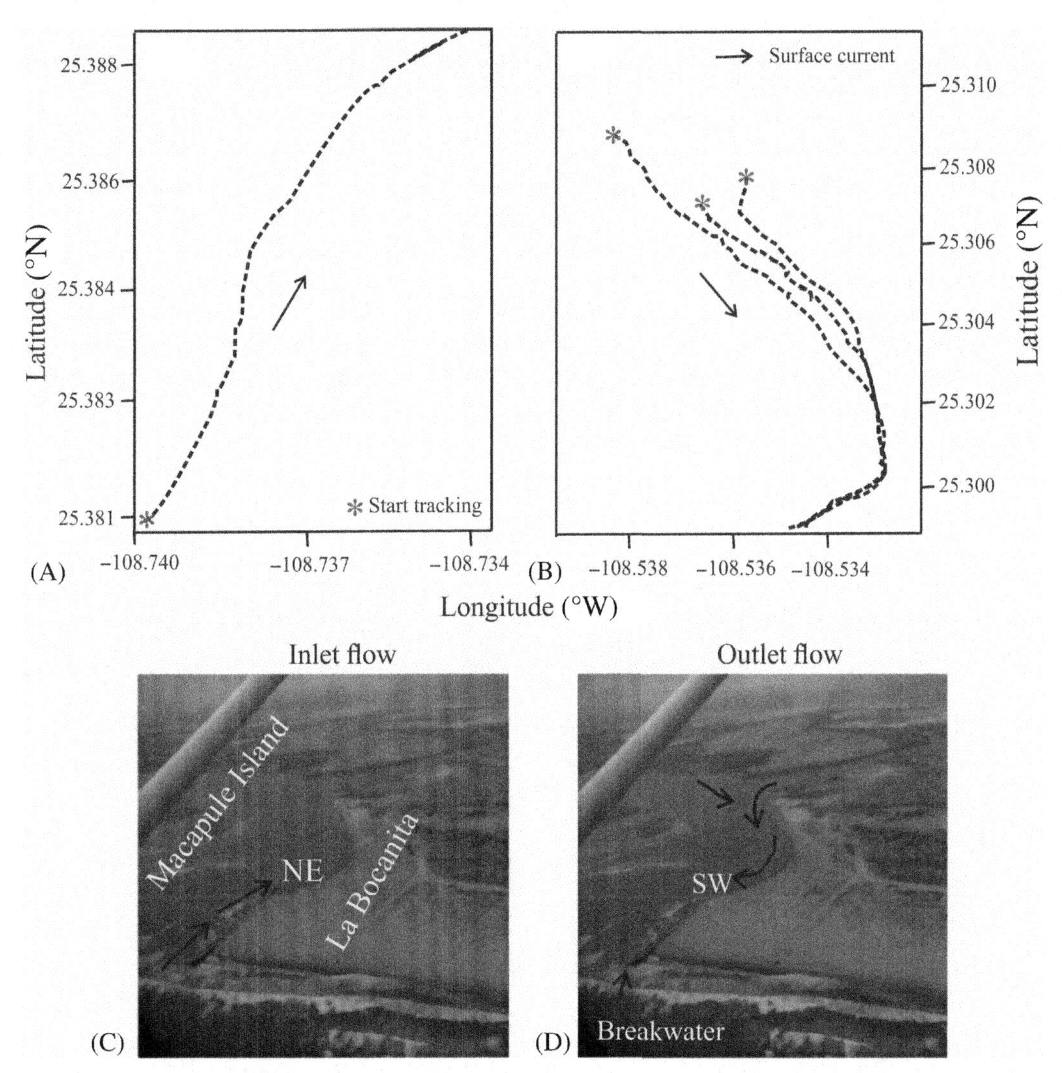

FIG. 6 Trajectories tracking of surface currents by drift bodies (A) and (B), inlet flow (C), and (D) outlet flow. *Modified from Espinosa-Carreón, T.L., 2013. Estudio Ambiental-Estudio de monitoreo de Procesos de Erosión en la Zona Costera de la Isla Macapule. SEMARNAT-CONANP. Informe técnico, 42 p.*

In Las Glorias Beach, the maximum measured velocities recorded were $0.89\,\mathrm{m\,s^{-1}}$, and the minimum velocities were $0.05\,\mathrm{m\,s^{-1}}$. The current speed increases in the navigation channel, in front of the La Bocanita, and produces eddies near the breakwater tip.

The physical dynamics determines the coastal transport of Macapule Island and La Bocanita (north of Las Glorias Beach). Coastal transport depends on the external waves present on the coast, but mainly on its angle of incidence, which is modified by refraction that in turn is caused by bathymetry because the waves propagate faster through deeper parts. The analysis of stream measurement data suggests that there is a parallel and near-shore current with intensity greater than $0.50\,\mathrm{m\,s^{-1}}$ at flow and at reflux greater than $1\,\mathrm{m\,s^{-1}}$, which is capable of eroding the shore and depositing the sediments where the intensity of the current is diminished. This causes the coast to be eroded, retracted, and the outside of the coast to grow (Sánchez-Lindoro et al., 2017).

During the period from November to March, the wind blows from the northwest, causing coastal transport toward the southeast while, in the summer, the wind blows from the southeast, causing coastal transport to the northwest. As you can see in Fig. 7, south of Macapule Island, one can observe the growth of a tongue or arrow of sand that grows the island toward the southeast and, for more evidence, the tongue of sand traps a small estuary. This is due to coastal transport from northwest to southeast as the southeasterly wave arrives at an angle greater than 5 degrees, which causes the lick of the wave to carry large amounts of sand into the system that surrounds the sandy tongue. The canal, which can be seen contiguous to the eastern coast of La Bocanita, passes through the south of the tongue and continues parallel to the coast of the island. It can also be observed that the waves do not break in the channel and the growth of Macapule Island to the south causes the speed of tidal currents to increase as

FIG. 7 Isla Macapule sand growing west-east. Bottom coastal line from 200 m. *Photograph taken from Leticia Espinosa.*

well as erosion in the western part of Las Glorias Beach in La Bocanita. When there is swell that breaks obliquely on this beach, this increases erosion by adding this factor of coastal transport to the currents.

6 VARIATION OF THE COASTLINE BY MEANS OF SATELLITE IMAGES AND PHOTOGRAPHS

6.1 Aerial Images

From Google Earth link were take images in differents years, and in 2006 and September 2009, two aerial tours were conducted to take photographs after the construction of the breakwater to know the differences in the coastline between the monitorings.

6.2 GPS Tracks With ATV

The beach contour was measured with GPS in an ATV that circulated over the coastlines, in Bajamar, approximately to the line of the mean sea level. Measurements were taken every 2 s with the GPS Garmin Etrex, capturing the position and then stored in an internal memory. This was captured in an electronic sheet Blue Chart America V8.5 and Garmin Map Source software was installed to download the GPSMAP data to the PC, with all the points of information in UTM coordinates. In addition, the contour of a 1994 topographic map, provided by INEGI 2004, was made and placed in layers to compare with the contour data obtained from the GPS track points with the ATV to measure the variations of the coastline.

6.3 Satellite Images

The satellite images of the land were obtained from 1999, 2003, 2007, 2008, 2010, and 2012 in order to carry out a comparative analysis between the images and then proceed to the evaluation of the coastal zone before, during, and after the construction of the breakwater. Through the satellite images, the coastlines were drawn, differentiating the panchromatic image. Subsequently, with the IDRISI 32 program, the study area was delimited and georeferenced, then later converted to a bitmap image (BMP). It was exported as an image to the Autocad 2008 program, and an image was placed in each layer, georeferenced, and placed on the same scale after delineating the coastlines of each one. Subsequently, areas of erosion and sedimentation were obtained for periods of time elapsed between each comparative image.

6.4 Littoral Transport Model and Resulting Erosion

In short, the construction of the breakwater partially halted the migration of the mouth, but in trying to avoid migration and by continuing the contribution of sand to the channel due to littoral transport (applying the Larras Model: 18 m^3/season (90 days)/m beach, 4 m^3/storm (3 days)/m beach), it reduced the mouth cross-section (from 2500 to 1000 m^2) and therefore caused an increase in the intensity of the currents that in turn caused an increase in beach erosion (37 m $year^{-1}$).

6.5 Digital Images From Google Earth Pro

From the digitization of the coastline, erosion was observed in La Bocanita before the construction of the breakwater (1999–2003), followed by a deposit or sedimentation from 2003 to 2008. Within La Bocanita, erosion was observed from 2008 to 2010. However, on the opposite side to the west of the Sinaloa River, from 1999 to 2003 there was erosion, from 2003 to 2008 sedimentation, and from 2008 to 2010, erosion. In the central part, constant erosion was observed (Fig. 8).

This can be separated into two processes, before and after the construction of the breakwater, which can be seen in Fig. 9A. Before the construction of the La Bocanita breakwater, there was already moderate erosion at Las Glorias Beach and accretion in La Bocanita, but especially at the eastern tip of Macapule Island. Between 2003 and 2007, there was accretion in Las Glorias Beach, but very large erosion in the eastern part of La Bocanita.

As you can see in Fig. 9B, the coastlines that are observed between 2003 and 2008, reflect an accretion to south of Macapule Island of 917 m^2, and along Las Glorias Beach. The Bocanita has an area of erosion of 1350 m^2 toward the dune zone and an average displacement (erosion) of the coastline of 289 m linear as well as an average erosion rate of 48 m $year^{-1}$. As for Isla Macapule, it has a growth area of 1692 m^2 to the southeast of the Bocanita and an average displacement (regression) of the coastline of 203 linear meters, resulting in an average erosion rate of 34 m $year^{-1}$.

The digitization of the images from 2008 to 2010 (Fig. 9C) showed along the coastline of Playa Las Glorias, an erosion of 2 km, which is equivalent to 1306 m^2, with an annual linear displacement of 25 m $year^{-1}$. Likewise, where the buildings are located at about 600 m has a very varied erosion behavior. La Bocanita presents an erosion area of 545 m^2 and an average coastline or erosion displacement of 70 linear meters, with an average erosion rate of 24 m $year^{-1}$ toward the dune zone. Macapule Island has an accretion area of 469 m^2 and an average coastline or accretion displacement of 317 linear meters, perpendicular to the beach, with an average accretion rate of 106 m $year^{-1}$ to the southeast of the Bocanita.

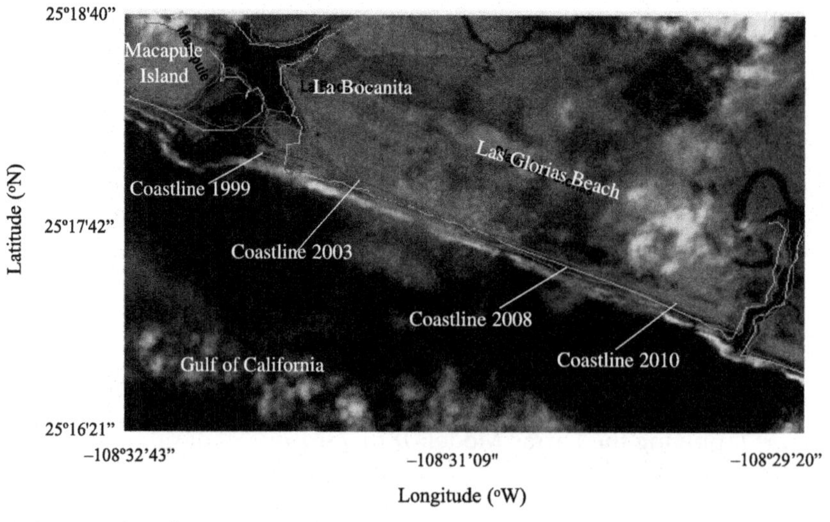

FIG. 8 Digitalization of satellite images, coastal line changes during 1999–2010.

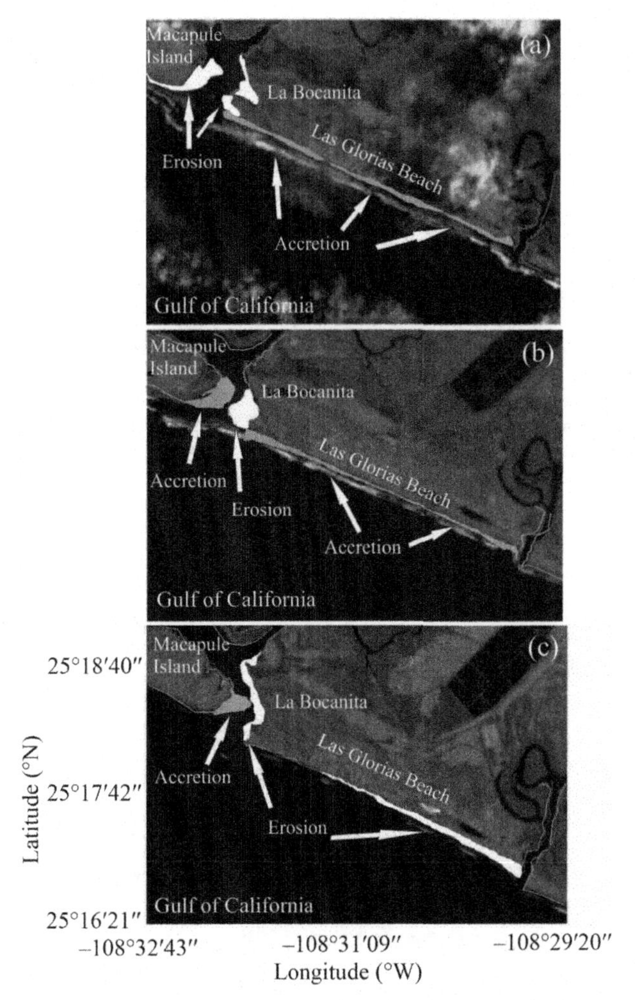

FIG. 9 Coastal lines in Playa Las Glorias and Isla Macapule: (A) in 1999 and 2003, (B) in 2003 and 2008, and (C) in 2008 and 2010.

A cut was made in the satellite images to observe in greater detail the coastline change in La Bocanita and Isla Macapule before, during, and after the construction of the spike (Fig. 10). Before 1997 (before the breakwater), the coastline could be considered with some stability, but in 2007 there is a great extension in La Bocanita that has been eroded (Fig. 10A). From 2007 to 2012, we can see the continuation of the deposit in the internal part of Macapule Island, but toward its border with the sea, a very marked erosion was observed. On the other hand, to the north of La Bocanita, the erosion continues, while next to the breakwater the accretion is observed (Fig. 10B). When comparing the coastlines in 1999 and 2012, the deposition in the inner part of Isla Macapule and the great extent of erosion in La Bocanita are evident (Fig. 10C).

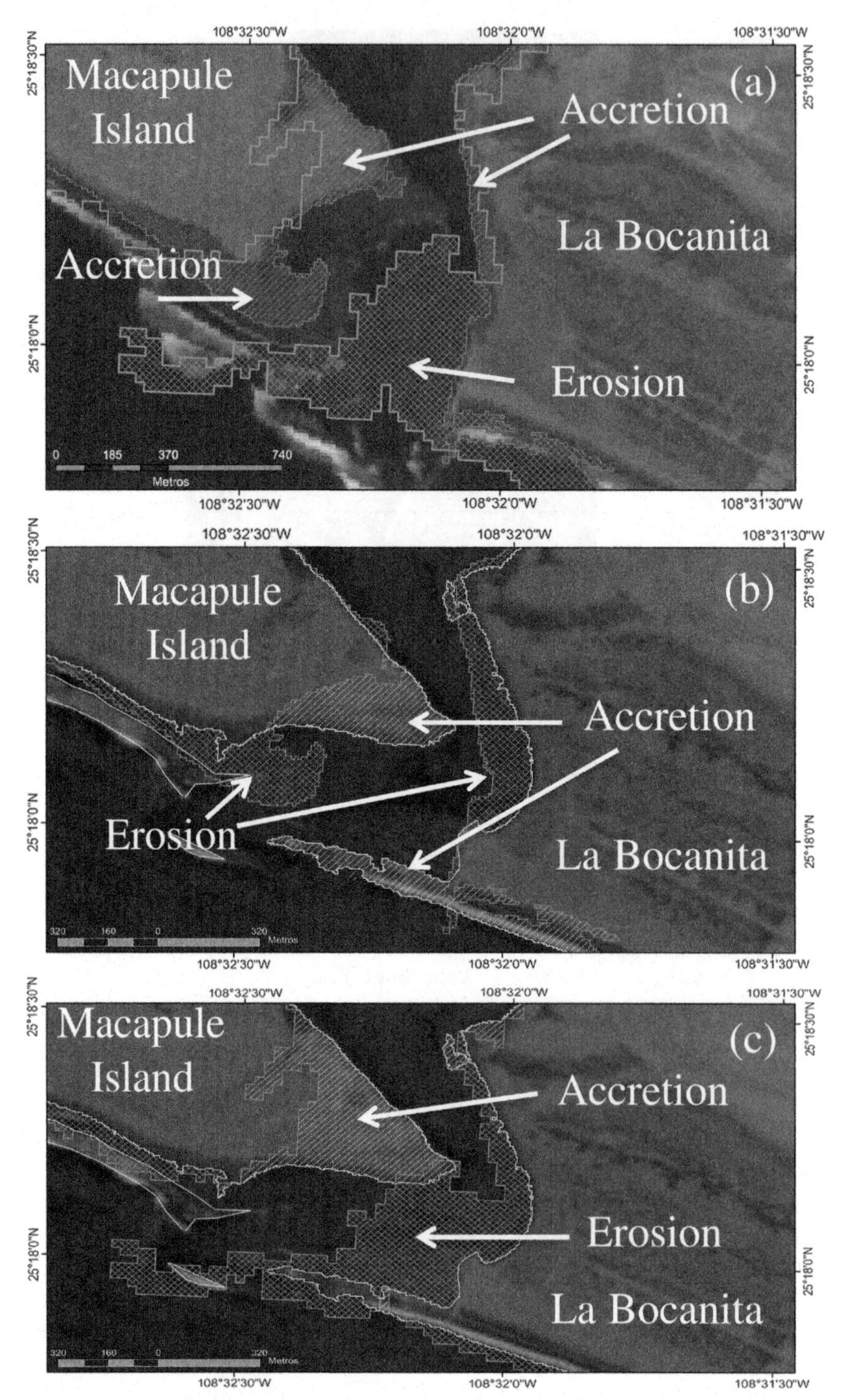

FIG. 10 Costal lines in La Bocanita and Isla Macapule: (A) in 1999 and 2007, (B) in 2007 and 2012, and (C) in 1999 and 2012. *Modified from Espinosa-Carreón, T.L., 2013. Estudio Ambiental-Estudio de monitoreo de Procesos de Erosión en la Zona Costera de la Isla Macapule. SEMARNAT-CONANP. Informe técnico, 42 p.*

6.6 Satellite Images

Naturally, the coastal dynamics of the beach are a process that presents deposition and an erosion, a behavior that is directly related to the currents, their speed, and angle of incidence. In the area of La Bocanita-Isla Macapule (southern part), the processes of retreat and advancement of the mouths that had naturally occurred, before the construction of the dams and the breakwater, oscillated around a base line during 50 years. However, this process has been significantly affected. From satellite images in 1999, 2007, 2008, and 2012, it is shown that the effect of the construction of the breakwater, mainly to the south of Isla Macapule and La Bocanita, has been modifying the patterns of currents. Due to the intense erosion processes in Playa Las Glorias (350 linear meters) and of acression in Isla Macapule (500 linear meters), goverment decision makers must be involved in generating management plans.

6.7 Erosion

There are four types of engineering works that affect the transport of sand and the stability of beaches: (1) Alteration of the production of sediments from the source, for example, the construction of dams in rivers and the protection of cliffs against erosion (Eurosión, 2005). (2) Breakwaters that extend from the coast and intercept coastal transport by reducing the supply of sand downstream from the site. (3) Coastal barriers comprising the breakwaters parallel to the coast, which are built offshore and intercept the waves before they reach the coast, including submerged breakwaters. For this reason, the littoral material accumulates in the protected area behind and the rate of transport of the breakwater is reduced. (4) Channels that drain across the shoreline. These channels create depths greater than normal and produce an accumulation of littoral material upstream of the littoral flow, and produce a reduction of downstream supply. Breakwaters generally complement dredging. In all cases, the reduction of coastal transport produces a recession of the beach, which is also facilitated because the waves continue to transfer sediments from the beach to and from the sea. Unless steps are taken to prevent these phenomena, beach erosion continues until stability is achieved for the new condition.

When erosion progresses and land is lost, the law says that from the highest tide up to 20 m inland is a federal maritime land area, so if the coastline advances, the land is lost as private property. If an extraordinary event, such as a hurricane, causes a landowner loses ground, he has 6 months to rebuild his land and leave the federal zone as it was. Otherwise, if he exceeds that time interval, he loses his land. Land reclaimed from the sea is a federal zone, and if it is to be used, it must be granted to ZOFEMAT (Federal Coastal Sea Zone)-SEMARNAT and paid for its use, but these legal matters are not part of this work.

The dunes are stable natural structures that are formed by the wind when the sand of the berm is dried; they serve to supply sand to the beach when there is erosion. Nothing should be built on the dunes so that their operation is free and they can recover the beaches. When building a house on sand, the foundations must include piles that must be calculated so that they hold the house by friction, even if they do not reach a hard and firm substrate; therefore, they should reach at least −4.00 NMM. In this way, even if the water passes under the house, it will not collapse.

When there is an erosion on one side of a breakwater or jetty due to the accumulation of sand on the other side of the breakwater, it is recommended to maintain the sand balance. Therefore, the sand must be transported to the other side using a two-way bypass system with Moyno type sand pumps to maintain the balance. Monitoring the beach (beach nourishment) is recommended to maintain the normal level and slopes. This method is widely used in advanced countries.

7 CONCLUSIONS

The construction of the breakwater definitely influenced the erosion that occurred in La Bocanita and affected the normal erosion and accretion process of Las Glorias Beach. The construction of the breakwater caused the growth toward the east of the Island Macapule and the growth to the east of the breakwater. A new cove was forme, ending the inner part of the breakwater that now is used as a shelter by fishermen.

The construction of the breakwater accelerated the erosion in the east part near the mouth of the River Sinaloa. The reconstruction of the beach is possible if that problem is treated as a state problem of Sinaloa, or treating it as a basin with a sedimentary imbalance, or as a municipality in which in one season the irrigation system was given priority, which caught the sediments. Now, it is necessary to remedy the problems of beach erosion caused by not having Integral Coastal Zone Management.

8 SOLUTIONS PROPOSED

8.1 Geotubes

The reconstruction of beaches is possible with the installation of a barrier of geotubes parallel to the beach, placed on antiscouring mantles. This would form a dam that would avoid the erosion of Las Glorias Beach and protect it, but perhaps the most valuable thing would be the recovery of the land lost by filling the sand dykes of the area behind the break. This method should achieve a smooth slope and a large berm to give security and stability to new terrain.

These geotubes, built with geotextiles, are very resistant to ultraviolet ray, and can be sandy in color. They are very friendly to the environment and also to the beachgoers' feet, which is very different with rock spikes or concrete elements such as cubes, drawers, tetrapods or acropods that end with the recreational use of the beach.

In addition to the damage to infrastructure that has caused erosion in Las Glorias Beach, the advance of the sea has destroyed more than 70 houses and restaurants in the last 15 years; the most serious impact is the loss of tourism potential. This site was in its moment a place of summer for American tourists, who since the beginning of the destruction of the beach have not returned. Local and regional tourists continue to visit this place, but not in the same amount or as often as they did before, all to the detriment of more than 30 restaurants established between Las Glorias and Boca Del Rio. Service providers and the residents of

FIG. 11 Geotubes as proposal for Playa Las Glorias recovery, an example of the protection of beaches in CC Villa, Villasur, Peru. *From Royal Duth and Tencate (2011).*

these two communities are clamoring for the government to solve the problem because they are afraid of a hurricane. For several years, researchers from the Interdisciplinary Research Center for Integral Regional Development and the Interdisciplinary Center for Marine Sciences, both from the National Polytechnic Institute, have been studying the problem and have proposed a solution. That is the construction of geotubes on the coastline Las Glorias, which can be an immediate solution. At the moment the geotubes are installed and filled, the sand immediately recovers the beach and stabilizes it, to recover part of the lost ground and to prevent erosion (Fig. 11). Tencate Geotubes were installed in CC Villa in Villa Sur, Perú. Geotubes are tube-shaped containers made of high-strength geotextiles and filled with sand from the same beach. They are placed as barriers to reduce or absorb wave energy (each bag is 50 m long, 3.0 m wide, and 2.5 m high). The structure of the geotubes is designed so that the waves drag the adjacent sand, and this is retained in the structure, generating the process of beach recovery. The placement of the geotubes has an estimated cost of more than $100 million pesos, but the task of management corresponds to the local authorities or to the same federal legislators. Geotubes can be an immediate solution because the moment the geotubes are installed and filled with sand, the recovery process of the beach immediately begins.

Carter (1988) mentions that man has maintained a very unstable and unbalanced relationship with the coast. Where man interacts with the coast, the natural system tends to become unbalanced and sooner or later, this has repercussions on its functioning. The current situation of its littoral provides local people and municipal authorities with a great potential for sustainable, diversified, well-planned development focused on maintaining the functions and environmental services that the coast provides. All this must take place under a vision of development and equity for the local population. Sustainability as part of coastal development is achieved at the intersection of the ecological, economic, and social spheres that represent the three components of sustainable development (Campbell and Heck, 1997). This development occurs when the management objectives and actions taken are simultaneously: (1) Ecologically viable through an integrated environmental management that maintains the integrity and functionality of the ecosystems, which does not exceed the carrying capacity of the ecosystems, maintains biodiversity, and contributes to the maintenance of global life systems on the planet. (2) Economically possible, ensuring growth with equity and efficiency in the use of resources and the economic improvement of the local population. (3) Socially desired, promoting participation, social mobility and social cohesion, cultural identity, and increased quality of life.

9 INTEGRAL MANAGEMENT OF THE COASTAL ZONE

9.1 Coastal Zone

There is a need to work on Integral Coastal Zone Management (ICZM) under an integral and holistic vision (Moreno-Casasola et al., 2005). The hydrological basins represent natural units of environmental management, responsible for the dragging of sediments and water to coastal areas. In the marine coast it is much more difficult to decide the units, because they are not clearly defined and they are not obvious. Since the 1970s, geomorphologists have identified a structure for coasts: they have found that the movement of coarse sediments is restricted to specific areas and only in extreme conditions does it move out of them.

These units are called coastal cells. They are defined as discrete units that operate in isolation from one another. The individual boundaries are not fixed and in extreme conditions there may be sediment movement between cells, which makes it difficult to trace them on paper. When sediment transport along the coast is reduced to almost zero, there is an edge or boundary between cells. This usually occurs on protrusions in estuaries or rivers where there is freshwater outflow. Therefore, in coastal management it is essential to take into account the quantities of sediments (i.e., entrances, movements, and exits of coarse and fine sediments by any process and their travel along the area where they move freely, or a coastal cell) and maintain the natural dynamics to make it ecologically viable. At this point it is worth making a difference. When reading coastal management text, emphasis is placed on coastal protection issues. This refers to all those works or structures, whether hard engineering (concrete, steel, etc.) or soft engineering (free or confined sand, creation of habitats, etc.) that will serve to protect the coastline as well such as settlements, properties, or activities that occur in it and the coast dynamics, both in times of great strength (hurricanes, storms, etc.) and in daily life.

9.2 Erosion

Erosion and land use affect almost all European, Asian, and American countries to a greater or lesser degree when there has been intense economic activity based on the coast (i.e., tourism, ports, etc.). There are countries such as Japan where more than half the coast has been transformed; this has had a strong impact on coastal processes and sediment supply. The situation of most of Mexico's coasts is very different from that of developed countries, but one must learn from their experiences. Most of our coasts are in natural conditions, that is, there have been few attempts at handling and protection. An example of this protection and the problems it brings are the breakwaters of the port of Veracruz and the environs of Puerto Progreso in Yucatan, which accumulate sand on one side and erode it on the other side. Therefore new ones have to be permanently added. Another example is the Coatzacoalcos boardwalk, Veracruz, which due to the coastal dynamics, the sediments are dragged causing a large accumulation of sediments.

9.3 Coastal Management in Mexico

The main coastal management that Mexico has been in the coastal lagoons, where it has opened mouths and continues to open them with few studies and predictive capacity. In some cases, when the economy allows it and the social demand is great, rocks are built. In Veracruz there is also the example of the Laguna Verde power plant. Due to the large amount of sediment transport by coastal currents and wind drag, constant dredger is necessary to keep the inlet channel open to water exchange. In this context within development planning is prevention, which takes a preponderant role on remediation. The option of a more natural coastline has future benefits because if there is still a free flow of sediment, more natural habitats may develop in the future and therefore the coast can respond more easily to changes in wind, swell, and sea level. This quickly leads to economic benefits by making coastal protection works unnecessary and reducing the economic costs of development. However, in the case of Sinaloa, the irrigation system has the priority and it has constructed dams in most of the rivers. However, the dams caught the sediments unbalancing the coastal zone from a lack of sediments, causing erosion in the beaches. Only through the integrated management of the coastal zone (ICZM) can these needs be taken into account as well as those of tourism and port and industrial developments.

That is, coastal economic development and the protection and conservation of coastal dynamics and the functioning of these ecosystems can be harmonized. Historically, coastal management during the last 30 years has evolved from few uses and land-based management to multiple uses and management covering not only land but also marine areas. The main efforts to manage the coast from a more holistic point of view began when, in developed countries, the coasts began to degrade due to mismanagement. In 1966 and 1972, Australia and the United States, respectively, were among the first countries to have a Coastal Zone Management Act. It was in 1992 at the United Nations Conference on Environment and Development that the concept and practice of the ICZM were introduced for the first time at the highest political level as an approach to sustainable coastal development. During this conference, the various countries signed several multilateral agreements such as Agenda XXI. Chapter 17 of this agenda stresses the importance of conserving the coasts and oceans. By signing this agreement, the coastal nations of the world committed themselves to developing and implementing comprehensive management programs and to promoting the sustainable development of coastal and marine areas under their jurisdiction. By 1993, 142 integrated coastal zone management efforts had been recorded in 57 countries (Sorensen et al., 1992). In Latin America, countries such as Belize, Costa Rica, Ecuador, Chile, and Brazil, among others, have already made progress in these efforts. In this sense,

Mexico is lagging behind because to date it does not have a coastal management law or coastal law. In 1999, there was an initiative by SEMARNAP, which published the strategy for the integral management of the coastal zone containing the recommendations of experts in the field (SEMARNAP, 2000). At present, the most notorious administrative advance is the inclusion of the term "coastal environments" and the theme of conservation of coastal environments within the General Direction of the Federal Maritime Zone and Coastal Environments.

In 2001, about 7600 km of the European coast were affected by erosion control projects, 80% of which have been carried out over the last 15 years. These projects use and combine a wide range of techniques, including "hard" engineering techniques that use concrete or breakwater to set the coastline and protect coastal assets and facilities. These techniques of breakwaters, walls, exempt breakwaters, etc., characterize more than 70% of the protected coastline in Europe. Engineering "soft" techniques that are used include the artificial regeneration of sand beaches or beach nourishment performed according to coastal transport processes, or using the natural environment such as sand, dunes, marshes, or vegetation to prevent erosion from reaching the inland (Erosión, 2005). The piers are effective only on a limited stretch of coast; then the erosion is accentuated downstream and requires the extension of the field of breakwaters, creating a domino effect.

Acknowledgment

The authors are thankful to reviewers whose suggestions and recommendations contributed to enhance the quality of this research. Baja Ferries support the travel of La Paz-Topolobampo-La Paz from ARJI. MZE had a fellowship of CONACYT. LEC have fellowship of COFAA and EDI. Support of projects: CONANP (SEMARNAT, Mexico) financed a research from erosion in Macapule Island. SENER: CEMIE-Oceano (G-LE1) "Detección de lugares de aprovechamiento de gradientes de temperatura con potencial energético de explotación en México para determinar la factibilidad de construcción de una planta OTEC", and IPN-SIP 20161087, 20170983. In this work have been taken some figures from Espinosa-Carreón (2013) financed from CONANP (SEMARNAT, Mexico).

References

Alcántar-Elizondo, R., 2007. Variabilidad espacio temporal del perfil de playa, en Playa Las Glorias. IPN-CIIDIR Unidad Sinaloa, Guasave. Tesis de Maestría. 64 p.

Anderson, L., Glover, A., Edgmon, D., Whipp, J., 2014. Coastal Erosion of the Gulf Coast and Its Effects in the Region. Media Nola, A project of Tulane University. USGS, Science for a Changing world.

Barnard, P.L., Hubbard, D.M., Dugan, J.E., 2012. Beach response dynamic of a littoral cell using a 17-year single-point time series of sand thickness. Geomorphology 139–140, 588–598.

Barnard, P.L., Short, A.D., Harley, M.D., Splinter, K.D., Vitousek, S., Turner, I.L., Allan, J., Banno, M., Bryan, K.R., Doria, A., Hansen, J.E., Kato, S., Kuriyama, Y., Randall-Goodwin, E., Ruggiero, P., Walker, I.J., Heathfield, D.K., 2015. Coastal vulnerability across the Pacific dominated by El Niño/Southern Oscillation. Nat. Geosci. 8, 801–807. https://dx.doi.org/10.1038/ngeo2539.

Barnard, P.L., Hoover, D., Hubbard, D.M., Snyder, A., Ludka, B.C., Allan, J., Kaminsky, G.M., Ruggiero, P., Gallien, T.W., Gabel, L., McCandless, D., Weiner, H.M., Cohn, N., Anderson, D.L., Serafin, K.A., 2017. Extreme oceanographic forcing and coastal response due to the 2015–2016 El Niño. Nat. Commun. https://dx.doi.org/10.1038/ncomms14365.

Campbell, C., Heck, W.W., 1997. An ecological perspective on sustainable development. In: Muschett, F.D. (Ed.), Principles of Sustainable Development. St. Lucie Press, Delray Beach, FL, pp. 47–68.

Carter, R.W.G., 1988. Coastal Environments. An Introduction to the Physical Ecological and Cultural Systems of Coastlines. Academic Press, Londres.

CONANP, 2010. Decreto 2 de agosto de 1978 por el que se establece una zona de refugio de aves migratorias y de la fauna silvestre, en las islas que se relacionan, situadas en el Golfo de California.

Espinosa-Carreón, T.L., 2013. Estudio Ambiental-Estudio de monitoreo de Procesos de Erosión en la Zona Costera de la Isla Macapule. SEMARNAT-CONANP. Informe técnico. 42 p.

Eurosión, 2005. In: Doody, P., Ferreira, M., Lombargo, S., Lucius, I., Misdorp, R., Niesing, H., Salman, A., Smallegange, M., Raventos, J.S., Roca, E., Fernández Bautista, P., Pérez, C. (Eds.), Vivir con Erosión Costera en Europa, Sedimentos y Espacio para la Sostenibildad. Ministerio de Medio Ambiente. Europan Comunities. Oficina de Publicaciones Oficiales de la Comunidades Europeas. ISBN 92-894-9918-4.

Fluviomarítima (Consultores Ingeniería) S.A., 1992. Definición del Plan Maestro de desarrollo turístico. FONATUR, México, D.F, Las Glorias.

García, E., 2004. Modificaciones al sistema de clasificación climática de Köppen. Instituto de Geografía-UNAM, CONABIO, México.

González-Ruelas, M.E., Navarro-Rodríguez, M.C., Carrillo, G., González Guevara, F.M., Flores-Vargas, R., 2005. Espigones en Bahía de Banderas, Jalisco. México *Aleph Zero* 40. Centro Universitario de la Costa, Campus Vallarta, Departamento de Ciencias, Universidad de Guadalajara, Jalisco.

Hansen, J.E., Elias, E., Barnard, P.L., 2013. Changes in surfzone morphodynamics driven by multi-decadal contraction of a large ebb-tidal delta. Mar. Geol. 345, 221–234.

Harley, M.D., Turner, I.L., Kinsela, M.A., Middleton, J.H., Muford, P.J., Splinter, K.D., Phillips, M.S., Simmons, J.A., Hanslow, D.J., Short, A.D., 2017. Extreme coastal erosion enhanced by anomalous extratropical storm wave direction. Sci. Rep. www.nature.com/cientificreports.

Hubbard, D.M., Dugan, J.E., Schooler, N.K., Viola, S.M., 2014. Local extirpations and regional declines of endemic upper beach invertebrates in southern California. Estuar. Coast. Shelf Sci. 150, 67–75.

Klemas, V., 2013. Remote sensing of coastal hazards. In: Finkl, C.W. (Ed.), Coastal Hazards. Springer Science + Busines Media, Dordrecht. https://dx.doi.org/10.1007/978-94-007-5234-4_2.

Massenlink, G., Castelle, B., Scott, T., Dodet, G., Suanez, S., Jackson, D., Floc'h, F., 2016. Extreme wave activity during 2013/2014 winter and morphological impacts along the Atlantic coast of Europe. Geophys. Res. Lett. 43, 2135–2143. https://dx.doi.org/10.1002/2015GL067492.

Montes de Oca, M., 1985. Topografía. Alfaomega, México.

Moreno-Casasola, P., Peresbarbosa-Rojas, E., Travieso-Bello, A.C., 2005. Manejo costero integral: el enfoque municipal. In: Moreno-Casasola, P., Peresbarbosa-Rojas, E. (Eds.), Manejo integral de la zona costera. Instituto de Ecología, A.C., ISBN 970-709-039-1.

Moshinsky, M.R., Jiménez, E.M., Vázquez, C.M.T., 2014. Atlas Climatológico de Ciclones Tropicales en México. CENAPRED Secretaría de Gobernación, ISBN: 970-628-633-0.

Orme, A., 1985. Understanding and predicting the physical world. In: Johnston, R.J. (Ed.), The Future of Geography. Methuen, London, pp. 258–275.

Patsch, K., Griggs, G., 2008. A sand budget for the Santa Barbara littoral cell, California. Mar. Geol. 252, 50–61.

Prasad, D.H., Kumar, N.D., 2014. Coastal erosion studies—a review. Int. J. Geo. 5, 341–345. https://dx.doi.org/10.4236/lig.2014.53033.

Sánchez-Lindoro, F.J., Jiménez-Illescas, A.R., Espinosa-Carreón, T.L., Obeso-Nieblas, M., 2017. Hydrodynamic model in the lagoon complex Navachiste, Guasave, Sinaloa, Mexico. Biol. Mar. Ocean. 52, 210–231.

SEMARNAP, 2000. Estrategia ambiental para la gestión integrada de la zona costera. Propuesta. Instituto Nacional de Ecología-SEMARNAP, México.

Sorensen, J.C., MeCreary, S.T., Brandani, A., 1992. Costas. Arreglos institucionales para manejar ambientes costeros. Coastal Resource Center United States Agency for International Development, Washington, DC.

Splinter, K.D., Davidson, M.A., Golshani, A., Tomlinson, R., 2012. Climate control on longshore sediments transport. Cont. Shelf Res. 48, 146–156.

Stockdon, H.F., Doran, K.S., Sallenger, A.H., 2009. Extraction of lidar-based dune-crest elevations for use in examining the vulnerability of beaches to inundation during hurricanes. J. Coastal Res. (53), 59–65.

Zayas-Esquer, M., 2010. Efectos en Las Glorias Beach causados por la construcción del espigón al modificar el transporte litoral. Tesis Maestría, IPN-CIIDIR Unidad Sinaloa 107 pp.

CHAPTER

10

An Innovative Technique of Tidal River Sediment Management to Solve the Waterlogging Problem in Southwestern Bangladesh

*Md Sharif Imam Ibne Amir**, *M. Shah Alam Khan*†

*School of Engineering and Technology, Central Queensland University, QLD, Australia
†Institute of Water and Flood Management, Bangladesh University of Engineering and Technology, Dhaka, Bangladesh

1 INTRODUCTION

The Bengal Basin is morphologically active and geologically the youngest drainage basin of the world (Amir et al., 2013a). This basin has been developed by alluvial sediment deposition from the Ganges-Brahmaputra-Meghna river system and forms the world's largest sediment distribution system (Mukherjee et al., 2009). The rivers of southwestern Bangladesh are part of this basin. The land level close to the river bank is raised by repeated sediment deposition; however the land between the two rivers remains low. These highly active rivers are characterized by active deposition of sediment, causing a significant reduction in their drainage capacity (IWM, 2009).

In the 17th century during the period of Zamindari or large landowners who also served as principal revenue agents for the government, tenant farmers had to pay a large portion of their income, usually a percentage of the crop, to the Zamindars (Amir, 2010). Because the income of the Zamindars depended largely on crop production, they built low earthen dikes around the tidal flats to prevent tidal intrusion as well as wooden sluices to drain off surplus rainwater and then cultivate indigenous varieties of flood-tolerant and saline-tolerant rice. Their tenants cultivated similar indigenous varieties of rice and reaped bumper harvests. The dikes were traditionally cut from November to July, allowing rivers to naturally flow and ebb over the floodplains during the rest of the year. After the harvest, the dikes and sluices were dismantled,

© 2019 Elsevier Inc. All rights reserved.
https://doi.org/10.1016/B978-0-12-810473-6.00011-X

and the people grazed cattle and fished in the tidal flood plains. Thus, the environment, ecosystem, and biosystem that evolved in the coastal area were in balance. The dikes were not sufficiently high and strong. Opening the sluice gates was not enough and the gates were weak. Therefore, they required considerable maintenance each year.

In 1951, the Zamindari system was abolished by the East Bengal State Acquisition and Tenancy Act, 1950, and the Zamindars were relieved of their power and authority. Many had been living in other countries and did not return to the area while some residing in the area left. T hose who remained were stripped of their power. As a result, there was no one to take the responsibility for the repair and maintenance of existing dikes or the construction of new ones. Gradually, the dikes were breached and overtopped by tides, becoming practically useless. There was nobody to manage the river basins in the age-old manner, leaving farmers unable to cultivate crops due to annual flooding.

Various attempts were made during that time to improve conditions in the coastal areas, but none were fruitful until the late 1960s and early 1970s. At that time, the Government of Bangladesh (GOB) constructed flood protection embankments and various types of drainage structures to safeguard urban and agricultural lands from damage due to frequent tidal inundation, monsoon flooding, and the intrusion of saline water. This river management technique was called polder. The polder was made up of encircled earthen embankments around depressions. The main tidal channels were kept outside the polder. There were gated hydraulic structures on the intersecting points of polder and channels. This man-made intervention yielded good results until the 1980s.

In the early 1980s, polders became a bane rather than a boon for the people, as rivers failed to maintain their natural courses. Tides deposited silt in the riverbeds rather than the floodplains for more than two decades, halting the natural flow of the rivers. The consequent dearth of land formation left floodplains inside the polders lower than riverbanks outside the polders. Rainwater, therefore, could not drain from the areas, leading to chronic waterlogging. Adding to the tragedy, the construction of the Farraka Barrage on the Ganges river and the unilateral diversion of its water by India beginning in 1975 caused the deterioration of the balanced (fresh water—tidal flow) ecosystem of the region. This was further aggravated by the construction of reservoirs on the upper catchment of all transboundary rivers of the southwestern region. During the dry period of the year (January to April), the area receives almost no upland freshwater flow. Under the changed hydrological situation, many tidal channels outside the polders started experiencing abnormal sedimentation, blocking the drainage paths of the polders. Prolonged waterlogging inside the polders was so severe in some cases that the people of the area had the only option of migration.

The Bangladesh Water Development Board (BWDB) was fully aware and concerned about the drainage congestion of the polders of southwestern Bangladesh. But the spatial extent of the problem and the prevailing hydrogeomorphological conditions of the area are so complex that a holistic and well-planned approach was needed. BWDB conducted six studies by engaging international and national consultants between 1986 and 1998. Besides this planning exercise, BWDB dredged some badly silted up channels for immediate relief of the drainage congestion problem. From 1996 to 1998, about 0.610 million cubic meters were dredged and 1.480 million cubic meters were reexcavated (manual and mechanical) to keep the main river system active. Dredging and reexcavation were done several times; however this faced siltation every time.

All the tidal floodplains were enclosed within polders and tidal intrusion was stopped into the polders by implementing a coastal embankment. Only surplus rainwater was allowed to drain out through hydraulic structures. This had one advantage and one disadvantage. While it enabled the creation of a perennial freshwater regime within the polders for agriculture to be practiced year round, it also denied the land the silt required to maintain the land level. The continued subsidence of the loose delta soil was not compensated. Subsequently, an indigenous water management practice was introduced in this area. This popular practice is known as tidal river management (TRM).

The TRM technique uses the full advantage of the natural tide movement in rivers. During high tide, sediment-borne water is allowed to enter into an embanked low-lying area, known as the tidal basin or beel, where the sedimentation takes place during the storage period. During ebb tide, the water flows out from the tidal basin with greatly reduced sediment load and eventually erodes the downstream riverbed. The natural movement of flood and ebb tide along the tidal basin and along the downstream part of the river maintains a proper drainage capacity in that river.

With Bangladesh being a deltaic country, the land in the plains has been formed by sediments carried down by the Ganges, Brahmaputra, and Meghna river systems. Depressions are formed by numerous causes such as subsidence of topsoil caused by creation of a vacuum below by decomposition of organic substances mixed with silt as well as subsidence caused by tectonic movement and nondestructive floods depositing sediment close to the riverbank. Such repeated deposits raise the level of land close to the riverbank. But the land between the two rivers remains low-lying. Such low-lying land is also known as a beel. A tidal basin is a beel or depressed low-lying area adjacent to the sediment-laden tidal rivers. There are several tidal basins in the Bhabodah area (southwestern Bangladesh) that are very useful for management of the sediment-laden tidal rivers.

The concept of TRM is, in fact, a natural water management process with very little human intervention. However, it needs strong participation and consensus with a great deal of sacrifice by the stakeholders for a specific period of 3–5 years, depending on the tidal volume and the area of the beel (Rahman, 2008). When a beel is selected for TRM, a potential location of the link canal is identified, considering tidal movement and sedimentation inside the beel and drainage of the other beels. This is done after consulting with local stakeholders and with the help of a numerical model study. Then sedimentation inside the tidal basin is allowed through that link canal. In almost all cases, most of the sedimentation takes place near the entrance of the link canal.

Sediment management inside the beel and maintaining the proper drainage capacity of the river through sequential operation of a potential beel for TRM by involving people's participation for sustainable drainage management are the two main objectives in this area. From field visits and monitoring results, it has observed that sedimentation inside the beel was not uniform in the Beel Kedaria and the East Beel Khuksia (EBK). So, the main objective was not attained by the TRM practice for the lack of technical effectiveness during the TRM operation. The present practice is that a one- or two-link canal is constructed that connects the tidal basin with the river. But most of the sedimentation takes place near the entrance of the link canal in almost all cases. Sediment management has been the most challenging yet important aspect of TRM in the study area (SMEC, 1997). People allow their land to be used for tidal basin operation without any compensation, hoping that the land will rise after 3 or 4 years. However, monitoring results and community consultation reveal that in almost all cases, sedimentation inside the tidal basin does not occur as expected (CEGIS, 2002a). This results in people's unwillingness to allow their land to be used for basin operation. Besides, social conflicts among various groups such as farmers, fisherman, landowners, etc., and institutional conflicts among government agencies, water management associations (WMA), and local government institutions (LGIs) have made the TRM practices unsuccessful (CEGIS, 2002b).

Therefore, it was important to find out the real reason behind the ineffectiveness of the TRM practices, considering all relevant social, institutional, technical, and economical aspects with a view to ensuring successful enhancement of the overall environmental condition. The main two objectives of the study were:

- To determine and analyze relevant social and institutional causes behind the problems with sediment management.
- To formulate and analyze different socially acceptable, technically feasible, and economically viable options for uniform sediment deposition inside the tidal basin.

2 STUDY AREA

The study area is located in the southwest region of Bangladesh and falls under the administrative jurisdiction of Jessore and Khulna. The area has morphologically active tidal rivers and channels and the main river system in the study area is the Mukteshwari-Hari river system (Fig. 1). The Mukteswari-Hari river forms a drainage route of about 40 km, meeting the Harihar-Upper Bhadra system near Ranai. The Teligati-Gengrail system receives the combined flow of the Harihar-Upper Bhadra and Mukteswari-Hari river systems. The Upper Sholmari, Lower Sholmari, and Lower Salta rivers are the main drainage channels for Beel Dakatia, Polder 27, and Polder 28. The Teligati-Gengrail system is the only main outlet for the drainage of the Bhabodah area. These river systems are deteriorating rapidly due to sedimentation (siltation) and causing drainage congestion in the adjacent areas. There are several channels in this study area that are important for drainage. However, most of the channels are not functioning properly due to sedimentation on the channel's bed. The channels that are functioning, mainly carry water from rainfall and runoff to the beels and rivers in the wet season. The study area has a huge number of beels (Fig. 2) that are very important for biodiversity, such as freshwater fish and birds. The beel area is about 45% of the total study area.

The Hari river bed is covered with fine sediment (mud), originating from the Bay of Bengal. The average grain size is typically less than 0.063 mm and the bed behaves as a cohesive sediment bed. Grain size distribution has been done for the Hari river at Ranai from the measured bed sample and it has been found that $D_{50}=0.017$ mm and $D_{90}=0.050$ mm. The maximum concentration is observed during the dry season in the month of May.

3 GENESIS OF TRM

3.1 The Initiative of Local People to Solve the Problem

The polders disconnected the floodplains from the rivers and caused sedimentation in the river beds. As a result, the Bhabodah area faced severe drainage congestion and waterlogging during the 1980s and 1990s. Huge river training and drainage improvement works were carried out during that time to solve these long-standing problems. However, the drainage

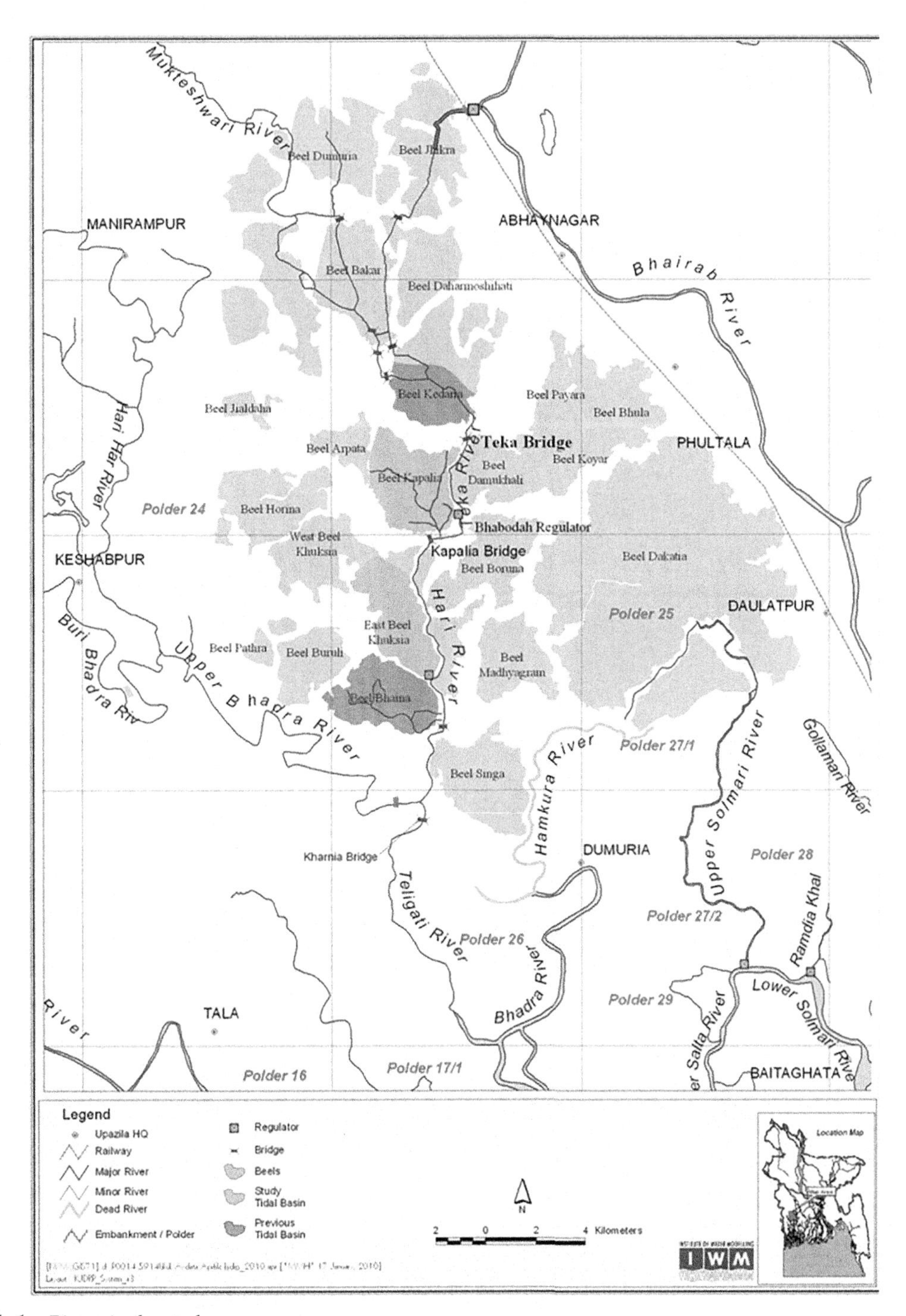

FIG. 1 Rivers in the study area.

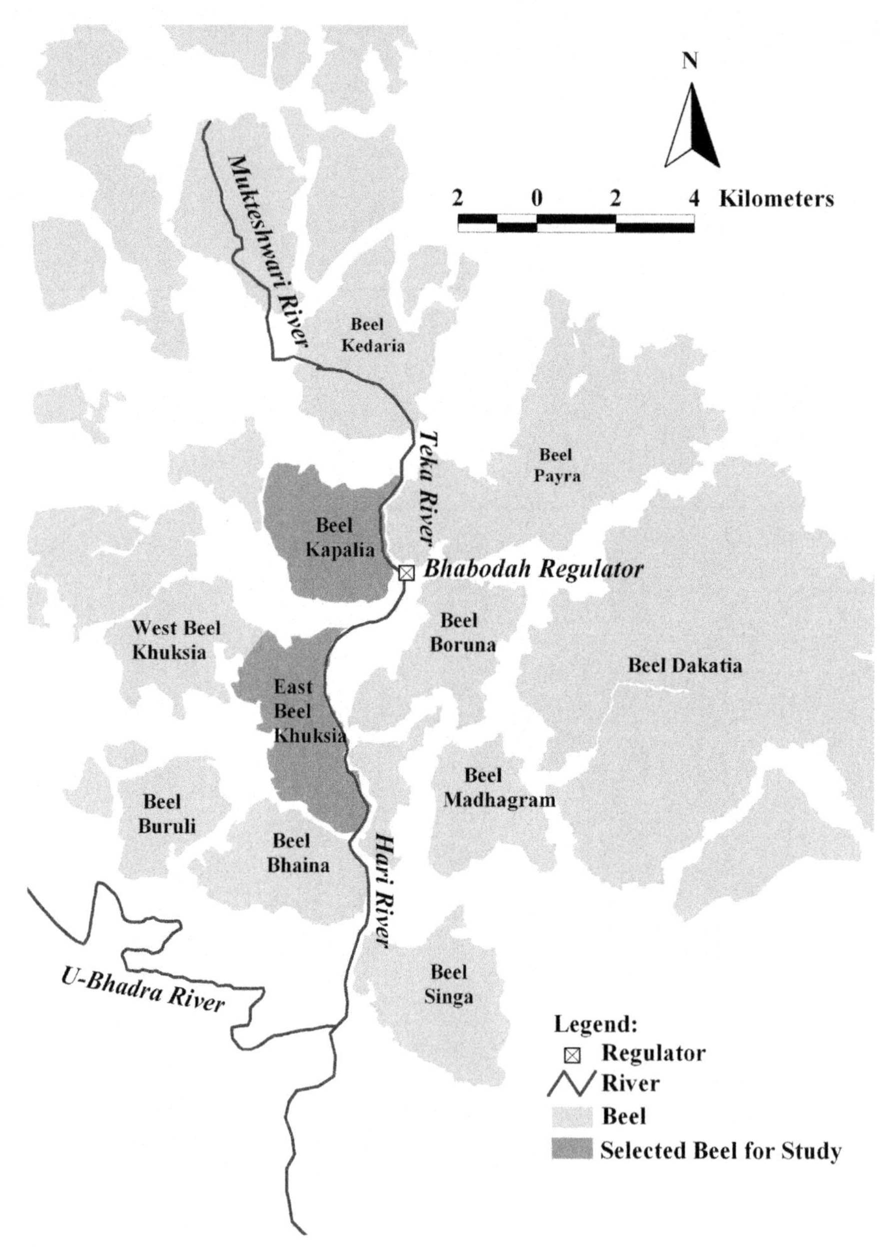

FIG. 2 Beels in the Bhabodah area.

congestion and waterlogging problems were not improved substantially. The people of the waterlogged area petitioned the authorities to solve the problem. As the authorities paid no heed to their grievances, the people themselves took the initiative to organize and mobilize the community, devising plans for solving the problem. From their own experience and observation, people identified the polders as the main cause of waterlogging and began to present their reasoned arguments for breaching or cutting away polders to allow tidal flows. Their logic was that if tidal waters can be allowed to flow freely, the navigability of the rivers would be restored, the enclosed lands would be free from waterlogging, alluvium would accumulate inside the polders, and as a result, the level of the land would rise. The first manifestation of this logic was seen in 1990. This concept is called the TRM technique.

In mid-September 1990, after prolonged hardship due to drainage congestion, the people of Beel Dakatia made public cuts in the polder embankment at four locations. The people made the public cut to relieve drainage congestion and improve the water quality of the beel. It was also hoped that sedimentation would occur in the beel, thereby raising the land level appreciably. The effect of the public cuts on sediment deposition was not as great as had been hoped for. In the area of cuts 1–3, no significant sedimentation was recorded. However, appreciable sedimentation was reported near the area of cut 4; however, that was not uniform (IWM, 2002). The sediment deposit inside the beel extended outward in a delta formation over an area of about 900 ha out of 18,000 ha (SMEC, 2002). At distances of more than 3.5 km, no significant sedimentation was observed. After the breaching of the embankment of Dakatia beel, the Hamkura river became a strongly flowing river 300 ft wide and 30 ft deep at the new highway bridge on the Khulna-Chuknagar Road. Therefore, Beel Dakatia operated successfully as a tidal basin and enlarged the downstream river. However, people in the beel experienced difficulties due to a lack of fresh water, reduced fish yields, death of trees due to salinity, and loss of dry season crops due to higher prevailing water levels (H&A, 1993).

3.2 Unplanned TRM Practices

The success in draining water from the Dakatia beel encouraged people in the adjacent waterlogged areas. They organized themselves and formed committees at different levels, then undertook initiatives to turn their waterlogged land into agricultural land again. The Madhukhalir beel and Patra beel are examples of such collective efforts. However, these efforts could not achieve the desired results at every stage due to a lack of proper organizational structure and planning. In the meantime, the Bhabodah area (Jessore zone) started to experience widespread waterlogging. The people of the area organized themselves, removed the accumulated silt from the exit of the Bhabodah sluice gates every year, and opened a narrow drainage channel. Each year, they retrieved more land for agricultural production. The people there learned from the Dakatia beel experience and tried the TRM concept on Bharter beel, Golner beel, Bahadurpur beel, and Magurkhali beel. The experiments proved successful.

The affected people cut the embankment to connect the Hari river with Beel Bhaina in October 1997. It was closed in December 2001. About 600 ha of land was raised by 1.0 m and the Hari River, downstream of the cut, revived for a length of 4 km with a more than 8 m depth. The deposited sediment volumes at different periods from the beginning of the operation in November 1997 to the closing of the basin in December 2001 are presented in Table 1.

It is evident from Table 1 that sediment deposition was relatively less at the beginning of the operation of the tidal basin. Sediment deposition is less during the monsoon in 2000 in comparison to that of the dry season. Sediment deposition was reduced considerably in the last year of operation (2011). So, the rate of sediment varies both temporally and spatially.

TABLE 1 Deposited Sediment Volume in the Beel Bhaina Tidal Basin

Period of Operation	Deposited Sediment Volume (Million m^3)
November 1997–March 2000	1.90
April 2000–June 2000	1.10
July 2000–December2000	1.75
January 2001–December 2001	1.73
Total period: November 1997–December2001	6.48

Higher sedimentation is observed close to the downstream opening of the tidal basin and decreases gradually to the farthest end of the beel.

Another issue is uneven sedimentation inside tidal basins. Such uneven sedimentation created drainage congestion in Beel Bhaina, especially in the northwestern part of the beel. Appropriate measures such as compartmentalization or rotation of the opening need to be undertaken so that a similar situation does not arise in Beel Kedaria (Rahman, 2008).

3.3 Institutionalization of TRM

Because TRM was an emerging successful practice, the Asian Development Bank studied the TRM options in greater detail in terms of technical feasibility as well as environmental and social impacts based on the feedback of the project beneficiaries and suggestions received from the stakeholders. It was found that the TRM is technically feasible and attractive from social and environmental points of view. In response to this situation, the GOB, with financial support from the Asian Development Bank, undertook the Khulna-Jessore Drainage Rehabilitation Project during 1994–2002. After implementation of the project, the prevailing drainage congestion was solved considerably and agricultural, social, and economical benefits were obtained. It was observed that benefits were sustained in the Khulna region where drainage management was solved by the construction of regulators, the dredging of drainage channels, and keeping up the practice of regular removal of silt at the downstream of the regulators. But in the Bhabodah and its adjacent area (Jessore part), sustainable drainage improvement was not achieved before the implementation of TRM. This study mainly concentrates on the Jessore part of the named Bhabodah area. The rotational tidal basin was proposed to share both the inconveniences and the benefits with adjacent beels. In this regard, nine beels, one after another, were selected to operate as a tidal basin (Rahman, 2008).

3.4 Planned Tidal Basin in Beel Kedaria

Monitoring results show that the Beel Kedaria tidal basin performed as an effective tidal basin for maintaining the design drainage capacity of the Hari River during its operation from January 2002 to January 2005. The analysis of the cross-section at Ranai of the Hari River indicates that the river was in the dynamic equilibrium condition in this river reach during the

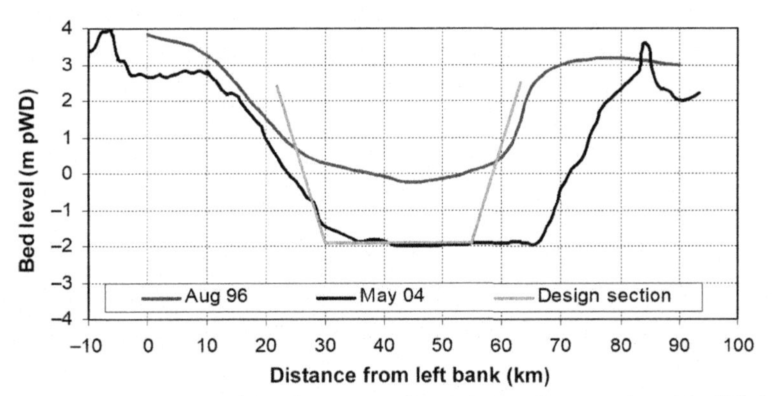

FIG. 3 Stability of drainage capacity of the Hari river at Ranai during the operation of Beel Kedaria tidal basin.

operation of the Beel Kedaria for TRM as the drainage capacity of the Hari River reached a stable condition with the small seasonal change. It is evident from Fig. 3 that the drainage capacity was also higher compared to the design one during the operation of Beel Kedaria.

The net silt deposition in the Beel Kedaria tidal basin since its operation from 2002 to May 2004 was about 0.49 million m^3 over an area of 524 ha. It is apparent that deposition took place almost over the whole area; however the deposition is not uniform over time and space.

3.5 Performance of Beel Bhaina and Beel Kedaria Tidal Basins

The Beel Bhaina tidal basin generated about 10 times higher tidal volume than that generated by the Beel Kedaria tidal basin. This higher tidal volume generated in Beel Bhaina was mainly due to the location of the basin. The Beel Bhaina tidal basin is located near the downstream of the Hari River where the tidal range was more than 1.0 m; it is about 0.15–0.20 m in the Kedaria tidal basin (IWM, 2002). The higher tidal range at the mouth of Beel Bhaina caused higher flow and flow velocity that led to the river bed erosion and siltation in the basin.

3.6 Waterlogging After Closing the Beel Kedaria Tidal Basin

The TRM operation in the Beel Kedaria basin was stopped by the landowners by closing the gates of the Bhabodah regulator. Consequently, siltation occurred along the 17 km stretch of the Teka-Hari river system. A severe waterlogging problem prevailed in the Bhabodah area from October 2005 to November 2006, due to the discontinuation of operations of the Beel Kedaria tidal basin for TRM. The area inundated due to drainage congestion was about 18,100 ha in September 2006. The inundated area includes agricultural land, homesteads, schools, colleges, and roads under the three Upazilas (Manirampur, Keshabpur, and Abhaynagar) of the Jessore District. The affected areas are 6120, 8980, and 3000 ha in Abhaynagar, Manirampur, and Keshabpur Upazila, respectively (IWM, 2007). Altogether, 313,045 people in these three Upazilas were affected due to drainage congestion. It was observed that sanitation, drinking water, and health were the urgent issues in all the affected areas. There was no scope for cultivation at all. Local communications were disrupted. There was great scarcity of food and drinking water, and all sanitary latrines were destroyed. It was observed that a large number of different social components were also affected due to waterlogging. The drainage congestion continued until the next TRM started in the EBK on Nov. 30, 2006 (Amir, 2010).

3.7 Planned Tidal Basin in the EBK

The EBK was an ongoing TRM basin and the TRM operation was started in December 2006. A considerable river-bed scouring occurred at the downstream reach of the Hari River due to the operation of the EBK tidal basin. The evolving cross-sections of the Hari River at Ranai are shown in Fig. 4. About 2m scouring was observed at Ranai and about 5km was observed downstream of the tidal basin during December 2006 and April 2007 due to TRM operation.

It was found that the river at this location is still adjusting with the tidal prism to reach a new equilibrium under a changed morphological condition. The conveyance of the Hari River at Ranai increased 2.6 times (from $125\,m^{8/3}$ to $338\,m^{8/3}$) after 5 months of TRM operation (IWM, 2007). It is also found that the drainage capacity of the Hari River at the downstream reaches of the basin increased from its design drainage capacity.

Deposition of sediment in the tidal basin is an important issue as it determines the lifetime of the tidal basin as well as the development of land for agricultural production. In order to assess the impact of TRM operation in terms of siltation inside the tidal basin, a bottom topography survey of the basin was carried out in February and May 2007. After processing the data, two digital elevation models (DEM) were prepared that show the actual deposition/erosion pattern inside the basin for a specified time period (Fig. 5). About 0.9 million

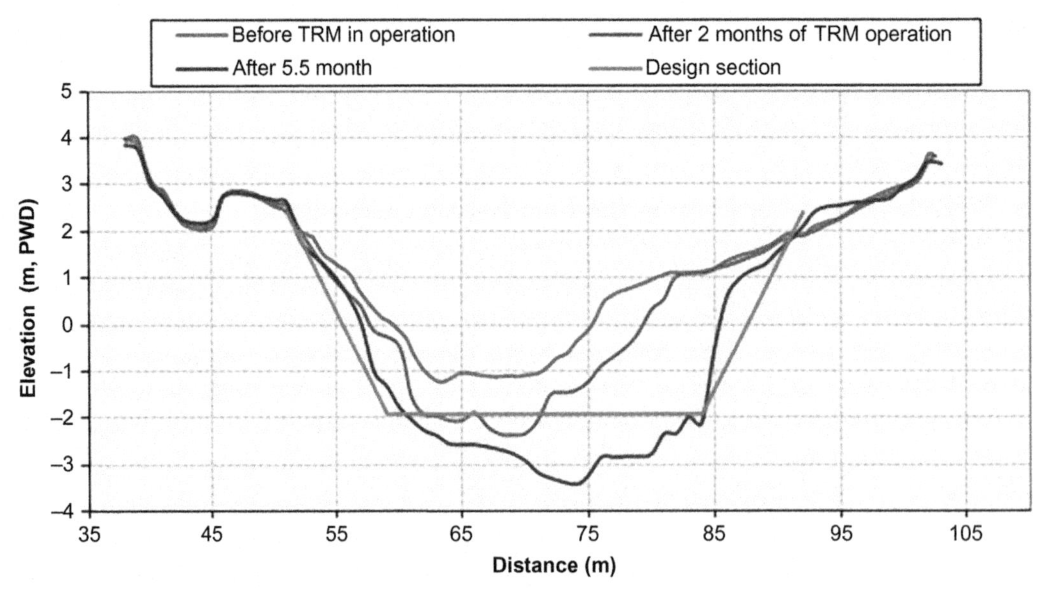

FIG. 4 Cross-section comparison of the Hari River at Ranai after 5 months of TRM operation.

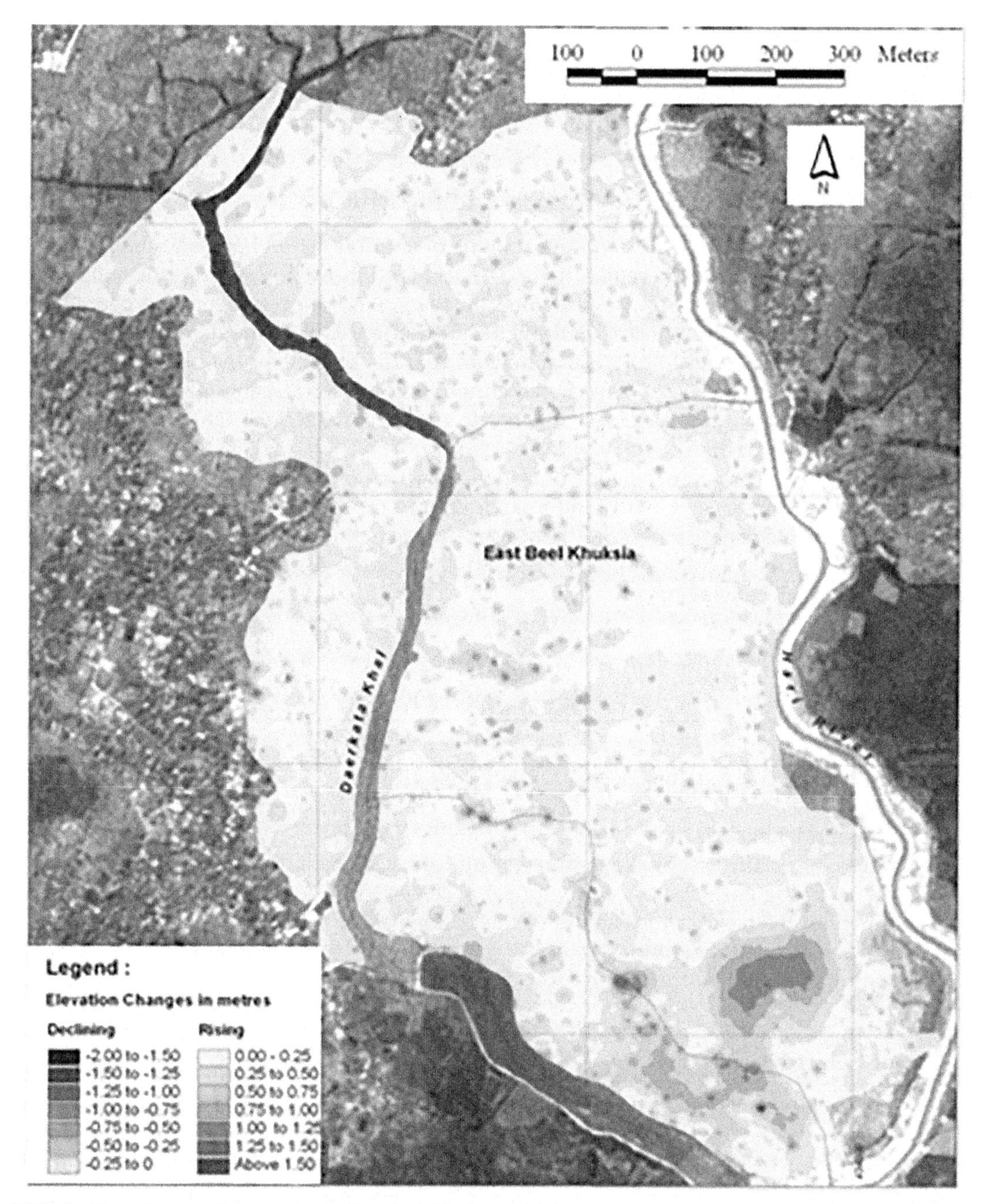

FIG. 5 Deposition and erosion in the East Beel Khuksia tidal basin.

m^3 siltation took place in the tidal basin during the 5-month operation of the basin. It is apparent that deposition occurred mostly near the downstream area (nearest to the cut point) of the basin. There are lots of fishing infrastructures (ghers) in the basin that create obstructions to uniform spreading of incoming silt over the basin area.

3.8 Lessons Learned From Beel Kedaria and the EBK

The region has been subjected to different kinds of environmental imbalance and natural disasters. So, the people of the region have been contending with the forces of nature for generations, and they know best what is good for them. On the other hand, the people, on the basis of their traditional wisdom and practical experience, have the democratic right to say something about technical projects that may profoundly affect their lives and livelihoods. Therefore, their ideas and views must be taken into account in decision-making processes. All the organizations and agencies related to river management should develop a people-oriented approach.

The success of TRM largely depends on proper selection of a beel. The selection of a tidal basin needs detailed surveys, hydraulic modeling, and morphological and environmental studies involving the beneficiaries. The TRM needs routine monitoring of the key hydromorphological indicators to collect information and knowledge to evaluate its performance, future planning, and management.

4 METHODOLOGY

To select a functional sediment management option, a well-organized approach was followed, which is given below.

- Selection of tidal basins.
- Generation of socially acceptable options for uniform sediment deposition inside the tidal basin by applying participatory rural appraisal (PRA) tools.
- Assessment of the technical feasibility of the options using numerical modeling.
- Economic analysis of the options.
- Finalization of an option for future TRM operation.

4.1 Beel Selection Criteria

Two suitable beels were selected for this study in such a way that would serve the research objectives of the study. One beel was selected at the upstream part of the Hari River and another at the downstream part. Accessibility to the beels and availability of secondary data were the other two considerations during the selection process. The selection process also involved extensive review of secondary data and literature and expert consulting with the local stakeholders, BWDB field officials, members of LGIs, and representatives of WMA/WMF in the field visits. Finally, the Beel Kapalia (BK) upstream and the EBK downstream were selected for this study.

4.2 PRA Techniques to Select Sediment Management Options

PRA is a family of approaches and methods to enable rural people to share, enhance, and analyze their knowledge of life and conditions in order to plan and to act (Ricaurte et al., 2014). PRA is an intensive and systematic but semistructured learning experience carried out in a community by a multidisciplinary team, which includes community members (Nguyen et al., 2013). Information from the field has been synthesized and analyzed to understand the perspectives and priorities of the local communities regarding sediment management (Alvarez-Guerra et al., 2010). Semistructured interviews, resource mapping, and focus group discussion (FGD) are effective PRA tools for collecting information regarding water management (Franzén et al., 2015; Hauck et al., 2013; Kraaijvanger et al., 2016; Maynard, 2015). Therefore, PRA tools and several stakeholder consultations were conducted for the collection of primary data and important information.

4.2.1 Semistructured Interview

Semistructured interviews were conducted with a fairly open framework that allowed for focused, conversational, two-way communication. In this technique, relevant topics were identified before the interview and most of the questions were created during the interview. This technique was used to collect qualitative data by setting up a situation that allowed a respondent to talk about their opinions on the particular subjects. The wording of the questions was not same for all respondents. Interviews were conducted at several places in the Bhabodah area. The interviewees were landowners, farmers, fishers, and members of the WMAs. Interviews were also conducted with the key informants from representatives of the LGIs, nongovernment organizations (NGOs), and the field offices of BWDB, Department of Fisheries, and the Department of Environment (DOE). The focus of these interviews was to collect information to understand the present problems regarding TRM operation, mitigation options, identification of major stakeholder groups, locations of community interactions, beels for study, and locations for resource mapping.

4.2.2 Resource Mapping

Resource mapping (RM) is a map to depict the resources, mainly natural rivers, channels, and beels available in the study area (Amir, 2010). RM normally covers the area of the entire study along with some adjacent areas. RM is often used as a base map at the time of planning as it enlists and visualizes almost all resources. It also acts as a documentation of the situation in the study area during the time of planning. RM was done in two villages in the EBK and BK. A brown paper and four-color marker pens were provided to the participants to draw the resource map. One person from a group of 15–20 people drew the map. Mistakes were corrected by the rest of the participants. The whole exercise took place in the open field. Two maps were prepared for two selected beels.

4.2.3 Focus Group Discussion

FGD is an efficient and effective tool for collecting information (Maynard, 2015). A group of 6–8 people who representated a much larger sector of a society or community was selected for FGDs. A homogeneous group of people discussed a topic of concern with the help of a facilitator in FGDs. The FGDs were conducted by the author himself, as a facilitator. Four FGDs, two in each beel, were conducted in four villages (Kakbadhal and Kalicharanpur in the EBK, and Kapalia and Monoharpur in BK). Farmer and fisher groups were the main source of primary information. Approximately 6–8 people were involved in each FGD and the average age of the people was 40. Most of the FGDs were done in the open field of the villages. The earlier prepared resource maps were used during the FGDs.

Finally, three socially acceptable sediment management options were identified for the TRM operation. The three identified options are described in the Result and Discussion section.

4.3 Technical Feasibility Using the Numerical Modeling

MIKE packages are the professional water-modeling software that is capable of simulating the flows of rivers, channels, estuaries, and other waterbodies (Amir et al., 2012, 2013a,b, c, 2018; IWM, 2009). In this study, the MIKE 21 Flow Model FM was chosen to assess the technical feasibility of the three sediment management options because of its flexibility, speed, accuracy, and popularity. The MIKE 21 Flow Model FM uses a depth-integrated flexible mesh approach. In this model, the transport of fine-grained material is averaged over depth and appropriate parameterization of the sedimentary processes is applied. The MIKE 21 Flow Model FM has six modules: hydrodynamic (HD), transport, ECO Lab, mud transport, particle tracking, and sand transport. The HD module is a prerequisite for the other five modules. The resulting flow and distributions of salts, temperature due to variety of forcing, and boundary conditions are calculated in the HD module (Delestre et al., 2016). The HD module extracts numerical solutions from the depth-integrated incompressible Reynolds averaged Navier-Stokes equations (two-dimensional (2D) shallow water equations), which consist of continuity, momentum, temperature, salinity, and density equations (Amir et al., 2013a). The cell-centered finite volume method was used to solve the model equations. A triangular grid was used in the horizontal plane. An explicit scheme was used for time integration.

Topographic DEM data were used for model bathymetry generation. Time series data of discharge, water level, and suspended sediment concentration were used for the model set-up, calibration, and validation. Major steps that were followed for the 2D HD model development were:

- Processing and analysis of DEM data.
- Specification of model domain.
- Generation of mesh.
- Generation of bathymetry.
- Specification of boundaries.
- Specification of model parameters.
- Specification of simulation period.
- Simulation of model.
- Calibration/validation of model.
- Options simulation.

4.4 Economic Analysis

Several civil works are needed for the operation of tidal basins for TRM operation. Construction of a peripheral embankment is essential to protect the homestead and crop lands from flooding. In the BK, the link canal will disrupt the present communication in the Bhabodah area. Therefore, one Bailey bridge is required to maintain the existing communication link in the area. At the end of the operation of the tidal basin, the canal needs to be closed by a closure dam after removing the Bailey bridge. A seasonal earthen cross dam is also required on the Hari river at the upstream off-take of the link canal to divert total tidal flow of the river to the tidal basin. Dredging or reexcavation of canals is needed for uniform silt deposition in the tidal basin. Besides, compensation is required for the landowner of the beel. The cost was calculated for all options based on the current schedule of rates of the BWDB.

5 DATA AND MODEL

5.1 Data Collection and Processing

The required data for the HD module of the MIKE 21 Flow Model FM are bathymetric data of rivers, the topographic data of beels, and the time series data of rainfall, evaporation, water level, and discharge. The extensive cross-sections (500 m interval) and surveyed topographic data were collected from IWM. The time series data of rainfall and evaporation were collected from the BWDB. Hourly discharge data for the same period were used as the upstream boundary for the HD model. Similarly, hourly water level data at the Ranai station on the Hari River and hourly discharge data of two locations of the EBK were used to calibrate the HD model.

Suspended sediment concentration data are required to develop and calibrate the mud transport model (MTM). Therefore, hourly suspended sediment samples were collected from IWM for one tidal cycle (13h) at two locations in the rivers and two locations in the beel.

The quality of the data was assured by visual checking and plotting hydrographs. Then, the data were prepared for the required format for the modeling purpose.

5.2 Model Setup

The MTM setup involves a geometrical description and specification of physical characteristics of the HD and sediment transport of the study area. The major components of the model setup include module selection, domain specification, mesh and bathymetry generation, boundary condition, HD, and mud transport parameter specification. Most of the components have to be specified in the Setup Editor (Fig. 6). There are three separate panes in the setup editor of the MIKE 21 Flow Model FM. The navigation tree, which is situated on the left, shows the structure of the model setup file. The corresponding editor is shown in the central pane of the setup editor when an item in this tree is selected. Validation error is shown in the bottom pane of the setup editor and errors can be minimized dynamically in this pane.

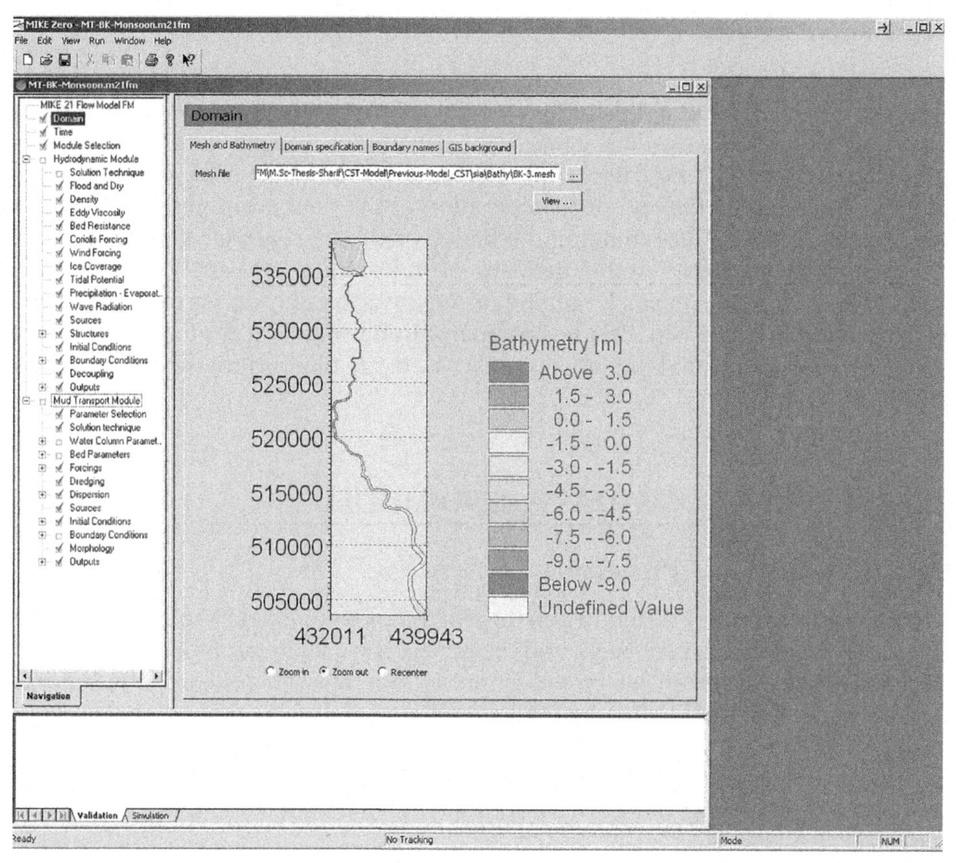

FIG. 6 Setup editor of MIKE 21 flow model FM.

The numerical model was developed integrating the main river system Teka-Hari-Teligati-Gengrail and the EBK and BK tidal basins. An explicit scheme was used for time integration. The model domain is subdivided into nonoverlapping triangular elements or cells (mesh/grid). Generation of suitable cells of the model is important for a reliable result. The total cell number of the model was 13,562 and 11,122 for the BK and EBK, respectively. The size of the cell was not uniform in the model. The minimum and maximum area of the cell was 120 and 20,000 m^2, respectively. The smaller cell was specified in the river system, the channels in the beel, and the connecting canals. The cell size of the numerical MTM for both beels is presented in Fig. 7. The bathymetry of the model was generated using the 87 measured cross-sections data at different locations of the four rivers (Teka, Hari, Telegati, and Gengrail River) and measured topographic data of 50 m resolution. Hourly water level data at the Kanchannagar station on the Gengrail River were used as the downstream boundary of the HD model. In the MTM, the downstream model boundary was defined by measured time series suspended sediment concentration and the upstream boundary was defined by constant concentration.

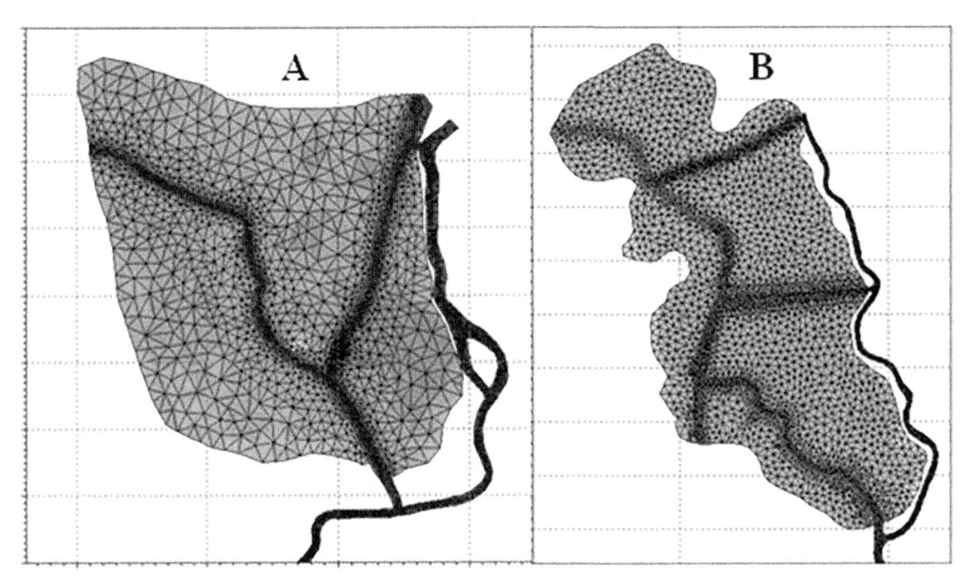

FIG. 7 The cell size of the cohesive sediment transport model for BK (A) and EBK (B).

The integrated HD and MTM were simulated simultaneously. The information of water levels and currents from the HD model is used to calculate the mud transport using a cell-centered finite volume method. The same mesh was used both for HD and MTM.

5.3 Model Calibration and Verification

This stage allowed fine tuning the numerical model by comparing simulated results with measured values. The goal was to obtain minimum differences between model output and field data. The model was calibrated against water level and discharge in the HD model. In the mud transport module, the model was calibrated against suspended sediment concentrations using settling velocity, bed roughness height, critical bed shear stresses, dispersion coefficient, and concentration at the open boundaries.

A comparison of measured and simulated discharge and sediment concentration data is shown in Figs. 8 and 9, respectively, indicating that a reasonably good calibration was achieved.

To check whether the calibrated model is an adequate representation of the physical system, simulated land elevation was verified with the measured data. The measured land surface was generated from land elevation measurements by a topographic survey inside the tidal basin. Good agreement was achieved between the measured and simulated data within different ranges of elevation inside the basin (Fig. 10).

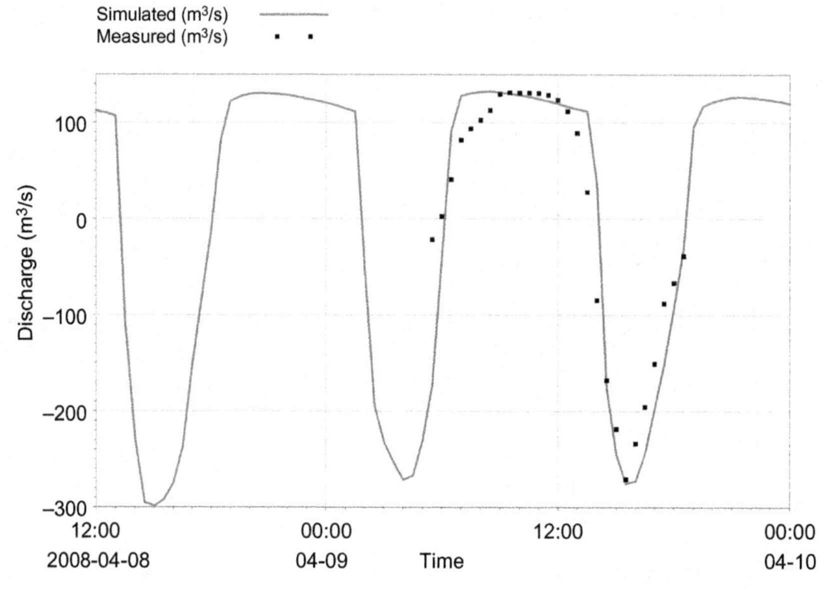

FIG. 8 Comparison of measured and simulated discharge in the Hari river near Dierkatakhali link canal of EBK.

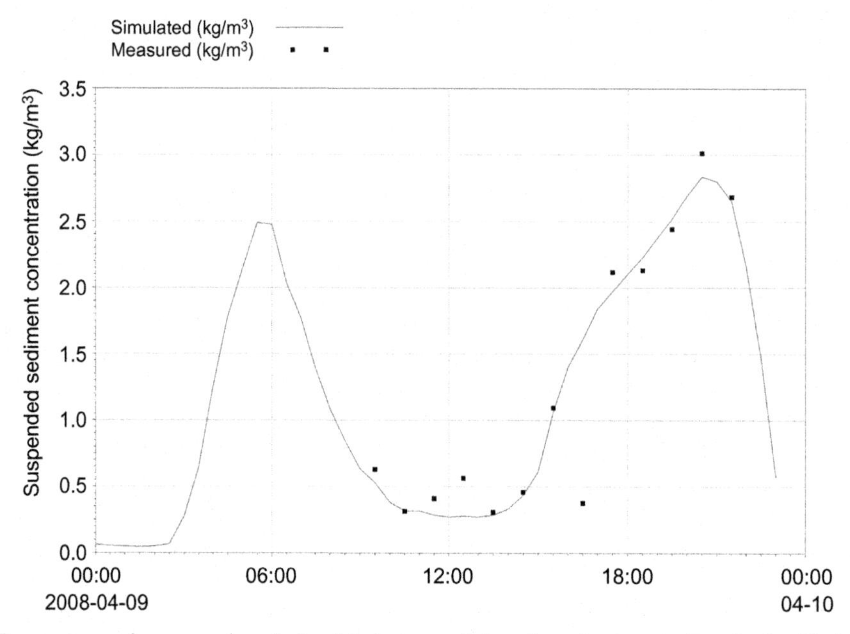

FIG. 9 Comparison of measured and simulated suspended sediment concentration in the Hari river near Dierkatakhali link canal of EBK.

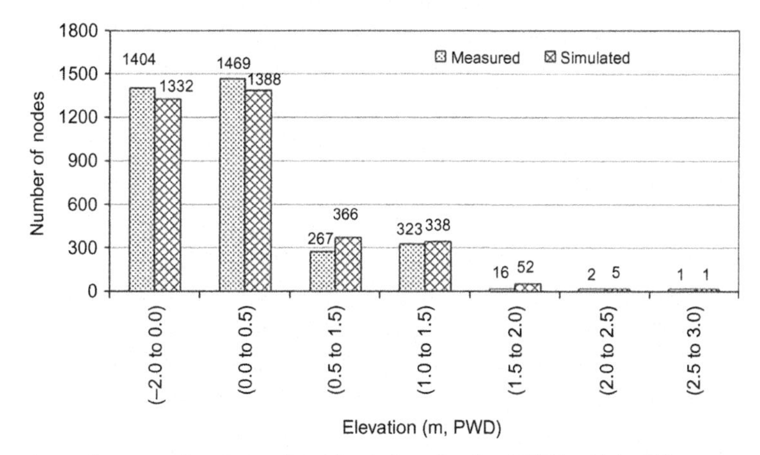

FIG. 10 Comparison of measured and simulated land elevation (m, PWD) within different ranges of elevations.

6 RESULTS AND DISCUSSION

6.1 Synthesis of Primary Data From PRA

Stakeholder consultations and semistructured interviews were carried out at the beginning of the study for collection of information regarding the present problem during TRM operation, identification of major stakeholder groups, location of community interactions, beels for the study, and locations for resource mapping. After stakeholder consultation, it was identified that farmers and fishers are the major stakeholder groups. It was also found that coordination among local stakeholders is required for effective sediment management.

Resource maps are very useful media for interaction with the local people. In all the FGDs in the EBK and BK, resource maps were used to understand the views of major stakeholder groups, farmers, and fishers. In the discussions, the main issue was to identify options for the uniform sediment deposition inside the tidal basin. Other important issues were also discussed, including coordination among local stakeholders, conflict between farmers and fishermen, compensation for the land, and alternate job opportunities. The brief descriptions are presented in the following sections.

6.1.1 Coordination Among Local Stakeholders

Field investigation from stakeholder consultation and semistructured interviews indicated that there was a lack of coordination among different stakeholders in the study area. Local stakeholders were also interested in establishing a mechanism for joint efforts of government agencies, LGIs, WMA, and NGOs for the sustainable water management in the study area. For interinstitutional coordination, BWDB might maintain contact with other government agencies such as the Local Government Engineering Department, DAE, and DOE, whose activities have an impact on water management in the study area. BWDB would liaise with LGIs and harmonize the works of LGIs and WMOs. BWDB can retain the responsibility of periodic and emergency maintenance works, but may accomplish them with the participation of WMOs and LGIs. Besides O&M works, BWDB would assist in strengthening of WMO both institutionally and technically. They may also carry out an information campaign and motivation work in close cooperation with WMOs and LGIs. It is required for arranging the training and capacity building of WMA and LGIs to ensure active participation of local people at all stages.

6.1.2 *Conflict Between Farmers and Fishers*

In many parts of the study area, the widespread practice of shrimp cultivation was taking place. For economic benefits, people often acquire the drainage canals and prepare fishing ghers without considering the drainage problems. Peripheral embankment of fishing ghers inside the tidal basin and fishing patta in the river and link canal should not be allowed during TRM operation because it restricts the smooth spreading of incoming sediment over the entire basin. To solve this problem, the BWDB should hand over the canals and lands that are under their control to WMA, letting them remove the illegal ghers and patta with the help of local initiatives and local administration. For the sustenance of the fishers, it is essential to develop public awareness of improved rice-fish culture and indigenous fish culture.

To avoid indiscriminate fishing and to save open water fisheries, the TRM basin may be well managed, organizing the present fishers through conservation and fish harvesting in a suitable manner. It would be wise to manage the expected increased fish resources through the WMA. Coordination is essential among the WMA, the Department of Fisheries, LGIs, NGOs, and the local community for integrated fisheries development.

6.1.3 *Compensation*

In the field investigation, the issue of compensation was discussed at a great length. Landowners demand compensation for their crops for the period of tidal basin operation. A significant number of participants, however, felt that it would be difficult to maintain such spirit for very long. Some suggested that the Union Parishad (UP) should be given the authority to collect taxes to be distributed among the affected households, according to certain predefined criteria. This, however, would require detailed field-level investigations and negotiations to develop cooperation among the different regions and local councils. Landowners whose land will be used for the designated basin should be compensated through payment of a fixed rent. The rate of compensation should be followed in accordance with market prices of food grains and average productivity of the land. Therefore, for the smooth and long-term operation of the tidal basin for TRM, a compensation mechanism should be introduced. This system would enable the landowners/farmers to manage alternative livelihoods during the operation of the tidal basin for TRM.

6.1.4 *Alternate Job Opportunity for Fishers and the Landless*

There are many landless and fisher people in the study area who are involved in farming and fisheries activities. When a beel is under TRM operation, these groups of people are jobless. So, alternate job opportunities are essential for the survival of these people. Many of these people were getting support from NGOs, but they have specific rules and regulations and may not be helpful for overall water resource development at all. Water management associations (WMAs), in this case, are not linked with greater socioeconomic developments and even with the rehabilitation activities of economic life. WMAs can play specific roles to solve the economic problems of the study area if WMAs activities and other NGO initiatives can be incorporated for the greater rehabilitation of the economic situation.

To mitigate the suffering of the affected landless people, it has been suggested to enroll them under the vulnerable group feeding (VGF) program. Once the area is demarcated, the appropriate authority should record the names and particulars of the affected landowners. Each should be given a VGF card to enable them to receive a fixed amount of food grain per month. The existing rules of the VGF program should be followed in implementing the program.

Support from government organizations and NGOs could be used to develop the technical knowledge to do other things such as handicrafts, livestock other than agriculture, and fisheries for the jobless. Many civil works are needed when a beel is selected for TRM operation. The jobless people should get priority for that construction work. Besides this, interest-free loans and/or credit from the government are essential for the jobless. Financial support/assistance from donor agencies will be very useful for the jobless.

People affected by the tidal basins should be given material and technical help for their resettlement, both in an economical and a professional sense. In assessing compensation, criteria such as land loss as a proportion of the absolute land size of the household should be considered. Households for which proportional land loss would be above a critical percentage (to be defined in light of the national standard) would be categorized as a separate group, either for a better compensation rate (through arbitration suits) or for specific resettlement programs.

6.2 Generation of Sediment Management Options

Three socially acceptable options for TRM operation were identified for the uniform sediment deposition inside the tidal basin through participatory approach. Dredging was considered for all three options. The general practice and three identified options of the TRM operation are described below. The general practice is presented as Option-0, which was useful to compare with the other options.

6.2.1 Option-0

The general practice of TRM operation was to construct one- or two-link canals to connect the tidal basin with the river. But in that case, most of the sedimentation took place in the vicinity of the link canal. Normally, sediment does not spread out in the areas far away from the canal. Therefore, dredging or reexcavation of the canals was considered for the other options to get the uniform sedimentation inside the tidal basin. Fig. 11 shows the schematization for Option-0 in the BK.

6.2.2 Option-1

In this option, each beel was divided into three compartments by considering an embankment around each compartment. Fig. 12 shows the schematization for Option-1 in the BK and EBK. To allow sedimentation in one compartment at a time, only that compartment was connected with the river by cutting an artificial canal, which is called a link canal. Each beel can be selected for TRM operation for 3 years period. So, in this option, each compartment

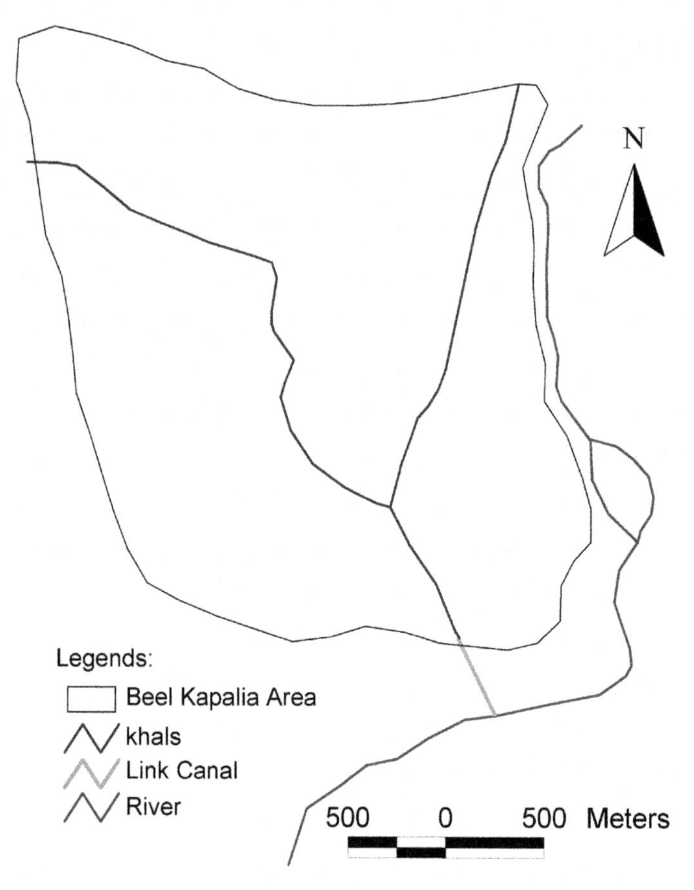

FIG. 11 Schematization for Option-0 in Beel Kapalia.

was connected with the river for 4 months period in every year out of 3 years. The compartments were devised based on three criteria: area of the compartment, existence of the channel in the compartment, and land topography of the beel. The areas of the compartments for both beels are given in Table 2.

6.2.3 Option-2

In this option, an embankment was considered along both banks of the main channel (left channel) in the tidal basin, thereby allowing sedimentation by cutting the embankment part by part gradually from downstream to upstream for both beels, as shown in Fig. 13. The total length of the main channel of the BK and the EBK is 3660 and 5760 m respectively. For the first year, the embankment was considered for a length of three quarters of the total length, that is, 2745 m for the BK (Fig. 13A1) and 4320 m for the EBK (Fig. 13B1) and sedimentation was allowed for the rest. For the second year, the embankment was removed in such a way that half the main channel was open for water flow and sedimentation (Fig. 13A2 and B2). In the third year, the embankment existed for only one quarter of the main channel

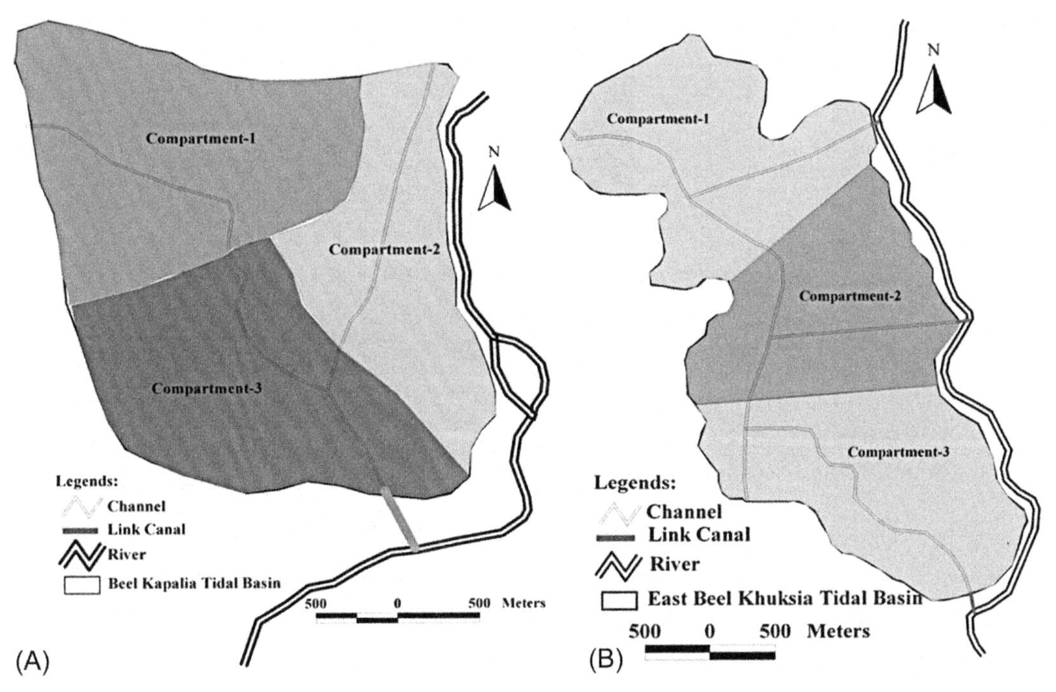

FIG. 12 Schematization of Option-1 for BK (A) and EBK (B).

TABLE 2 Area of Three Compartments in Each Beel

Name of the Beel	Area (ha)		
	Compartment-1	Compartment-2	Compartment-3
EBK	265	242	274
BK	254	169	235

and the other three quarters were allowed for sedimentation (Fig. 13A3 and B3). During the fourth year, sedimentation was allowed for the full channel without any embankment (Fig. 13A4 and B4).

6.2.4 Option-3

In this option, all the existing major channels of the beel were connected with the river by link canals, that is, allowing sedimentation in the whole basin at the same time (Fig. 14). In the BK, there are two channels that are nearer to the Hari River and those channels were connected with the river by constructing two link canals (Fig. 14A). Similarly, three channels in the EBK were connected with the Hari River with three link canals (Fig. 14B).

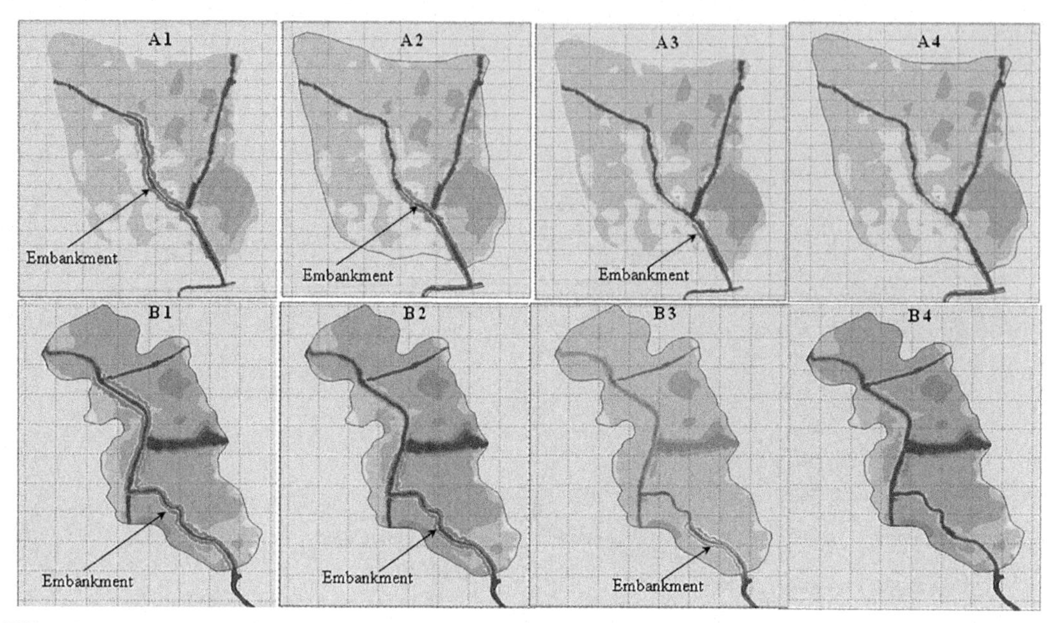

FIG. 13 Different stages of Option-2 for BK (A1–A4) and EBK (B1–B4).

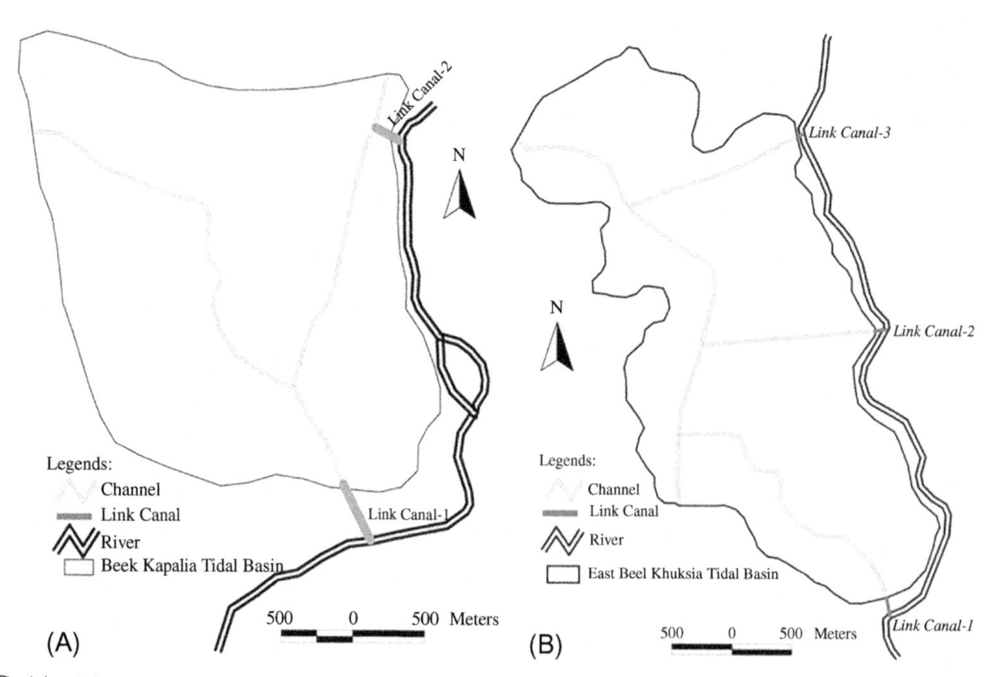

FIG. 14 Schematization of Option-3 for BK (A) and EBK (B).

6.3 Option Simulation Results

Options identified from the field were analyzed for their technical feasibility using an MTM developed by the Danish Hydraulic Institute (DHI) Water and Environment. To set the options in the model, the size of the grid was taken to be very fine. For all the options, total bed thickness was changed and net deposition in the beels was calculated.

The MTM was simulated for 4 years for all three options. Continuous 4-year model simulation of the tidal river is quite complex and time consuming. Initially, simulation was done in the dry period. Then the monsoon flow was simulated with the updated bed level. In this way, prediction for a year was obtained for all options. Then, the model was simulated for different time periods, 1–4 years, with dredging of canals. Changes of net deposition in the tidal basin were computed for the next 3 years after considering the changes that occurred in the previous year.

6.3.1 Option-0

This option was a business-as-usual option or a traditional sediment management practice. The simulation time was 1 year for Option-0. The simulated net deposition pattern in the tidal basin for this option is presented in Fig. 15. It is clear from the figure that most of the sedimentation took place in the channel and the immediate vicinity of the channel. Sediment couldn't spread out in the areas far away from the canal. Fig. 15 also indicates a very nonuniform sedimentation in the basin.

6.3.2 Option-1

The simulated net deposition pattern inside the tidal basin by the operation of TRM in different periods with dredging of channels for Option-1 is shown in Fig. 16. In Fig. 16, the results (predicted net deposition pattern inside the tidal basin) from Option-1 for 1 year, 3 years, and 4 years operation on the BK (Fig. 16A–C) and the EBK (Fig. 16D–F) are presented, respectively. The figures show that for both the beels, the sedimentation is increasing and distributes from canal to far away from year 1 to year 4. Fig. 16A–D shows that in 1 year, the net sediment deposition around three entry points is high, which is around 1 m. In the rest of the area, the net sediment deposition is mostly within 0.1–0.2 m. After 3 years, that sediment deposition starts to spread all over the tidal basin (Fig. 16B–E). Finally, within a 4-year period, the sedimentation spreads over the whole area and ranges from 0.3 to 1.0 m (Fig. 16C–F).

6.3.3 Option-2

In Fig. 17, the predicted net deposition pattern inside the tidal basin for Option-2 for 6 months, 1 year, 2 years, 3 years, and 4 years operation on the BK (Fig. 17A–E) and the EBK (Fig. 17F–J) are presented, respectively. The impact of embankment on both sides of the major canal is clearly viewed in Fig. 17. The figure shows that sedimentation is progressing year by year with the removal of the embankment. At the end of the fourth year, the sedimentation is spread over the total area. Though the net sediment deposition is high (above 1 m) around the main canal and at the remotest point from the main channel, the net sediment deposition is 0.4–0.5 m, which is shown in Fig. 17E. A similar result was found for Beel Khukshia in Fig. 17J, but here the net sediment deposition is from 0.3 m to above 1 m.

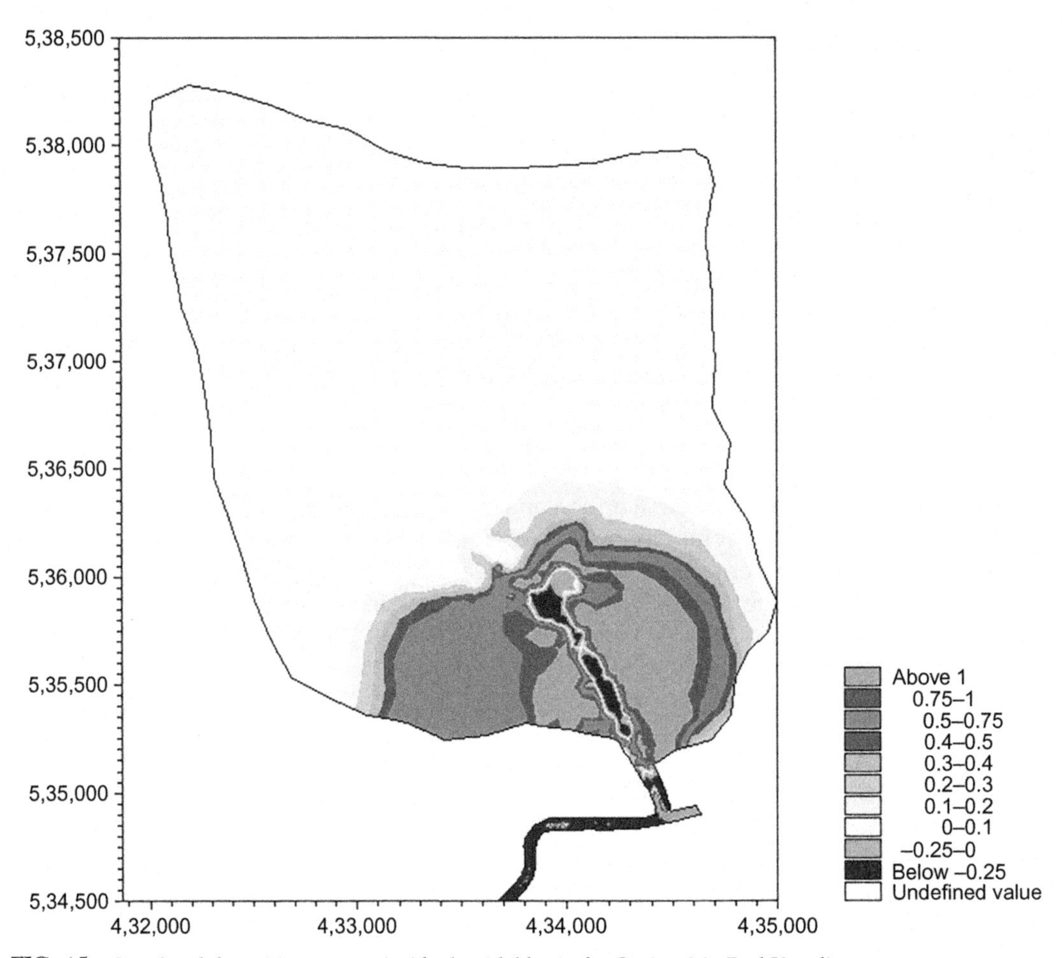

FIG. 15 Simulated deposition pattern inside the tidal basin for Option-0 in Beel Kapalia.

Therefore Fig. 17A–J show that sedimentation is increasing year by year for both the beels and propagates along the main canal throughout the area within a 4-year period. Though the sedimentation is not fully uniform, that is in an acceptable limit.

6.3.4 Option-3

The predicted net deposition pattern for different periods inside the tidal basin for Option-3 is presented in Fig. 18. This figure shows that the response of sedimentation in Option-3 is comparatively slower than Option-1 and Option-2. Though at the end of the fourth year, reasonable sedimentation is found in both tidal basins. However, the sedimentation in the EBK is higher than the BK (Fig. 18E and J).

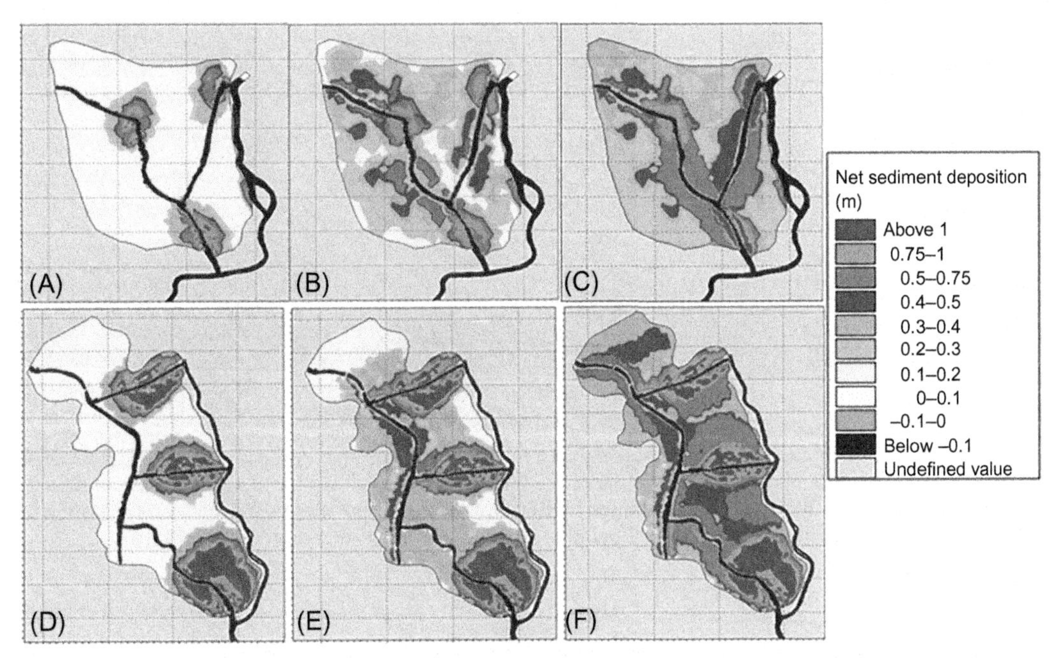

FIG. 16 Predicted net deposition pattern inside the tidal basin for Option-1 after the operation of TRM in different periods with dredging of channels; 1 year operation for BK (A), 3 year operation for BK (B), 4 year operation for BK (C), 1 year operation for EBK (D), 3 year operation for EBK (E), 4 year operation for EBK (F).

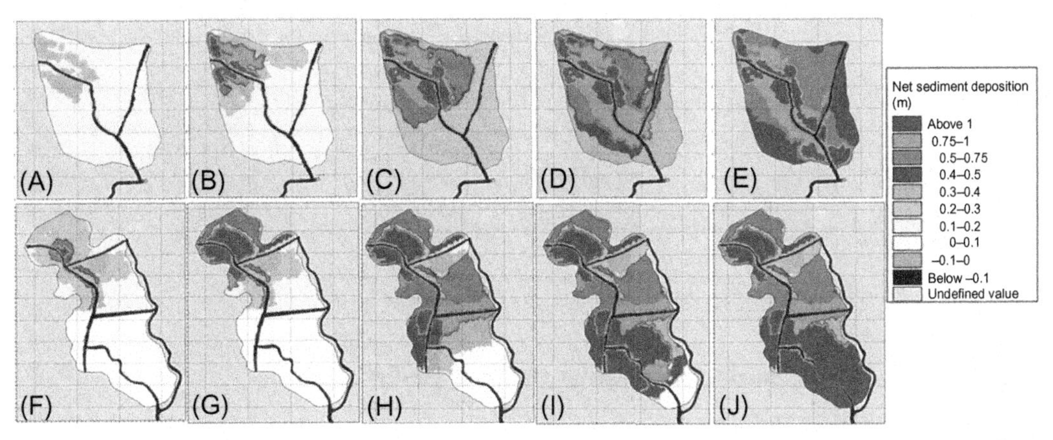

FIG. 17 Predicted net deposition pattern inside the tidal basin for Option-2 after the operation of TRM in different periods with dredging of channels; 6 months operation for BK (A), 1 year operation for BK (B), 2 years operation for BK (C), 3 year operation for BK (D), 4 years operation for BK (E), 6 months operation for EBK (F), 1 year operation for EBK (G), 2 years operation for EBK (H), 3 year operation for EBK (I), 4 years operation for East BK (J).

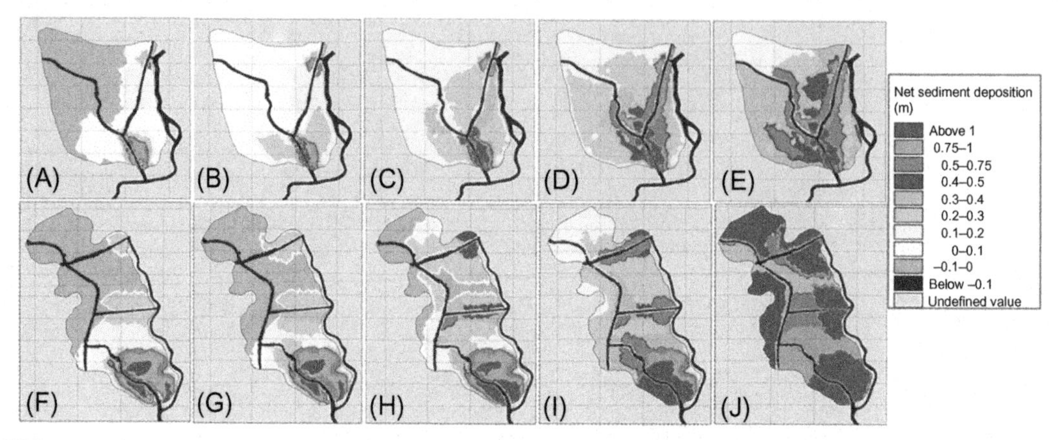

FIG. 18 Predicted net deposition pattern inside the tidal basin for Option-3 after the operation of TRM in different periods with dredging of channels; 6 months operation for BK (A), 1 year operation for BK (B), 2 years operation for BK (C), 3 year operation for BK (D), 4 years operation for BK (E), 6 months operation for EBK (F), 1 year operation for EBK (G), 2 years operation for EBK (H), 3 year operation for EBK (I), 4 years operation for East BK (J).

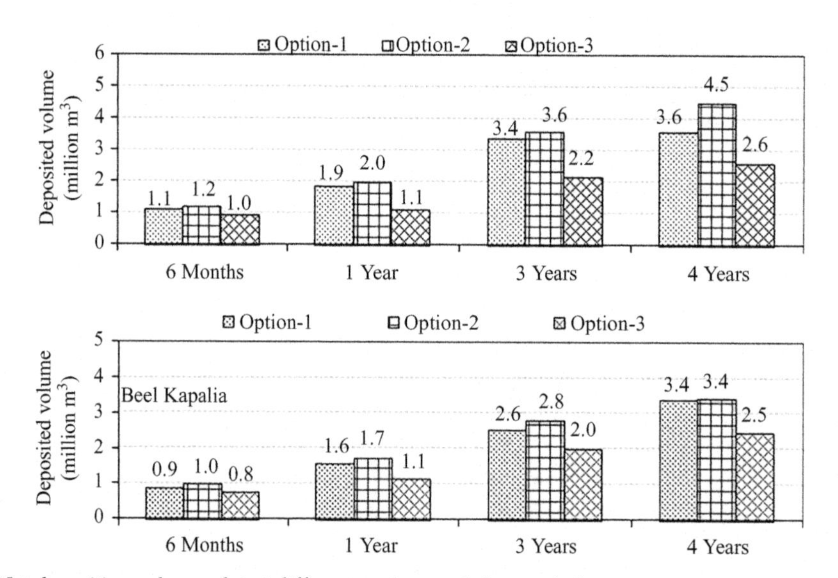

FIG. 19 Net deposition volume plot at different options and time period.

6.4 Comparison of Simulation Results of Two Beels

6.4.1 Net Deposition Volume

A comparison of net deposition volume in the EBK and the BK for the three options is given in Fig. 19. The net deposition volume after 4 years under Option-1, Option-2, and Option-3 is 3.58 million m^3, 4.51 million m^3, and 2.61 million m^3, respectively, in the EBK and 3.40 million m^3, 3.43 million m^3, and 2.45 million m^3, respectively in the BK. As seen from the figure, the

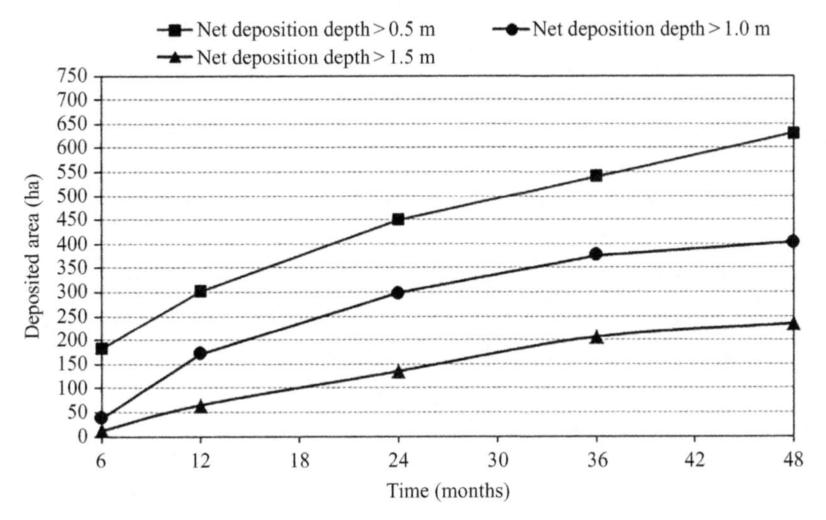

FIG. 20 Deposited area plot for Option-1 of East Beel Khuksia.

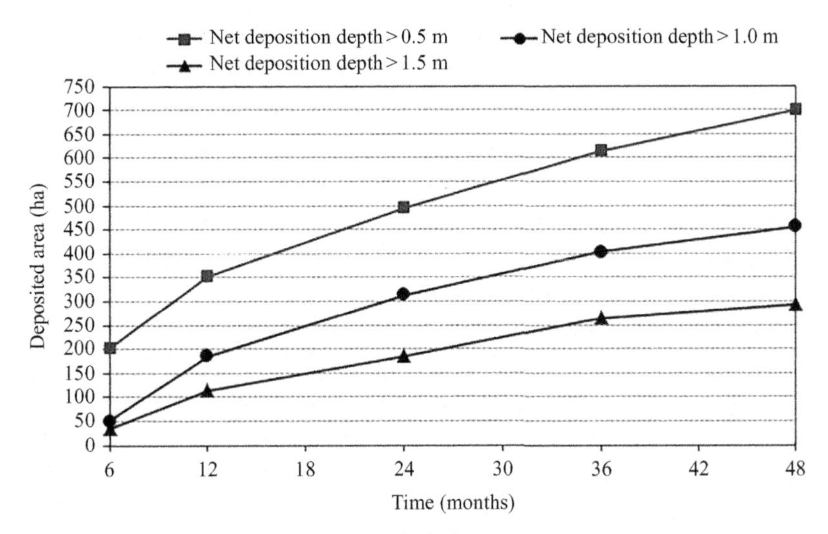

FIG. 21 Deposited area plot for Option-2 of East Beel Khuksia.

maximum net deposition occurred for Option-2 and the minimum net deposition occurred for Option-3 in both beels. Net deposition volume is higher in the EBK than the BK because the EBK is located at the downstream of the Hari River. This means that the beels located in the downstream of the river where tidal influence is stronger are more suitable for TRM operation.

6.4.2 Deposition Area

From the simulated results, six plots (Figs. 20–25) of deposited area versus time were prepared for each option in both beels. The plots were prepared for three levels of deposition: net

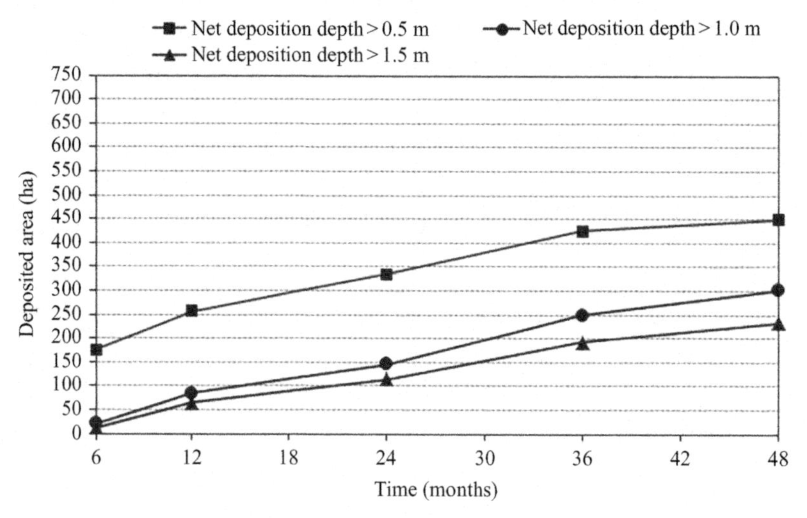

FIG. 22 Deposited area plot for Option-3 of East Beel Khuksia.

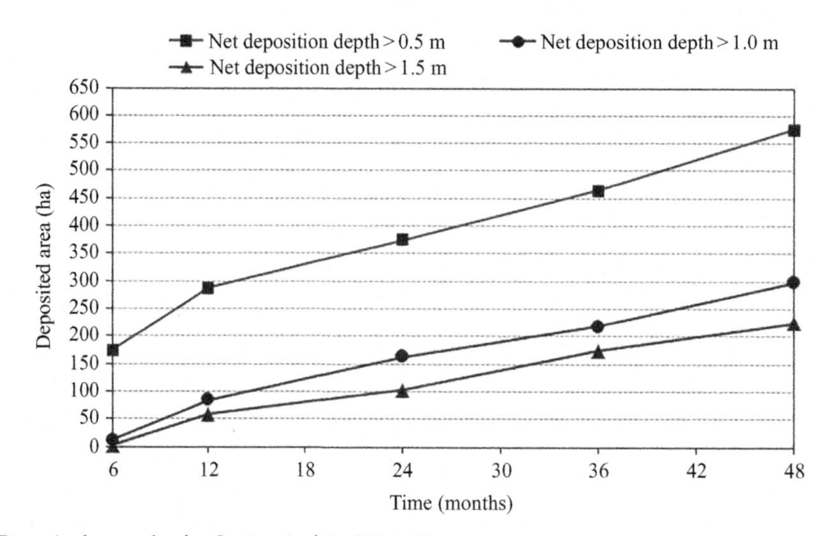

FIG. 23 Deposited area plot for Option-1 of Beel Kapalia.

deposition greater than 0.5m, net deposition greater than 1.0m, and net deposition greater than 1.5m. It is seen that maximum and minimum net deposition occurred for Option-2 and Option-3, respectively.

Figs. 20 and 21 for Option-1 and Option-2 in the EBK show that for deposition depth greater than 1.0 and 1.5m, sediment deposition does not increase significantly after almost 36 months. But further deposition will occur under 48 months in areas where the net deposition depth is greater than 0.5m. In Option-3 (Fig. 22), all areas where the net deposition depth is greater than 0.5, 1.0, and 1.5m continue to increase, even after 48 months. Similar situations were observed from Figs. 23–25 for the BK.

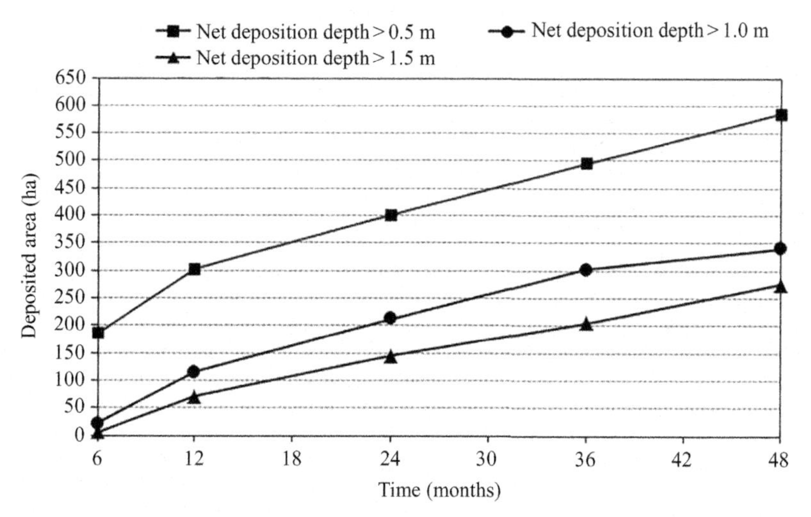

FIG. 24 Deposited area plot for Option-2 of Beel Kapalia.

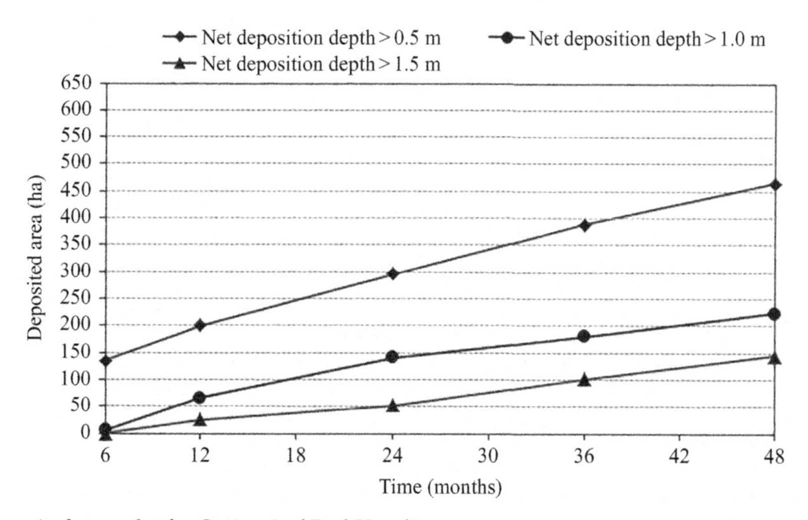

FIG. 25 Deposited area plot for Option-3 of Beel Kapalia.

6.5 Cost for Different Options

The total estimated cost of the three options for the two beels is presented in Fig. 26. The cost was based on current schedule rates of the Bangladesh Water Development Board. The total estimated cost for the three options are Tk.28,58,48,912, Tk.21,34,55,375, and Tk.35,58,37,393, respectively, in the EBK and Tk.20,79,89,120, Tk.16,16,72,991, and Tk.25,21,70,405, respectively, in the BK. Thus, Option-2 is the cheapest option for both beels.

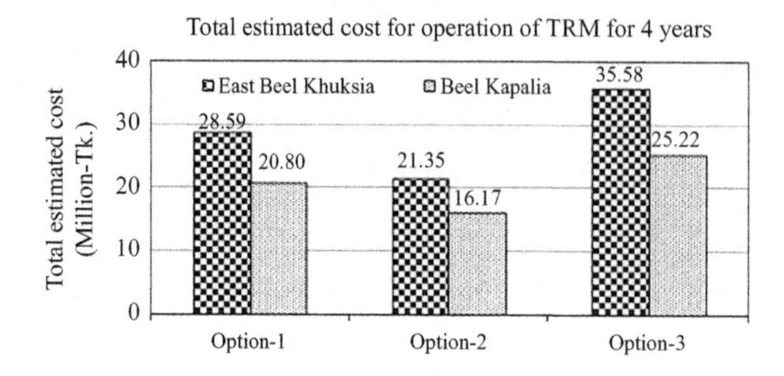

FIG. 26 Total estimated cost for the three options in the two beels.

6.6 Stakeholder Consultation for Option Selection

Stakeholder consultation was carried out to finalize the acceptable option(s) for sediment management inside the tidal basin. Discussions of technical feasibility and economic analysis were held with different groups, including farmers, fishermen, day laborers, and traders. After detailed consideration of the different aspects of sediment management, including the estimated cost of options, most stakeholders agreed that Option-2 is the preferred option.

7 CONCLUSION AND DISCUSSION

7.1 Conclusion

The TRM is an ecotechnological concept designed to solve the waterlogging problem while at the same time improving the environment. By implementing the concept, the natural environment was restored and the ecology of the wetlands was conserved. Raising the land consistently inside a beel and maintaining the proper drainage capacity of the river are the two main objectives of TRM operation. From field visits and monitoring results, it was observed that sedimentation inside the beel was not uniform in the Beel Kedaria and the EBK. This happened due to the lack of social and technical limitations during TRM operation.

In this study, three sediment management options for uniform sediment deposition in the tidal basins were identified through FGDs and consultation with the local stakeholders. In Option-1, each beel was divided into three compartments separated by an embankment around the compartment and allowing sedimentation in the compartments one after another. In Option-2, embankments were constructed along both banks of the main channel through the beel and sedimentation in the basin was allowed by cutting the embankment part by part gradually from upstream to downstream. In Option-3, all existing channels were connected with the river at the same time by constructing link canals, that is, allowing sedimentation in the whole basin at the same time. The technical feasibility of these options was assessed by an MTM using the modeling system of MIKE 21 Flow Model FM.

In Option-1, after 1 year without dredging of canals and with dredging of canals, the net sediment deposition volumes were 1.63 and 1.88 million m^3, respectively, in the EBK, and 1.02 and 1.56 million m^3, respectively, in the BK. Simulation results indicate more sedimentation deposition inside the tidal basin if dredging after 4 months. Besides, more uniform deposition is observed after dredging. Without dredging, sedimentation normally takes place in the canals and can't spread out in the area far away from the canal/channel. It is the main barrier of uniform sediment deposition in the beel. In addition, problems always arise regarding disposing of the dredged spoil. It would be better to store the dredged spoil in the remote portion of the beel where sedimentation is less. Most of the stakeholders opine that dredging or reexcavation of canals and channels inside the tidal basin should be done with the participation of the local stakeholders using mechanical dredgers or manual reexcavation every 4 months.

The net deposition volume was higher in the EBK than in the BK because the EBK is located at the downstream of the Hari river. Similar results were obtained for the other two options. The net deposition volume after 4 years in Option-1, Option-2, and Option-3 are 3.58, 4.51, and 2.61 million m^3, respectively, in the EBK and 3.40, 3.43, and 2.45 million m^3, respectively, in the BK. So, maximum deposition was observed for Option-2.

The total estimated costs for the three options were BDT28,58,48,912, BDT21,34,55,375, and BDT35,58,37,393, respectively, in the EBK and BDT20,79,89,120, BDT16,16,72,991, and BDT25,21,70,405, respectively, in the BK. This indicates that minimum cost was required for implementing TRM in Option-2 for both beels.

Stakeholder consultation was carried out to finalize the acceptable option(s) for sediment management inside the tidal basin. Discussion was held with different groups, including farmers, fishers, day laborers, and traders. Technical feasibility and economic analysis were presented in the discussion. People allow their land to be used for tidal basin operation hoping that the land will rise after a certain period. But their previous experience from the Beel Kedaria and the EBK was not good because sedimentation inside these two beels was not uniform. In Option-1 and Option-2, the sedimentation pattern inside the tidal basin was uniform. However, at least a 2 million m^3 tidal prism is required for sustainability of the Hari river. If the area of the tidal basin is more, then the tidal prism will increase. In Option-1, the area of the tidal basin is one-third of its original area. Most of the stakeholders opine that in Option-1, the minimum tidal prism will not be attained. In Option-2, the area of the tidal basin is not minimized. Therefore, there was no question for the generation of the required tidal prism for Option-2. In Option-3, the sedimentation pattern was not uniform. Thus, after detailed discussions and consideration of different aspects of sediment management, including the cost estimates for different options, it was apparent that Option-2 is acceptable by most stakeholders in the area.

7.2 Recommendations

Based on the present study, the following recommendations can be made:

- The models were simulated for a period of 4 years due to time constraints. Further simulation for a longer period may be carried out to determine the actual required lifetime of a tidal basin.
- Water level and discharge data at several locations can be collected for the most accurate HD calibration.

- To calibrate and verify the model, more detailed data of sedimentation in the tidal basin are required. Therefore, long-term sediment concentration measurement programs may be undertaken to support similar studies in the future.
- To assess the effectiveness of the TRM operation in terms of sedimentation inside a tidal basin, accurate measurement of sedimentation is essential. From direct field measurement, it is possible to quantify the sedimentation volume. This will also help to investigate the sediment distribution over the entire basin area.
- Parts of the beel away from the channels were not raised in all the options. In those parts, new canals may be constructed. The technical feasibility of this can be assessed by using the present MTM.
- Similar studies may be carried out in other areas of the coastal region to investigate the regional variability.

References

Alvarez-Guerra, M., Canis, L., Voulvoulis, N., Viguri, J.R., Linkov, I., 2010. Prioritization of sediment management alternatives using stochastic multicriteria acceptability analysis. Sci. Total Environ. 408 (20), 4354–4367.

Amir, M.S.I.I., 2010. Socio-technical Assessment of Sediment Management Options in Tidal Basins in Southwestern Bangladesh. (M.Sc.). Bangladesh University of Engineering and Technology, Bangladesh.

Amir, M.S.I.I., Khan, M.M.K., Rasul, M.G., Sharma, R.H., Akram, F., 2012. In: A review of downscaling and 1-D/2-D hydraulic model for flood studies.Paper Presented at the Central Region Engineering Conference, 10–12 August, Rockhampton, Australia.

Amir, M.S.I.I., Khan, M.S.A., Khan, M.M.K., Rasul, M.G., Akram, F., 2013a. Tidal river sediment management—a case study in southwestern Bangladesh. World Acad. Sci. Eng. Technol. 7 (3), 175–185.

Amir, M.S.I.I., Khan, M.M.K., Rasul, M.G., Sharma, R.H., Akram, F., 2013b. Automatic multi-objective calibration of a rainfall runoff model for the Fitzroy Basin, Queensland, Australia. Int. J. Environ. Sci. Dev. 4 (3), 311–315.

Amir, M.S.I.I., Khan, M.M.K., Rasul, M.G., Sharma, R.H., Akram, F., 2013c. Numerical modelling for the extreme flood event in the Fitzroy Basin, Queensland, Australia. Int. J. Environ. Sci. Dev. 4 (3), 346–350.

Amir, M.S.I.I., Khan, M.M.K., Rasul, M.G., Sharma, R.H., Akram, F., 2018. Hydrologic and hydrodynamic modelling of extreme flood events to assess the impact of climate change in a large basin with limited data. J. Flood Risk Manag. 11, S147–S157. https://dx.doi.org/10.1111/jfr3.12189.

CEGIS (Center for Environmental and Geographic Information Services), 2002a. Monitoring and Integration of the Environmental and Socio-economic impacts of implementing the Tidal River Management option to solve the problem of drainage congestion in KJDRP area. Final Report, Part A: Monitoring, Bangladesh Water Development Board.

CEGIS (Center for Environmental and Geographic Information Services), 2002b. Monitoring and Integration of the Environmental and Socio-economic impacts of implementing the Tidal River Management option to solve the problem of drainage congestion in KJDRP area. Final Report, Part D: WMA-O&M Plan.

Delestre, O., Abily, M., Cordier, F., Gourbesville, P., Coullon, H., 2016. Comparison and validation of two parallelization approaches of FullSWOF_2D software on a real case. In: Gourbesville, P., Cunge, A.J., Caignaert, G. (Eds.), Advances in Hydroinformatics: SIMHYDRO 2014. Springer Singapore, Singapore, pp. 395–407.

Franzén, F., Hammer, M., Balfors, B., 2015. Institutional development for stakeholder participation in local water management—an analysis of two Swedish catchments. Land Use Policy 43, 217–227.

H&A (Haskoning and Associates), 1993. Second Coastal Embankment Rehabilitation Project. vol. I. Bangladesh water Development Board. Executive Summary and Main Report.

Hauck, J., Görg, C., Varjopuro, R., Ratamäki, O., Jax, K., 2013. Benefits and limitations of the ecosystem services concept in environmental policy and decision making: some stakeholder perspectives. Environ. Sci. Policy 25, 13–21.

IWM (Institute of Water Modelling), 2002. Special Monitoring of Rivers and Tidal Basins for Tidal River Management. Final Report, Volume 1: Main Report, Bangladesh Water Development Board.

IWM (Institute of Water Modelling), 2007. Monitoring the effects of Beel Khuksia TRM Basin and Dredging of Hari River for drainage improvement of Bhabodah area. Final Report, Bangladesh Water Development Board.

IWM (Institute of Water Modelling), 2009. Mathematical modelling study for planning and design of Beel Kapalia Tidal Basin. Final ReportBangladesh Water Development Board.

Kraaijvanger, R., Almekinders, C.J.M., Veldkamp, A., 2016. Identifying crop productivity constraints and opportunities using focus group discussions: a case study with farmers from Tigray. NJAS Wagening. J. Life Sci. 78, 139–151. https://dx.doi.org/10.1016/j.njas.2016.05.007.

Maynard, C.M., 2015. Accessing the environment: delivering ecological and societal benefits through knowledge integration—the case of water management. Appl. Geogr. 58, 94–104.

Mukherjee, A., Fryar, A.E., Thomas, W.A., 2009. Geologic, geomorphic and hydrologic framework and evolution of the Bengal basin, India and Bangladesh. J. Asian Earth Sci. 34 (3), 227–244.

Nguyen, Q., Hoang, M.H., Öborn, I., van Noordwijk, M., 2013. Multipurpose agroforestry as a climate change resiliency option for farmers: an example of local adaptation in Vietnam. Clim. Chang. 117 (1), 241–257.

Rahman, M.R., 2008. Investigation into replicability of good practices in flood management. Final Report, Institute of Water and Flood Management, Bangladesh University of Engineering and Technology.

Ricaurte, L.F., Wantzen, K.M., Agudelo, E., Betancourt, B., Jokela, J., 2014. Participatory rural appraisal of ecosystem services of wetlands in the Amazonian Piedmont of Colombia: elements for a sustainable management concept. Wetl. Ecol. Manag. 22 (4), 343–361.

SMEC, 1997. Khulna-Jessore Drainage Rehabilitation Project, Sediment and River Morphology Study. Final Report.

SMEC, 2002. Khulna-Jessore Drainage Rehabilitation Project, Feasibility Study Report for Overall Drainage Plan. Final Report, Bangladesh Water Development Board.

CHAPTER

11

Coastal and Marine Biodiversity of India: Challenges for Conservation

C. Raghunathan, R. Raghuraman, Smitanjali Choudhury

Zoological Survey of India, Andaman and Nicobar Regional Centre, Port Blair, India

1 INTRODUCTION

From bacteria to baleen whales, our planet is home to tens of millions of different life forms; biologists can only guess at the true number of species. Oceans and the major seas cover 70.8% or the Earth's 362 million km^2, with a global coastline of 1.6 million km. Coastal and marine ecosystems are found in 123 countries around the world. Marine ecosystems are strongly connected through a network of surface and deep-water currents, and they are among the most productive ecosystems in the world. Coastal and marine ecosystems include near-shore coastal areas, open-ocean marine areas, and sand dune areas, where freshwater and seawater mix. Marine systems extend from the low-water mark, that is, a depth of 50 m, to the high seas, and coastal systems stretch from the coastline to depths less than 50 m. The Indian Ocean accounts for 29% of the global ocean area, 13% of the marine organic carbon synthesis, 10% of the capture fisheries and 90% of the culture fisheries, 30% of the coral reefs, and 10% of the mangroves. It has 246 estuaries draining a hinterland greater than 2000 km^2, in addition to coastal lagoons and backwaters. Being landlocked in the north, and with the largest portion of it lying in the tropics, the Indian Ocean is a region of high biodiversity with one of the countries in the region, India, rated as one of the megabiodiversity centers of the world. In the current context of international trade and intellectual property regimes, it is important for all the Indian Ocean countries to understand their marine biodiversity. India is one among 17 megabiodiversity countries and represents four biodiversity hotspots among the reported 32 richest and highly endangered ecoregions of the world. In terms of the marine environment, India has a coastline of about 8000 km. The exclusive economic zone (EEZ) of the country has an area of 2.02 million km^2, comprising 0.86 million km^2 on the west coast, 0.56 million km^2 on the east coast, and 0.6 million km^2 around the Andaman and Nicobar Islands. Adjoining the continental regions and the offshore islands is a wide range of coastal ecosystems such

https://doi.org/10.1016/B978-0-12-810473-6.00012-1
© 2019 Elsevier Inc. All rights reserved.

as estuaries, lagoons, mangroves, backwaters, salt marshes, rocky coasts, sandy stretches, and coral reefs, which are characterized by unique biotic and abiotic properties and processes.

Among the Asian countries, India is perhaps the only one that has a long record of inventorying coastal and marine biodiversity, going back at least two centuries. However, there is so much diversity in space, time, and taxonomical diversity that it is almost impossible to review all the records and reports. The synthesis of what is known about the coastal and marine biodiversity of India that is attempted in this overview relies mainly on systematic accounts, records, and reports of two major institutions concerned with surveying and inventorying fauna and flora—the Zoological Survey of India and the Botanical Survey of India—as well as other research organizations such as the Central Marine Fisheries Research Institute and the National Institute of Oceanography.

1.1 Marine Algae

Marine macro algae (seaweed) from Indian coasts have been fairly well studied for several decades. Oza and Zaidi (2001) listed 844 species (including forms and varieties) a decade ago. However, a recent report identifies a total of 936 species of marine algae from different areas of India (Rao, 2010). The most abundant among them are rhodophytes (434 species in 136 genera) followed by chlorophytes (216 species in 43 genera), phaeophytes (191 species in 37 genera), and xanthophytes (three species in one genus). The estimated total standing crop of seaweed in the intertidal and shallow waters of the Indian coast is 91,345 tons wet weight. In deep water, the amount is 75,373 tons, which includes 6000 tons of agar-yielding seaweed (Roy and Ghosh, 2009). Red algae (*Gelidiella acerosa, Gracilaria edulis, G. crassa, G. foliifera,* and *G. verrucosa*) are used for the manufacture of agar, and brown algae (*Sargassum* spp., *Turbinaria* spp. and *Cystoseira trinodis*) are used for alginates and seaweed liquid fertilizers. The bulk of the harvest is from the natural seaweed beds of the Gulf of Mannar Islands. The data collected by the Central Marine Fisheries Research Institute (CMFRI) on seaweed landings in Tamil Nadu from 1978–2000 reveal that the quantity (dry weight) exploited annually during this period was 102–541 tons for *Gelidiella acerosa*, 108–982 tons for *Gracilaria edulis*, 2–96 tons from *G. crassa*, 3–110 tons for *G. foliifera*, and 129–830 tons for *G. verrucosa* (Kalimuthu et al., 1991; Ramalingam, 2000).

1.1.1 Diatoms

Diatoms are the dominant component of phytoplankton in all the Indian estuaries and coastal waters where detailed inventories of floristic composition and seasonal changes are available. Among the estuaries of the east coast, the phytoplankton composition has been studied in detail only from the Hooghly, Rushikulya, Godavari, Cooum, Ennore, Adyar, and Vellar. A total of 102 species of diatom belonging to 17 families is known from the east coast, with the largest diversity being found in the family Naviculaceae (21 species) and the families Chaetoceraceae and Coscinodiscaceae (11 species each) (Fig. 1). A total of around 200 species of diatoms are reported from Indian marine environments. The diatom diversity along the west coast is relatively higher, with 148 species under 22 families. The family Naviculaceae is dominant with 22 species, followed by Biddulphiaceae (16 species), Lithodesmiaceae (15 species), and Thalassiosiraceae (12 species) (Fig. 2). The families

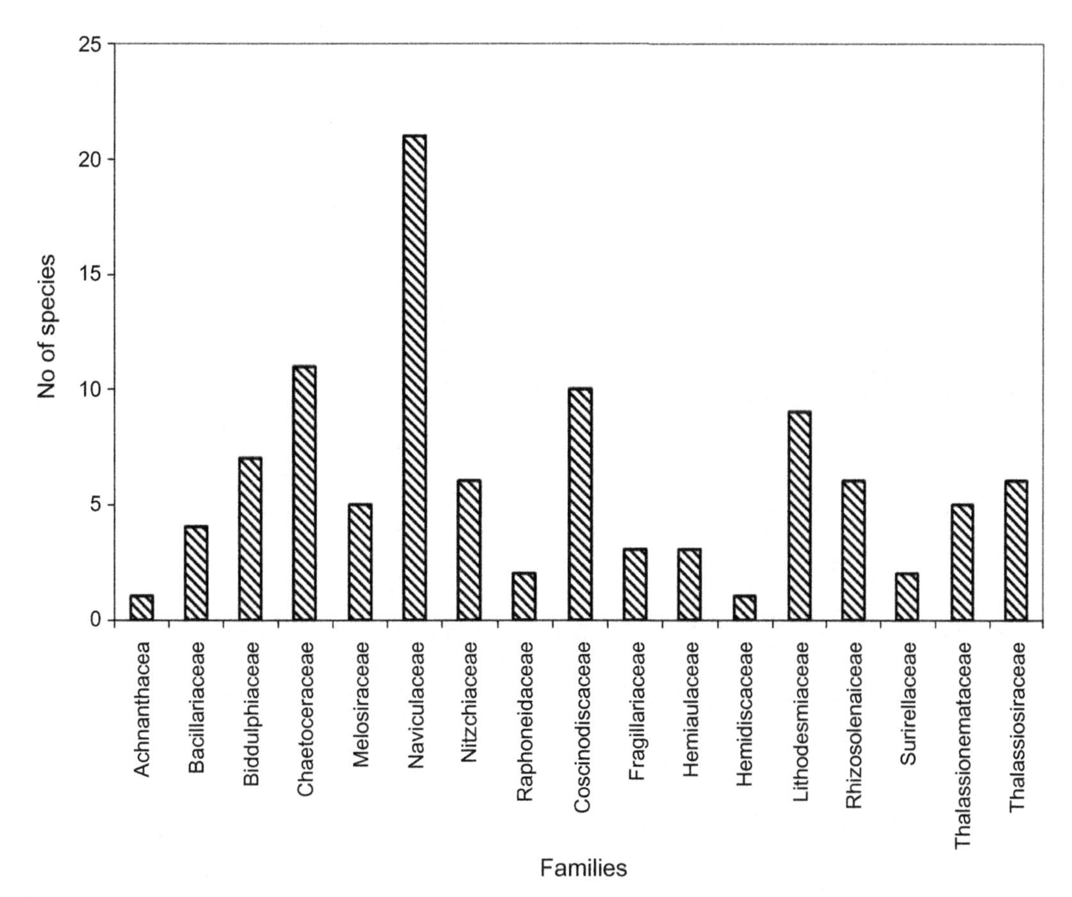

FIG. 1 Species diversity of diatoms along the east coast.

Bacillariaceae, Biddulphiaceae, Chaetoceracae, Naviculaceae, Thalassiosiraceae, Thalassionemataceae, and Rhizosoleniaceae are the most cosmopolitan in distribution. Of the few groups of marine organisms, planktonic algae appear to have been more completely catalogued (Venkataraman, 1939; Sournia et al., 1991). Their compilation suggests that the number of pennate diatoms in the world's oceans could range from 500 to 784 and that of centric diatoms from 865 to 999. Not more than 25% of these diatom species have been recorded from Indian waters.

DINOFLAGELLATES

The dinoflagellate species diversity in the east coast estuaries is relatively low (15 species in seven families) compared with the west coast estuaries (76 species from 10 families). The family Dinophyceae is dominant with 18 species, followed by Peridiniaceae and Ceratiaceae, with 13 and 10 species, respectively. Unlike the diatoms, the estimated number of dinoflagellate species in the marine environment varies from 1000 to 2000. Compared with these numbers,

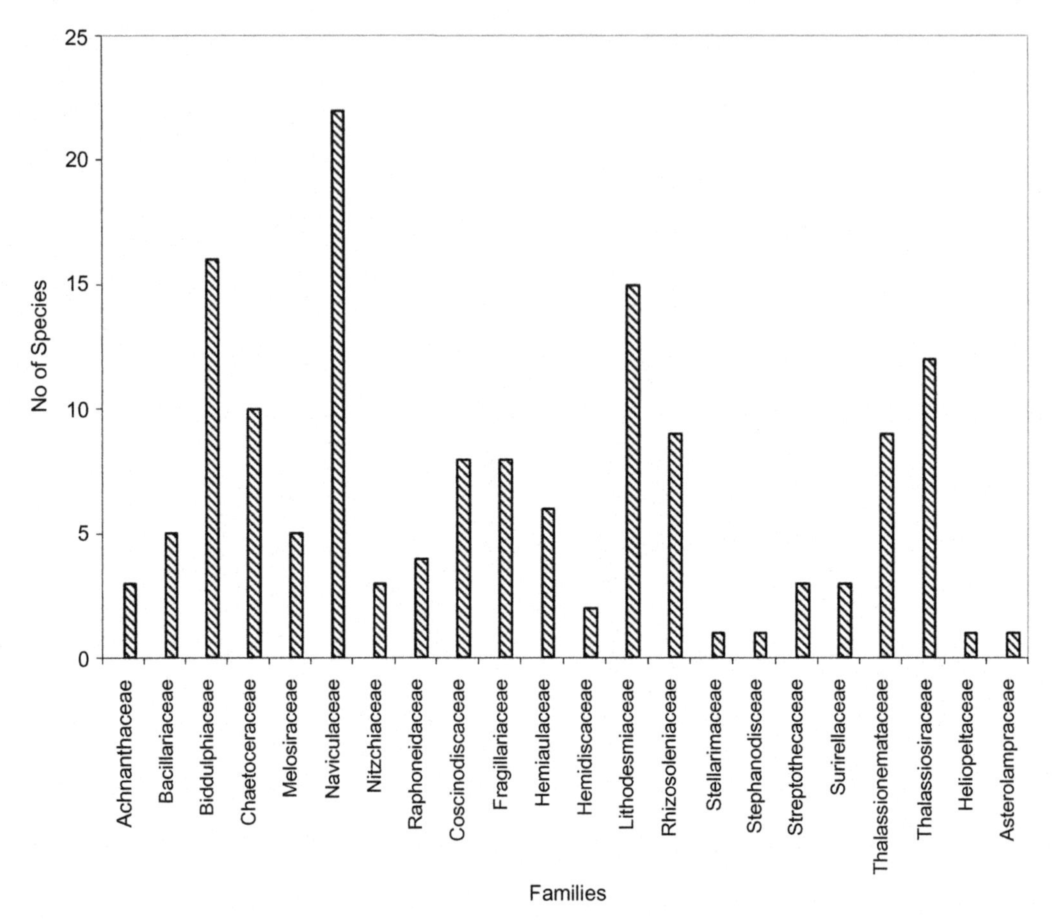

FIG. 2 Species diversity of diatoms along the west coast.

the current inventory of dinoflagellates in the Indian waters appears to be too small. A total of 90 species of diiflagellates are reported from Indian waters.

1.2 Seagrasses

The global composition of seagrass ranges from 0.1% to 0.2% of the aquatic flora. Seagrasses provide a habitat for *Dugong dugon*, a wide variety of marine organisms both plant and animal; these include meiofauna and flora, benthic flora and fauna, epiphytic organisms, plankton, and fish. Seagrass meadows account for 15% of the ocean's total carbon storage. The ocean currently absorbs 25% of global carbon emissions. It is estimated that seagrasses are capable of binding about 1000 g of carbon per square meter each year. Sixty species of seagrass have been described from the world's oceans with 12 genera in five families; the greatest diversity is found in tropical regions (Spalding et al., 2003; Short et al., 2007). Fourteen species of seagrass under six genera are known from Indian seas (Table 1). They are often found in

TABLE 1 Distribution of Seagrasses in Five Different Regions of the Indian Coast

Sl. No.	Species	Coromandel/ East coast	Palk Bay and Gulf of Mannar	West Coast	Andaman and Nicobar Islands	Lakshadweep Islands
1	*Enhalus acoroide* (L.f.) Royle	–	+	–	+	+
2	*Halophilia beccarii* Asch.	+	+	+	–	–
3	*Halophilia decipiens* Osten.f.	–	+	–	–	–
4	*Halophilia ovalis* (R.Br.) Hook.f. subsp. *ovalis*	+	+	+	+	–
5	*Halophilia ovalis* Hook.f. subsp. *ramamurthiana*	+	–	–	–	–
6	*Halophilia ovata* Gaud	+	+	+	+	+
7	*Halophilia stipulacea* (Forsk.) Asch.	–	+	–	–	–
8	*Thalassia hemprichii* (Ehrenb.) Asch.	–	+	–	+	+
9	*Cymodocea rotundata* Ehren. and Hempr. Ex Asch	–	+	+	+	+
10	*Cymodocea serrulata* (R.Br.) Asch. and Magnus	–	+	+	+	+
11	*Holodule pinifolia* (Miki) Hartog	+	+	+	+	–
12	*Halodule uninervis* (Forsk.) Asch.	+	+	+	+	+
13	*Holodule wrightii* Asch.	+	+	–	–	–
14	*Syringodium isoetifolium* (Asch.) Dandy	–	+	–	+	+
	Total	7	13	7	9	7

Courtesy: Status of sea grasses of India; Kannan, L., Thangaradjou, T., Anantharaman, P., 1999. Seeweed Res. Utiln. 21 (1&2), 25–33.

association with coral reef areas. Eleven species are known from the Palk Bay and 13 species occur in the Gulf of Mannar Biosphere Reserve. *Thalassia* and *Syringodium* are dominant in coral reef areas and coral rubble whereas the others are distributed in muddy and fine sandy soils. Along the west coast, only the *Halophila* and *Halodule* species are cosmopolitan in distribution; *Cymodocea* spp. and *Syringodium isoetifolium* occur as very small patches at the southernmost end of Thiruvananthapuram. Nine species, among which *Thalassia hemprichiii* and *Cymodocea rotundata* are dominant, occur in the Andaman and Nicobar Islands. Seven species are known from the Lakshadweep Islands, with *Thalassia hemprichi* being dominant.

1.3 Mangroves

Mangroves occur in 112 countries, primarily in the tropical regions of the world, over an area of 1,89,399 km^2. One hundred species of true mangroves have been described so far around the world. The Indian mangroves cover about 4827 km^2, with about 57% of it along the east coast, 23% along the west coast, and the remaining 20% in the Andaman and Nicobar Islands. The mangrove formations are of three types: deltaic, backwater-estuarine, and insular. The deltaic mangroves occur mainly along the east coast, the backwater-estuarine type along the west coast, and the insular type in the Andaman and Nicobar Islands. A total of 69 mangrove species belonging to 42 genera and 22 families are known from India (Table 2). While several of them are cosmopolitan in distribution, five—*Aegialitis rotundifolia, Heritiera fomes, H. kanikensis, Rhizophora annamalayana,* and *R. stylosa*—are restricted to the east coast and one, *Lumnitzera littorea,* is present only in the Andaman and Nicobar Islands (Kathiresan, 1999). Mangroves serve as abodes for several faunal communities and in India, mangroves harbor 2359 faunal species (Table 3). Twenty species of mangrove fall under the rare, endemic, and restricted categories (Table 4).

1.4 Protozoa

The known number of protozoan species from Indian seas is 2577, equivalent to about 8% of the total world protozoan fauna. Among them, 52% are free-living and the remaining are parasitic species. Out of seven protozoan phyla, only one, Labyrinthomorpha, has not yet been reported from India.

TABLE 2 Mangrove Diversity in India

Sl. No.	State	Area (km^2)	No. of Species
1	Andhra Pradesh	353	31
2	Goa	17	Not known
3	Gujarat and Daman and Diu	1047	12
4	Karnataka	3	29
5	Kerala	5	27
6	Maharashtra	186	26
7	Orissa	221	60
8	Tamil Nadu and Pondicherry	39	24
9	West Bengal	2152	57
10	Andaman and Nicobar Islands	615	44

Courtesy: Bhatt, J.R., Ramakrishna, Sanjappa, M., Remadevi, O.K., Nilaratna, B.P., Venkataraman, K., 2013. Mangroves of India: Their Biology and Uses. Zoological Survey of India, pp. 640.

TABLE 3 Mangrove-Associated Fauna in India

Sl. No.	Faunal Element	No. of Species
1	Prawns	55
2	Crabs	138
3	Mollusks	308
4	Insects	711
5	Fish	546
6	Amphibians	13
7	Reptiles	85
8	Birds	433
9	Mammals	70

TABLE 4 Rare, Endemic, and Restricted Mangroves in India

Sl. No.	Species	Category
1	*Acanthus ebracteatus*	Restricted to Andaman and Nicobar Islands
2	*Aegialitis rotundifolia*	Confined to West Bengal, Orissa, and Andhra Pradesh
3	*Aglaia cuculata*	Restricted to West Bengal and Orissa
4	*Brownlowia tersa*	Restricted to West Bengal, Orissa, and Andhra Pradesh
5	*Heritiera fomes*	Restricted to West Bengal and Orissa
6	*Heritiera kanikensis*	Endemic to Bhitarkanika
7	*Lumnitzera littorea*	Restricted to Andaman and Nicobar Islands
8	*Merope angulata*	Confined to West Bengal and Orissa
9	*Nypa fruticans*	Restricted to West Bengal and Andaman and Nicobar Islands
10	*Phoenix paludosa*	Restricted to Andaman and Nicobar Islands and Andhra Pradesh
11	*Rhizophora annamalayana*	Endemic to Pitchavaram
12	*Rhizophora stylosa*	Confined to Orissa
13	*Scyphiphora hydrophyllacea*	Restricted to Andaman and Nicobar Islands and Andhra Pradesh
14	*Sonneratia apetala*	Rare in several areas
15	*Sonneratia griffithii*	Restricted to West Bengal, Orissa, and Andaman and Nicobar Islands
16	*Tylophora tenuis*	Restricted to West Bengal and Orissa
17	*Urochondra setulosa*	Endemic to Gujarat
18	*Thespesia populacea*	Restricted to West Bengal and Orissa
19	*Xylocarpus makongensis*	Restricted to West Bengal, Orissa, and Andaman and Nicobar Islands
20	*Xylocarpus mollucensis*	Restricted to Andaman and Nicobar Islands

1.4.1 *Foraminifera*

Foraminiferans are eukaryotic unicellular organisms with the general characteristics of protists. Their exoskeleton is commonly made of calcium carbonate while some species have agglutinated shells made up of sediments or shells of dead organisms. Their small size, sensitivity to small changes in the environment, and ability to preserve these changes in their hard part make them very useful in palaeoclimatic reconstruction and environmental monitoring (Rana et al., 2007). It has been estimated that the total number of extant foraminiferan species is approximately 10,000. Foraminifera are among the relatively well-studied groups, with the earliest descriptions of new species dating back to the 18th century (Von Fichtel and Von Moll, 1798). The major part of the work on this group has been done along the east coast of India by several scientists: Sarojini (1958), Bhatia and Bhalla (1959) Subba Rao and Vedantam (1968), Bhalla (1970), and Gnanamuthu (1943). Comparatively less work has been done on the west coast of India and the Arabian Sea. Mention may be made of the work of Antony (1968), Siebold (1974), Chaudhury and Biswas (1954), and Rao (1970, 1971). With the compilation of all the available literature, a total of around 329 foraminiferan species were recorded from Indian waters.

1.5 Porifera

The phylum Porifera represents the most primitive under the metazoan group (Müller, 1995), and they are commonly known as sponges. They are sessile and attached to a substrate during the living state. They are the only multicellular animals with a cellular grade of organization (Bergquist, 1978) and do not form any tissues. They show an amazing variety of colors, shapes, and sizes from flat cushions to elaborate branching or cup-shaped forms, from tiny crusts measured in mm to giant shapes in meters, but essentially they share a simple body plan. Numerous tiny pores, or ostia, on the surface allow water to enter and circulate through a series of canals where plankton and organic particles are filtered out and eaten. This network of canals and a relatively flexible skeletal framework give most sponges a characteristic spongy texture. The sponge fauna of the Indian region was initially documented by Carter (1887). Of the approximately 15,000 sponge species, most occur in marine environments while aonly about 1% of the species inhabits freshwater. The sponges under the class Calacrea are an evolutionarily as well as economically important group. Calcarean sponges first appeared during the Cambrian and their diversity was greatest during the Cretaceous period. Some of the calcareous sponges (Pharetronids) have contributed significantly to reef-building throughout different periods of the earth's history. So far, only 400 of those are Calcarean sponges identified from world's oceanss. In Indian waters, only 10 species of Calcarean sponges were reported and all the 10 species of calcarean sponges are protected and listed under the Schedule III category of the Wildlife (Protection) Act, 1972. India represents a total of 476 species of Porifera from the marine habitats.

1.6 Cnidaria

The global estimates of cnidarian diversity range between 9000 and 12,000 species. In India, 178 species of Hydrozoa, 30 species of Scyphozoa, five species of Cubozoa, and about 1117 species of Anthozoa have been reported so far. Because not all groups of cnidarians have received adequate attention from Indian taxonomists, the above figures cannot be taken as final.

Annandale (1907a,b, 1915, 1916), Leloup (1934), and Menon (1931a,b) carried out pioneering work. Comprehensive accounts are available only for siphonophores (Daniel, 1985), scyphomedusae (Chakrapany, 1984), and scleratinian corals (Pillai 1991).

1.6.1 *Siphonophora*

Siphonophora comes under the hydrozoan class, is abundant in the Indian seas, and constitutes an important part of the marine plankton. The siphonophores from the Indian Ocean have been studied by several people—Leloup (1934) and Daniel and Daniel (1963); Patriti (1970); Totton (1954); Alvarino (1974); Rengarajan (1974); and Daniel (1966, 1974a). A comprehensive accounting of the siphonophora shows that there are 116 valid species, of which one variety and three doubtful species are known from the Indian Ocean; 89 of these occur in the Indian seas. A total of 178 species of hydrozoan are reported from Indian waters.

1.6.2 *Anthozoa: Scleractinia & Octocorals*

Studies on the taxonomy of Indian coral reefs were carried out in India as early as 1847 by Rink in the Nicobar Islands and later, in 1888, by Thurston in the Gulf of Mannar region. Brook (1893) recognized eight species of *Acropora* from Rameswaram, of which three species were described as new species. Subsequent contributions to the inventory of coral species were made by Alcock (1893, 1898), Gardiner (1903–1906), Matthai (1924), Gravely (1927a,b,c), and Sewell (1935). India represents a total of 627 species of scleractinian corals under 99 genera and 19 families against the reported 1498 species of the world. Among the four major reef areas of India, the Andaman and Nicobar Islands are the richest in coral species diversity whereas the Gulf of Kachchh is the poorest. The Lakshadweep Islands have a larger number of species compared with the Gulf of Mannar. Among the deep water (ahermatypic) corals, so far 720 species belonging to 110 genera and 12 families have been reported from around the world, of which 227 species belonging to 71 genera and 12 families have been reported from the Indian Ocean region (Cairns and Kitahara, 2012; Venkataraman and Satynarayana, 2012). However, the effort made to inventory the deep-water corals has been meager, and as a result, only 44 species are known from Indian seas.

Octcorals are a major subclass of marine cnidaria of the class Anthozoa, including the orders Alcyonacea (soft corals and gorgonians), Pennatulacea (sea pens and sea pansies), and Helioporacea (blue corals). A total of about 415 species of octocorals have been reported from the Indian seas. Venkataraman et al. (2004) reported 27 species of gorgonians belonging to eight families and 19 genera from India. Of these, 12 species of gorgonians belonging to four families and nine genera have been reported from the northeast coast of India (Thomas et al., 1995). Recently, Yogesh Kumar et al. (2014) reported 51 species belonging to 25 genera, eight families, and three suborders. Of these, 44 species belonging to 24 genera and seven families are new to India.

1.6.3 *Cubozoa*

Cubozoans are box jellyfish distinguished by their cube-shaped medusae. They resemble basic jellyfish but they can swim quite rapidly, maneuver with great agility, and have good vision despite not having a brain. Some species have tentacles that can reach up to 3 m in length. *Chironex fleckeri*, *Carukia barnesi*, and *Malo kingi* are the most venomous creatures in the world. Although the venomous species of box jellyfish are almost entirely restricted to the tropical Indo-Pacific, various species of box jellyfish can be found widely in tropical and subtropical oceans, including the Atlantic and east Pacific, with species as far north as California and the Mediterranean. A few other fossils that may be cubozoans have been found in the Jurassic Solnhofen Limestone of Bavaria, Germany (www.ucmp.berkeley.edu). In the world's oceans, 45 species of cubozoans have been recorded (WoRMS, 2016), although only five species have been reported from Indian waters (Ramakrishna and Sarkar, 2003; Haldar and Choudhury, 1995).

1.6.4 Scyphozoa

The Scyphozoans are members of the phylum Cnidaria with large solitary marine invertebrates; they are known as true jellyfish. The term Scyphozoa was first used by Lankester in 1881. Jellyfish inhabit every major oceanic region of the world and are capable of withstanding a wide range of salinities and temperatures. Most jellyfish live in shallow coastal waters, but a few species inhabit depths of 4600m. The inverted bell-shaped morphology of many of the species that make up this class is referred to as "medusoid." Most scyphozoan species are capable of swimming by use of the muscular mesoglea. The adult jellyfish drifts in the water with limited control over its horizontal movement. A total of 189 species of scyphozoans belonging to three orders, 18 families, and 61 genera were recorded from the world's oceans (WoRMS, 2016); India contributes only 30 species (Kramp, 1961; Ramakrishna and Sarkar, 2003; Haldar and Choudhury, 1995).

1.7 Ctenophora

Ctenophores are variously known as comb jellies, sea gooseberries, sea walnuts, or Venus's girdles. They are biradially symmetrical, acoelomate organisms that resemble cnidarians. They are largely planktonic, exclusively marine animals found throughout the world's oceans. Ctenophores are distinguished from all other animals by having colloblasts that capture prey by squirting glue on them, although a few ctenophore species lack these. These animals have three main cell layers and no intermediate jelly-like layer. Many ctenophores, like various other planktonic organisms, are bioluminescent. Two species of fossil ctenophores have now been found from the Late Devonian period in the famous Hunsrückscheifer slates of southern Germany (Stanley and Stürmer, 1983, 1989). A total of 197 species of ctenophores have been recorded from the world's oceans (Mills, 2014; WoRMS, 2016) while India represents only 19 species (Ramakrishna et al., 2003; Prasade et al., 2015).

1.8 Platyhelminthes

The flatworms come under Phylum Platyhelminthes and comprise a very diverse group of worms, with 12,821 species described (WoRMS, 2016). The parasitic flatworms, such as tapeworms and liver flukes, are included within this large phylum. They are bilaterally symmetrical, triploblastic, they lack an anus, and they have no body cavity other than the gut. Platyhelminthes are divided into three classes: Turbellaria, a free-living marine species; Monogenea, ectoparasites of fish; Trematoda, internal parasites of humans; and other species. The most common marine flatworms belong to a different group called polyclads (or Polycladida). Triclads and polyclads are all free-living (i.e., they are not parasites).

Polyclads are found in most marine habitats, usually on the sea floor among algae, corals, or rocky reefs. Platyhelminths have practically no fossil record. A few trace fossils have been reported that were probably made by platyhelminths. Newman and Cananon (2003) reported 1500 species of polyclads in the world. Recently, studies conducted by the Zoological Survey of India have found 38 species of polyclad from the Andaman and Nicobar Islands (Venkataraman et al., 2015a). A total of 46 species of polyclads have been reported in India.

The subclass Digenea (Class Trematoda) is associated with marine fish of the Indian region. Madhavi (2011) has enumerated 662 species of digenean trematodes belonging to 200 genera and 32 families reported from Indian marine fish. In addition, a total of 122 species of class Monogenea and two species of cestoda are reported from India.

1.9 Nemertea

The phylum Nemertea is also known as "ribbon worms" or "proboscis worms." They are typically marine, benthic animals and are found burrowed in mud, sand, or other sediments, among rocks, or associated with algae or other plant masses. The smallest are a few millimeters long; more are less than 20 cm, and several exceed 1 m. The longest animal ever found, at 54 m long, may be a specimen of *Lineus longissimus* (Ruppert et al., 2004). A few live in the open ocean while the rest find or make hiding places on the bottom. Most nemerteans are carnivores, feeding on annelids, clams, and crustaceans. A few species are scavengers and some live commensally inside the mantle cavity of mollusks. The Middle Cambrian fossil Amiskwia from the Burgess Shale has been classed as a nemertean based on a resemblance to some unusual deep-sea swimming nemerteans (Schram, 1973). Traditional taxonomy divides the phylum in two classes, Anopla and Enopla, with two orders each. A total of 1362 species have been described and grouped into 285 genera (Kajihara et al., 2008; WoRMS, 2016). India harbors only sixspecies of Nemertea from the entire coastal and marine habitat (Shrinivaasu et al., 2011, 2015a; Shynu et al., 2015).

1.10 Gastrotricha

The gastrotrichs, commonly referred to as hairybacks, are a group of microscopic (0.06–3.0 mm), worm-like, pseudocoelomate animals that are characterized by a meiobenthic life style. In marine habitats they are mainly interstitial whereas in fresh waters, they are a ubiquitous component of periphyton and benthos. In marine sediments, gastrotrich density may reach 364 individuals/10 cm^2; typically they rank third in abundance following Nematoda and Copepoda (Harpacticoid), although in several instances they have been found to be the first or the second most abundant meiofaunal taxon. The phylum is cosmopolitan with about 765 species grouped into two orders: Macrodasyida, with 310 strap-shaped species, all but two of which are marine or estuarine; and Chaetonotida with 455 tenpin-shaped species, three-fourths of which are freshwater. Recently, a total of 497 species of gastrotricha were reported from the world's oceans (WoRMS, 2016). A total of 61 species (45 species of Macrodasyida and 16 species of Chaetontida) are reported from Indian waters (Rao, 1991, 1993; Priyalakshmi et al., 2007; Priyalakshmi and Menon, 2012).

1.11 Rotifera

The rotifers are microscopic aquatic animals commonly known as wheel animals. They are pseudocoelomate animals. They were first described by Rev. John Harris in 1696. They may be sessile or sedentary and some species are colonial. They are found in marine, brackish, and fresh waters throughout the world, excluding the Antarctic. Most rotifers are around 0.1–0.5 mm long, although their size can range from 50 μm to more than 2 mm (Howey, 1999). Rotifers eat particulate organic detritus, dead bacteria, algae, and protozoans. They eat particles up to 10 μm in size. A total of about 172 species of rotifers have been described from the world (WoRMS, 2016) while India contributes 47 species (George et al., 2011; Anandakumar and Thajuddin, 2013).

1.12 Nematoda

The individuals under the phylum Nematoda are called nematodes or roundworms. They are bilaterally symmetrical and worm-like organisms. Near the body wall but under the epidermal cells are muscle cells; they run in the longitudinal direction only by the process of contraction. Recently, it has been claimed that nematodes are one of the three major radiations that have produced most of the world's multicellular species (May, 1988; Gaston, 1991). Lambshead (1993) estimated that there may be as many as 1×108 nematode species in the deep sea, but the number of described species of nematodes is only about 20,000, of which 6833 species are free-living marine organisms (Gerlach, 1980; WoRMS, 2016). Six species and one genus of free-living marine nematodes belonging to two orders and seven families were recorded from Indian waters of the Pichavaram mangrove of India's Southeast Coast (Chinnadurai and Fernando, 2007). A total of 356 species of marine nematodes are reported from Indian waters (Annapurna et al., 2012; Sivaleela and Venkataraman, 2015; Dutta et al., 2016).

1.13 Mollusca

The number of molluscan species recorded from various parts of the world is between 80,000 and 150,000. The approximate number of marine mollusk species is 46,315 including opisthobranch. Studies on Indian mollusks were initiated by the Asiatic Society of Bengal (1784) and the Indian Museum, Kolkata (1814). Benson, in 1830, was the first author to publish a scientific paper on the Mollusca. The beginning of the 20th century was the most productive and significant period in the history of Indian malacology. In India, 5100 species of Mollusca have been recorded from freshwater (22 families, 53 genera, and 183 species), land (26 families, 140 genera, and 1487 species) and marine (242 families, 591 genera, and 3400 species) habitats (Subba Rao, 1991, 1998; Subba Rao and Dey, 2000; ZSI, 2011). The Andaman and Nicobar Islands are one such area, having more than 1000 species from the marine region (Subba Rao and Dey, 2000). The Gulf of Mannar and Lakshadweep have 428 and 424 species, respectively (Venkataraman et al., 2004). Eight species of oyster, two species of mussel, 17 species of clam, six species of pearl oyster, four species of giant clam, one species of window-pane oyster, other gastropods such as the Sacred Chank, *Trochus*, and *Turbo*, and 15 species of cephalopod are exploited from the Indian marine region. On the basis of studies from the

Zoological Survey of India, it was revealed that 2000 species of mollusks (except opisthobranchs) have been reported from Indian marine environments.

1.13.1 *Opisthobranchia*

The opisthobranchs, or sea slugs, are shell-less mollusks. About 3736 species of opisthobranch have been reported from around the world. The earlier works on opisthobranchs were initiated during the 1880s (Alder and Hancock, 1864). The first report on opisthobranchs was made by Eliot (1906) and it deals with 42 species belonging to 10 families. A total of 260 opisthobranch species have been reported from Indian waters, with the greatest number, 200 species, having been recorded from the Andaman and Nicobar Islands (Ramakrishna et al., 2010). However, recent reports from the Zoological Survey of India indicated 388 species from Indian waters.

1.14 Annelida

1.14.1 *Polychaeta*

In the phylum Annelida, the Polychaeta have received considerable attention since 1909. Polychaetes form an important component in the marine food chain, especially for demersal fish. The number of polychaetes worldwide is estimated to be 11,513 species excluding Echuira. Surveys of this group actually started with Southern's (1921) work, "Polychaeta of Chilka Lake," and continued with the study of the littoral fauna of Krusadai Island in the Gulf of Mannar by Gravely (1927a) and by Fauvel (1930) (119 species under 22 families). A perusal of the literature shows that most of the records pertaining to this group are either from the Madras coast or the Gulf of Mannar (Aiyar, 1931; Aiyar and Alikunhi, 1940; Gravely, 1942; Alikunhi, 1946, 1947; Ganapati and Radhakrishna, 1958; Ghosh, 1963; Krishnamoorthi, 1963). The Central Marine Fisheries Research Institute has listed 200 species under 46 families in their catalogue of types and reference collections. From the collections of the Zoological Survey of India and the Indian Museum, Fauvel (1932) described 300 species under 30 families and in his later monograph (Fauvel 1953), he raised the total to 450 species. Hartman (1974), dealing with polychaetes of the Indian Ocean, recorded 244 species, of which 116 are considered new to the region. The catalogue of the polychaetous annelids from India lists 522 species.

1.14.2 *Archiannelida*

Pioneering studies on the archiannelids of India were made by Aiyar and Alikunhi (1944) and Alikunhi (1946, 1948a,b) along the Madras coast, from which two species of *Polygordius*, to species of *Protodrilus*, and four species of *Saccocirrus* were described as new to science. Rao and Ganapati (1968) recorded 15 species of archiannelids from the beach sands along the Waltair Coast. Thus, compared to the vast stretch of Indian coast, the investigations hitherto carried out on Archiannelida are quite limited and any further intensive surveys of the fauna in other areas are quite likely to yield interesting results. The world records of Archiannelida hitherto fell under five families, 18 genera, and more than 90 species, of which about 20 species are reported from Indian coasts.

1.14.3 *Echiura*

Commonly recognized as the "spoon worm" are the inhabitants of the shores of polar, temperate, and tropical seas. The first echiurans described appear to be Thalassema (= Lumbricus) thalassemum (Pallas, 1766). Echiurans are specialized for the formation of U-shaped burrows. A total of 197 species have been reported from the world's oceans (WoRMS, 2016). In India, 48 species have been reported from the Gulf of Kachchh, the Gulf of Cambay, the Gulf of Mannar, Lakshadweep, the Andaman and Nicobar Islands, Kerala, West Bengal, and the Odisha coasts. Twelve species of Ehiurans are endemic to India (Maiti and Maiti, 2011).

1.14.4 *Oligochaeta*

The marine oligochaete fauna of India is poorly known, and most of the species have been recorded from the littoral zones of small freshwater bodies such as ponds, tanks, pools, and ditches from all over the country. The Enchytraeidae (pot worms) occur in terrestrial, littoral, and marine habitats, being abundant in acidic soils with high organic content. Only 3% of the enchytraeid species of the world have so far been reported from this region, mainly from Orissa.

1.15 Arthropoda (Crustacea)

The global crustacean species diversity is estimated to be 150,000, of which 40,000 species have been described so far. Of the 2818 species of Crustacea and 38 species of Chelicerata that have been reported from India, marine species (94.85%) contribute the most. In India, as many as 93 species of stomatopods (four families and 26 genera), 38 species of lobsters (four families, 11 genera), 120 species of hermit crabs (three families, 40 genera), 705 species of brachyuran crabs (28 families, 270 genera), and 439 species of shrimps and prawns (Penaeidea and Caridea) under seven families, 19 genera and 159 species of the Caridea (15 families, 56 genera) have been recorded. Apart from these, 541 species of copepods, 74 species of thoracica, and 125 species of ostracods have also been recorded.

1.15.1 *Copepoda*

Copepods are the most widely studied group among the marine zooplankton. There are approximately 210 described families, 2280 genera, and more than 11,246 species from around the world. Important contributions to the systematics of copepods from Indian waters are those of Sewell (1929), Krishnaswamy (1950, 1952), Krishnaswamy (1953), and Pillai (1967). Very few papers dealing with the marine Harpacticoida of India and the neighboring seas have been published so far (Sewell, 1924, 1940). Studies on sand-dwelling forms are even fewer: those of Krishnaswamy (1953, 1956a,b, 1957) provide an account of 17 sand-dwelling harpacticoids under five families. A total of 106 species belonging to 23 families are known from the east coast estuaries (Fig. 3). The calanoids are dominant among them, distributed in 16 families, followed by the harpacticoids (five families) and cyclopoids (two families). The diversity in the west coast estuaries is relatively high, with 179 species in 31 families (Fig. 4). The calanoids are dominant with 20 families. Though the numbers of families of harpacticoids and cyclopoids are the same (six families), the latter are more diverse, with 22 species, compared with seven species of harpacticoids. The compilation of the available literature and the recent studies of the Zoological Survey of India reveal a total of 541 species of copepods from Indian waters.

1.15.2 *Ostracoda*

The Ostracoda are one of the crustacean groups and are generally small, ranging in length from 0.1 to 32 mm; they are commonly known as seed shrimp. Ecologically, marine ostracods can be a part of zooplankton or part of the benthos living on or inside the upper layer of the sea floor. They contain calcified valves and became the main attribute for the fossil records, which includes the last 425 million years. A total of 33,000 species have been described, of which 5930 are accepted species (Brandão, 2016). A total of 125 species of ostracods are reported in Indian waters, comprising 65 species from the east coast, 39 species from the west coast, 22 species from the Andaman and Nicobar Islands, and one species each from Lakshadweep and Gulf of Mannar (George and Nair, 1980; Rao, 1989; Jain, 1981; Sivaleela and Venkataraman, 2009; NIO, 2016).

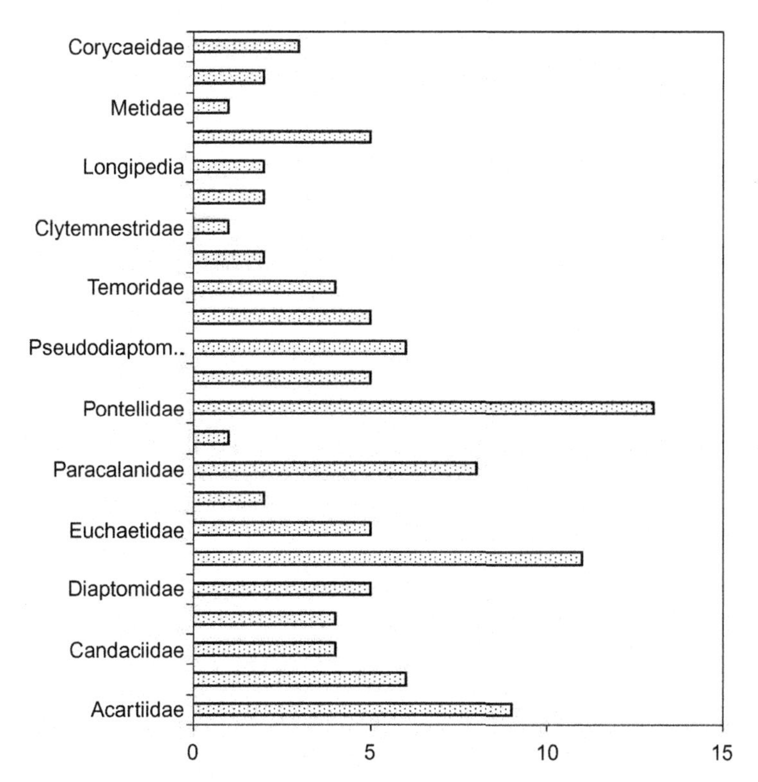

FIG. 3 Diversity and distribution of copepod families along the east coast of India.

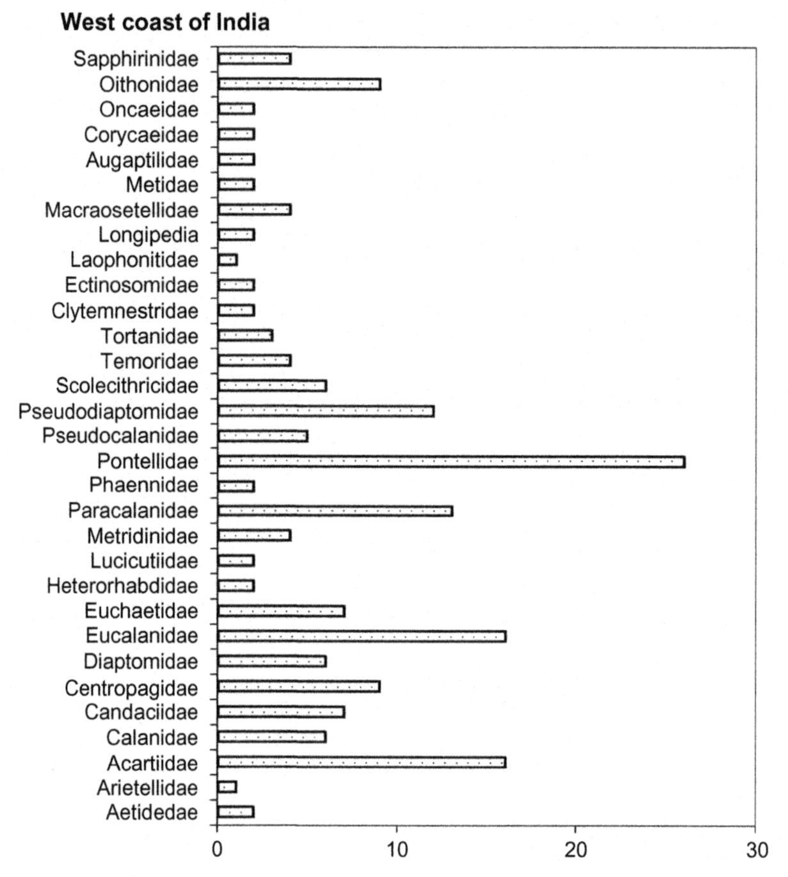

FIG. 4 Diversity and distribution of copepod families along the west coast of India.

1.15.3 *Branchiura*

Our knowledge of this group in the Indian region is rather scanty. It was not until 1951, when Ramakrishna contributed to our knowledge of the Indian species of argulids that are parasitic on fish, that the group received adequate attention. Ramakrishna described five species of the genus *Argulus*, of which three were new to science.

1.15.4 *Thoracica*

Our information on the diversity of the Thoracica under the infra class Cirripedia of the Indian coast is far from complete, with only 36 species having been recorded so far. Even this record (Weltner, 1894; Borradaile, 1903a,b; Annandale, 1907a,b, 1913, 1914; Nilsson Cantell, 1938; Panikkar and Aiyar, 1937; Daniel, 1971, 1974b) is rather sketchy and provides low geographical coverage. A total of 74 species of Thoracica are recorded from Indian waters.

1.15.5 *Malacostraca: Mysidacea*

Mysids are shrimp-like animals with a shield-shaped carapace in which the first three segments are fused. The total number of species worldwide is known to be 1139 under 120 genera (WoRMS, 2016). Mysidaceans, with a total number of about 84 species, are known so far only from the works of Tattersall (1922) and Pillai (1964, 1968, 1973).

1.15.6 *Cumacea*

A total of 1631 species of marine cumaceans have been reported (WoRMS, 2016). Cumacean species are also little known, except for the studies of Calman (1904), Kemp (1916), and Kurian (1954, 1965). Before these studies, 23 species of the Bodotriidae, three species of the Disstylidae, four species of the Nannastaeidae, and one species of the Camylaspididae were known from the Indian region. A recent study reveals that a total of 88 cumacean species under eight genera and three families are found in the Indian seas (Chatterjee and Pesic, 2010; Petrescu and Chatterjee, 2011).

1.15.7 *Tanaidacea*

Tanaidacea is an order of crustaceans thar includes benthic macroinvertebrate species. A total of 1,371 marine tanaidacean species are described worldwide (WoRMS, 2016). The order Tanaidacea comprises four suborders: Anthracocaridomorpha, Apseudomorpha, Neotanaidomorpha, and Tanaidomorpha. The first is represented by a fossil species, but the others are represented by a total of 1052 species (Apseudomorpha, 457 species under 93 genera; Neotanaidomorpha, 45 species under four genera; and Tanaidomorpha, 550 species under 120 genera) (Blazewicz-Paszkowycz et al., 2012). Our knowledge of Tanaidacea is rather poor from the Indian region. Chilton (1923) contributed a paper dealing with a species belonging to this group from Chilka Lake.

1.15.8 *Isopoda*

Very little is known about the marine isopods compared with the terrestrial isopods of India. In 1923, Chopra contributed a monumental monograph on the bopyrid isopods of Indian Macrura, wherein 33 species pertaining to 13 genera were described from the Andaman Islands, Ganges delta, Madras, and other areas. Chopra (1923) contributed another paper on the bopyrid isopods of Indian Macrura. The collection included 12 species pertaining to seven genera, collected mostly from the Andaman and Nicobar Islands, Ganges delta, Gulf of Mannar, and Bombay. The contributions on marine woodborers made from 1963 to 1968 by various Indian authors indicate that there are six species of the genus *Sphaeeroma* and nine species of *Limnoria* in Indian waters. A total of 300 species of isopod are reported from Indian waters.

1.15.9 *Amphipoda*

Studies on the amphipods of the Indian and neighboring waters were carried out by the zoologist Giles (1885, 1887) from Bengal. His subsequent works raised the number to 27. Bernard (1935) reported amphipods from collections made from the Travancore, Cochin, and Bengal coasts. In the last half of the previous century, Sivaprakasam (1968a,b, 1970a,b), in a series of contributions, enriched our knowledge about the amphipods of the east coast of

India and listed 61 species. Surya Rao (1974) provided a detailed account of the intertidal gammarid amphipods from the Indian coast and listed 132 species in 54 genera; it has since been upated to 167 species.

1.15.10 *Euphausiacea*

The earliest accounts of Indian euphasids are provided by Wood Mason and Alcock (1891) and Alcock and Anderson (1893, 1894). Tattersall (1911) gave an account from the Indian Ocean. Twenty-three species of euphasid have been recorded so far from the Laccadive and Maldive islands and two species have been recorded from adjoining regions along the southwest coast of India.

1.15.11 *Stomatopoda*

Kemp (1913) published a monograph on Indo-Pacific stomatopods, covering the 139 species and varieties known then. Ghosh (1975, 1976) and Tiwari and Ghosh (1975) contributed a series of papers highlighting the present knowledge of the Stomatopoda in Indian waters. Our information about the Stomatopoda of India is, however, far from complete as only 93 species have been reported so far.

1.15.12 *Decapoda: Macrura*

The Decapoda as a whole have received good attention from scientific workers compared with other groups. The earliest to contribute was De Man (1908), who did a collection from the brackish water ponds of Lower Bengal. The contributions of Kemp on the decapod crustaceans of the Indian Museum were published in 24 parts in the *Records of the Indian Museum.* Alcock (1901, 1906) contributed a comprehensive catalogue of the penaeid prawns of India. Although large numbers of species of prawns and lobsters are known to occur along the Indian coast, work on these species is very limited. Over 17 families, 67 genera, and 439 species have been recorded as commercially important from around the world. A total of 55 species of commercial shrimps and prawns have been recorded in India while a total of 110 species are available in Indian waters. The east coast of India contributes about 24.5% to the country's production, and the west coast contributes 75.3%.

1.15.13 *Brachyura*

The earliest works on the crabs of the Indian seas are those of Milne-Edwards (1834–1837), Henderson (1893), and De Man (1887). The first comprehensive study of the crabs of the west coast was that of Borradaile (1900, 1902a,b), Borradaile (1903a,b). Kemp (1915) dealt with 38 species under six families collected from Chilka Lake and in 1923 dealt with crabs collected from the mouth of the Hooghly River. Chopra (1930, 1931, 1933a,b), described in seven parts, dealt with crabs of the groups Hymenosomatidae, Dromiacea, Oxystomata, Oxyrhyncha, Brachyrhyncha, and Potamonidae. Many other Indian authors added to the earlier works, raising the total carcinological fauna known to more than 250 species. About 254 species of crab belonging to 120 genera under 24 families have been recorded from along the west coast of India. The family Leucosiidae is represented by the highest number of species, 20, followed by the subfamily Thalamitae of the family Portunidae (19 species). The family Xanthidae is represented by 10 subfamilies, of which the subfamily Zosiminae is represented by 14 species. A total of 705 species of brachyurans are reported from Indian waters.

1.15.14 Anomura

In 2009, a total of, 1069 anomuran species were documented by De Grave et al. and a recent study lists 2815 species worldwide (WoRMS, 2016). Sarojini and Nagabhushanam (1972) have given a detailed account of the porcellanids from the Waltair coast, and a recent checklist prepared by Prakash et al. (2013) indicates that 30 valid porcellanid species belonging to 11 genera are found in Indian waters. Reddy and Ramakrishna (1972) listed 20 species in the families Paguridae and Coenobitidae. Studies on the anomuran crabs are far from complete, and more studies on this group are needed. A total of 120 anomurans are reported from Indian waters.

1.15.15 Chelicerata (Pycnogonida)

The Pycnogonida are commonly known as sea spiders and are found in intertidal to abyssal depths up to 7000 m around the world's oceans, from the tropics to the polar seas. They are seen as a cryptive epibenthic community on other invertebrates and algae and sometimes in coastal regions. The pycnogonids were first discovered from the Antarctic waters (Eight, 1835). The studies on Indian pycnogonids were started during the 1870s (Mason, 1873). The oldest fossil record of pycnogonids was recorded during the Upper Cambrian period (Waloszek and Dunlop, 2002). They resemble spiders and other arachnids in aspects of their morphology and anatomy, with small bodies and relatively long, hinged legs (Dunlop and Arango, 2005). A total of 1330 species of pygnogonid under four orders, 15 families, and 85 genera were recorded from the world's oceans (Bamber, 2011) while 36 species of pycnogonids belonging to 16 genera in nine families were recorded from Indian waters (Daniel and Sen, 1975; Krapp, 1996; Satheesh and Wesley, 2012).

1.16 Sipuncula

The Sipuncula or Sipunculida is commonly called the "peanut worm" because some have the general shape of shelled peanuts. They are bilaterally symmetrical, un-segmented marine worms. The first species, *Sipunculus vulgaris*, of this phylum was described in 1827 by the French zoologist Henri Marie Ducrotay de Blainville. The studies on Indian sipunculates were initiated in 1903 at the Minicoy Islands of Lakshadweep (Shipley, 1903). Sipunculids are relatively common and live in shallow waters, either in burrows or in discarded shells like hermit crabs. Some bore into solid rocks to make a shelter for themselves. A total of 147 species have been recorded from the world's oceans (WoRMS, 2016) while a total of 41 species are recorded from Indian waters (Haldar, 1985, 1991; Cutler, 1977; Cutler and Cutler, 1979; Saiz et al., 2015).

1.17 Tardigrada

The individuals under the phylum Tardigrada are known as water bears or moss piglets. They are cosmopolitan in distribution. The tardigrades are bilaterally symmetrical micrometazoans with four pairs of lobopod legs terminating in claws. Tardigrades were first described by German zoologist Johann August Ephraim Goeze in 1773 (Ontario, 2010). Of the three orders of the phylum Tardigrada, the only one found in marine, freshwater, and high-altitude mountain habitats is the Heterotardigrada. They range in size from 50 to 250 mm, but some giants can reach 1700 mm. Tardigrades can withstand 1000 times more radiation than other organisms, and are able to survive in extreme environments that would kill almost any other animal. Tardigrades are the first known animal to survive in space (Simon and Matt, 2014). About 202 species under five families and 20 genera have been described from the world (Zhang, 2011; WoRMS, 2016) while only eight species under four families and five genera are reported from Indian waters (Rao, 1971, 1972, 1975).

1.18 Phoronida

The Phoronids are commonly known as horseshoe worms. These organisms exclusively occur in marine environments. They live in upright chitin tubes made by their own to protect soft bodies. Phoronids are found throughout the world's oceans between the intertidal zone and about 400 m down, except Antarctica (Temereva et al., 2000). Phoronids consist of a crown of tentacles (Lophophore) at the upper end of the stalk to filter-feed and the stalk descends into the tube. Phylum Phoronida is comprised of two genera and 11 species (Emig, 2014). Only three species of phoronids are reported from Indian waters and the distribution was documented from the Digha coast in West Bengal and the Andaman Islands (Venkataraman et al., 2015b).

1.19 Bryozoa

Byozoans are colonial invertebrates originally called "Polyzoa" but renamed "Bryozoa" by Ehrenberg's term (Mayr, 1968; Beatty and Blackwelder, 1974). They comprise about 16,000 fossil species. Bryozoans are commonly found in high diversity and greater abundance in shallow water regions worldwide. Almost all bryozoans are colonial, composed of anywhere from a few to millions of individuals. Each individual, or zooid, is enclosed in a sheath of tissue, the zooecium, in which many species secrete a rigid skeleton of calcium carbonate. All the 6148 species of Bryozoa are recorded under three classes, four orders, 187 families, and 808 genera (Bock and Gordon, 2013; WoRMS, 2016). India represents a total of 272 species from the marine and coastal habitat (Satyanarayana Rao, 1998; Soja, 2006; Mankeshwar et al., 2015; Shrinivaasu et al., 2015b).

1.20 Entoprocta

The phylum Entoprocta is also called as Kamptozoa. They may by sessile, solitary, or colonial. Entoprocta resembles hydroids and bryozoans, as these organisms have thin stalks and branched tips. Phylum Entoprocta includes around 181 species classified under four families such as Loxosomatidae, Loxokalypodidae, Pedicellinidae, and Barentsiidae (WoRMS, 2016). Marine species are found throughout the world's oceans. The studies on the Indian Entoprocta started in the early 19th century (Annandale, 1908, 1912). Later, this phylum was the least studied from Indian waters. Only 4four species of Entoprocta are updated as being from Indian waters (Venkataraman et al., 2015b).

1.21 Brachiopoda

Brachiopoda is a primitive minor phyla in the animal kingdom. They are commonly called lamp-shells and, like the bivalve (Mollusca), they have two shells, with the upper shell smaller than the lower one. They are mostly sedentary, aquatic bottom dwelling and restricted to marine. Primarily they are filter feeders that mostly feed on algae, small planktonic organisms, and detritus. They are distributed from intertidal to greater depths, but are often found between 5 and 80 m. Among the Brachiopods, the genus Lingula is considered as a "living fossil" and is the oldest of all living genera of animals (Goodnight et al., 1970). However, about 250 million years ago during the great Permian extinction, 95% of the species vanished (Raup, 1979). It is estimated that only 1% of brachiopod species are extant against their diversity date back to Ordovician (450 million years ago) period. About 419 of the existing brachiopod species belonging to 116 genera can be found in the world's oceans (WoRMS, 2016). About 30,000 species of brachiopods became extinct in the Devonian era (Hyman, 1959). A total of eight species of brachiopods belonging to five genera, four families, and two orders under a single class Rhynchonellata were recorded from Indian waters (Sundaresan, 1968; Ramamoorthi et al., 1973; Raghunathan and Jothinayagam, 2007; Samanta et al., 2014).

1.22 Echinodermata

Plancus and Gualtire made the first report on Indian echinoderms from Goa in 1743. Subsequently, the accounts of Müller (1849), Luetken (1865, 1872), and Marktanner-Turneretscher (1887) added a few new species from the Bay of Bengal. India has 765 species (Crinoidea: 13 families, 43 genera, and 95 species. Astereroidea: 20 families, 81 genera, and 180 species. Ophiuroidea: 15 families, 67 genera, and 150 species. Echioidea: 28 families, 79 genera, and 150 species. Holothuroidea: 14 families, 62 genera, and 160 species) recorded so far, and about 257 species are known from the Andaman and Nicobar Islands (Sastry 1998; James 1986). The total Indian echinoderm database has been updated to 777 species. Economically, only the Holothuroidea were exploited on a commercial scale for export. Twelve holothurian species belonging to the genera *Actinopyga*, *Bohadschia*, *Holothuria*, *Stichopus*, and *Thelenota* are known to be of commercial importance in India. However, only three species, *Bohadschia marmorata*, *Holothuria scabra*, and *H. spinifera*, are being exploited to a large extent in the Gulf of Mannar. All the holothurians are now included in Schedule I of the Wildlife (Protection) Act 1972.

1.23 Chaetognatha

Members of Phylum Chaetognatha (bristle jaws) are commonly known as arrow worms. The first *Chaetognath* was described in 1778 by the Dutch worker Martin Slabber. Most of them are benthic and attached to algae, dead shells, and hard rocks. Chaetognaths appear to have originated in the Cambrian period. Complete body fossils have been formally described from the Lower Cambrian Maotianshan shales of Yunnan, China, that is, *Eognathacantha ercainella* (Chen and Huang, 2002) and *Protosagitta spinosa* (Hu, 2005), and the Middle Cambrian Burgess Shale of British Columbia (*Oesia disjuncta*) Walcott (Szaniawski, 2005). Chaetognaths rank second in terms of abundance after copepods in marine zooplankton and are cosmopolitan in distribution. A total of 131 living species of chaetognaths were recorded from the world's oceans (Thuesen and Pierrot-Bults, 2016) while India has 44 species (Srinivasan, 1971; Casanova and Nair, 1999, 2002).

1.24 Hemichordata

The phylum Hemichordata represents the marine deuterostome. They are divided into three extant classes: Enteropneusta, Pterobranchia, and Planctophaeroidea. Another class, that is, Graptolithina, under this phylum was recorded as extinct (Cameron et al., 2000). The first hemichordate, that is, *Ptychodera flava*, was discovered by Eschscholtz in 1825. Hemichordata fossils appear in the Lower or Middle Cambrian. The body has three parts—prosome, mesosome, and metasoma—and each has a coelomic cavity or paired cavities whereas chordates have but one coelom pair. The prosome is the proboscis, the mesosome is the collar, and the metasome contains the pharynx, gonads, and gut. Pterobranchs are filter-feeders, mostly colonial, living in a collagenous tubular structure called a coenecium (Sato et al., 2008). A total of 130 species have been recorded from around the world while only 14 species are known from Indian marine waters (Dhandapani, 1998; Venkataraman et al., 2002)

1.25 Chordata

This phylum includes two subphyla, Cephalochordata and Tunicata. A total of 3087 species of protochordates are reported globally. Out of these, 2912 belong to the Ascidiacea, 30 species under the subphylum Cephalochordata, 82 species in Thalicea, and 67 species described under the class Larvacea of Tunicata. A total of 515 species of protochordates under 63 genera and 18 families were reported from Indian waters. The subphylum Tunicata is divided into the class Ascidiacea (sea squirts), which are sessile or benthic and attached to the substratum of a coral reef, the class Thaliacea (salps), and the class Larvacea, which is planktonic.

1.26 Fish

The histories of Indian ichthylogy may be found in Day (1875–1878, 1889) and Whitehead and Talwar (1976). Among the books published on Indian fish, Francis Day's (1875–1878) treatise *The Fishes of India* is particularly important. The publications "Commercial Sea Fishes of India" by Talwar and Kakkar (1984) and "Fishes of the Laccadive Archipelago" by Jones and Kumaran (1980) are noteworthy works regarding the fish faunal resources of India. Fish constitute about half the total number of vertebrates. The estimated number of living fish species of the world is close to 28,000. Day (1889) described 1418 species of fish under 342 genera from British India. Talwar (1991) described 2546 species of fish belonging to 969 genera, 254 families, and 40 orders. The distribution of marine fish is rather wide, and some genera are common to the Indo-Pacific and Atlantic regions. The number of fish of the coastal and marine ecosystems of India is more than 2618 species: 154 species belong to Chondricthyes (cartilaginous fish) and more than 2275 species belong to Actinopterigii (bony fish). The

Lakshadweep Islands have a total of 603 species of fish (Jones and Kumaran 1980). Recent studies have updated the database of Indian fish to 2629 species (Ramakrishna et al., 2010). More than 1000 species are found in the Andaman and Nicobar Islands, and about 538 species are found in the Gulf of Mannar Biosphere Reserve. Fish found in the coral reef ecosystems of India including groups such as the damselfish (more than 76 species), butterfly fish (more than 40 species), parrot fish (more than 24 species), sea bass, groupers, and fairy basslets (more than 57 species), cardinal fish (more than 45 species), jacks and kingfish (more than 46 species), wrasses (more than 64 species), comb-tooth blennies (more than 58 species), gobies (more than 110 species), surgeon fish, tangs, and unicorn fish (more than 40 species) (Ramakrishna et al., 2010). Cryptic and nocturnal species that are confined primarily to caverns and reef crevices during daylight periods constitute another 20%.

1.27 Reptiles

A total of 32 species of marine reptiles have been reported from Indian seas, including 24 species of sea snake belonging to the family Hydrophiidae, five species of sea turtles, and the saltwater crocodile, *Crocodylus porosus*. All the species of sea snakes, four species of turtles, and the crocodile are also found in the marine environment of the Andaman and Nicobar Islands. Studies on sea turtles in the coastal waters of India and their nesting grounds were neglected until Smith (1931), who focused our attention on these giants among the sea reptiles. The leatherback sea turtle *Dermochelys coriacea* is the sole representative of the family Dermochelyidae and is a rare species. The remaining four species, namely the green turtle (*Chelonia* mydas), the olive ridley (*Lepidochelys olivacea*), the hawksbill (*Eretmochelys imbricata*), and the loggerhead (*Caretta caretta*), belong to a single family, Cheloniidae.

1.28 Seabirds

The marine ecosystem offers a feeding and breeding ground for a number of birds. Although there is not much diversity among seabirds, a number of seabirds are found regularly in marine and estuarine ecosystems. Some of the pelagic seabirds that have been reported, notably boobies (Sulidae), shearwaters (Procellariidae), and terns (Sternidae), rarely nest in the Andaman and Nicobar Islands. Smaller numbers of waders and other seabirds such as sandpipers, oystercatchers, turnstones, and plovers are also found on or near coral reefs. Egrets and herons are also widespread, often feeding across reef flats at low tide. Pelicans and flamingos have been recorded in the Gulf of Kachchh. Birds of prey, including ospreys and sea eagles, are likewise occasional visitors to the marine region. A total of 123 species of waterfowl and 85 species of terrestrial birds were reported from the Gulf of Kachchh Marine Park area in 2002. A total of 187 species were recorded from the Gulf of Mannar Marine National Park during 1985–1988, of which 84 were aquatic and the remaining terrestrial. At Manali and Hare islands, 23 species of migratory birds were found to oversummer each year. Uncommon waders in India, such as *Calidris canuta, Calidris tenuirostris, Numenius auquata, Numenius phaeopus*, and *Limosa lapponica* are regular winter visitors to this area (Balachandran, 1995).

1.29 Marine Mammals

Marine mammals include representatives of three major orders, namely Cetacea (whales, dolphins and porpoises), Sirenia (manatees and dugong), and Carnivora (sea otters, polar bears and pinnipeds). The order Cetacea consists of two suborders, namely Mysticeti (baleen whales) and Odontoceti (toothed cetaceans). The suborder Mysticeti is represented by four families with 14 species while the suborder Odontoceti is represented by 10 families with 73 species. A total of 130 marine mammal species have been recognized from around the the world (Jefferson et al., 2008). The Indian seas support 25 species of marine mammal, including the families Delphinidae, Physeteridae, Kogiidae, Ziphiidae, Phocoenidae, and Platanistidae (Kumaran, 2002; Vivekanandan and Jeyabaskaran, 2012). Of the 25 cetacean species, five are baleen whales and the rest are odontocetes, including members of the families Delphinidae, Physeteridae, Kogiidae, Ziphiidae, Phocoenidae, and Platanistidae (Kumaran 2002). A majority of these are oceanic forms, and occasionally a few individuals get stranded on the shore. Stranding and sighting records suggest that only 20 cetacean species and one sirenian are found in Indian seas (Vivekanandan and Jeyabaskaran, 2012). The Sea Cow *Dugong dugon* occurs in the near-shore waters of the Gulf of Mannar, the Gulf of Kachchh, and the Andaman and Nicobar Islands. All the marine mammal species reported from Indian waters are protected under the Wildlife (Protection) Act 1972.

1.29.1 *Marine Biodiversity in India and the World*

The current inventory of the coastal and marine biodiversity of India is summarized in Table 5. It depicts a total of 18,135 species from the faunal and floral communities reported from the seas around India. The data reveal that India contributes 7.33% to global marine biodiversity. Further intensive explorations and taxonomical studies may bring out several more species in India. The species composition of flora and fauna and species diversity of fauna are illustrated in the graph (Figs. 5–6). The pictures of marine fauna are shown in Plate 1.

1.29.2 *Marine Protected Fauna*

To protect nature and wild animals, 730 areas (including National Parks, Wildlife Sanctuaries, Community Reserves, and Conservation Reserves) have been declared as protected areas in India. Seventeen biosphere reserves have also been declared to protect entire ecosystems. According to a series of notifications issued under the Wildlife (Protection) Act 1972 by the Ministry of Environment, Forests, and Climate Change, Government of India, so far 1191 species (Table 6) of marine fauna belonging to six phyla (Porifera, Coelenterata, Arthropoda, Mollusca, Echinodermata, fish, reptiles, and mammals) are listed under different categories (Schedules I, II, III, IV) as their natural populations are being depleted.

2 CHALLENGES

The major challenges for marine ecosystems are waves and storms, particularly cyclones. Cyclonic disturbances develop during October–November along the coast. These cyclones have sustained winds with speeds ranging from 65 to 120 km per hour. High-speed winds cause extreme wave action that kills many fauna and flora while also breaking coral into

TABLE 5 Marine Biodiversity of India Compared With Global Marine Biodiversity

Taxon/Group	India	World
BACTERIA		2058
SEAWEED		
Cyanophyta	9	1000
Chlorophyta	216	3032
Phaeophyta	191	1600
Rodhophyta	434	8174
Xanthophyta	3	
Bacillariopyta (diatoms)	200	5000
SEAGRASSES	14	60
MANGROVES	69	100
Other Protoctista		23000
PROTOZOA	2577	
Dinophyceae (dinoflagellates)		2000
Euglenophyceae		250
Cilophora		2668
Bicoecea		
Chrysophyceae		500
Dinomastigota		4000
Radiolaria		550
Protista	532	
Foraminifera	329	8989
PORIFERA	476	8339
CNIDARIA		
[a]Actiniaria	75	1109
Hydrozoa	178	3662
Scyphozoa	30	189
Cubozoa	5	45
Anthozoa (only Scleractinian Coral and Octocoral)	1042	5912
[a]Staurozoa	1	48
CTENOPHORA	19	197
PLATYHELMINTHES		

(Continued)

TABLE 5 Marine Biodiversity of India Compared With Global MarineBiodiversity—cont'd

Taxon/Group	India	World
Polyclads	44	1005
Monogenea	122	2494
Digenea	662	5308
Cestoda	2	1767
DICYEMIDA	6	122
NEMERTEA (Ribbon Worms)	6	1362
GASTROTRICHA	61	497
ROTIFERA	47	172
CEPHALORHYNCHA		
[a]Kinorhyncha	10	188
NEMATODA	356	6833
[a]GNATHOSTOMULIDA	4	101
MOLLUSKS (Except opisthobranchs)	2000	42579
Opisthobranchia	388	3736
ANNELIDA		
Polychaeta (Except echuira)	522	11513
Archianellida	20	90
Echiura	43	176
ARTHROPODA		
Chelicerata (Pycnogonida)	36	1330
[a]Merostomata	2	4
Crustacea		
Copepoda	541	11246
Ostracoda	125	5930
[a]Branchiopoda	4	100
[a]Branchiura	4	48
Malacostraca		
Mysidacea	84	1139
Cumacea	88	1631
Tanaidacea	3	1371
Isopoda	300	7884

TABLE 5 Marine Biodiversity of India Compared With Global MarineBiodiversity—cont'd

Taxon/Group	India	World
Amphipoda	167	8165
Euphausiacea	23	86
Stomataopoda	93	480
[a]Thoracica	74	1302
Decapoda		
Prawn, Shrimp and Lobsters	477	
Brachyura	705	5899
Anomura	120	2815
Insecta		
[a]Halobatinae	10	51
SIPUNCULA	41	147
TARDIGRADA	8	202
PHORONIDA	3	11
BRYOZOA	272	6148
ENTOPROCTA	4	181
BRACHIOPODA	8	419
ECHINODERMATA	777	7000
CHAETOGNATHA (Arrow Worms)	44	131
HEMICHORDATA	14	130
CHORDATA		
Cephalochordata	6	30
Tunicata	515	3057
Pisces	2618	18194
Reptiles	32	74
Birds	211	
Mammalia	33	130
[a]FUNGI		1369
Total	18135	247129

[a] *Desription not provided in the text of this article.*

Courtesy: WoRMS, 2016. World Register of Marine Species. http://www.marinespecies.org/aphia.php?p=taxdetails&id=1839 (Accessed 08 September 2016); Bouchet P., 2006. The magnitude of marine biodiversity. In: Duarte, C.M. (Ed.), The Exploration of Marine Biodiversity, Scientific and Technological Challenges. Fundacion BBVA Bilbao; Groombridge, B., Jenkins, M.D., 2000. Global Biodiversity: Earth's Living Resources in the 21st Century. World Conservation Press, Cambridge, IUCN, 2015. http://www.iucnredlist.org, No reliable data available for the blank area.

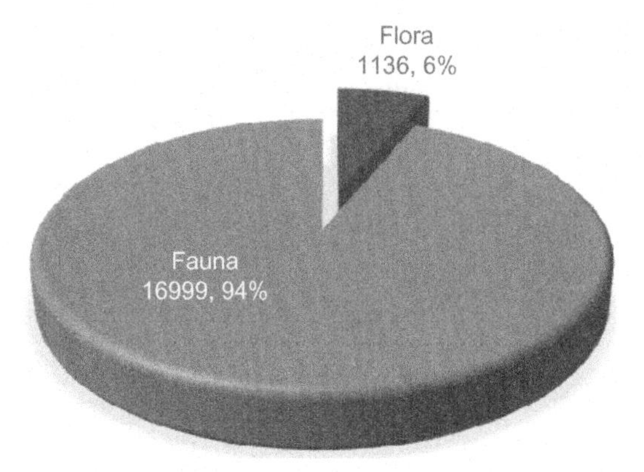

FIG. 5 Species composition of flora and fauna of India.

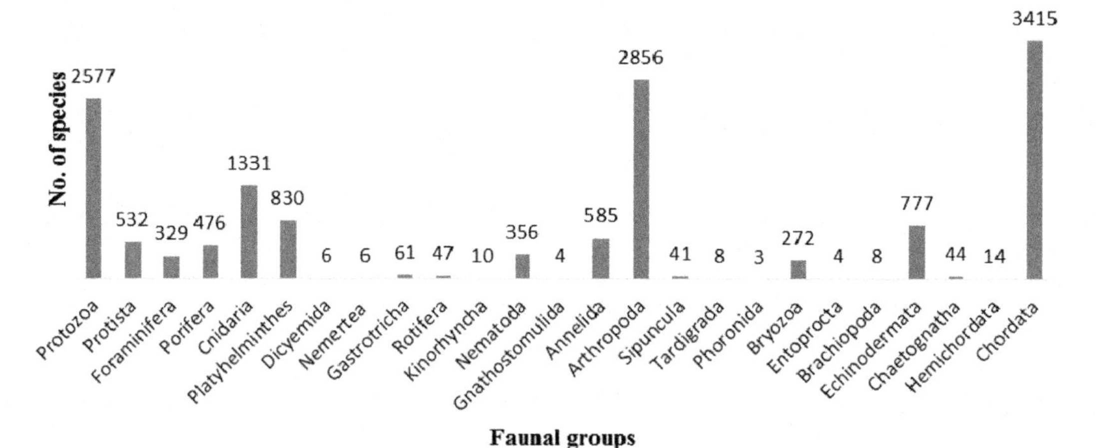

FIG. 6 Faunal diversity of India.

rubble. Sometimes, large amounts of sand and other material may be dumped onto the coral reef. Also, freshwater runoff kills many fauna and flora in semienclosed bays and lagoons by lowering salinity and depositing large amounts of sediments and nutrients.

Varied anthropogenic activities that area a cause for concern over and above natural disturbances include runoff and sedimentation from development activities (projects); eutrophication from sewage and agriculture; the physical impact of maritime activities; dredging, collecting, and destructive fishing practices; pollution from industrial sources, golf courses, and oil refineries; and the synergistic impacts of anthropogenic disturbances. The amount of sediments and chemicals the runoff carries to the sea has profound effects on the fertilization of eggs of marine species. Likewise, the quality of runoff water can affect the metamorphosis of the larvae of many species. Oil pollution induces mortality, decreases fecundity and fails recruitment. India has three megacities as well as many small, medium, and major ports and industries around the 8000 km coast. The enactment of the Water Pollution Act in 1974 and the Environment Protection Act in 1986 has helped in regulating the disposal of waste from

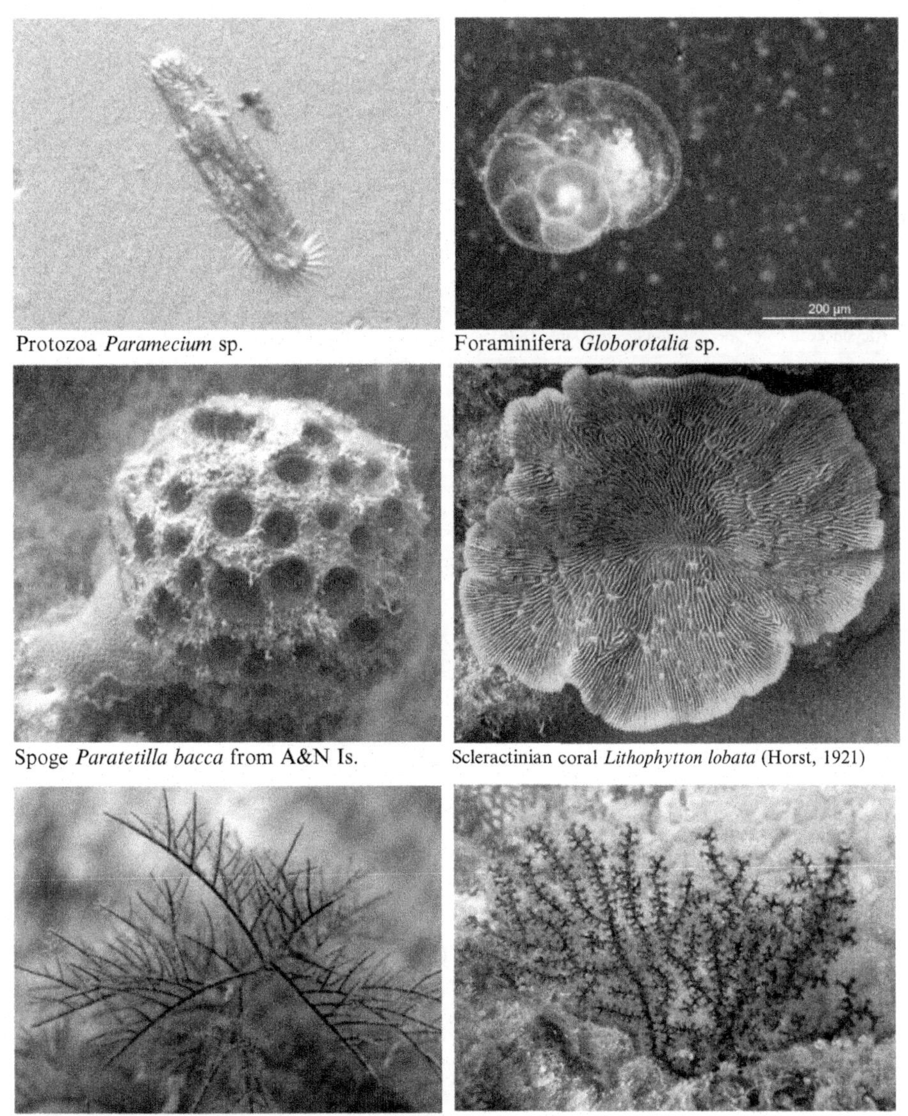

Protozoa *Paramecium* sp.

Foraminifera *Globorotalia* sp.

Spoge *Paratetilla bacca* from A&N Is.

Scleractinian coral *Lithophytton lobata* (Horst, 1921)

Antipatherian coral *Cladopathes plumosa* Brook, 1889

Gorgonid *Acanthogorgia spinosa* Hiles, 1899

PLATE 1 Marine fauna in Andaman and Nicobar Islands, India.

(Continued)

Alcyonacean coral *Sarcophyton latum* (Dana, 1846)

Organi pipe coral *Tubipora musica* Linnaeus, 1758

Sea pen *Scytallium martensi* Kolliker, 1870

Blue coral *Heliopora coerulea* Pallas 1766

Scyphozoa *Cassiopea andromeda* (Forsskål, 1775)

Fire coral *Millipora dichotoma* (Forsskål, 1775)

Hydrozoa *Pennaria disticha* Goldfuss, 1820

Cubozoa *Tamoya* sp.

PLATE 1—CONT'D

Ctenophore *Ceonoplana* sp.

Polyclad *Pseudoceros tristratus* Hayman, 1959

Namertean *Baseodiscus quinquelineatus* (Quoy & Gaimard, 1833)

Rotifera colonies

Chiton *Cryptoplax larvaeformis* (Burrow, 1815)

Gasropod *Cassis cornuta* (Linnaeus, 1858)

Bivalve *Tridacna gigas* (Linnaeus, 1758)

PLATE 1—CONT'D

(Continued)

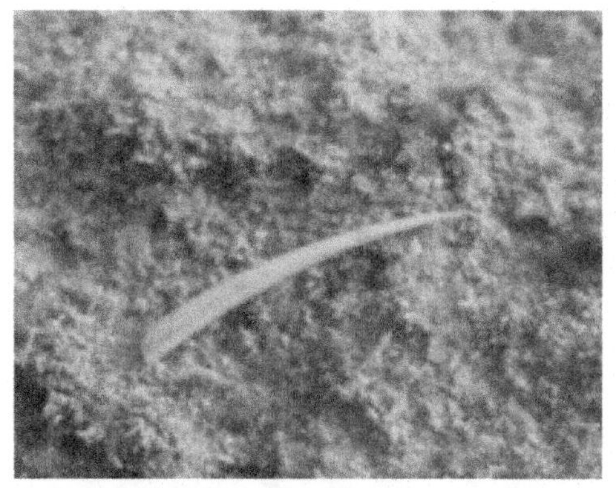
Tusk shell *Dentalium octangulatum*

Cephalopod *Octopus* sp.

Nudibranch *Doriprismatica atromarginata* (Cuvier, 1804)

Polychaete *Spirobranchus* sp.

Ehiura *Bonellia* sp.

Brachyuran crab *Neopetrolisthes maculatus* (H. Milne Edwards, 1837)

PLATE 1—CONT'D

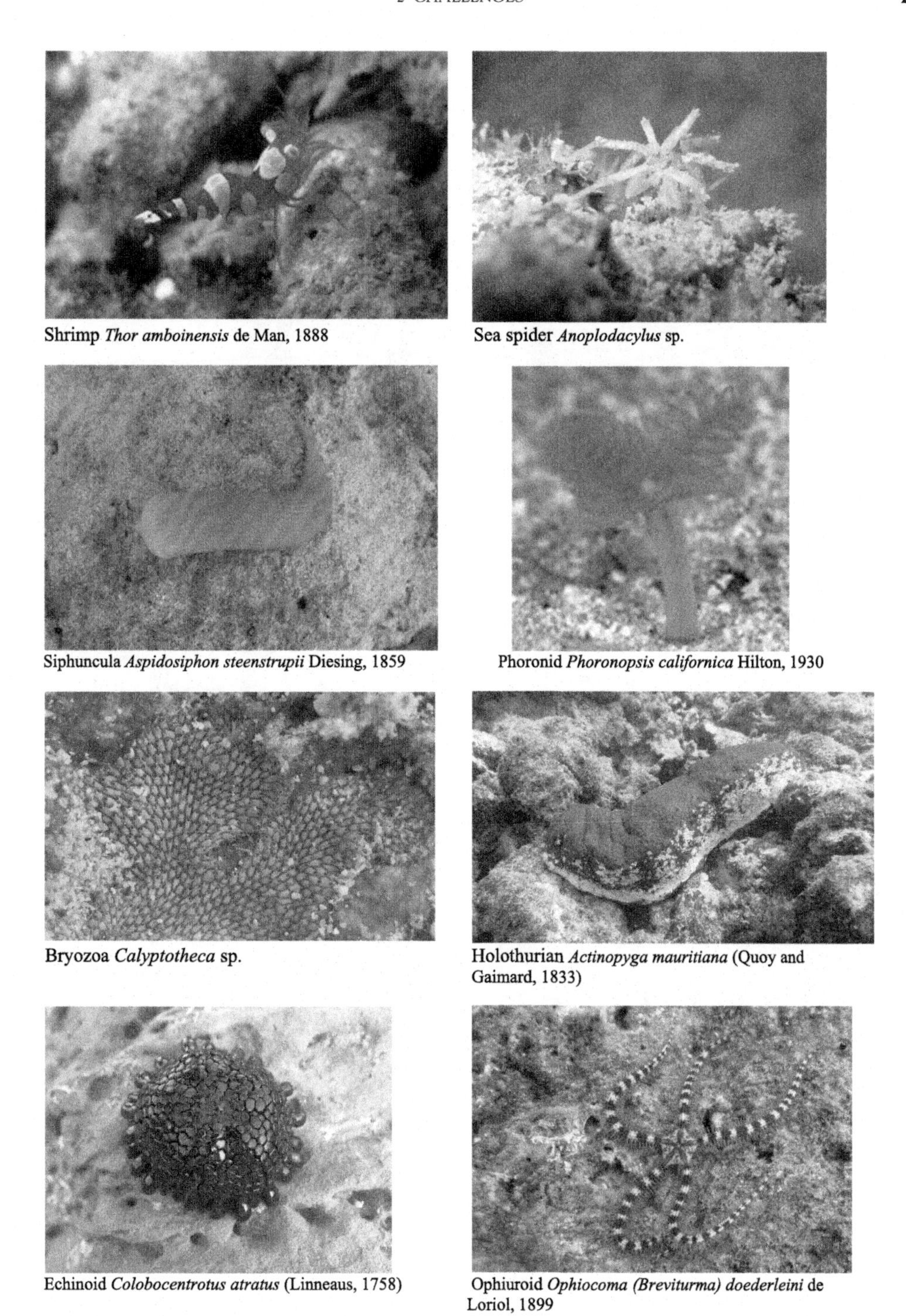

Shrimp *Thor amboinensis* de Man, 1888

Sea spider *Anoplodacylus* sp.

Siphuncula *Aspidosiphon steenstrupii* Diesing, 1859

Phoronid *Phoronopsis californica* Hilton, 1930

Bryozoa *Calyptotheca* sp.

Holothurian *Actinopyga mauritiana* (Quoy and Gaimard, 1833)

Echinoid *Colobocentrotus atratus* (Linneaus, 1758)

Ophiuroid *Ophiocoma (Breviturma) doederleini* de Loriol, 1899

PLATE 1—CONT'D

(Continued)

Asteroid *Fromia indica* (Perrier, 1869)

Crinoid *Cenometra bella* (Hartlaub, 1890)

Hemichordate *Balanoglossus*

Ascidian *Phallusia julinea* Sluiter, 1915

Reef fish *Synanceia verrucosa* Bloch & Schneider, 1801

Reef fish *Pomacanthus semicirculatus* (Cuvier,1831)

PLATE 1—CONT'D

Laticauda colubrina (Schneider, 1799)

Chelonia mydas (Linnaeus, 1758)

Crocodylus porosus (Schneider, 1801)

Egretta sacra (Gmelin, 1789)

Dugong dugon (Müller, 1776)

Tursiops aduncus (Ehenberg, 1833)

PLATE 1—CONT'D

TABLE 6 Protected Marine Faunal Species in India

S. N.	Protected Faunal Group	No. of Protected Species	IW(P)A, 1972
1	*Porifera*		
	Calcareous sponges (All species)	10	Schedule III
2	*Cnidaria*		
	Reef building corals (All Scleractinians)	627	Schedule I
	Black coral (All Antipatharians)	10	Schedule I
	Organ Pipe Coral (Tubipora musica)	1	Schedule I
	Fire Coral (All Millepora species)	5	Schedule I
	Sea Fan (All Gorgonians)	221	Schedule I
3	*Arthropoda*		
	Robber Crab (Crustacea)	1	Schedule I
	Horseshoe Crab (Merostomata)	2	Schedule IV
4	*Mollusca*		Schedule I
	Gastropod and Bivalve	24	Schedule I&IV
5	*Echinodermata*		
	Sea Cucumber (All Holothurians)	188	Schedule I
6	*Fishes*		
	Elasmobranchs (Sharks and Rays)	10	Schedule I
	Sea horse (All Syngnathidians)	50	Schedule I
	Giant Grouper	1	Schedule I
7	*Marine Reptiles*		
	Marine Turtles	5	Schedule I
	Saltwater Crocodile	1	Schedule I
8	*Marine Mammals*	33	Schedule I& II
	Total	1191	

industries. These measures have resulted in the reduction of pollution loads of the coastal waters, to a certain extent. Major industries such as fertilizer, petrochemical, agrochemical, and chemicals are mainly located along the coasts. Sometimes, the anthropogenic threats are the prime regulators to initiate as well as develop natural threats. Some of the threats are categorically interpreted below.

Climate change: It is one of the most important and dominant factors to change the oceanic temperatures, acidity, and patterns of water currents, eddies, and fronts. With the alteration of the ocean's physiological condition, it used to change in vulnerable habitats of a wide group of floral and faunal communities especially mangroves, coral reefs, sea weed, sea grass etc. The coral reef ecosystem is most susceptible one due course of climatic alteration or changes with an alarming rate of mortality. The rise in sea level can be seen as another future threat toward coral reef biodiversity as the enhanced water level deeper reefs will be unable to get sustainable environments. This will change the patterns of coastal erosion as well as affect intertidal ecosystems. Temperature is the prime factor for the development pattern of marine faunal and floral communities. Corals are well-known stenothermic organisms that are very sensitive to temperature ranges. The rise of temperature both in the atmosphere andon the sea surface can result in coral bleaching, which usually leads to subsequent death. The global emergence of the greenhouse effect is the cause for temperature increases of around 2°C in the tropics, which can lead to a doubling of CO_2. These may cause destruction for the scleractinian corals (Fig. 7).

Invasive species: Invasion of nonnative or invasive species is a major natural threat to native species. Being a nonnative species, they occupy the habitat of a native species and absorb or use the local nutrient cycle, which may cause species competition that leads to the death or destruction of the natives (Fig. 8).

Nutrient loading: Coastal areas the most supportive location for the sustainable development of the human population. The domestic waste, runoff from agricultural land, and other chemical wastes flow to marine water without proper treatment, which increases the ratio of nitrogen and phosphorus. *Eutrophication* is the prime destructor of an entire marine biodiversity.

UV radiation: The globe is well protected with the cover of a wide range of gaseous layers. Ozone is one of the layers that protects the Earth from the sun's ultraviolet (UV) radiation. Over thecourse of time, it has been noticed that ozone depletion has caused increased UV radiation. The productivity of phytoplankton is reduced due to the radiation in surface waters, which will affect the entire marine ecosystem, especially the coral reef ecosystem.

Carbonate mineral saturation: Carbonate mineral saturation can affect coral calcification. High levels of dissolved CO_2 in sea water will equilibrate the increased partial pressure of the CO_2 in the atmosphere and reduce $CaCO_3$ precipitation. The reduction will reduce the growth rate of the corals. Scleractinain corals need a healthy sustainable saline condition, as they are stenohaline creatures. Differentiation in salinity level by rainfall, runoff, or sewage affects the reef growth rate.

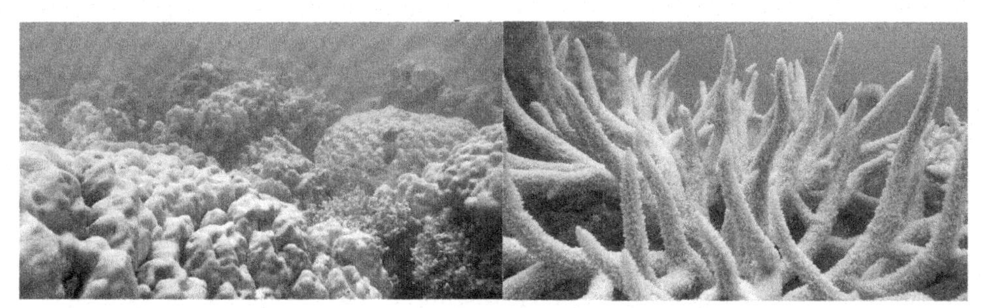

FIG. 7 The coral bleaching in the Andaman and Nicobar Islands.

FIG. 8 *Carijoa riisei* from Andaman and Nicobar Islands.

Deforestation and mining: Deforestation and mining at the coastal areas as well as hundreds of kilometers inland as well as construction of dams for hydroelectricity generation or for irrigation purposes have led to great increases in sediments. The residue loads destroy the coastal ecosystems such as coral reefs, mangroves, seagrass, and seaweeds.

Habitat alteration: The floral and faunal components face the effects of habitat alteration due to fishing gear, especially bottom trawling. The destruction can be seen in coral reef communities and is also associated with a wide range of faunal groups due to the alteration.

Overfishing and exploitation: At present, in the 3651 fishing villages situated along the 8129 km coastline of India, about one million are occupied full time in marine capture fisheries. Several types of net fishing have also been responsible for overexploitation of marine resources. The use of fish traps made of long-lasting materials with small mesh sizes results in the capture of prereproductive juveniles, affecting future populations. This can also lead to the death of fish when traps become dislodged during storms but continue to capture fish, which eventually starve. Fishing operations with the latest technologies are causing damage to the marine living resources. Shrimp trawlers probably have the highest rate of by catch bringing in up to 90% more of "trash fish". Random capture techniques destroy immature fish and other nontargeted marine species (Fig. 9).

Tourism: This is one of the great resources of economic development of several areas, especially marine and coastal tourism. Uncontrolled tourism may be seen as the destructive driver of coastal and marine biodiversity. Coral reef ecosystems with a wide variety of faunal communities, mangrove ecosystems, etc., are the most attractive areas for he tourists. The nonecotourism strategy may cause great harm to those ecosystems (Fig. 10).

Chemical Stress: Toxic chemical compounds contribute to algal blooming. Severe impacts can be seen in coastal resources, local economies, and public health due to this effect. The blooming may cause the alteration of water colors. As a result of this blooming, the rate of accumulation, decomposition, and displaced biomass will be altered. This may cause destructive effects for the scleractinian corals.

Pollution: Marine pollution and litter are among the most destructive ecological threats toward marine life. This includes a wide range of sediments, nutrients, plastics, hazardous and radioactive substances, heavy metals, discarded fishing gear, microbial pollution, medical waste, domestic waste trace elements such as carcinogens, endocrine-disruptors,

FIG. 9 Fishing nets affect coral reefs and associated organisms in the Andaman and Nicobar Islands.

FIG. 10 Marine tourism at the Andaman and Nicobar Islands.

and info-disruptors. The acute and chronic effects of these molecules directly as well indirectly damage the physiological status of the organisms and the marine environment as a whole (Fig. 11).

Natural calamities: Storms and wave energies may be seen as typhoons, quakes, and tsunamis. These are the natural catastrophic events that may cause the destruction of marine biodiversity (Figs. 12–13).

Oil spills and anchor: Oil spills from boat and ships damage marine biodiversity by altering the entire ecosystem and causing massive poisoning. Sometimes, boats anchor in reef

FIG. 11 Plastic pollution in Andaman and Nicobar Islands.

FIG. 12 Exposed reef at North Andaman due to earthquake.

FIG. 13 Impact of tsunami on littoral forest ecosystem.

ecosystems without bothering the fragile life of reef biodiversity (Fig. 14). However, this may cause the destruction of coral reefs and associated marine animals.

Bacterial effects: Several cyanobacterial species initiate the diseases of marine animals. The diseases are the most destructive agents for the devastation of marine animals. Red Band Disease (microbial manifestation, Black Overgrowing Cyanophyta (cyanophytic manifestation), Black Aggressive Band, Lethal Orange Disease (bacterial pathogen manifestation), Fungal Infection (FI), Yellow Band Disease (YBD also Yellow-Blotch Disease), Dark Spot Disease (DSD), Skeleton Eroding Band (SEB), White Syndromes (WS), Epizoism, Peyssonnelia, Hyperplasia. and Neoplasia. etc.. can cause serious damage to scleractinian corals.

FIG. 14 Anchors at reef biodiversity.

3 CONCLUSION

The world's biodiversity is estimated at 1.75 million species, excluding microbial species (Heywood and Watson, 1996); according to another estimate it falls within a range of from 5 to 120 million species (Reaka-Kudla, 1998). Approximately 340,000 marine species are known, compared with more than 1.5 million known terrestrial species. This indicates only a lack of knowledge about marine species rather than greater terrestrial diversity (Alder, 2003). It is thus evident that most of the world's marine species are unknown and may remain so, given a 500-year projection to describe all species based on the current knowledge of 1.36 million animal species and the possibility of 10 times this number actually existing (Heywood and Watson, 1996). In the context of the Indian seas, it is also evident that even many of the better-known taxonomic groups remain poorly studied. We may of course never know the full extent of biodiversity in any of the world's oceans, and the rate at which we improve our understanding (Keesing and Irvine, 2005) is likely to be lowest in the Indian seas. The impacts of climate change will alter coastal marine ecosystems, affecting the range of species and their ecology at a rate faster than we are recording them (Keesing and Irvine, 2005). It is evident that comprehensive taxonomic coverage of the marine biota of any region remains a monumental task (Griffiths, 2005) beyond the capacity of existing local taxonomic expertise, given the existing taxonomic impediments. It would be quite appropriate to carry out studies systematically rather than describe new species in a haphazard and opportunistic manner as and when they are discovered, as remarked by Griffiths (2005).

Acknowledgments

The authors are grateful to the Dr. Kailash Chandra, Director, Zoological Survey of India, Kolkata, for facilities and encouragement.

References

Aiyar, R.G., 1931. Proc. 18th Indian Sci. Congr. p. 244.
Aiyar, R.G., Alikunhi, K.H., 1940. Rec. Indian Mus. Calcutta 42, 89–107.

Aiyar, R.G., Alikunhi, K.H., 1944. Proc. Nat. Inst. Sci. 10 (1), 113–140.
Alcock, A., 1893. J. Asiat. Soc. Beng. 62 (2), 138–149.
Alcock, A., 1898. Mem. Indian. Mus., 1–29.
Alcock, A., 1901. A Descriptive Catalogue of the Indian Deep-Sea Crustacea, Decapoda: Macrura and Anomura in the Indian Museum, Being a Revised Account of the Deep-Sea Species Collected by the Royal Indian Marine Survey Ship, "Investigator" Calcutta, India, pp. 1–286.
Alcock, A., 1906. Trustees of the Indian Museum (Crustacea), Calcutta. pp. 1–55.
Alcock, A., Anderson, R.S., 1893. Ann. Mag. Nat. Hist. 3 (7), 1–27. 278–292.
Alcock, A., Anderson, R.S., 1894. J. Asiat. Soc. Beng. 43, 141–185.
Alder, E., 2003. A world of neighbours: UNEP's Regional Seas Programme (UNEP). www.unep.ch/seas/Library/-neighbours.pdf.
Alder, J., Hancock, A., 1864. Trans. Zool. Soc. Lond. 5, 117–147.
Alikunhi, K.H., 1946. Curr. Sci. 15, 140.
Alikunhi, K.H., 1947. Proc. Nat. Inst. Sci. India 13 (3), 105–127.
Alikunhi, K.H., 1948a. J. Roy. Asiat. Soc. Bengal 14 (1), 17–25.
Alikunhi, K.H., 1948b. Proc. Nat. Inst. Sci. India 14 (8), 373–383.
Alvarino, A., 1974. J. Mar. Biol. Assoc. India 14 (2), 713–722.
Annandale, N., 1907a. Proc. Asiat. Soc. Bengal NS III (2), 79–81.
Annandale, N., 1907b. Rec. Indian Mus. 1, 197–205.
Annandale, N., 1908. The fauna of brackish ponds at Port Canning, lower Bengal: Part VII: further observations on the Polyzoa, with the description of a new genus of Entoprocta. Rec. Indian Mus. 2, 24–32.
Annandale, N., 1912. The occurrence of Entoprocta in Indian waters. Records of the Indian Museum. .
Annandale, N., 1913. Rec. Indian Mus. 9, 227–236.
Annandale, N., 1914. Rec. Indian Mus. 10, 273–280.
Annandale, N., 1915. Mem. Indian Mus. 5, 65–114.
Annandale, N., 1916. Mem. Asiat. Soc. Bengal 5, 18–24.
Anandakumar, N., Thajuddin, N., 2013. Physico-chemical properties, seasonal variations in species composition and abundance of micro zooplankton in the Gulf of Mannar, India. Ind. J. Mar. Sci. 42 (3), 383–389.
Annapurna, C., Vijaya Bhanu, C., Srinivasa Rao, M., Sivalakshmi, M., Cooper, L.M.G., Rao, Y.K., 2012. Free-living nematodes along the continental slope off northeast coast of India. J. Mar. Biol. Assoc. India 54 (2), 52–60.
Antony, A., 1968. Studies on the shelf water Foraminifera of the Kerala coast of India. 4, Bulletin Deep Marine Biological Oceanography, University of Kerala, India, pp. 1–154.
Balachandran, S., 1995. J. Bombay Nat. Hist. Soc. 92 (3), 303–313.
Bamber, R.N., 2011. Class Pycnogonida Latreille, 1810. In: Zhang, Z.Q. (Ed.), Animal Biodiversity: An Outline of Higher-Level Classification and Survey of Taxonomic Richness.In: Zootaxa, vol. 3148. pp. 100–111.
Bergquist, P.R., 1978. Sponges. Hutchinson, London, 268 pp.
Bernard, K.H., 1935. Rec. Indian Mus. 37, 279–319.
Beatty, J., Blackwelder, 1974. Names of invertebrate phyla. Syst. Zool. 23 (4), 545–547. https://dx.doi.org/10.2307/2412472.JSTOR2412472.
Bhalla, S.N., 1970. Foraminifera from the Marina beach sands, Madras and faunal provinces of the Indian Ocean. J. Foraminiferal Res 21, 156–163.
Bhatia, S.B., Bhalla, S.N., 1959. Recent Foraminifera from beach sands of Puri, Orissa. J. Palentological Soc. India 4, 78–81.
Blazewicz-Paszkowycz, M., Bamber, R., Anderson, G., 2012. PLoS One 7 (4), e33068. https://dx.doi.org/10.1371/journal.pone.0033068.
Bock, P., Gordon, D., 2013. WoRMS Bryozoa: World List of Bryozoa (version 2013-03-04). In: Roskov, Y., Kunze, T., Paglinawan, L., Orrell, T., Nicolson, D., Culham, A., Bailly, N., Kirk, P., Bourgoin, T., Baillargeon, G., Hernandez, F., De Wever, A. (Eds.), Species 2000 & ITIS Catalogue of Life, 2013 Annual Checklist. In: Species 2000, Reading, UK. DVD.
Borradaile, L.A., 1900. Proc. Zool. Soc. London, 568–596.
Borradaile, L.A., 1902a. The fauna and geography of the Maldives and Laccadive archipelagoes, the account of the work carried on and of the collections made by an expedition during the years 1899 and 1900 by J. Stanley Gardiner, marine crustaceans. I On varieties 1 (2), 191–208.
Borradaile, L.A., 1902b. The fauna and geography of the Maldives and Laccadive archipelagoes, being the account of the work carried on and of the collections made by an expedition during the years 1899 and 1900 by J. Stanley Gardiner, marine crustaceans. II Portunidae 1 (3), 237–271.

Borradaile, L.A., St. Gardiner, J., 1903a. Marine crustaceans. IV. Some remarks on the classification of the crabs. In: The Fauna and Geography of the Maldive and Laccadive Archipelagoes.vol. 1, no. 4, pp. 424–429.
Borradaile, L.A., 1903b. St. Gardiner, J. (Ed.), The Fauna and Geography of the Maldive and Laccadive Archipelagoes. p. 431.
Brandão, S., 2016. Ostracoda. In: Brandão, S.N., Angel, M.V., Karanovic, I., Parker, A., Perrier, V., Yasuhara, M. (Eds.), World Ostracoda Database. Accessed through: World Register of Marine Species at http://www.marinespecies.org/aphia.php?p=taxdetails&id=1078. (Accessed 9 September 2016).
Brook, G., 1893. The genus *Madrepora*. In: Catalogue of the Madreporarian Corals in the British Museum (Natural History).I, pp. 1–212.
Calman, W.T., 1904. Rep. Ceylon Pearl Oysters Fish. Gulf of Mannar 2, 159–180.
Cairns, S.D., Kitahara, M.V., 2012. ZooKeys 227, 1–47. https://dx.doi.org/10.3897/zookeys.227.3612.
Cameron, C.B., Garey, J.R., Swalla, B.J., 2000. Evolution of the chordate body plan: new insights from phylogenetic analyses of deuterostome phyla. Proc. Natl. Acad. Sci. U. S. A. 97 (9), 4469–4474.
Carter, H.J., 1887. Report on the Marine Sponges, chiefly from King Island, in the Mergui Archipelago, collected for the Trustees of the Indian Museum, Calcutta, by Dr. John Anderson, F.R.S., Superintendent of the Museum. J. Linnean Soc. Zool. 21 (127–128), 61–84, pls 5–7.
Casanova, J.P., Nair, V.R., 1999. A new species of the genus Sagitta (Phylum Chaetognatha) from the Agatti lagoon (Laccadive Archipelago, Indian Ocean) with comments on endemism. Indian J. Mar. Sci. 28 (2), 169–172.
Casanova, J.P., Nair, V.R., 2002. A new species of Sagitta (Chaetognatha) from a Laccadive lagoon (Indian Ocean) having fan-shaped anterior teeth: phylogenetical implications. J. Nat. Hist. 36, 149–156.
Chakrapany, S., 1984. Studies on marine invertebrates. Scyphomedusae of the Indian and adjoining seas. Ph.D. thesis, University of Madras 206 pp.
Chatterjee, T., Pesic, V., 2010. Cah. Biol. Mar. 51, 289–299.
Chaudhury, A., Biswas, B., 1954. Recent perforate Foraminifera from Juhu beach, Bombay. Micropaleontology 8 (4), 30–32.
Chen, J.Y., Huang, D.-Y., 2002. A possible lower Cambrian Chaetognath (arrow worm). Science 298 (5591), 187.
Chilton, C., 1923. Mem. Indian. Mus. 5, 877–895.
Chinnadurai, G., Fernando, O.J., 2007. Meiofauna of Mangroves of the southeast coast of India with special reference coast of India with special reference to the free-living nematode assemblage. Estuar. Coast. Shelf Sci. 72, 329–336.
Chopra, B., 1923. Rec. Indian Mus. 25, 411–550.
Chopra, B., 1930. Rec. Indian Mus. 32, 413–429.
Chopra, B., 1931. Rec. Indian Mus. 33, 303–324.
Chopra, B., 1933a. Rec. Indian Mus. 35 (1), 25–32.
Chopra, B., 1933b. Rec. Indian Mus. 35 (1), 77–86.
Cutler, E.B., 1977. The bathyal and abyssal Sipuncula. Galathea Rep. 14, 135–156. 13 figs.
Cutler, E.B., Cutler, N.J., 1979. Madagascar and Indian Ocean Sipuncula. Bull. Mus. Natn. Hist. Nat. Paris 4 (1 N4), 941–990. 21 figs.
Daniel, R., 1966. Ann. Mag. Nat. Hist. 9 (13), 689–692.
Daniel, A., 1971. J. Mar. Biol. Assoc. India 13 (1), 82–85.
Daniel, R., 1974a. Mem. Zool. Surv. India 15 (2), 865–868.
Daniel, A., 1974b. Proc. Indian Natl. Sci. Acad. 38 (3and4), 179–189.
Daniel, R., 1985. Siphonophora. Zoological Survey of India. 440 pp.
Daniel, R., Daniel, A., 1963. A. J. Mar. Biol. Assoc. India 5 (2), 185–220.
Daniel, A., Sen, J.K., 1975. Studies on the Pycnogonids from the collection of the Zoological Survey of India, Calcutta, together with notes on their distribution in the Indian Ocean. J. Mar. Biol. Assoc. India 17 (2), 160–167.
Day, F., 1875–1878. The Fishes of India: Being a Natural History of the Fishes Known to Inhabit the Seas and Fresh Waters of India, Burma and Ceylon. Reprinted by William Dawson and Sons Ltd, London 778 pp.
Day, F., 1889. The Fauna of British India, Including Ceylon and Burma. Fishes, Vol. II. Taylor and Francis, London, 509 pp.
De Man, J.G., 1887. Zool. Jahrb. (Syst.) 2, 639–689.
De Man, J.G., 1908. The fauna of brackish ponds of Port Canning, lower Bengal. Part X. Decapod Crustacea with an account of small collection from brackish water near Calcutta and in the Dacca District, Eastern Bengal. Rec. Indian Mus. 2, 211–231.
Dhandapani, P., 1998. Hemichordata. In: Faunal Diversity in India. Zoological Survey of India, Kolkata, pp. 406–409.
Dunlop, J.A., Arango, C.P., 2005. Pycnogonid affinities: a review. J. Zool. Syst. Evol. Res. 43, 8–21.

Dutta, T.K., Miljutin, D.M., Chakraborty, S.K., Mohapatra, A., 2016. Cyathoshiva amaleshi gen. n. sp. n. (Nematoda: Cyatholaimidae) from the coast of India. Zootaxa 4126 (4), 577–586.

Eliot, C., 1906. On the nudibranchs of South India and Ceylon with special reference to the drawings of Kelaart and the collection belonging to Alder and Hancock preserved in the Hancock Museum at New Castle-on-Tyne North. Proc. Zool. Soc. London 111, 636–691. 997–1008.

Eight, J., 1835. Description of a new animal belonging to the Arachnides of Latreille, discovered in the sea along the shores of the New South Shetland Islands. Boston J. Nat. Hist. 1, 203–206.

Emig, C.C., 2014. Phoronida. In: Emig, C.C. (Ed.), World Phoronida database. Accessed through: World Register of Marine Species athttp://www.marinespecies.org/aphia.php?p=taxdetails&id=1789. (Accessed 7 August 2014).

Fauvel, P., 1930. Bull. Madras Govt. Mus. I (2), 1–72.

Fauvel, P., 1932. Mem. Indian Mus. XII (1), 1–262.

Fauvel, P., 1953. Fauna of India including Pakistan, Ceylon, Burma and Malaya. Annelida, Polychaeta. 507 The Indian Press, Allahabad.

Ganapati, P.N., Radhakrishna, Y., 1958. Andhra Univ. Mem. Oceanogr. 2, 210–237.

Gardiner, J.S., 1903–1906. Gardiner, J.S. (Ed.), The Fauna and Geography of the Maldive and Laccadive Archipelagoes. Cambridge. vol. 2, no. 1. p. 1679.

Gaston, K.J., 1991. The Magnitude of Global Insect Species Richness. Conserv. Biol. 5, 283–296.

George, J., Nair, V.R., 1980. Planktonic ostracods of the Northern Indian Ocean. Mahasagar-Bull. Natl. Inst. Oceanogr. 13 (1), 29–44.

George, G., Sreeraj, C.R., Dam Roy, S., 2011. Brachionid diversity in Andaman waters. Ind. J. Mar. Sci. 40 (3), 454–459.

Gerlach, S., 1980. Development of marine nematode taxonomy up to 1979. Veroffentlichungen desInstituts fur Meeresforschung in Bremerhaven 18, 249–255.

Ghosh, A., 1963. J. Mar. Biol. Assoc. India 5, 239–245.

Ghosh, H.C., 1975. Crustaceana 28 (1), 33–36.

Ghosh, H.C., 1976. Rec. Zool. Surv. India 71, 51–55.

Giles, G.M., 1885. J. Asiat. Soc. Beng. 54, 69–71.

Goodnight, C.J., Goodnight, M.L., Gray, P., 1970. General Zoology. Oxford & IBH Publishing Co, New Delhi, p. 564.

Griffiths, C.L., 2005. Indian J. Mar. Sci. 34 (1), 35–41.

Gnanamuthu, C.P., 1943. The Foraminifera of the Krusadi Island (in the Gulf of Mannar). Bull. Madras Govt. Mus. N S (N H) 1 (2), 1–21.

Gravely, F.H., 1927a. Bull. Madras Govt. Mus. N S (N H) 1 (1), 41–51.

Gravely, F.H., 1927b. Bull. Madras Govt. Mus. 1 (1), 123–124.

Gravely, F.H., 1927c. Bull. Madras Govt. Mus. N S (N H) 1 (2), 123–128.

Gravely, F.H., 1942. Bull. Madras Govt. Mus. N S (N H) 5 (2), 104.

Haldar, B.P., 1985. In: Echiura and Sipuncula.State of Art Report on Estuarine Biology, Workshop on Estuarine Biology, Berbampore, Odissa, 19–22 February1985. No. 9, 13 pp.

Haldar, B.P., 1991. Sipunculans of the Indian coast. Memoirs Zool. Surv. India 17, 1–169.

Haldar, B.P., Choudhury, A., 1995. Hooghly Matla Estuary. Medusae: Cnidaria. Estuarine Ecosystem Series. Part-2. Zool. Surv. India, 9–30.

Hartman, O., 1974. J. Mar. Biol. Assoc. India 16 (2), 609–644.

Henderson, J.R., 1893. Trans. Linn. Soc. Lond. Zool. 5 (2), 325–458.

Heywood, V.H., Watson, R.T. (Eds.), 1996. Global Biodiversity Assessment. Cambridge University Press, New York.

Howey, R.L., 1999. Welcome to the wonderfully weird world of rotifers. Micscape Magazine. http://www.microscopyuk.org.uk/mag/indexmag.html?; http://www.microscopyuk.org.uk/mag/artnov99/rotih.html. (Accessed 10 May 2012).

Hu, S.-X., 2005. Taphonomy and palaeoecology of the Early Cambrian Chengjiang Biota from Eastern Yunnan, China. Berliner Paläobiologische Abhandlungen 7, 1–197.

Hyman, L.H., 1959. The Invertebrates: Smaller Coelomate Groups. vol. 5 McGraw- Hili Book Company, Inc, New York, p. 783.

Jain, S.P., 1981. Checklist of Ostracoda from India-1. Cenozoic Ostracoda (Marine). J. Palaeontol. Soc. India 25, 88–105.

James, D.B., 1986. In: James, P.S.B.R. (Ed.), Recent Advances in Marine Biology. Today and Tomorrow's Printers and Publishers, New Delhi, pp. 569–591.

Jefferson, T.A., Webber, M.A., Pitman, R.L., 2008. Marine Mammals of the World: A Comprehensive Guide to Their Identification. Academic Press, Oxford, UK 592 pp.
Jones, S., Kumaran, M., 1980. In: Fishes of the Laccadive Archipelago: The Nature Conservation and Aquatic Sciences Service. 760 pp.
Kajihara, H., Chernyshev, A.V., Sichun, S., Sundberg, P., Crandall, F.B., 2008. Checklist of nemertean genera and species (Nemertea) published between 1995 and 2007. Species Diversity 13, 245–274.
Kalimuthu, S., Kaliaperumal, N., Ramalingam, J.R., 1991. J. Mar. Biol. Assoc. India 33 (1 & 2), 170–174.
Kathiresan, K., 1999. Mangrove Atlas and Status of Species in India. Project report, Ministry of Environment and Forests, Government of India, New Delhi. 235 pp.
Kemp, S., 1913. An account of the crustacean Stomatopoda of the Indo-Pacific region based on the collection in the Indian Museum. Mem. Indian Mus. 4, 10217.
Kemp, S., 1915. Mem. Indian Mus. 5, 199–325.
Kemp, S., 1916. Rec. Indian Mus. 12 (8), 386–405.
Keesing, J., Irvine, T., 2005. Coastal Biodiversity of Indian Ocean: the known, the unknown and the unknowable. Indian J. Mar. Sci. 34 (1), 11–26.
Kramp, P.L., 1961. Synopsis of the Medusae of the World. J. Mar. Biol. Assoc. UK 40, 1–469.
Krapp, F., 1996. Anoplodactylus sandromagni n.sp. from Kerala, South India (Pycnogonida, Arthropoda). Bollettino del Museo civico di Storia natural di Verona 20, 521–529.
Krishnamoorthi, B., 1963. J. Mar. Biol. Assoc. India 5, 97–102.
Krishnaswamy, S., 1950. Rec. Indian Mus. 48, 117–120.
Krishnaswamy, S., 1952. Rec. Indian Mus. 50, 324.
Krishnaswamy, S., 1953. J. Madras Univ. 23B (l&2), 61–75.
Krishnaswamy, S., 1956a. Rec. Indian Mus. 54, 23–28.
Krishnaswamy, S., 1956b. Rec. Indian Mus. 54, 29–32.
Krishnaswamy, S., 1957. Studies on the Copepoda of Madras. Ph.D. thesis, University of Madras.
Kumaran, P.L., 2002. Curr. Sci. 83 (10), 1210–1220.
Kurian, C.V., 1954. Rec. Indian Mus. 52 (2–4), 275–311.
Kurian, C.V., 1965. J. Mar. Biol. Assoc. India 2, 630–633.
Lambshead, P.J.D., 1993. The first differentiation of roundworms from horsehair worms, though. Oceanis 19 (6), 5–24.
Leloup, E., 1934. Siphonophores calycophorides de l'océan. Atlantique tropical et austral. Bull. Mus. R. Hist. Nat. Belg. 10 (6), 1–87.
Luetken, C., 1865. Kritiske Bemaerkninger om forskillige Seostjermer (Asterider) med Beskrivelse af nogle nye aster. Vidensk. Meddr dansk naturh. Foren 1864, 194–230.
Luetken, C., 1872. Ophiuridarum novarum vel minus cogniterum descriptiones nouvelles. Overs. K. danske Vidensk. Selsk. Forh. 77, 75–158.
Madhavi, R., 2011. Checklist of digenean trematodes reported from Indian marine fishes. Syst. Parasitol. 78, 163–232.
Maiti, P.K., Maiti, P., 2011. Biodiversity: Perception, Peril and Preservation. PHI Learning Pvt. Ltd., p. 542.
Marktanner-Turneretscher, G., 1887. Annln Naturh Mus Wein 2, 291–316.
Mankeshwar, M., Kulkarni, A., Apte, D., 2015. Diversity of Bryozoans of India with new records from Maharashtra. In: Venkatraman, K., Sivaperuman, C. (Eds.), Marine Faunal Diversity in India Taxonomy. Ecology and Conservation. Elsevier, USA, pp. 95–106.
Mason, J., 1873. On *Rhopalorhynchus* Kroyeri, a new genus and species of Pycnogonida. J. Asiat. Soc. Bengal. 42 (2), 171–175.
Matthai, G., 1924. Mem. Indian Mus. 8, 1–59.
May, R., 1988. How many species are there on earth? Science 241, 1441–1449.
Mayr, E., 1968. Bryozoa versus Ectoprocta. Systematic Zool. 17 (2), 213–216. https://dx.doi.org/10.2307/2412368. JSTOR2412368.
Menon, M.G.K., 1931a. Bull. Madras Govt. Mus. (NH) 3 (2), 1–32.
Menon, K.S., 1931b. Rec. Indian Mus. 33, 489–516.
Mills, C.E., 2014. Phylum Ctenophora: List of all valid species names. https://faculty.washington.edu/cemills/Ctenolist.html.
Milne-Edwards, H., 1834–1837. Historie naturelle des Crustaces. Parts I–II (1834–1837). 532 pp.
Müller, J., 1849. Über die Larven un die Metamorphose der Holothurien. 1849, Müller's Archiv, Berlin, pp. 364–399.
Müller, W.E.G., 1995. Molecular phylogeny of Metazoa (animals): monophyletic origin. Naturwiss. 82, 321–329.

Newman, L.J., Cananon, L.R.G., 2003. The World of Polyclads. CSIRO Publishing. 97 pp.
Nilsson Cantell, C.A., 1938. Mem. Indian Mus. 13, 1–81.
NIO, 2016. http://www.nio.org/userfiles/classification.pdf. as on 27/06/2016 Prepared by Nair, V. and Stephen, R.
Ontario, B., 2010. http://www.eoearth.org/view/article/156414.
Oza, R.M., Zaidi, S.H., 2001. A Revised Checklist of Indian Marine Algae. CSMCRI, Bhavnagar. 296 pp.
Pallas, P.S., 1766. Lumbricus echiurus. Miscellania Zoologica. Hagae Comitum, pp. 146–151.
Panikkar, N.K., Aiyar, R.G., 1937. Proc. Indian Acad. Sci. 6 (5), 284–336.
Patriti, G., 1970. Marsielle Fac ser suppl 10, 285–303.
Petrescu, I., Chatterjee, T., 2011. Zootaxa 2966, 51–57.
Pillai, C.S.G., 1991. Zool. Surv. India 1, 41–47.
Pillai, N.K., 1964. J. Mar. Biol. Assoc. India 6 (1), 1–40.
Pillai, N.K., 1968. J. Zool. Soc. India 20, 6–24.
Pillai, P.P., 1967. J. Mar. Biol. Assoc. India 13, 162–172.
Pillai, N.K., 1973. Mysidacea of the Indian Ocean. International Indian Ocean Expedition. In: Handbook of Zooplankton Collection.vol. 6. pp. 1–126.
Prakash, S., Ajith Kumar, T.T., Khan, S.A., 2013. Checklist of the Porcellanidae (Crustacea: Decapoda: Anomura) of India. Check List 9 (6), 1514–1518.
Prasade, A., Apte, D., Kale, P., Oliveira, O.M.P., 2015. Vallicula multiformis Rankin, 1956 (Ctenophora, Platyctenida): first record from the Indian Ocean. J. Biodiversity. https://dx.doi.org/10.15560/11.1.1544.
Priyalakshmi, G., Menon, N.R., 2012. Kerala- an abode of taxonomically challenging permanent meiofauna, Gastrotricha. J. Mar. Biol. Assoc. India 54 (2), 61–64.
Priyalakshmi, G., Menon, N.R., Todaro, M.A., 2007. A new species of Pseudostomella (Gastrotricha: Macrodasyida: Thaumastodermatidae) from a sandy beach of Kerala, India. Zootaxa 1616, 61–68.
Raghunathan, C., Jothinayagam, J.T., 2007. Occurrence of 'Living Fossil' Lingula translucida Dall (Brachiopoda, Lingullidae) along Krishnapatnam Coast of Bay of Bengal. Seshaiyana 15 (1), 3–6.
Ramakrishna, Sarkar, J., 2003. On the Scyphozoa from East Coast of India, including Andaman & Nicobar Islands. Rec. Zool. Surv. India 101 (Part 1–2), 25–56.
Ramakrishna, Sarkar, J., Alukdar, S.T., 2003. Marine invertebrates of Digha coast and some recomendations on their conservation. Rec. zool. Surv. India 101 (Part 3–4), 1–23.
Ramakrishna, Sreeraj, C.R., Raghunathan, C., Sivaperuman, C., Yogesh Kumar, J.S., Raghuraman, R., Immanuel, T., Rajan, P.T., 2010. Guide to Opisthobranchs of Andaman and Nicobar Islands. 196 pp.
Ramalingam, J.R., 2000. Golden Jubilee Celebrations Souvenir 2000, Mandapam R C of CMFRI, Mandapam Camp. pp. 81–83.
Ramamoorthi, K., Venkataramanujam, K., Srikrishnadhas, B., 1973. Mass mortality of Linqula anatine (Lam) (Brachiopoda) in Porto Novo waters. Curr. Sci. 42 (8), 285–286.
Rana, S.S., Nigam, R., Panchang, R., 2007. Relict benthic Foraminifera in surface sediments off central east coast of India as indicator of sea level changes. Indian J. Mar. Sci. 36 (4), 355–360.
Rao, G.C., Ganapati, P.N., 1968. Proc. Indian Acad. Sci. Sec. B (1) 67, 24–29.
Rao, K.K., 1970. Foraminifera of Gulf of Cambay. J. Bombay Nat. Hist. Soc. 66 (3), 584–596.
Rao, G.C., 1971. Gastrotricha in the interstitial fauna of the Bay Islands. J. Biol. Ecol. 1, 42–44.
Rao, G.C., 1972. Occurrence of the interstitial tardigrade Parastygarctus higginsi Renud-Debyser, in the intertidal sands of Andaman Islands. Curr. Sci. 41, 845–846.
Rao, G.C., 1989. Ecology of the meiofauna of sand and mud flats around Port Blair. J. Andaman Sci. Assoc. 5 (2), 99–107.
Rao, G.C., 1991. Meiofauna. In: Fauna of Lakshadweep. State Fauna Series, vol. 2. Zool. Surv. India, pp. 63–81.
Rao, G.C., 1993. Littoral meiofauna OF Little Andaman. Rec. Zool. Surv. India, Occ. Pap. 155, 26–44.
Rao, M.U., 2010. Nat. Symp. Mar. Plants Parangipettai, 5–6.
Raup, D.M., 1979. Size of the Permo-Triassic bottleneck and its evolutionary implication. Science 206, 217–218.
Roy, A., Ghosh, A., 2009. Ocean. Sheuli Chatterjee, Sa Explorer's Institute, Kolkata. 256 p.
Reaka-kudlA, M.L., 1998. The global biodiversity of coral reefs: a comparison with rain forests. In: Reaka-Kudla, M.L., Wilson, D.E., Wilson, E.O. (Eds.), Biodiversity II: Understanding and Protecting Our Biological Resources. Joseph Henry Press, Washington, DC, pp. 36–50.
Reddy, K.N., Ramakrishna, G., 1972. Rec. Zool. Surv. India 66 (1–4), 19–30.
Rengarajan, K., 1974. J. Mar. Biol. Assoc. India 16, 280–286.

Ruppert, E., Fox, R., Barnes, R., 2004. Invertebrate Zoology: A Functional Evolutionary Approach, seventh ed. Thomson-Brooks/Cole, Belmont, CA.

Saiz, J.I., Bustamante, M., Tajadura, J., Vijapure, T., Soniya, S., 2015. A new subspecies of Phascolion Théel, 1875 (Sipuncula: Golfingiidae) from Indian waters. Zootaxa 3931 (3), 433–437.

Samanta, S., Choudhury, A., Chakraborty, S.K., 2014. New record of a primitive brachiopod benthic fauna from North-East coast of India. Int. J. Cur. Res. Acad. Rev. 2 (3), 70–73.

Sarojini, D., 1958. Studies on littoral Foraminifera in the Bay of Bengal. M.Sc. thesis, Andhra University.

Sarojini, R.A., Nagabhushanam, R., 1972. Rec. Zool. Surv. India 66, 249–272.

Sastry, D.R.K., 1998. Echinodermata. In: Faunal Diversity in India. Zoological Survey of India, Kolkata, pp. 398–403.

Satheesh, S., Wesley, S.G., 2012. Occurrence of sea spider Endeis mollis Carpenter (Arthropoda: Pycnogonida) on the test panels submerged in Gulf of Mannar, Southeast coast of India. Arthropods 1 (2), 73–78.

Sato, A., Bishop, J.D.D., Holland, P.W.H., 2008. Developmental biology of pterobranch hemichordates: history and perspectives. Genesis 46 (11), 587–591. https://dx.doi.org/10.1002/dvg.20395.

Satyanarayana Rao, K., 1998. Bryozoa. In: Faunal Diversity in India. Zoological Survey of India, Kolkata, pp. 371–377.

Schram, F.R., 1973. Pseudocoelomates and a nemertine from the Illinois Pennsylvanian. J. Paleontol. 47, 985–989.

Sewell, R.B.S., 1924. Mem. Indian Mus. 5, 771–852.

Sewell, R.B.S., 1929. Mem. Asiat. Soc. Beng. 9, 133–205.

Sewell, R.B.S., 1935. Mem. Asiat. Soc. Beng. 9, 461–540.

Sewell, R.B.S., 1940. Rep. John Murray Exped. 7, 117–382.

Shipley, A.E., 1903. Gardiner, J.S. (Ed.), Fauna and Geography of the Maldive and Laccadive Archipelagoes.In: vol. 1, pp. 131–140.

Siebold, I., 1974. Rev. Esp. Micropaleontol. 7, 175.

Simon, Matt, 2014. Wired (magazine). http://www.wired.com/wiredscience/2014/03/absurd-creature-week-waterbear/.

Sivaprakasam, T.E., 1968a. J. Mar. Biol. Assoc. India 14, 34–51.

Sivaprakasam, T.E., 1968b. J. Mar. Biol. Assoc. India 14, 274–282.

Sivaprakasam, T.E., 1970a. J. Mar. Biol. Assoc. India 16, 81–92.

Sivaprakasam, T.E., 1970b. J. Mar. Biol. Assoc. India 16, 93–96.

Sivaleela, G., Venkataraman, K., 2015. Free-living Marine Nematodes of Tamil Nadu Coast, India. Records of the Zoological Survey of India, Occasional Paper No. 336, pp. 1–40.

Sivaleela, G., Venkataraman, K., 2009. Meiofauna of Gulf of Mannar. Rec. Zool. Surv. India 109 (4), 25–33.

Smith, M.A., 1931. Fauna of British India, etc. Reptiles and Amphibia 1. Loricata. Testudines. Taylor and Francis, London, p. 185.

Soja, L., 2006. Taxonomy, bionomics and biofouling of bryozoans from the coast of India and the Antarctic waters. Ph.D. thesisCochin University of Science and Technology. 336 pp.

Short, F., Carruthers, T., Dennison, W., Waycott, M., 2007. Global seagrass distribution and diversity: a bioregional model. J. Exp. Mar. Biol. Ecol. 350, 3–20.

Sournia, A., Chrétiennot-Dinet, M.J., Ricard, M., 1991. Marine phytoplankton: How many species in the world's oceans? J. Plankton Res. 13, 1093–1099.

Southern, R., 1921. Mem. Indian Mus. 5, 563–659.

Srinivasan, M., 1971. Two new records of meso- and bathy-planktonic chaetognaths from the Indian seas. J. Mar. Biol. Assoc. India. 13 (1) 130–132.e.

Shrinivaasu, S., Venkatraman, K., Mohanraju, R., 2011. Baseodiscus Hemprichii Ehrenberg (1831) (Phylum Nemertea) New Distributional Record from Andaman and Nicobar Islands. India. Rec. Zool. Surv. India 111, 1–4 Part-1.

Shrinivaasu, S., Venkatraman, K., Venkatraman, C., 2015a. Checklist of Nemerteans with a New Record in Indian Coastal Waters. In: Venkataraman, K., Raghunathan, C., Mondal, T., Raghuraraman, R. (Eds.), Lesser Known Marine Animals of India. Zool. Surv. India, Kolkata, pp. 173–175 (Chapter 10).

Shrinivaasu, S., Venkatraman, C., Rajan, R., Venkataraman, K., 2015b. Marine Bryozoans of India. In: Venkataraman, K., Raghunathan, C., Mondal, T., Raghuraraman, R. (Eds.), Lesser Known Marine Animals of India. Zool. Surv. India, Kolkata, pp. 1–550.

Shynu, S.P., Shibu, S., Jayaprakas, V., 2015. First record of nemertean Lineus mcintoshii (Nemertea: Anopla: Heteronemertea) from the Indian coast. Mar. Biodiversity Rec. 8, e25.

Spalding, M., Taylor, M., Ravilious, C., Short, F., Green, E., 2003. The distribution and status of seagrasses. In: Green, E.P., Short, F.T. (Eds.), World Atlas of Seagrasses. University of California Press, Berkeley, CA, pp. 5–26.

Stanley, G.D., Stürmer, W., 1983. The first fossil ctenophore from the Lower Devonian of West Germany. Nature 303, 518–520.

Stanley, G.D., Stürmer, W., 1989. A new fossil ctenophore discovered by X-rays. Nature 327, 61–63.

Subba Rao, N.V., 1991. Mollusca. In: Animal Resources of India. Zoological Survey of India, pp. 125–147.

Subba Rao, N.V., 1998. Mollusca. In: Faunal Diversity in India. Zoological Survey of India, pp. 104–117.

Subba Rao, N.V., Dey, A., 2000. Rec. Zool. Surv. India. Occ. Paper No. 187. 323 pp.

Subba Rao, M., Vedantam, D., 1968. Distribution of Foraminifera in shelf sediments off Visakhapatnam. Nat. Inst. Sci. Ind. Bull. 38, 491–501.

Surya Rao, K.V., 1974. Proc. Indian Natn. Sci. Acad. 38, 190–205.

Sundaresan, D., 1968. Brachiopod larvae from west coast of India. Proc. Ind. Acad. Sci. 68, 59–68.

Szaniawski, H., 2005. Cambrian chaetognaths recognized in Burgess Shale fossils. Acta Palaeontol. Pol. 50 (1), 1–8.

Talwar, P.K., 1991. Pisces. In: Animal Resources of India: Protozoa to Mammalia. 1. Zoological Survey of India, Kolkata, pp. 577–630.

Talwar, P.K., Kakkar, R.K., 1984. Handbook of Commercial Sea Fishes of India. 997 Zoological Survey of India, Kolkata.

Tattersall, W.M., 1911. Trans. Linn. Soc. Lond. 15 (1), 119–136.

Tattersall, W.M., 1922. Rec. Indian Mus. 24, 445–504.

Temereva, E.N., Malakhov, V.V., Yakovis, E.L., Fokin, M.V., 2000. Phoronis ovalis (Phoronida, Lophophorata) in the White Sea: the first discovery of phoronids in the Arctic Basin. Dokl. Biol. Sci. 374, 523–525.

Thomas, P.A., George, M.A., Lazarus, S., 1995. J. Mar. Biol. Assoc. India 37, 134–142.

Thuesen, E.V., Pierrot-Bults, A., 2016. WoRMS Chaetognatha: World list of Chaetognatha (version 2016-07-01). In: Roskov, Y., Abucay, L., Orrell, T., Nicolson, D., Kunze, T., Flann, C., Bailly, N., Kirk, P., Bourgoin, T., De Walt, R.E., Decock, W., De Wever, A. (Eds.), Species 2000 & ITIS Catalogue of Life, 28th July 2016. Species 2000: Naturalis, Leiden, The Netherlands. ISSN 2405-8858. Digital resource at www.catalogueoflife.org/col.

Tiwari, K.K., Ghosh, H.C., 1975. Proc. Zool. Soc. Calcutta 26, 33–37.

Totton, A.K., 1954. Disc. Rep. Cambridge 27, 161.

Venkataraman, G., 1939. A systematic account of some south Indian diatoms. Proc. Indian Acad. Sci. 10, 293–368.

Venkataraman, K., Satynarayana, C.H., 2012. Corals Identification Manual. 136 pp.

Venkataraman, K., Srinivasan, M., Satyanarayana, C.H., Prabhakar, D., 2002. Faunal Diversity of Gulf of Mannar Biosphere Reserve. In: Conservation Area Series 15. Zool. Surv. India, Kolkata, pp. 1–77.

Venkataraman, K., Jeyabaskaran, R., Raghuram, K.P., Alfred, J.R.B., 2004. Bibliography and checklist of corals and coral reef associated organisms of India. Rec. Zool. Surv. India Occ. Paper 226. 648 pp.

Venkataraman, K., Raghunathan, C., Raghuraman, R., Dixit, S., 2015a. Fascinating Seaslugs and Flatworms of Indian Seas. Director, Zool. Surv. India, Kolkata, pp. 1–149.

Venkataraman, K., Raghunathan, C., Raghuraman, R., Mondal, T., 2015b. Horse-shoe worm Phoronida. In: National Workshop on Lesser Known Marine Animals of India, 11–13 June 2015. Director, Zoo. Surv. India, Kolkata.

Vivekanandan, E., Jeyabaskaran, R., 2012. Marine Mammal Species of India. Central Marine Fisheries Research Institute, Kochi 228 pp.

Von Fichtel, L., Von Moll, J.P.C., 1798. Testacea microscopica, aliaqua minuta ex generibus Arpmauta et Mautilus and naturam pieta et descripta, 2nd edition vol. 7. p. 123. 1803.

Waloszek, D., Dunlop, J.A., 2002. A larval sea-spider (Arthropoda: Pycnogonida) from the upper Cambrian 'orsten' of Sweden and phylogenetic position of pycnogonids. Palaeontology 45 (3), 421–446.

Weltner, W., 1894. Zwei neue Cirripediea aus dem Indischen Ocean. S.B. Ges Natur Fr, Berlin.

Whitehead, P.J.P., Talwar, P.K., 1976. Bull. Br. Mus. Nat. Hist. (Sr.) 5, 1–189.

Wood Mason, J., Alcock, A., 1891. Ann. Mag. Nat. Hist. 7, 1–19. 186–202, 258–272.

WoRMS, 2016. World Register of Marine Species. http://www.marinespecies.org/aphia.php?p=taxdetails&id=1839. (Accessed 8 September 2016).

Yogesh Kumar, J.S., Raghunathan, C., Geetha, S., Venkataraman, K., 2014. New species of soft corals (Octocorallia: Alcyonacea) on the coral reef of Andaman and Nicobar Islands. Int. J. Integrative Sci., Innov. Technol. 1 (3), 8–11 (eISSN 2278-1145).

Zhang, Z.Q., 2011. Animal biodiversity: an introduction to higher-level classification and taxonomic richness. Zootaxa 3148, 7–12.

Zoological Survey of India (ZSI), 2011. Protected Species of Mollusca.

Further Reading

Annandale, N., Kemp, S., 1915a. Mem. Indian Mus. 5, 55–63.
Annandale, N., Kemp, S., 1915b. Mem. Indian Mus. 5, 519–558.
Barman, R.P., 1998. Pisces. In: Faunal Diversity in India. Zoological Survey of India, Kolkata, pp. 418–426.
Bhatt, J.R., Ramakrishna, Sanjappa, M., Remadevi, O.K., Nilaratna, B.P., Venkataraman, K., 2013. Mangroves of India: Their Biology and Uses. Zoological Survey of India, p. 640.
Borradaile, L.A., 1906. St. Gardiner, J. (Ed.), The Fauna and Geography of the Maldive and Laccadive Archipelagoes, p. 431.
Bouchet, P., 2006. The magnitude of marine biodiversity. In: Duarte, C.M. (Ed.), The Exploration of Marine Biodiversity, Scientific and Technological Challenges. Fundacion BBVA, Bilbao.
Browne, E.T., 1905. Rep. Govt. Ceylon Pearl Oyster 4, 131–166.
Browne, E.T., 1906. Trans. Linn. Soc. London (Zool.) 10, 163–187.
Browne, E.T., 1916. Trans. Linn. Soc. London (Zool.) 17, 169–210.
Daniel, A., 1975. J. Mar. Biol. Assoc. India 16 (2), 182–210.
Devanesan, D.W., Varadarajan, S., 1939. Curr. Sci. 8 (4), 157–159.
Groombridge, B., Jenkins, M.D. (Eds.), 2000. Global Biodiversity: Earth's Living Resources in the 21st Century. World Conservation Press, Cambridge.
Haldar, B.P., 1977. Newsl. Zool. Surv. India 3 (3), 120–123.
Haldar, B.P., 1985b. Second National Seminar on Marine Intertidal Ecology, Waltair. Abstract 26.
Harmer, S., 1915. Siboga Expedition Monographs. 28a 565 pp.
Jaume, D., Boxshall, G.A., 2008. Hydrobiologia 595, 225–230.
Johnson, p., 1964. Ann. Mag. Nat. Hist. 7 (13), 331–335.
Johnson, P., 1969. J. Bombay Nat. Hist. Soc. 66 (1), 43–46.
Johnson, P., 1971. J. Bombay Nat. Hist. Soc. 68 (3), 596–608.
Kaladharan, P., Kaliaperumal, N., 1999. Naga 22 (1), 11–14.
Kannan, L., Thangaradjou, T., Anantharaman, P., 1999. Seeweed Res. Utiln. 21 (1 & 2), 25–33.
Laidlaw, F.F., 1902. Fauna and Geology of the Maldive and Laccadive Archipelagoes 1, 282–312.
Madhu, R., Madhu, K., 2007. J. Mar. Biol. Assoc. India 49 (2), 118–126.
Madhu Pratap, M., 1979. Indian J. Mar. Sci. 8 (1), 1–8.
Mammen, T.A., 1963. J. Mar. Biol. Assoc. India 5 (1), 27–61.
Mammen, T.A., 1965. J. Mar. Biol. Assoc. India 5 (1), 1–57.
Myers, N., Mittermeier, R.A., Mittermeier, C.G., Da Fonseca, G.A.B., Kent, J., 2000. Nature 403, 853–858.
Nair, K.K., 1945. Proc. Indian Sci. Congr. 3, 97. 32nd Sess.
Nair, K.K., 1954. Bull. Cent. Res. Inst. Univ. Travancore 2 (1), 47–75.
Nair, R.V., 1967. Proc. Symp. Indian Ocean Bull. N I S I 38, 747–752.
Nair, R.V., 1971. J. Mar. Biol. Assoc. India 13 (2), 226–233.
Nair, R.V., 1973. I O B C Hand Book 5, 87–96.
Nair, R.V., 1974. J. Mar. Biol. Assoc. India 16 (3), 721–730.
Nair, R.V., 1975. Mahasagar 8 (1&2), 81–86.
Nair, V.R., Achuthankutty, C.T., Nair, S.S.R., Madhupratap, M., 1981. Indian J. Mar. Sci. 10 (3), 270–273.
Nair, R.V., Rao, T.S.S., 1973. In: Zeitschel, B. (Ed.), The Biology of the Indian Ocean. Chapman and Hall, London, pp. 1–549.
Nair, R.V., Selvakumar, R.A., 1979. Magasagar Bull. Nat. Inst. Oceanogr. 12, 17–25.
Nayar, K.N., 1959. Bull. Madras Govt. Museum (N H) VI (3), 59.
Nayar, K.N., 1966. Proceedings of the Symposium on Crustacea. MBAI. 1, pp. 133–168.
Prasad, P.R., 1956. Indian J. Fish. III, 1–42.
Prashad, B., 1919. Rec. Indian Mus. 16, 399–402.
Prashad, B., 1935. Rec. Indian Mus. 37, 39–43.
Prashad, B., 1936. Rec. Indian Mus. 38, 231–238.
Raghunathan, C., Raghuraman, R., Choudhury, S., Venkataraman, K., 2014. Diversity and distribution of sea anemones in India with special reference Andaman and Nicobar Islands. Rec. Zool. Surv. India 114 (2), 269–294.
Rao, T.S.S., 1958a. Andhra Univ. Mem. Oceanogr. 2, 137–146.
Rao, T.S.S., 1958b. Andhra Univ. Mem. Oceanogr. 2, 164–167.

Rao, T.S.S., Ganapati, P.N., 1958. Andhra Univ. Mem. Oceanogr. 2, 147–163.

Rao, G.C., 1975. The interstitial fauna in the intertidal sands of Andaman and Nicobar group of islands. J. Mar. Bioi. Ass. India 17 (2), 116–128.

Renganathan, T.K., 1986. Studies on the ascidians of south India. Ph.D. thesis, Madurai Kamaraj University 249 pp.

Rink, H.J., 1847. Die Nikobarischen Inselin. Eine geographische Skizee, mit spezieller Berucksichtigung der Geograhie. H. G. Klein, Kopenhagen.

Ruppert, E.E., Barnes, R.D., 1994. Invertebrate Zoology, Sixth Edition Saunders College Publishing Harcourt Brace and Company, Orlando, FL. 1100 pp.

Salinas, J.I.S., 1993. J. Nat. Hist. 27 (3), 535–555. https://dx.doi.org/10.1080/00222939300770301.

Sreeraj, C.R., Raghunathan, C., 2011. J. Mar. Biol. Assoc. UK 4, e73. https://dx.doi.org/10.1017/S1755267211000819.

Sreeraj, C.R., Raghunathan, C., 2013. Proc. Int. Acad. Ecol. Environ. Sci. 3 (1), 36–41.

Stephen, A.C., Edmonds, S.J., 1972. The Phyla Sipuncula and Echiura. Trustees of the British Museum (Natural History), London.

Sudarsan, D., 1963. Indian J. Fish. 8 (2), 364–382.

Thomas, P.A., Alfred, J.R.B., Das, A.K., Sanyal, A.K., 1998. Porifera. In: Faunal Diversity in India. ENVIS Centre, Zoological Survey of India, Kolkata, pp. 27–36.

Varadarajan, S., 1934. Curr. Sci. 8, 3–6.

Varadarajan, S., Chacko, P.I., 1943. Proc. Nat. Inst. Sci. India 9, 245–248.

Venkataraman, K., 1998. In: Alfred, J.R.B., Sanyal, A.K., Das, A.K. (Eds.), Faunal Diversity in India. Zoological Survey of India, Kolkata, pp. 391–395.

CHAPTER

12

Implication and Management of Coastal Salinity for Sustainable Community Livelihood: Case Study From the Indian Sundarban Delta[☆]

*Rajarshi DasGupta**, *Rajib Shaw*[†], *Mrittika Basu*[‡]

*Institute for Global Environmental Strategies (IGES), Hayama, Kanagawa, Japan

[†]Keio University, Japan

[‡]United Nation's University Institute for Advanced Studies (UNU-IAS), Tokyo, Japan

1 INTRODUCTION

The history of coastal development in the Asian megadeltas predominantly narrates the reclamation of marshy swamps into agricultural establishments, thereby expanding human settlements toward the coast. Over the past several decades, agriculture, especially rice cultivation, has played a dominant role in these megadeltas by shaping the local economy, catering to regional food demands, and providing income opportunities to isolated rural communities. However, with recent cyclonic storms over the north Indian Ocean and intensified surges (e.g., Cyclone Sidr in 2007; Cyclone Nargis in 2008; Cyclone Aila in 2009; Cyclone Phailin in 2013, etc.), sustainability of coastal agriculture remains at a stake. In particular, prolonged surge water flooding over the low alluvial plains and the consequent salinization of coastal agricultural lands have led to a massive discontinuation of community livelihood in

[☆] This chapter is derived from first author's doctoral dissertation, archived in the Kyoto University library as Dasgupta Rajarshi (2016): "Enhancing Coastal Communities' Disaster and Climate Resilience in Mangrove Rich Indian Sundarban."

https://doi.org/10.1016/B978-0-12-810473-6.00013-3

© 2019 Elsevier Inc. All rights reserved.

the Ganges-Brahmaputra-Meghna (GBM) Delta in India and Bangladesh, the Ayeyarwady Delta in Myanmar, and the Mekong Delta in Vietnam (Nhan et al., 2012; DasGupta and Shaw, 2015; SeinnSeinn et al., 2015). It is further expected that the unfavorable soil and water salinity scenario will continue to be aggravated by a number of rapid and slow onset coastal hazards, led by climate change-induced sea level rise and rainfall variability in coastal regions (Knutson et al., 2010; Redfern et al., 2012; Dasgupta et al., 2014).

Irrespective of its confinement to relatively small geographical areas, unfavorable water and soil salinity are being compounded as an immediate threat to coastal agriculture, faster and more intensely than any other climate variables. Empirical studies have already indicated that each unit (dS/mL) rise of salinity roughly corresponds to a 12% reduction in rice yields, along with structural damage resulting in poor market values (Hanson et al., 1999; Redfern et al., 2012). As for example, Abedin et al. (2012) mentioned that, on average, Bangladesh losses 0.2 Mmt of rice every year due to salinity intrusion in coastal areas. The adverse impact of salinity is further hypothesized by a number of researchers as a determining factor behind the remarkable outward migration of agricultural workers from the Asian megadeltas (Basu and Shaw, 2013; Dasgupta et al., 2014; Basu et al., 2015). Needless to say, adapting to such an adverse salinity scenario remains highly imperative for coastal managers, policy planners, and coastal agricultural communities at large.

Since the realization of climate change adversity and its predicted impacts, adaptation has occupied the central theme of agricultural sustainability because of the greater susceptibility and contribution to domestic production in the developing countries (Smit and Skinner 2002; Bradshaw et al., 2004). Yet, despite extensive research on agricultural adaptation options against climate variables such as temperature and rainfall (e.g., Deressa et al., 2009), the understanding of adaptive responses against unfavorable salinity scenarios is incomplete at the present (Connor et al., 2012). Considering the above, this chapter presents the results of an empirical study conducted over the salinity-impacted coastal plains of the lower Gangetic delta in India (*aka Indian Sundarban delta*) involving local farmers who suffered severe livelihood discontinuation from Cyclone Aila in 2009. The objectives of this chapter are threefold. First, it narrates the implications of unfavorable salinity scenarios in the study area and the adverse consequences in rice production following Cyclone Aila (2009). Second, the chapter narrates the findings from an empirical study conducted to identify the potential adaptive and coping options against the rising salinity scenario. Finally, it further narrates the results of a pilot study to identify farmers' willingness and intention to embrace specific adaptive/coping actions.

2 DELINEATION OF THE STUDY AREA

The lower Gangetic Delta, also known as the Indian Sundarban Delta, is an archipelago of 103 low-lying, densely populated deltaic islands at the confluence of the River Ganges-Brahmaputra-Meghna and the Bay of Bengal (Fig. 1). The area is geographically bounded between 21°32′ to 22°40′ northern latitude and 88°05′ to 89°00′ eastern longitude. It spreads over nearly 10,000 sq. km within the geographical territory of the state of West Bengal in India. Historically, the entire delta was covered by dense mangrove forests. However, over the

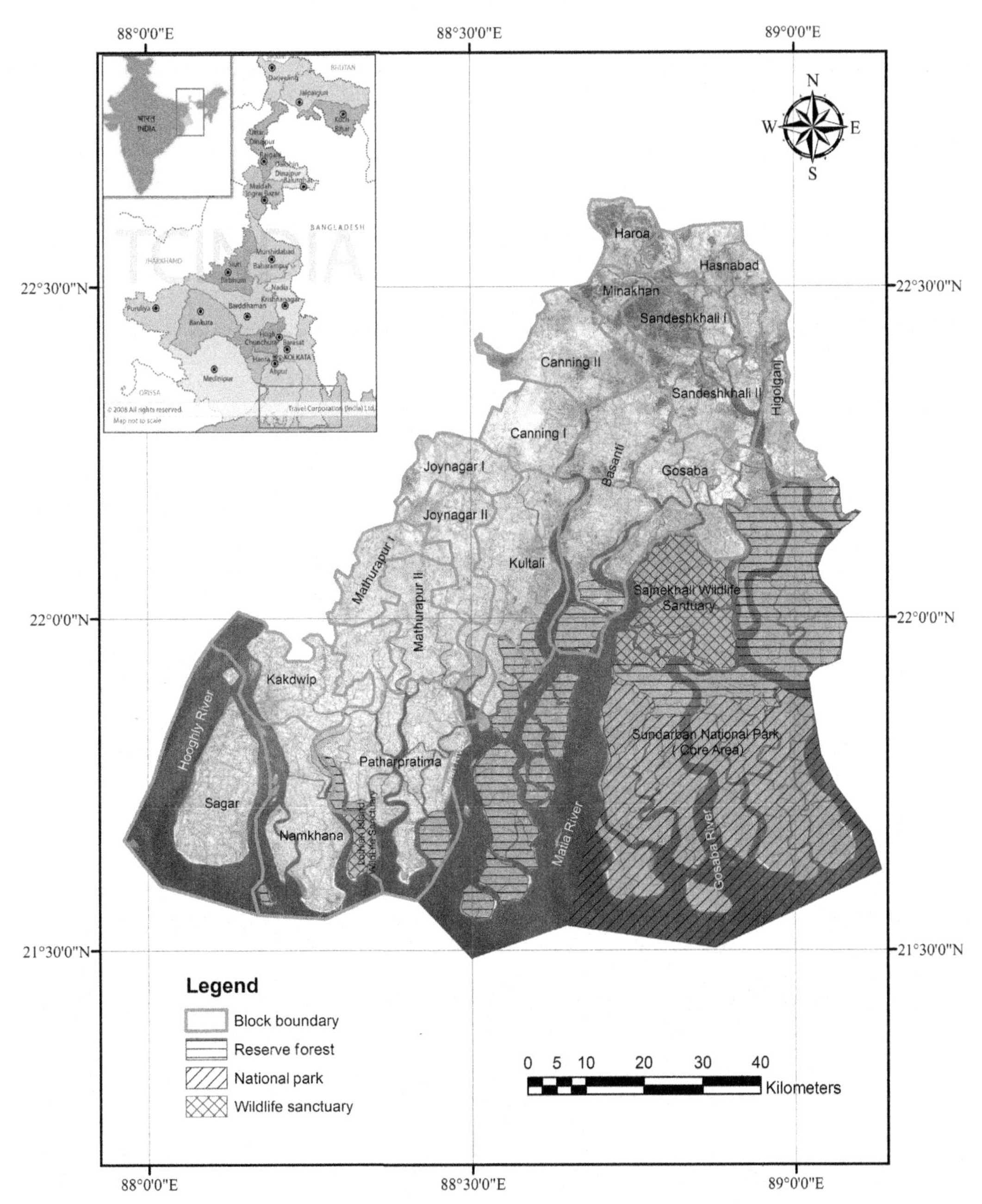

FIG. 1 Location Map of the Indian Sundarban Delta on a recent Landsat OLI satellite image, along with the protected mangrove area and reclaimed islands marked within the yellow block boundaries.

years, more than 54 islands were reclaimed for the purpose of agriculture by establishing a nearly 3500km long earthen embankment to prevent daily tidal inceptions. Presently, the delta has two distinct landscapes that include large protected mangrove forests in the eastern part while the majority of the western part consists of reclaimed agricultural land with rural settlements. Nearly half the reclaimed area, measuring about 2482.33 sq. km. is used for coastal agriculture, mainly producing rice as a major harvest (SBR, 2008). As per the official records, 60%–65% of the 4.37 million dwellers are directly involved in agriculture while indirect dependency may reach well up to 95% (District Human Development Report, 2009). However, about 54% of the communities are landless, which implies a considerably high number of "agricultural labors" who hold no land. Apart from rice, the other important cash crops are chili, cotton, watermelon, cucumber, betel leaves, etc.

2.1 Impact of Cyclone Aila on Coastal Agriculture

In this section, we attempt to provide a landscape-scale analysis of the impact of salinity on the Indian Sundarban Delta, with particular emphasis on the storm surge and flooding caused by Cyclone Aila in 2009. Information provided in this section is derived from various local government sources through a careful screening of a number of unpublished reports. Cyclone Aila, a severe cyclonic storm with maximum sustained wind speeds of approximately 110km/hour, hit the Indian Sundarban Delta on May 26, 2009. The storm overlapped with the morning retreating tide on the day just before a new-moon night, generating a storm surge of more than 3m above the high tide line (HTL). This was higher than the maximum height of most of the earthen embankments protecting the low-lying islands of the delta. As per the official damage assessment report, nearly 613.785 km of earthen embankments were either breached or completely washed away by the powerful surges, leading to instant flooding of the agricultural lands along with the rural settlements. The storm resulted in severe damage to agricultural lands, completely destroying the livelihood of the communities (Fig. 2). Direct loss of agricultural products and revenue was estimated as 1472ha of rice cultivation, 748ha of betel leaves cultivation, and 2151ha of vegetable cultivation, costing nearly INR 12 crores (approximately $20 million). Indirect losses due to the salinization of agricultural plots and the loss of freshwater ponds may reach up to INR 125 crores ($26.3 million). Importantly, because Aila occurred in late May, there was no particular damage to rice cultivation itself. However, it completely destroyed the crop potential for the next couple of years through extensive salinization of fertile agricultural lands (DasGupta and Shaw 2015). An internal administrative report from the district administration suggests that nearly 1390 sq. km of agricultural land was flooded and salinized due to Aila (Office of the District Magistrate, South 24 Parganas 2010). In the post-Aila scenario, most of these agricultural lands could not be cultivated due to excessive salinity, although the official report suggests that, in some areas, 20% of the regular agriculture was possible with the help of extensive irrigation. Due to the process of salinization, an estimated 226,345 families were without employment for the next couple of years. Not surprisingly, it also led to massive outward migration and deterioration of local law and order. In addition, Aila also caused vital damage to the rural agricultural facilities, particularly the freshwater ponds. The same administrative report mentioned that nearly 160,000 ponds were affected by saline water while the local

FIG. 2 (*Top*) Aerial photographs of flooded agricultural land after Cyclone Aila. (*Bottom*) Ground images of flooded agricultural lands in the aftermath of Cyclone Aila. (Top): *Courtesy: Indian Air Force.* (Bottom): *Courtesy: District Disaster Management Authority, South 24 Parganas, Government of West Bengal.*

government could retrieve only 60,000 ponds the following year. The cultivable fields could not be desalinized and the situation became even worse because of a paucity of rain in June 2009. As revealed by the local farmers, the unfavorable salinity scenario lasted until 2012 with crop production reduced to less than half the normal yield.

3 MATERIAL AND METHODS

Apart from the immediate stress posed by the storm, it is widely acclaimed that the regional trend of climate-induced sea level rise will adversely aggravate the salinity scenario. In particular, changes in inland water and soil salinity have been identified as one of the major determinants that can jeopardize the traditional agricultural practices in the Sundarban Delta (Abedin et al., 2012; Dasgupta et al., 2014). Therefore, it remains imperative that the communities as well as the local government prepare a robust anticipatory adaptation strategy that can sustain the traditional agro-based livelihood.

Agricultural adaptation, in simpler terms, essentially denotes the methods and practices that maintain agricultural objectives such as productivity, price, and market sustainability under external stressors. However, being a complex multilayer system, it requires systematic intervention at different levels of operations (e.g., farm level or institutional level) (Risbey et al., 1999; Smit and Skinner, 2002). Nevertheless, researchers argue that, irrespective of proper institutional arrangements, sustainability of adaptive actions largely depends on the farmers' intention to adapt, their ability to apprehend the threats,a and a multitude of other factors such as technical and financial capacity, social capital, and an institutional support system (Smit and Skinner, 2002; Deressa et al., 2009; Feola and Binder 2010; Basu et al., 2015). Considering the above, the study was essentially designed to keep the farmers or the rural producers at the core of the planning process.

The methods applied for this study essentially rely on the conventional participatory rural appraisal (PRA) tool that includes six focus group discussions (FGDs), a transect walk across salinity-affected areas, and a follow-up questionnaire survey of 126 farmers. We essentially relied on selective sampling techniques and participation preferences were given to those who had suffered livelihood discontinuation in Cyclone Aila. Site selection for the FGDs was chosen from the villages of Gosaba, Pathar Pratima, Kultali, and Sagar Blocks (see Fig. 1 for the location of the blocks), which sustained massive disruption of coastal agriculture. During the FGDs, enquiries were typically made regarding the farmer's coping/adaptive methods in response to high soil salinity and other environmental stressors. The discussion summarized a number of adaptive/coping actions that the farmers proposed based on their individual experiences, shared learning from the local NGOs, and local agricultural research agencies. Thereafter, a list of agricultural coping/adaptive actions was finalized from the above discussions. Physical (visible) adaptation measures were also subsequently verified by transect walks in the affected areas.

In the second stage of the research, 126 farmers were surveyed in Gosaba, Sagar, Kultali, and Pathar Pratima through a questionnaire, typically designed on the outcome of the above-mentioned FGDs. The sample population of farmers was later classified based on their landholding, that is, *Class I or Agricultural labors* (*n* = *35*) (no landholding or landholding of less than a bigha), *Class II or Marginal Farmers* (possess land measuring within 1–3 bighas or 1 acre) (*n* = *41*), and *Class III or Small farmers* (*n* = *50*), (land holding greater than 3 bighas but less than 10 bighas).[1] The questionnaire survey was typically intended to identify the variation

[1] As per the prevailing definition of the provincial government of West Bengal, farmers having less than 1 ha of land are broadly defined as marginal farmers whereas farmers cultivating land measuring between 1 and 2 ha are defined as small farmers (Adhikari et al., n.d.). However, as per this definition, the classification of farmers in the study area and the interpretation of their perception would be a gross generalization. This is because the local unit of land is a bigha (1 bigha equals one-third of an acre or 0.1338 hectares) and farmers having one acre of land are considered to be significantly dependent on agriculture whereas the definition does not mention agricultural laborers who often are landless or cultivate a tiny portion of land. Official figures suggest that nearly 85% of the farmers in the Indian Sundarban Delta are considered marginal or small farmers that do not allow the proper data representation for a microlevel perception analysis. Considering the above, we made this simple classification method to better understand the micro-level adaptation intentions.

between incurred agricultural damages and choices of adaptive actions among these three groups of cultivators. The questionnaire survey essentially relied on a five-point Likert scale-based prioritization technique, where "1" denotes least priority/significance and "5" denotes highest priority/significance. Using this classification system, farmers were asked to classify each of the adaptive/coping actions based on their individual perception. The data obtained from the questionnaire survey were first converted into frequencies and thereafter, statistical hypothesis testing was conducted over the observed frequencies for each of the threats and adaptive actions. The Null hypothesis (H_0) for this study was framed such that there is no statistically significant difference in adaptation intentions among these three groups, and conversely the alternative hypothesis was farmed. In the majority of cases, the χ^2 test was conducted for hypothesis testing. However, in some cases, due to the nonfulfillment of the test conditions, the Freeman Halton Extension of the Fisher Exact Test was adopted ($P < .05$).The observed P values were later used to accept or discard the null hypothesis.

4 RESULTS AND DISCUSSIONS

4.1 Individual Loss Assessment

During the FGDs, local farmers mentioned that because the landfall of Aila coincided with the no-crop season, direct damage to agriculture was limited, hence it didn't affect investments at large. However, as mentioned earlier, the main problems identified by the farmers are the post-Aila unfavorable soil and inland water salinity scenario and the heavy yield loss in consecutive years, which eventually led to a major livelihood crisis. In general, the identified issues during and after the cyclone revolve around crop yield loss, permanent land loss, high residual salinity in agricultural land, loss of livestock, financial loss, and loss of life. In the follow-up questionnaire, 100% of the agricultural labors and nearly 90% of the marginal and small farmers reported yield loss while more than 70% of respondents mentioned the occurrence of high residual soil salinity in their agricultural land (Fig. 3). It is important to

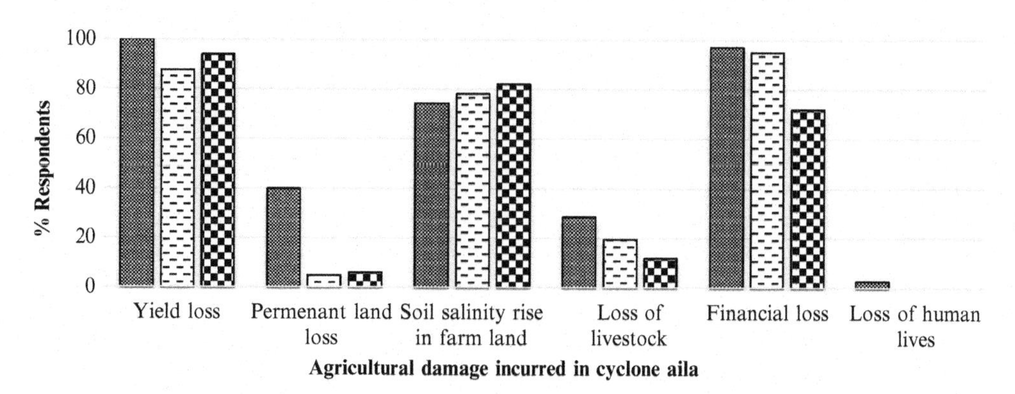

FIG. 3 Agricultural damage incurred in Cyclone Aila.

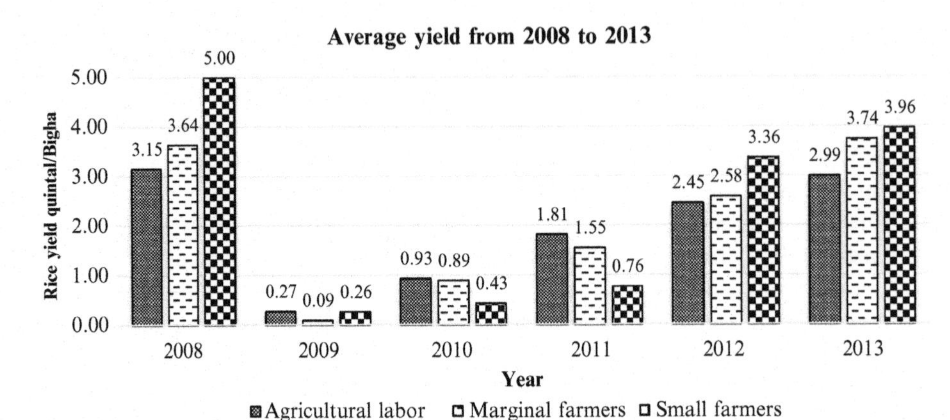

FIG. 4 Reported rice yields in Quintal/Bigha from 2008 to 2013.

note that the yield loss and consequent loss of invested capital are the cumulative consequences of high residual soil salinity that led to massive discontinuation of agriculture. In order to explore the issue in detail, the questionnaire typically inquired about the chronological crop yield since 2008. It was revealed that during 2008–2009 (before Cyclone Aila), the average crop yield (for the Aman Paddy) was nearly five quintal/bigha (approximately 37,000 kg/ha) for small farmers and 3.64 quintal/bigha (approximately 27,200 kg/ha) and 3.15 quintal/bigha (approximately 23,500 kg/ha) for marginal farmers and agricultural labors, respectively. However, in 2009–2010, due to extensive salinization of agricultural lands, virtually no crops were grown. Since 2010–2012, the scenario improved slowly although, as Fig. 4 suggests, crop yields did not recover fully until 2013.

4.2 Farmer's Understanding About the Drivers of Unfavorable Salinity

The community-level understanding of the existing drivers of unfavorable soil and inland water salinity, as revealed during the FGDs, remains primarily linked to the extensive saline water flooding during Cyclone Aila. However, farmers also highlighted other auxiliary factors such as lack of irrigation facilities, reduction of fresh water supply, drowning of tube wells, poor seed variety, and lack of rainfall during recent years which made the problem of unfavorable salinity even severe. It is important to mention that the Indian Sundarban Delta has largely missed out on the benefits of the "Green Revolution" in India. Therefore, rice cultivation still follows primitive practices, and is characterized by lack of mechanized farms. The majority (approximately 80%) of the agricultural land remains nonirrigated. Except for the northern interior blocks, mono-crop rice cultivation, particularly *Aman Rice* (monsoonal rice grown during the monsoon and harvested in the postmonsoon season), predominates the agricultural production of the delta. A good amount of annual rainfall (about 1600 mm/year) ensures a favorable physical environment by driving the additional salinity from the cultivable lands. However, respondents mentioned the lack of rainfall and its erratic behavior in recent years, among the important factors that disrupt their traditional livelihood practices. In particular, they mentioned about the late arrival of monsoon in previous years, which severely disrupted their traditional crop calendar.

4.3 Farmers' Adaptation Intentions Against Perceived Agricultural Stressors

Since Cyclone Aila, there has been a growing interest among the local communities to cope or adapt to extreme environment. As a result, communities have undergone several small-scale agricultural transformations to sustain a normal living from their limited land, water, and economic resources. Local and international NGOs have also supported the communities by developing several pilot studies of ameliorative agricultural practices that can sustain agriculture under the external stressors. Nevertheless, in most cases, it has been done with only a segment of the population in the pockets of the delta. Therefore, these measures have not been tested on a landscape scale, nor they could be so easily generalized considering the diversity of the farmers, diverse landholding and income-level. In order to bridge this gap, the study principally aimed to conduct a microlevel perception analysis to identify and prioritize the existing agricultural adaptation/coping options (i.e., which option is suited to whom), and further attempted to develop a sectorial guideline by which adaptive practices can be systematically infused within the existing social and economic scenarios.

From the FGDs and following the transect walks, the study could identify a total of 11 potential adaptive/coping mechanisms that the farmers believed to have local applicability or have practiced in the aftermath the Aila. The identified adaptation/coping options can be broadly classified under three behavioral adaptation measures, seven technical adaptation measures, and one institutional adaptation option. Table 1 provides a detailed analysis of the identified adaptive measures with subsequent illustrations on the specific advantages and disadvantages with respect to the existing social and geomorphological features of the delta. Pictorial references of some of the practiced adaptation measures are provided in Fig. 5. In short, five potential adaptive options were repeatedly mentioned by the group of surveyed farmers. These five actions are cultivation of salinity resilient paddy species, development of local irrigation, crop/flood insurance, soil and water conservation structures, diversification of livelihood, and outward migration. These five adaptive/coping options were mostly practiced by the communities in the aftermath of Cyclone Aila. It is, however, imperative to understand that each of the adaptive actions has a specific relevance in order to achieve agricultural sustainability in the study area. Yet, the choice of adaptation is largely influenced by the availability of technical resources, knowledge, awareness of the farmers, training, landholdings, and financial capacity. Therefore, in the follow-up questionnaire survey, an attempt was made to understand an individual's intensions to adapt, assuming that a farmer's decision to adapt is a complex function of several variables, including landholding, income, technical knowledge, etc.

4.4 Farmer's Intentions to Adapt

Out of the 11 mentioned adaptation/coping actions (see Table 1), eight show statistically significant variation among the three groups of famers while the adaptation intension for the rest of the three adaptive measures does not show any significant difference in intention to adapt among the three groups of farmers ($P < .05$). Difference in intention has been observed for the following adaptation options, namely, changing cropping pattern, intercropping, crop and flood insurance, dual use of agricultural land, construction of irrigation facilities, soil and water conservation structure, diversification of livelihood, and migration to other places. On

TABLE 1 Agricultural Adaptation Options and Their Applicability in the Indian Sundarban Delta

Adaptation Options	Description	Adaptation Type	Advantages	Disadvantages	Requirements
Change of seed sowing time	This intends to marginally reschedule the traditional agricultural calendar based on new weather patterns. This particular adaptation option is principally useful for monsoon variability or consistent late arrival of monsoons, as mentioned in the above discussions	Behavioral adaptation (farm level)	No major investments are required	Weather uncertainties increase the margin of error, and most importantly, the process is not irreversible	• Requires close monitoring of local weather as well as weather information dissemination among farmers • Market compatibility is always an issue with this particular adaptation option. Late arrival of crops in the market may cause lower prices or vice versa
Change of cropping pattern	Change in cropping patterns refers to the change in the proportion of area under different crops at two different points in time. In the Sundarban, the majority of the agricultural fields are used for mono-crop cultivation. It is possible to cultivate other crops, especially cash crops during the existing no-crop seasons that are not very water-intensive	Technical adaptation (farm level)	Economic advantages for local farmers, especially from cultivating two crops	Major investments required with provision for local irrigation, soil development. etc	• Farm mechanization is the primary requirement; however, in order to do this, farmers also require significant financial capital • Agricultural loan and lean season crop incentives by the local government can be a suitable way to promote this adaptation
Salinity resistance paddy species	Salinity-resistant rice species are specially engineered varieties of rice that can grow in high salinity. For example, while a normal variety grows under a salinity level less than 6 dS/meter, moderate and high salt-tolerant species can grow within a salinity level of 6–8 dS/m or higher	Technical adaptation (farm level)	Comparatively higher yield in the existing salinity scenario	Nonavailability of good quality seeds in the local market	• Salinity resistant rice varieties have significant potential in the Indian Sundarban Delta. However, the major requirements are the quality of seeds, seeds storage facilities, etc. Major local experimental varieties include *Luneshree*, *Bhutnath*, and *Sumati*, which showed promising results

Homestead gardening	Homestead garnering is an auxiliary income opportunity especially aimed at securing individual food security in case of complete discontinuation of farm level agriculture	Behavioral adaptation (individual level)	Auxiliary support, especially securing individual food security	No profitability and commercial production	No specific requirements
Inter cropping	Intercropping is a multiple cropping practice involving growing two or more crops in proximity. This type of adaptation is especially applicable when the yield from a particular crop is unsatisfactory. Especially in the case of the Indian Sundarban, rice cultivation can be combined with other crops/vegetables that do not compete with rice for physical resources	Technical adaptation (farm level)	Substantially increased yield, better pest control, and profitability	Requires scientific monitoring and assessment such as soil testing etc	• Providing technical guidance to local farmers is a prerequisite because the two varieties than can be grown together should not compete on the resources • Farmers also require some financial capital to conduct intercropping practices in their fields
Crop and flood insurance	Crop and flood insurance is a risk transfer mechanism adopted by farmers and others to protect themselves against the loss of their crops due to floods and cyclones. At present, there are some group insurance mechanisms existing in the Indian Sundarban. However, these are not for individuals because the determination of loss and damage is collectively based	Economic adaptation (institutional level)	Financial compensation in case of disaster damage	This measure is solely aimed to manage economic loss. It has significant limited applicability in terms of slow onset hazards such as salinity	• Local governments and financial institutions such as cooperative banks need to develop some scheme to protect a farmer's financial interests • Requires awareness camps to share the benefits of crop and flood insurance

(Continued)

TABLE 1 Agricultural Adaptation Options and Their Applicability in the IndianSundarban Delta—cont'd

Adaptation Options	Description	Adaptation Type	Advantages	Disadvantages	Requirements
Dual use of agricultural land	Dual use of agricultural land is a special intercropping pattern when rice and fish are cultivated in tandem. Cultivation of fish is done by deepening the rice filed (sometimes with impermeable layering) and the excavated soil is used to heighten the agricultural land. This process can suitably manage the water demand for rice cultivation as well as the use of water for pisiculture	Technical adaptation (farm level)	Diversification of income, especially due to cultivation of fish	Massive one-time investment that the majority of the farmers may not bear	• This adaptation measure is applicable only when the farmers have significant land • This measure also requires technological guidance from agricultural agencies
Construction of irrigation facility	Because the majority of the agricultural land in the Indian Sundarban is void of formal irrigation facilities, construction of a canal system for irrigation seems to be imperative for enhancing agricultural productivity. Particularly, this measure can promote two-crop cultivation in the Indian Sundarban delta	Infrastructural adaptation (institutional level)	Possibilities for two-crop cultivation	Considering the lack of freshwater availability, any centralized irrigation facilities may lead to massive groundwater pumping in coastal areas	Irrigation water budgeting for coastal areas and development of irrigation facilities. However, considering the local hydrological and topographical scenarios, centralized irrigation may attract a huge budget
Soil and water conservation structure	This is mostly an agricultural adaptation option attached to the farm level portion. Farmers with significant landholdings sacrifice a small portion to collect rainwater by creating ponds or tanks	Infrastructural adaptation (farm level)	Decentralized irrigation with generally higher productivity	Loss of fertile land	• No specific requirements, although availability of land is a major constraint

Diversification of livelihood from agriculture	Under the local context, it typically includes horticulture, ornamental fisheries, goats, ducks, etc. The local government provides several schemes to promote this alternative livelihood in the delta.	Behavioral adaptation (individual level)	Diversification of income generation	No proper markets within proximity	• Technical guidance from the local agricultural authorities • Existence of local markets and demands
Migration to other place	Migration is an extreme adaptation that was extensively practiced after Cyclone Aila. Many young farmers/agricultural laborers capable of delivering physical labor migrated to cities all across India. Although, to a major extent, it helped their families survive under an adverse economic scenario, it can be classified as an extreme adaptation measure that has little role to play in agricultural sustainability	Behavioral adaptation (individual level)	Continuing economic support to family members	Gradual depopulation of the delta	• No specific requirements

FIG. 5 (A) A community weather station especially intended for farmers and fishermen (in Kultali Block), (B) Dual use of agricultural land for rice and fish, developed by an NGO (in Kultali Block), (C) Intercropping with rice and vegetables (in Patharpratima Block) (D) Localized irrigation facilities by sacrificial of a portion (25%) of agriculture land (in Gosaba Block). *Source: Author, 2012–2013.*

the contrary, the results indicate a uniform perception over the adaptive actions such as changes in seed sowing time, salinity-resistant paddy cultivation, and homestead gardening. All three groups of farmers favored the cultivation of salinity-resistant paddy species (see Fig. 6). Understandably, as an immediate means to survive, the majority of the surveyed farmers put cultivation of salinity-tolerant rice species as the most favored adaptation option. In addition, many of them, particularly the small farmers, previously mentioned having cultivated salinity-resistant species. Importantly, farmers mentioned about six local indigenous, salt-tolerant rice species that were long lost due to the introduction of high yielding varieties. Consequently, farmers currently depend on external sources for salt-tolerant varieties that often do not come with seed certification. Hence, the major hindrances for a sustainable yield

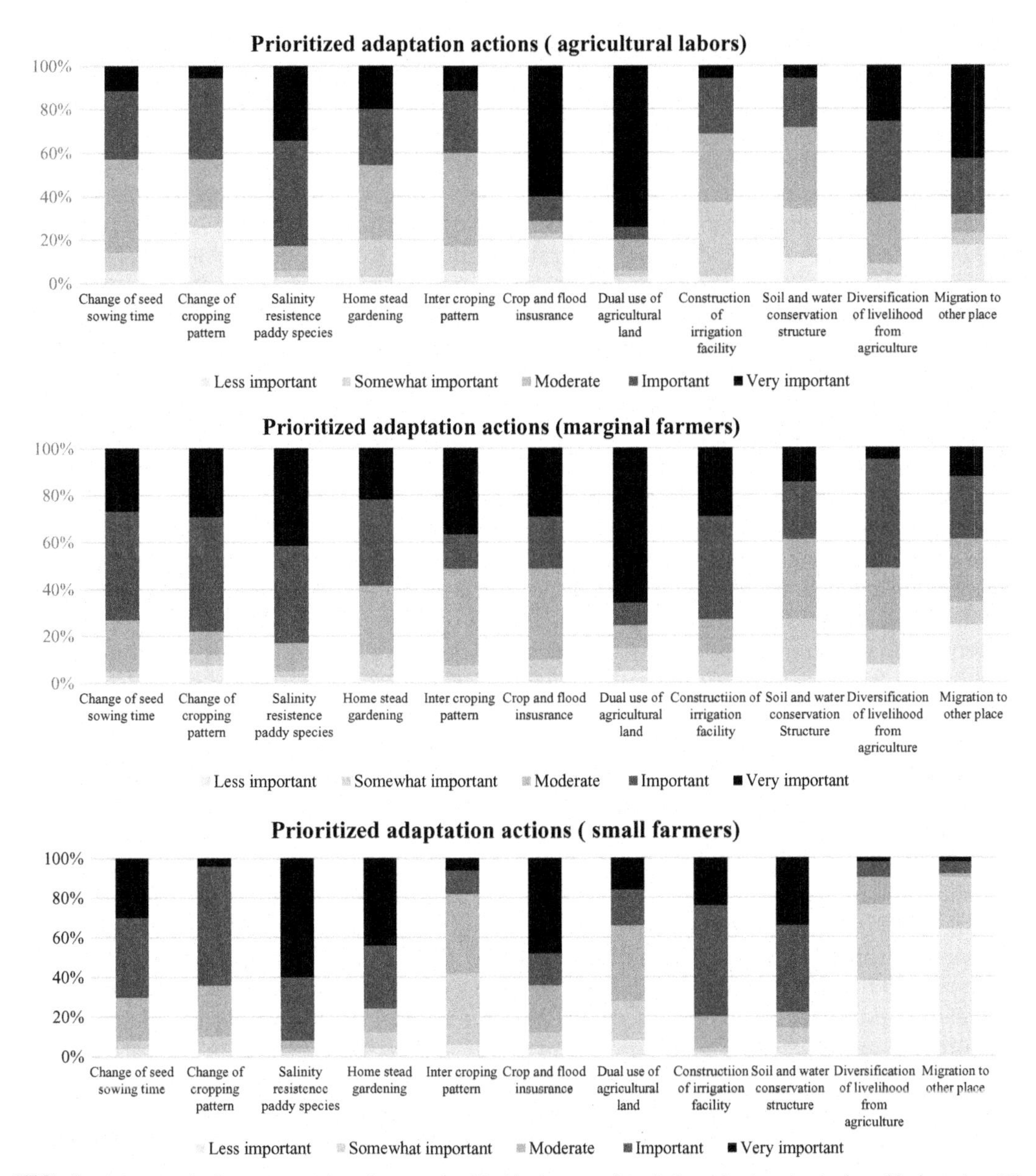

FIG. 6 Prioritized adaptation actions (represented in % of respondents). Sample size: Agricultural Labors ($n=35$), marginal farmers ($n=41$), and small farmers ($n=50$).

are the unavailability of good quality seeds and the lack of seed storage facilities (for preservation of good quality seeds).

As Fig. 6 suggests, the most favored adaptations for small farmers are construction of irrigation facilities, soil and water conservation structures, and cultivation of salinity-resilient rice varieties. Understandably, the combination of these three would make agricultural yields sustainable while other provisions such as *Boro crop* can also be exploited. On a broader scale, these prioritized options indicate the unavailability of adequate fresh water resources to sustain agriculture. As revealed by the farmers during the FGDs, the most feasible way to tackle salinity is to drain it with fresh water, for which robust irrigation facilities are an essential prerequisite. However, as mentioned, given the complex topography of the Indian Sundarban Delta, centralized irrigation facilities are virtually impossible. Decentralized zone-based irrigation is probably the best option under the existing circumstances. However, that may also lead to massive pumping of ground water because the river water is essentially saline. Therefore, although provisioning of irrigation facilities is a suitable adaptive option to increase productivity (from mono-crop to multicrop), it remains unlikely under the mentioned adversities. Consequently, the creation of localized soil and water conservation structures through the augmentation of rainwater can be explored as a viable alternative. On an individual basis, as depicted in Fig. 5, farmers develop small water conservation structures (water storage pit) within their agricultural land by sacrificing a portion (25%) of land. This essentially caters to the excess water requirements to drain out the salinity in the post-Aila period, and, therefore, remains as a local solution to the problems of residual salinity.

Similar to the small farmers, marginal farmers also mentioned the construction of irrigation facilities and salinity-tolerant rice varieties as the most intended adaptation option. However, they also put additional priority on the dual use of agricultural land, probably due to the apprehension of increasing income from the combination of rice and fish from limited land. On the other hand, most of the agricultural labors prioritized cultivation of salt-tolerant species, the dual use of agricultural land, and migration to different places. Following Aila, almost 65% of the agricultural laborers migrated to different parts of the country, leaving their families behind. While this can be argued as a suitable agricultural adaptation/coping measure in order to survive under extreme adversity, the possibilities of mass outward migration of agricultural laborers cannot be ruled out.

In addition to the above discussion, "crop and flood insurance" is also a desired adaptive action that remains imperative to manage sudden crop damage, as in the case of Aila. In this regard, it is important to mention that, at present, the government insurance scheme for disasters is primarily aimed at the village *panchayat* level. For example, if the local administrators, based on the stipulated guidelines, declare a *panchayat* as disaster affected, dwelling villagers including farmers can be compensated for the loss of crop under the National Agriculture Insurance Scheme (NAIS) or "*Rashtriya Krishi Bima Yojana (RKBY).*" The scheme is a nationwide plan to support farmers against adverse natural calamities and each state is required to reach to the level of *Gram Panchayat* as the unit of insurance in a maximum period of 3 years. The premium has been designated as 1.5%–3.5% of the sum assured on food crops. However, it can be argued that special cases such as Sundarban, where cyclones/floods affect more on an individual basis, this needs to done more on the individual level rather the village *panchayat* level.

Diversification of livelihood (especially development of an alternative livelihood) is one of the pressing agendas of the local government and also can be considered as a suitable adaptation option. The numbers of alternative livelihood schemes such as ornamental fisheries, horticulture, sericulture, goatery, piggery, etc., have been promoted by the local government under various alternative livelihood schemes. However, the lack of marketing provisions for the end products coupled with the lack of demand in the local/rural markets (e.g., ornamental fish or even horticulture products) is a significant barrier for upscaling these measures. On the other hand, we find that technological measures such as crop diversification, intercropping or the dual use of agricultural land have significant potential in the Indian Sundarban; however, their scopes remain vastly underutilized. The majority of the observed farm level adaptations are heavily supported by local NGOs. Although NGO lead approaches have been well accepted by the communities, the transformation of technical knowhow is vastly confined to small pockets, hindering the large-scale implementation of these adaptive actions. Additionally, upscaling these measures also requires some capital investments. Hence, the scope of these adaptative options is largely restricted under the current financial and technical capacities of the farmers.

5 CONCLUSION AND THE WAY FORWARD

Agricultural productivity is fundamental to the continual survival of the Asian megadeltas and their communities. Especially because of the topographical complexity of the megadeltas, several interlinked components of human sustainability such as food security, income generation, and poverty reduction are intrinsically linked with the agricultural productivity of the low-lying Asian megadeltas. With more than 65% of the population directly depending on agriculture, the Indian Sundarban Delta is no exception. In addition, without agricultural productivity, the possibilities of overexploitation of already vulnerable coastal resources such as mangroves and aquatic resources remain exceptionally high. Therefore, it remains imperative to develop a long-term, sustainable coastal agricultural management strategy that is resilient to future changes. The goal of this study was to identify and understand the scope and applicability of some of the potential adaptation options that communities can trust to maintain their agricultural and economic productivity. The study also addressed the gaps and challenges in improving agricultural productivity under the current environmental stressors, including the most pressing threats of unfavorable soil and inland water salinity. Additionally, the study attempted to understand an individual's intentions and ability to embrace specific adaptive/coping options.

In summary, the major finding of this study can be started that the cultivation of good quality salinity-resilient paddy species along with the provisioning of fresh water resources are among the immediate adaptation needs of the communities. For this purpose, localized irrigation facilities should be planned to supplement the traditional rain-fed rice cultivation. The local and provincial government should, therefore, explore the possibilities of rainwater harvesting and developing location-specific irrigation facilities. On the other hand, fulfilling the existing agricultural infrastructural deficits such as construction of a seed conservation center, a seed certification mechanism, and soil testing laboratories remain imperative. Nevertheless, for the longer run, we observed some of the prospective adaptive actions, such as dual use of agricultural land and intercropping, which can be considered for upscaling through proper institutional support. However, as observed during the PRA exercises, the main drawback for fostering these potential adaptive actions is the financial and technical incapacities of the farmers. Considering the above, it remains imperative to develop an adaptation fund for the Indian Sundarban Delta, possibly under the guidance of the National Bank for Agriculture and Rural Development (NABARD), which has been appointed as the nodal agency for utilizing the climate change adaptation fund formed under the Ministry of Environment, Forests, and Climate Change. In addition, possible upscaling of favorable agricultural adaptation actions can be considered broadly under the Integrated Coastal Zone Management (ICZM) plan which calls for sustainable and integrated management of human and coastal resources.

Acknowledgment

The authors would also like to thank all the district administrative and agricultural officials for their active support and help during the field survey in the Indian Sundarban. The financial support for necessary field work was provided by the GCOE-ARS and COHHO programs of Kyoto University which is also thankfully acknowledged.

References

Abedin, M.A., Habiba, U., Shaw, R., 2012. Health: impacts of salinity, arsenic and drought in South-western Bangladesh. In: Environment Disaster Linkages. Community, Environment and Disaster Risk Management, vol. 9. Emerald Group Publishing Limited, Bingley, UK, pp. 165–193 (Chapter 10).

Adhikari B., Bag M.K., Bhowmick M.K., Kundu C. (n.d.): Status Paper on Rice in West Bengal, Rice Research Station, Govt. of West Bengal.

Basu, M., Shaw, R., 2013. Water scarcity and migration: an Indian perspective. In: Water Insecurity: A Social Dilemma. Community, Environment and Disaster Risk Management, vol. 13. Emerald Group Publishing Limited, Bingley, UK, pp. 187–211.

Basu, M., Hoshino, S., Hashimoto, S., 2015. Many issues, limited responses: Coping with water insecurity in rural India. Water Resour. Rural Dev. 5, 47–63.

Bradshaw, B., Dolan, H., Smit, B., 2004. Farm-level adaptation to climatic variability and change: crop diversification in the Canadian prairies. Climat. Change 67 (1), 119–141.

Connor, J.D., Schwabe, K., King, D., Knapp, K., 2012. Irrigated agriculture and climate change: the influence of water supply variability and salinity on adaptation. Ecol. Econ. 77, 149–157.

DasGupta, R., Shaw, R., 2015. An indicator based approach to assess coastal communities' resilience against climate related disasters in Indian Sundarbans. J. Coast. Conserv. 19 (1), 85–101.

Dasgupta, S., Kamal, F.A., Khan, Z.H., Choudhury, S., Nishat, A., 2014. River Salinity and Climate Change: Evidence From Coastal Bangladesh. World Bank, Washington, DC.

Deressa, T.T., Hassan, R.M., Ringler, C., Alemu, T., Yesuf, M., 2009. Determinants of farmers' choice of adaptation methods to climate change in the Nile Basin of Ethiopia. Glob. Environ. Change 19 (2), 248–255.

District Human Development Report, 2009. Government of West Bengal. South 24 Parganas District Human Development Report.

Feola, G., Binder, C.R., 2010. Toward an improved understanding of farmers' behaviour: The integrative agent-centred (IAC) framework. Ecol. Econ. 69 (12), 2323–2333.

Hanson, B., Grattan, S.R., Fulton, A., 1999. Agricultural Salinity and Drainage. University of California Irrigation Program, University of California, Davis.

Knutson, T.R., Mcbride, J.L., Chan, J., Emanuel, K., Holland, G., Landsea, C., Held, I., Kossin, J.P., Srivastava, A.K., Sugi, M., 2010. Tropical cyclones and climate change. Nat. Geosci. 3, 157–163.

Nhan, D.K., Phap, V.A., Phuc, T.H., Trung, N.H., 2012. Rice production response and technological measures to adapt to salinity intrusion in the coastal Mekong delta. In: Mekong Program on Water, Environment and Resilience (MPOWER).

Office of the District Magistrate, South 24 Parganas, 2010. Internal Report on the Damage and response of Cyclone Aila. (unpublished).

Redfern, S.K., Azzu, N., Binamira, J.S., 2012. Rice in Southeast Asia: facing risks and vulnerabilities to respond to climate change. In: Building Resilience for Adaptation to Climate Change in the Agriculture Sector, 23,p. 295.

Risbey, J., Kandlikar, M., Dowlatabadi, H., Graetz, D., 1999. Scale, context, and decision making in agricultural adaptation to climate variability and change. Mitigat. Adapt. Strateg. Glob. Change 4 (2), 137–165.
SBR, 2008. Sundarban Atlas. In: Sundarban Biosphere Reserve.
SeinnSeinn, M.U., Ahmad, M.M., Thapa, G.B., Shrestha, R.P., 2015. Farmers' Adaptation to Rainfall Variability and Salinity through Agronomic Practices in Lower Ayeyarwady Delta, Myanmar. J. Earth Sci. Climat. Change 6 (2), 1.
Smit, B., Skinner, M.W., 2002. Adaptation options in agriculture to climate change: a typology. Mitigat. Adapt. Strateg. Glob. Change 7 (1), 85–114.

Further Reading

Adger, W.N., Dessai, S., Goulden, M., Hulme, M., Lorenzoni, I., Nelson, D.R., Wreford, A., 2009. Are there social limits to adaptation to climate change? Climat. Change 93 (3&4), 335–354.
District Disaster Management Authority, 2009. A brief report on 'AILA' and Review on Disaster Risk Management Programme. Unpublished Internal report.
INCAA, 2010. Climate Change and India: A 4 x 4 Assessment published by Indian Network for Climate Change Assessment (INCCA). Ministry of Environment and Forests, Government of India.
Le Dang, H., Li, E., Nuberg, I., Bruwer, J., 2014. Understanding farmers' adaptation intention to climate change: a structural equation modelling study in the Mekong delta, Vietnam. Environ. Sci. Policy 41, 11–22.
Roessig, J.M., Woodley, C.M., Cech Jr., J.J., Hansen, L.J., 2004. Effects of global climate change on marine and estuarine fishes and fisheries. Rev. Fish Biol. Fish. 14 (2), 251–275.

CHAPTER

13

Fostering Climate Change Mitigation Through a Community-Based Approach: Carbon Stock Potential of Community-Managed Mangroves in the Philippines

Dixon T. Gevaña, Juan M. Pulhin, Maricel A. Tapia

Department of Social Forestry and Forest Governance, College of Forestry and Natural Resources, University of the Philippines, Los Baños, Philippines

1 INTRODUCTION

Mangroves play an important part in the ecosystem. They enrich coastal waters, yield commercial forest products, protect coastlines from erosion, and support coastal fisheries. They have unique characteristics, being a true ecotone of land and sea. The term *mangrove* denotes two different concepts, according to Lugo and Snedaker (1974) and Alongi et al. (2005). Biologically, it is a group of salt-tolerant plants belonging to 9 orders, 20 families, 27 genera, and roughly 70 species. These plants can cope with changes in water and sediment salinity by evolving both xeromorphic and halophytic characteristics. Geographically, it refers to complex plant communities that fringe tropical and subtropical (32°N to 32°S) shores and are delimited by major ocean currents and a 20°C isotherm of seawater in winter. Mangrove plants are also well adapted to natural stressors such as high temperature, high salinity, anaerobic sediments, and extreme tides. However, because they live close to their tolerance limits, they are sensitive to other disturbances such as those brought about by human activities (FAO, 2007). In some countries, mangrove stands are on the brink of complete collapse

https://doi.org/10.1016/B978-0-12-810473-6.00014-5

© 2019 Elsevier Inc. All rights reserved.

for being utilized as sewage disposal sites and aquaculture pond development (Kathiresan and Bingham, 2001; Gevaña et al., 2015).

Records show that 50% of the world's mangrove forests were lost over the past half century (FAO, 2007). This reflects the immense damage to the vital ecological and economic benefits they provide. According to Costanza et al. (1998), global mangroves have an estimated worth of about \$180,900,000,000, with an average monetary value of \$10,000 ha^{-1} yr^{-1}. Despite these huge price equivalents, deforestation persists. Mangrove sites are usually favored as locations for settlement and industrial development. Furthermore, many governments have encouraged the development of agriculture, shrimp and fish farming, and salt and rice production in mangrove areas. All these have resulted in massive fragmentation, degradation, and pollution.

The Philippines is no exception to the above trends as the area of its mangrove forests declined by as much as 60% over the past eight decades (Garcia et al., 2014). From the early record of 400,000 ha (Brown and Fischer, 1920; Chapman, 1976; Primavera, 2000), the remaining mangrove cover today is estimated at 153,577 ha, with fairly extensive cover left on the island of Palawan at 41,830 ha (FMB, 2011).

Deforestation hampers the mangrove's ability to mitigate climate change. The decrease of forest biomass entails a loss of forest capacity to sequester atmospheric carbon. By some estimates, a healthy and well-protected mangrove stand can trap as much as 1 GtC ha^{-1} (Donato et al., 2011), a value that is double the capacity of other tropical forest stands. Allowing deforestation and forest degradation therefore can lead to serious carbon emission problems.

Local communities play a pivotal role in rehabilitating mangrove areas. They can be potential partners and stewards for more conscious management of the coastal resources. In the Philippines, the government has adopted the community-based forest management (CBFM) approach to manage most of the coastal forest areas. Such an approach has helped elevate the interests, roles, and responsibilities of local communities in managing mangrove resources. Unlike those in upland communities, the studies on Philippine CBFM in mangrove areas are very limited, especially in the context of climate change mitigation. This chapter therefore analyzes the achievements of CBFM as far as improving mangrove carbon stocks is concerned. Some issues and challenges on the sustainable implementation of CBFM on mangrove sites are also identified.

2 PHILIPPINE MANGROVES

The Philippines harbors rich tropical resources, hence it is known as the *Pearl of the Orient Sea*. Its verdant and blue coastal ecosystem is the fourth longest in the world with a length of 36,289 km. With the nation having about 7107 islands, mangroves are very common along coasts and estuaries. Philippine mangroves have six distinct formations: (a) *Rhizophora*-dominated stand along rivers and intertidal mudflats; (b) *Avicennia*-dominated stand at inundated beaches and mudflats; (c) *Sonneratia*-dominated at subtidal sediments; (d) *Rhizophora stylosa* stand along rocky coralline substrates; (e) *Nypa*-dominated stand along brackish rivers and lagoons; and (f) mixed trees, shrubs, and thorny bushes in elevated coasts (Fig. 1). Based on the listings of Fernando and Pancho (1980), these formations harbor 40 species of major

FIG. 1 Common mangrove formations in the Philippines: (A) *Rhizophora* along river; (B) *Avicennia* stand; (C) *Sonneratia* stand; (D) *R. styolsa* on rocky or corraline sediments; (E) *Nypa* stand; and (F) mixed species in elevated coast.

and minor mangroves that belong to 16 families, and as much as 30 species of mangrove associates (primarily shrubs and vines).

3 KEY INSTITUTIONAL ARRANGEMENTS AND POLICIES

Presidential Decree No. 705 of 1975, or the *Revised Forestry Code of the Philippines*, provides the foundation for determining the appropriate forest management systems in the country. This policy defines *mangrove* as a type of forest that thrives on tidal flats and seacoasts and those that extend through streams where the water is brackish. Section 16 of this legislation declares that those mangrove stands of at least 20m wide are owned by the state, hence they cannot be privately possessed. However, Section 13 placed an exemption on mangrove stands that are not needed for shore protection, and allowed them to be converted to aquaculture ponds.

The ensuing massive conversion of mangrove areas to fish ponds has led to tremendous loss in mangrove cover. For this reason, the government passed policies to seriously protect the remaining mangrove cover. These include the Republic Act (RA) 7161 or an *Act of Incorporating Certain Sections of the National Revenue Code* in 1991, (RA) 7586 or *National Integrated Protected Areas System Act* (NIPAS) of 1992, and (RA) 8550 or the *Philippine Fisheries Code of 1998*. By virtue of Section 71 of RA 7161, the government bans commercial cutting for all mangrove species. Moreover, Section 2 of RA 7586 had further placed the mangrove as an initial component in the list of protected areas, hence land use conversion was not allowed. Lastly, Section 94 of RA 8550 stipulates that abandoned fishponds (previously mangrove areas) shall be reverted back to mangrove stands through reforestation.

The focus of mangrove management policies and programs over the past four decades is protection and rehabilitation. Recognizing the vital role of local communities in pursuing this direction, the Department of Environment and Natural Resources (DENR) has placed a number of implementing rules and regulations to effectuate these mangrove policies. These include the following:

- DENR Administrative Order (DAO) 76 (1987): Local communities and fishpond leasers are required to establish mangrove buffer zones of: (a) 50 m fronting seas and oceans, and (b) 20 m along riverbanks.
- DAO 34 (1987): Guidelines on Environmental Clearance Certificate (strict permitting system that applies to fishpond development over mangrove areas),
- DAO 123 (1989): Local mangrove planters are awarded a 25-year tenure through the *Community Forestry Management Agreement (CBFMA)*, hence domestic mangrove use, the establishment of *Rhizophora* and *Nypa* plantations, and aquasilviculture are allowed.
- DAO 15 (1990): (a) Mangrove Stewardship Contracts (similar to DAO 123) are given to local communities and fishpond leasers, stipulating therein all the rights, roles, and responsibilities to conserve mangrove resources; (b) abandoned fishponds are required to be reverted back to mangrove forests through reforestation, (c) tree cuttings are banned in fishpond leased areas; and (d) conversion of thickly vegetative areas is prohibited.

Executive Order (EO) 263 or *Community-Based Forest Management* (CBFM) and DAO 10 (1998) or *Guidelines on the Establishment and Management of CBFM Projects with Mangrove Areas* have also provided opportunities for local communities to have legal access, management, and utilization rights (to some extent, that is, for domestic or noncommercial purposes) over mangrove forests.

4 MANGROVE VALUE

More than half the country's 1500 towns and 42,000 villages are intimately dependent on marine ecosystems for food and other benefits (Primavera, 2000). A study conducted by Carandang et al. (2013) on small community-managed mangroves (4426 ha) in Puerto Princesa, Palawan, values the annual direct use of mangroves to about $567,148.4 per hectare. This is reflective of the wide benefits that mangroves provide, namely, marine catches (fish, shrimp, and mollusks), timber, fuelwood, nipa thatching (*Nypa fruiticans*), and recreation.

Keeping a good mangrove cover has also been recognized as a mitigation strategy against natural disasters. Mendoza and Alura (2001) associated the significant uprooting of coconut trees with the lack of mangrove cover in a coastal site in Samar Province, Philippines. They underscored the interconnectedness of coastal ecosystems wherein mangroves work synergistically with seagrass, beach forest, and coral as tidal and erosion buffers. The findings of Macintosh (2010) on Japanese mangroves significantly reducing wave force by as much as 70%–90% confirm the previous studies. Further, Harada et al. (2002) emphasized that mangroves are more effective wave barriers than a concrete seawall in the event of a tsunami. Mazda et al. (1997) also reported that a 6-year-old mangrove belt (1.5 km width) can significantly reduce sea waves by 20 times its force.

5 CARBON STOCK CAPACITY OF COMMUNITY-MANAGED MANGROVES

Similar to the uplands, community forestry has become the model for forest management in coastal areas (Gilmour and Fisher, 1991; Pulhin, 2000; Walters, 2004). Governments need not singlehandedly manage forest resources as communities are recognized to be effective partners toward this end. Community-based reforestation and management are enthusiastically promoted by government, nongovernment, and aid agencies to cultivate a sense of stewardship over forest resources among communities (Pulhin et al., 2007). At the same time, this presents an opportunity to sustain the forest's vital function to sequester carbon.

Among the tropical forest ecosystems, mangroves have the largest carbon stock. A healthy mangrove stand can have 2–4 times the capacity of other forest ecosystems, sequestering as much as 1023 $tC\,ha^{-1}$ (FAO, 2010; Donato et al., 2011). The bulk of the carbon sequestered is stored in below-ground peat or sediment (Gevaña and Im, 2016). Thus, community-based mangrove management offers win-win solutions in strengthening local commitment on forest conservation and sustaining the carbon sequestration service from mangroves. This is demonstrated by two cases presented below: (a) San Juan in Batangas Province; and (b) Banacon Island in Bohol Province.

5.1 San Juan, Batangas

San Juan is situated at the southeastern tip of Batangas Province, forming part of one of the world's most megadiverse marine biodiversities, the Verde Island Passage Biodiversity Corridor (Fig. 2). It is geographically located along 13.49.6″ north and 121.23.8″ east, covering a total land area of 27,340 ha. The current population of San Juan is about 108,500 with an annual growth rate of 2.7%. Fishing and agriculture are the major livelihood activities, particularly of the coastal communities. Patches of mangrove forests are estimated at about 100 ha. Common mangrove species that thrive here include *Rhizophora mucronata*, *Rhizophora*

FIG. 2 Satellite image of San Juan, Batangas, Philippines (Google Earth Ver. 7.1.2.2019) and its typical mangrove vegetation types: (A) *Rhizophora*-dominated stand; (B) mixed species stand.

apiculate, *Xylocarpus granatum*, *Aegiceras corniculatum*, *Sonneratia alba*, *Avicennia marina*, *Avicennia marina* var. *rumphiana*, *Bruguiera parviflora*, *Ceriops tagal* and other associated species such as *Nypa fruiticans*, *Cocos nucifera*, and *Caesalpinia nuga*.

The community-based mangrove management in San Juan exemplifies the *comanagement* approach between local communities and the municipal local government unit (LGU). Local communities regard their mangroves as essential in sustaining the local fish catch and minimizing the detrimental impacts of flooding and tidal surges during the monsoon months. The local government shares this view and installed complementary actions by institutionalizing a collaborative program called *Bantay-Dagat* (Sea Patrol), a group of local community volunteers that is deputized and trained by the LGJ to conduct sea patrols and stop illegal fishing and mangrove-cutting activities. The *comanagement* proved to be effective and later opened opportunities for research and livelihood support from agencies such as the Conservation International or CI-Philippines (an international environmental NGO). CI-Philippines has conducted a vulnerability assessment and adaptation study in San Juan and other neighboring coastal municipalities along the Verde Island Biodiversity Corridor to help local communities better adapt to climate change. Further, the Bureau of Fisheries and Aquatic Resources (BFAR), a government agency mandated to assist in fishery development, has extended livelihood support to selected coastal communities through fingerlings distribution (e.g., *Chanos chanos*) and conducting training to boost aquaculture development.

The community-managed mangrove stands of San Juan have two distinct floristic zones: old growth *Rhizophora*-dominated stands along riverbanks, and old growth mixed-species stands at shallow inundated mudflats (Gevaña et al., 2008). Their carbon stocks are assessed at 115 tC ha^{-1} and 142 tC ha^{-1}, respectively, with a market value of about \$2954[1] $ha^{-1}$and \$3648 ha^{-1}. These values certainly merit serious conservation.

5.2 Banacon Island, Bohol

Banacon Island is lauded as one of Asia's largest community-initiated mangrove plantations (Fig. 3). Located in the province of Bohol, the island is home to at least 300 households (Gevaña and Pampolina, 2009). It has an area of about 660 ha, of which more than 80% is mangrove forest. It is located at 10°03′30″ to 10°15′30″ north and 124°03′30″ to 124°14′30″ east, forming part of the unique and eco-diverse protected marine sanctuary of Danajon Double Barrier Reef (Pichon, 1977).

In the early 1950s, Banacon Island was devoid of intact mangrove cover (Walters, 2004). Sandbars and reefs were the commonly seen features in the area, with just a few strips of heavily deforested stands. Local residents were engaged in massive cutting of *Sonneratia*, *Rhizophora*, and *Avicennia* trees for charcoal and fuelwood, which were then sold in the nearby city of Cebu. Realizing the economic gains from fuelwood, some planted mangroves in their backyard in the hope of high returns once harvested and sold to the market. This initiative was inspired by *Eugenio Paden*, a local who developed a dense planting method[2] of raising *Rhizophora stylosa* propagules on sandbars and shallow mudflats. Exhibiting ease in

[1]Carbon price based on \$7 per 1 tCO_2e ha^{-1} (Ecosystem Marketplace, 2011).

[2]Direct field planting of *R. stylosa* propagules with a distance of 0.5 m × 0.5 m.

FIG. 3 Satellite image of Banacon Island, Bohol, Philippines (Google Earth Ver. 7.1.2.2019. Typical mangrove vegetation types are: (A) *Rhizophora stylosa* plantation, and (B) *Avicennia* and *Sonneratia*-dominated natural stand (Gevaña et al. 2015).

harvesting and planting propagules, other residents have then followed suit. The practice later became popular and passed through generations, even with the unexpected issuance of cutting ban policies by the national government.

Stricter policies to avoid mangrove deforestation were passed through RA 7161 and RA 7586. However, this legislation did not totally stop illegal harvesting of mangroves in Banacon. In 1998, another attempt was done to address illegal cutting. This was though the CBFM agreement which is a tenurial program of the government. Managed cutting of mangrove plantations was then allowed as part of the provisions of EO 263 and DAO 1998–10. In 2004, Banacon planters were organized into a people's organization[3] (PO) called *Banacon Fisherfolks and Mangrove Planters Association*, or *BAFMAPA*, after which a CBFMA was awarded to them. Of the 300 households, 100 became PO members. Less extractive methods of harvesting, such as thinning or selection cutting, were implemented with the condition that it should be strictly for household consumption, such as fuelwood and poles for seaweed farms and house construction. The increase in the area of the mangrove plantation created a strong interest among the communities to commercially market mangrove products for additional income. DENR, however, remains rigid on its cutting ban policy, as this stipulation in RA 7161 is still in force.

Despite uncertainties on harvesting rights, planting efforts continued, resulting in thick and extensive mangrove forests. This accomplishment was noticed and recognized both locally and abroad. In 1981, the BAFMAPA received the *Likas Yaman Award* or the Natural Resources Award from the DENR for the exemplary performance in coastal reforestation. In 1991, the group also received the prestigious *Outstanding Tree Farmer Award* from the Food and Agriculture Organization (FAO). Such distinctions have stimulated interest from private

[3]Prior to issuance of their tenure right, the local community must be organized and legally registered as a people's organization.

and other government institutions to partner in plantation development. One of these was Kanepackage Philippines Inc., or KPG (an international corrugated box production company), which has funded a 200 ha plantation project as part of its corporate social responsibility (CSR). The plantation is a potential source of carbon credits in the future.

Camacho et al. (2011) assessed the carbon stocks of mangrove plantations in Banacon Island at 370 tC ha^{-1} while its natural *Avicennia* and *Sonneratia*-dominated stands contain at least 145 tC ha^{-1}. The carbon stocks of these stands have equivalent prices of at least \$9500 ha^{-1} and \$3725 ha^{-1}, respectively.

6 CHALLENGES AND WAY FORWARD

Notwithstanding the growing appreciation of a community-based approach in mangrove management, a number of challenges remain to be addressed. These include (1) unclear tree harvesting rights, (2) ill-founded motivation for reforestation, (3) species-site mismatch, and (4) poor coastal management planning.

6.1 Unclear Tree Harvesting Rights

The CBFM Agreement, as a rule, should provide the local communities with utilization rights over timber they raised in the designated production zones (Pulhin and Tapia, 2015; Larson and Pulhin, 2012). However, such a privilege does not apply to mangroves because there is a higher-level policy, that is, RA 7161, which prohibits commercial cutting in mangrove forests. Thus, many local communities are disillusioned to participate in reforestation projects because they cannot realize the financial benefits from the trees they planted. One way to address this problem is by compensating nonharvesting with payments for environmental service such as a carbon-offset project (e.g., REDD Plus and Reforestation/Afforestation Clean Development Mechanism). Through this, the local community can realize income benefits for each ton of carbon saved and sequestered by their mangroves.

6.2 Ill-Founded Motivation for Reforestation

In Negros Island, Philippines, Walters (2004) noted the ill-founded motivations of local communities in participating in a mangrove reforestation project. First, the local community sees their participation as an opportunity to expand their claims over the open intertidal spaces they planted. A de facto[4] ownership is likely to be acquired by the planters, hence securing access and utilization rights over the plantation they have grown. There is also a greater chance that other members will allow planters to convert their plantations into aquaculture ponds or settlement areas. Further, local communities are compelled to plant because this is required by the DENR and local government. Many reforestation programs were not sustained because the local communities see their role as mere providers of labor rather than stewards.

[4]Members of the local community recognize the access and management rights of an individual over the mangrove area he/she planted, despite the absence of a government-issued tenure certificate.

In order to address this issue, community organizing and awareness campaigns are needed to clarify a community's roles and responsibilities in mangrove management and increase awareness on the myriad benefits they can get in committing to conservation work. These can help elevate the level of participation toward achieving empowerment and self-reliance.

6.3 Species-Site Mismatch

Samson and Rollon (2008) conducted an extensive assessment of the growth and survival of monoculture plantations in Southern Luzon, Central Visayas, and Mindanao. They reported that reforestation sites planted with pure *Rhizophora* spp. yielded dismal outcomes, with high mortality and poor growth performance. Such a result was linked to the weak adaptability of species to their nonnatural habitat. Addressing this issue will require proper guidance on future rehabilitation work. It also calls for research and development programs on ecological species-site matching of mangrove species, which DENR and other concerned research institutions such as the academe could spearhead.

6.4 Poor Coastal Management Planning

Mangrove deforestation is reflective of poor coastal land use management planning. The experience of Manila (the national capital) exemplifies this case, where the eventual peeling off of the mangrove cover was observed to favor industrial port development and land reclamation. The name Manila was coined after a mangrove shrub called *nilad* (*Scyphiphora hydrophyllacea*), which used to be the predominant vegetation along its scenic bay. Courtney and White (2000), Yao (2001), and Primavera and Esteban (2008) stressed that the lack of clear land use zonation has resulted in poor community-based mangrove management. Without proper demarcation of the protection and production zones, the local community tends to either overprotect or overutilize mangroves, hence inciting conflicts between the desire to conserve and the need to cut. A well-designed land use management plan is vital in managing these two important objectives. The plan must also consider the varying interests and demands of mangrove stakeholders, particularly the local communities on whom the important task of conservation rests.

7 CONCLUSION AND RECOMMENDATIONS

Mangrove forests around the world have undergone similar trends of degradation and conversion to give way to more favored economic activities, such as settlement, industrial development, and agricultural and fisheries production. The severe loss in the area of mangroves also translates to a reduction in the billion-dollar services that this ecosystem provides, including climate change mitigation through carbon sequestration.

The Philippines' mangrove forests are an important source of livelihood for a significant portion of the coastal population. It is documented as an effective mitigation strategy for natural disasters. Further, it has the largest stock of carbon among tropical forest ecosystems.

However, the country's mangrove forests were not spared from massive deforestation, suffering a 60% decline in a span of almost a century. Hence, several policies have been passed and institutional arrangements set in place to appropriately manage and protect what remains of this resource.

The cases above highlight the central role of communities in mangrove protection and rehabilitation. While their motivation is primarily economic, what is obvious is that this goal can be integrated with ecological imperatives. In so doing, the carbon stocks of mangroves are enhanced, and this provides another opportunity for the communities to earn through the carbon market. Rehabilitating the mangrove ecosystem of the Philippines through community-based approaches therefore presents a huge potential for carbon sequestration, and a valuable pathway for complying with the Intended Nationally Determined Contributions (INDC) of the country submitted to the United Nations Framework Convention on Climate Change. This is congruent to the country's philosophy of pursuing climate change mitigation as a function of adaptation.

Mangrove forests are a government-owned resource, and the latter's issuance of stricter policies on mangrove utilization runs contrary to the primary motive of the communities to engage in its management and rehabilitation. Use rights bestowed through tenurial instruments, such as CBFMA, are overshadowed by stronger and more encompassing policies such as RA 7161. Prudence must be exercised on the part of the government to ensure that communities are not overburdened with responsibilities and shortchanged in terms of benefits. Furthermore, while stringency in implementing laws that protect the environment is laudable, there should be consistency and widespread dissemination to ensure that illegal activities will not proliferate, or communities would not participate in mangrove reforestation with the wrong expectations.

Considering the above, mechanisms should be in place to equitably reward the communities for their efforts in mangrove rehabilitation and management. At the core of this is the promotion of sustainable sources of livelihood to the poor and natural resource-dependent coastal communities. Basic needs of local communities such as fuelwood for cooking and boiling water should be taken into account by allocating areas for a community woodlot. Clear delineation of areas for protection and production through coastal planning would lend a good start toward this objective. Furthermore, reforestation is a major strategy in rehabilitating this mangrove ecosystem. Planting, however, should be done in a scientific way combined with indigenous and local knowledge so that mortalities would be avoided and resources and labor are not wasted.

References

Alongi, D., Clough, B., Robertson, A., 2005. Nutrient-use efficiency in arid-zone forests of the mangroves *Rhizophora stylosa* and *Avicennia marina*. Aquat. Bot. 82, 121–131.

Brown, W., Fischer, A., 1920. Philippine mangrove swamps. In: Brown, W.H. (Ed.), Minor Products of Philippine Forests I. Bureau of Printing, Manila, pp. 9–125. Bureau Forestry Bull. No. 22.

Camacho, L., Gevaña, D., Carandang, A., Camacho, S., Combalicer, E., Rebugio, L., Youn, Y., 2011. Tree biomass and carbon stock of a community-managed mangrove forest in Bohol, Philippines. For. Sci. Technol. 7 (4), 161–167.

Carandang, A., Camacho, L., Gevaña, D., Dizon, J., Camacho, S., de Luna, C., Pulhin, F., Paras, F., Peras, R., Rebugio, L., 2013. Economic valuation for sustainable mangrove ecosystems management in Bohol and Palawan, Philippines. For. Sci. Technol. 9, 118–125. https://dx.doi.org/10.1080/21580103.2013.801149.

Chapman, V., 1976. Mangrove Vegetation. J. Cramer, New York, p. 477.
Costanza, R., d'Arge, R., deGroot, R., Farber, S., Grasso, M., Hannon, B., 1998. The value of the world's ecosystem services and natural capital. Ecol. Econ. 25, 3–15.
Courtney, C.A., White, A.T., 2000. Integrated coastal management in the Philippines: testing new paradigms. Coast. Manag. 28, 39–53.
Donato, D., Kauffman, J., Kurnianto, S., Stidham, M., Murdiyarso, D., 2011. Mangroves among the most carbon-rich forests in the tropics. Nat. Geosci. 4, 293–297.
Ecosystem Marketplace, 2011. Back to the Future: State of the Voluntary Carbon Markets 2011. Bloomberg New Energy Finance, New York, NY, p. 93.
FAO, 2007. The World's Mangrove: 1980–2005: A Thematic Study Prepared in the Network of the Global Forest Resource Assessment 2005. FAO Paper No. 153, Food and Agriculture Organization, Rome.
FAO, 2010. Global Forest Resources Assessment 2010. FAO Forestry Paper No. 163, Food and Agriculture Organization, Rome, Italy, p. 378.
Fernando, E., Pancho, J., 1980. Mangrove trees of the Philippines. Sylvatrop Philipp. For. Res. J. 5 (1), 35–54.
FMB, 2011. Forestry Statistics 2011. Forest Management Bureau, Department of Environment and Natural Resources, Quezon City, Philippines.
Garcia, K., Malabrigo, P., Gevaña, D., 2014. Philippines' mangrove ecosystem: status, threats and conservation. In: Hakeem, et al., (Ed.), Mangrove Ecosystem in Asia: Current Status, Challenges and Management Strategies. Springer, New York, pp. 81–94.
Gevaña, D., Pulhin, F., Pampolina, N., 2008. Carbon stock assessment of a mangrove ecosystem in San Juan, Batangas. J. Environ. Sci. Manag. 11 (1), 15–25.
Gevaña, D., Pampolina, N., 2009. Plant diversity and carbon storage of a rhizopora stand in Verde passage, San Juan, Batangas, Philippines. J. Environ. Sci. Manag. 12 (2), 1–10.
Gevaña, D., Im, S., 2016. Allometric models for *Rhizophora stylosa* Griff. in dense monoculture plantation in the Philippines. Malaysian Forester 79 (1 and 2), 39–53.
Gevaña, D., Camacho, L., Carandang, A., Camacho, S., Im, S., 2015. Landuse characterization and change detection of a small mangrove area in Banacon Island, Bohol, Philippines using maximum likelihood classification method. For. Sci. Technol. 11 (4), 197–205.
Gilmour, D., Fisher, R., 1991. In: Victor, M., Lang, C., Bornemeir, J. (Eds.), Evolution in community forestry: contesting forest resources. Community forestry at crossroads: Reflections and future directions in the development of community forestry.Proceedings of an International Seminar, 17–19 July, Bangkok. RECOFTC Report No. 16, pp. 27–44.
Harada, K., Imamura, F., Hiraishi, T., 2002. In: Experimental study on the effect in reducing tsunami by the coastal permeable structures.Final Proceedings of the International Offshore and Polar Engineering Conference, USA, pp. 652–658.
Kathiresan, K., Bingham, B., 2001. Biology of mangroves and mangrove ecosystems. Adv. Mar. Biol. 40, 81–251.
Larson, A.M., Pulhin, J.M., 2012. Enhancing forest tenure reform through responsive regulations. Conserv. Soc. 10 (2), 103–113.
Lugo, A., Snedaker, S., 1974. The ecology of mangroves. Annu. Rev. Ecol. Syst. 5, 39–65.
Macintosh, D.J., 2010. Coastal Community Livelihoods: implication of intact ecosystem services. Paper presented at the 2010 Katoomba Meeting XVII held June 23–24, 2010 in Hanoi, Vietnam. http://www.ecosystemmarketplace.com/documents/acrobat/k17/Don%20Macintosh.pdf. [(Accessed 8 May 2013)].
Mazda, Y., Magi, M., Kogo, M., Hong, P., 1997. Mangrove on coastal protection from waves in the Tong King Delta, Vietnam. Mangrove Salt Marshes 1, 127–135.
Mendoza, A., Alura, D., 2001. In: Stott, D.E., Mohtar, R.H., Steinhard, G.C. (Eds.), Mangrove structure on the eastern coast of Samar Island, Philippines.Sustaining the Global Farm. Selected Papers from the 10th International Soil Conservation Organization Meeting held May 24–29, 1999 at Purdue University and the USDA-ARS National Soil Erosion Research Laboratory, pp. 423–425.
Pichon, M., 1977. Physiology, morphology and ecology of the Double Barrier reef of North Bohol (Philippines). In: Proceedings of the Thirds International Coral Reef Symposium, Miami, USA, pp. 261–267.
Primavera, J., 2000. Development and conservation of the Philippine mangroves: institutional issues. Ecol. Econ. 35, 91–106.
Primavera, J., Esteban, J., 2008. A review of mangrove rehabilitation in the Philippines: successes, failures and future prospects. Wetl. Ecol. Manag. 16 (3), 173–253.

Pulhin, J., 2000. In: Community forestry in the Philippines: paradoxes and perspectives in development practice.Paper presented in the 8th Biennial Conference of the International Association for the Study of Common Property (IASCP), Bloomington, Indiana, USA, p. 28.

Pulhin, J.M., Inoue, M., Enters, T., 2007. Three decades of community-based forest management in the Philippines: emerging lessons for sustainable and equitable forest management. Int. For. Rev. 19 (4), 865–883.

Pulhin, J.M., Tapia, M.A., 2015. Devolving bundles of rights or bundles of responsibilities? Impacts of forest tenure reform in the Philippines. Econ. Dev. J. 5 (2), 41–48.

Samson, M., Rollon, R., 2008. Growth performance of planted mangroves in the Philippines: revisiting forest management strategies. Ambio 37 (4), 234–240.

Walters, B., 2004. Local management of mangrove forests in the Philippines: successful conservation or efficient resource exploitation. Hum. Ecol. 32 (2), 177–195.

Yao, C., 2001. Community-based Forest Management: For Banacon Planters, Tenure Remains Elusive. The Online Magazine for Sustainable Seas. Available from:www.oneocean.org. [(Accessed January 2013)].

CHAPTER

14

Importance of Seagrass Management for Effective Mitigation of Climate Change

R. Ramesh, K. Banerjee, A. Paneerselvam, R. Raghuraman, R. Purvaja, Ahana Lakshmi

National Centre for Sustainable Coastal Management (NCSCM), Ministry of Environment, Forest and Climate Change, Anna University Campus, Chennai, India

1 INTRODUCTION

Human activities have resulted in increased atmospheric carbon dioxide (CO_2) concentrations from 280 ppmv in the preindustrial era to 408 ppmv (CO_2 Earth) currently; this in turn has led to a positive radiative forcing of climate. Currently, climate change is an important concern, and several governments worldwide are giving top priority toward mitigating climate change. Climate change mitigation refers to any actions or efforts taken to reduce or prevent the long-term risks of climate change on human life and property by reducing the sources or enhancing the sinks of greenhouse gases emissions. As a part of India's Intended Nationally Determined Contribution (INDC), an additional carbon sink of 2.5–3 billion tons of CO_2 equivalent is expected to be created through additional forest and tree cover by 2030. The National Mission for Green India (GIM) aims at restoring India's green cover as well as wetlands and other critical habitats, along with carbon sequestration as a cobenefit (MoEF, 2013). More than 0.1 mha of wetlands alone are to be restored as part of this program.

India is in a transition to a low-carbon economy, adopting several atmospheric CO_2 removal strategies. Among them, biosequestration by coastal ecosystems has gained much attention as this is known for its large biological carbon pools. Carbon accumulation in vegetated coastal sediments provides long-term storage of organic carbon, referred to as "blue carbon" (Mcleod et al., 2011), whereas storage in living biomass takes place over shorter timescales. With a coastline of about 7500 km, India has a fairly large area under coastal

Coastal Management
https://doi.org/10.1016/B978-0-12-810473-6.00015-7

© 2019 Elsevier Inc. All rights reserved.

wetlands. Among various blue carbon ecosystems, seagrass meadows provide high-value ecosystem services such as supporting fisheries and other habitats, regulating water quality, nutrient cycling, sediment stabilization, provisioning of fodder, green manure, medicines, and aesthetic values (Waycott et al., 2009). However, seagrasses at present are under direct threat from a host of anthropogenic influences as well as by the impacts of climate change and ecological degradation (Orth et al., 2006; Duffy, 2006; Waycott et al., 2009). Rapid loss of seagrass beds interrupts the important linkages between seagrass meadows and other habitats (Heck et al., 2008), which in turn creates a long-term impact on themselves. Hence, there is an urgent need to increase the resilience of this important ecosystem to ensure its survival into the future.

In this chapter, we look at the status of seagrasses in India, their importance, and their role in mitigating climate change in terms of the quantities of carbon that they sequester. It is clear that seagrass ecosystems need to be managed well for the varied ecosystem services they provide and specifically, for their contribution to climate change mitigation. Methods of seagrass management, including techniques for restoration of seagrass meadows, are discussed as it is essential to increase the area under seagrass meadows for the above mentioned reasons.

2 SEAGRASSES AND THEIR DISTRIBUTION IN INDIA

Seagrasses are considered one of the most important coastal habitats, as they support a wide range of keystone and ecologically important marine species from different trophic levels (Orth et al., 2006). They are marine-flowering plants that thrive fully submerged in shallow oceanic and estuarine habitats, colonizing soft substrates, especially in wave-sheltered conditions (Barbier et al., 2011). Global coverage of seagrass is estimated to be $3.45 \times 10^5 km^2$ (UNEP-WCMC and Short, 2016), which represents about 0.1%–0.2% of the ocean floor (Fourqurean et al., 2012; Greiner et al., 2013).

In India, the total seagrass cover is estimated to be 517 km^2 (Geevarghese et al., 2017) with 14 reported species. The overall distribution of seagrass meadows in India occurs from the intertidal zone to a maximum depth of 15 m with varying species diversity. Recent estimates suggest that 471.25 km^2 of seagrass meadows (Fig. 1) are distributed along the mainland coast of India, the Andaman and Nicobar Islands (Ganguly et al., 2017), and the Lakshadweep. The Gulf of Mannar and Palk Bay along the southeast coast of India comprise the largest seagrass meadows in India (Jagtap et al., 2003), covering an area of 76–85.5 and 320 km^2, respectively (Umamaheswari et al., 2009; Mathews et al., 2010; Ganguly et al., 2017). The meadows of Palk Bay are more luxuriant due to ideal topography and sediment texture, extend up to 9–10 km from the shore (Mathews et al., 2010; Geevarghese et al., 2017), and are rich in biodiversity. A substratum with sand, silt, and mud with thin layers of sand in Palk Bay/Gulf of Mannar supports the growth and establishment of seagrasses (Thangaradjou and Kannan, 2005, 2007). The Ramsar site os Chilika Lagoon in Odisha State also has seagrass meadows that have expanded from 20 km^2 to 80 km^2 after the opening of the new bar mouth (Kumar and Patnaik, 2010; Priyadarsini et al., 2014; Singh et al., 2015). Both Geevarghese et al. (2017) and Samal (2014) found changes in the areal distribution pattern of seagrasses during different seasons. On the west coast of India, the Gulf of Kachchh Marine National Park has 17–24 km^2 of

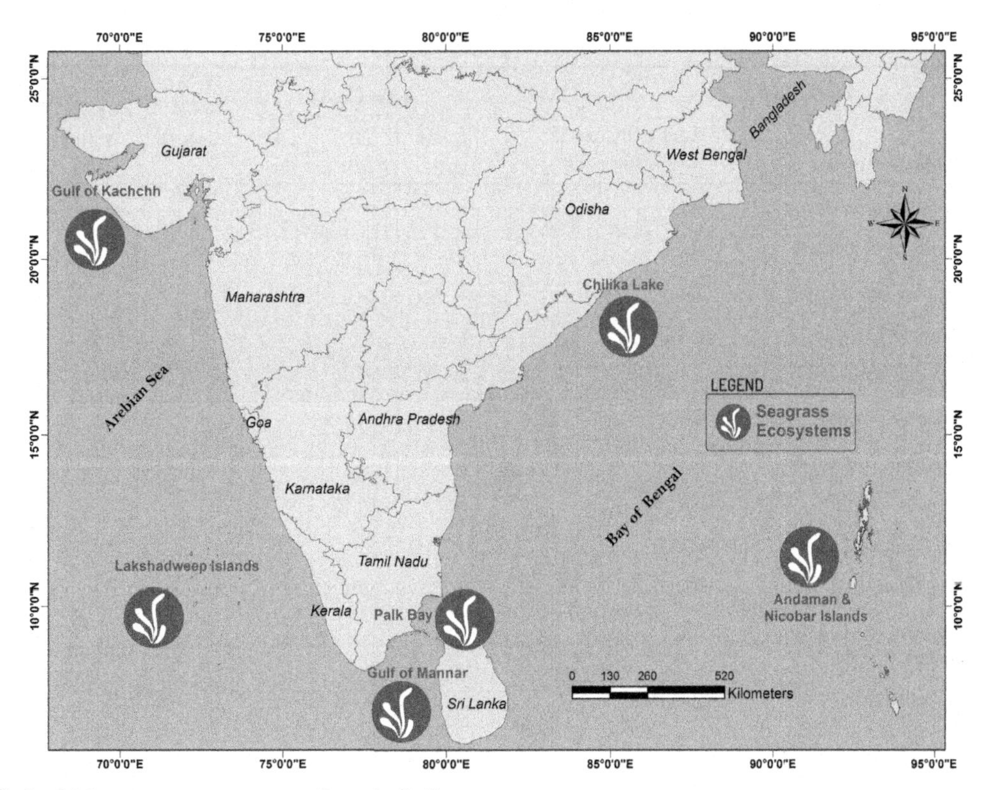

FIG. 1 Major seagrass ecosystems along the Indian coast.

seagrass beds distributed mainly in Bhural, Mundika, and Sikka reefs as well as Pirotan Island (Kamboj, 2014; Geevarghese et al., 2017). Very sparse/rare patches of seagrass meadows have been reported from the other coastal states of India such as Kerala, Karnataka, Goa, Maharashtra, Andhra Pradesh,and West Bengal (Jagtap et al., 2003). Apart from the mainland coast, the atolls of Lakshadweep (12–25 km^2) and the islands of the Andaman and Nicobar archipelago (8.3–29 km^2) have seagrass meadows (Nayak and Bahuguna, 2001), which were found to be reduced in extent during the recent observations by Geevarghese et al. (2017).

Of the 72 globally existing species of seagrass descending from four evolutionary lineages (Short et al., 2011), 14 species belonging to six genera are known to occur in peninsular India and the archipelagos (Ganguly et al., 2017). Species diversity in major seagrass beds in India follows the order Gulf of Mannar > Palk Bay > Andaman and Nicobar Islands > Lakshadweep > Gulf of Kachchh = Chilika Lagoon. A detailed overview of the seagrass species distribution in India is given in Table 1, which indicates the wide heterogeneous species distribution.

Both the Gulf of Mannar and Palk Bay have 13 tropical seagrass species compared to the 19 species recorded from insular Southeast Asia (Short et al., 2007). *Thalassia* and *Cymodocea* are the dominant seagrass genera in the Gulf of Mannar and Palk Bay (Mathews et al., 2010). A salinity regime change in the Chilika lagoon after 2000 introduced new species such as *Haloduleuninervis*, *Halodule pinifolia*, and *Halophila ovate* (Patnaik, 2003; Kumar and Patnaik,

TABLE 1 Distribution of Major Seagrass Meadows in India

			East Coast			West Coast	Islands	
No	**Seagrass Species**	**Reef (R)/Coastal (C)/Estuary (E)**	**GoM (14)**	**PB (13)**	**CH (5)**	**GoK (5)**	**A and N (9)**	**LK (8)**
1	*Cymodocea rotundata*	R,C	x	x			x	x
2	*Cymodocea serrulata*	R,C	x	x			x	x
3	*Enhalus acoroides*	R,C	x	x			x	
4	*Halodule pinifolia*	R,C,E	x	x	x		x	x
5	*Halodule uninervis*	R,C,E	x	x	x	x	x	x
6	*Halodule wrightii*	R	x					
7	*Halophila stipulacea*	R,C	x	x				
8	*Halophila beccarii*	R,C,E	x	x	x	x		
9	*Halophila decipiens*	R,C	x	x				x
10	*Halophila ovalis*	R,C,E	x	x	x	x	x	
11	*Halophila ovalis sub sp: ramamurthiana*	R,C	x	x				
12	*Halophila ovata*	R,C,E	x	x	x	x	x	x
13	*Syringodium isoetifolium*	R,C	x	x			x	x
14	*Thalassiahemprichii*	R,C	x	x		x	x	x

GoM, Gulf of Mannar; *PB*, Palk Bay; *CH*, Chilika Lagoon; *GoK*, Gulf of Kuchchh; *A and N*, Andaman and Nicobar Islands; *LK*, Lakshadweep Islands; "x" indicates presence.

K. Ravikumar, R. Ganesan, A new subspecies of halophilaovalis (R. Br.) J.D. Hook. (Hydrocharitaceae) from the eastern coast of peninsular India, Aquat. Bot. 36 (4) (1990) 351–358; T.G. Jagtap, Distribution of seagrasses along the Indian coast, Aqua. Bot. 40 (1991) 379–386; L. Kannan, T. Thangaradjou, P. Anantharaman, Status of seagrasses of India, Seaweed Res. Util. 21 (1&2) (1999) 25–33; V. Nair, Status of Flora and Fauna of Gulf of Kachchh. National Institute of Oceanography, Goa, 2002; A.K. Patnaik, Phyto-diversity of Chilika Lake, Orissa, India. PhD Thesis, Utkal University, 2003, pp. 1–105; T. Thangaradjou, R. Sridhar, S. Senthilkumar, S. Kannan, Seagrass resource assessment in the Mandapam coast of the Gulf of Mannar biosphere reserve, India. Appl. Ecol. Environ. Res. 6 (1) (2007) 139–146; R. Umamaheswari, S. Ramachandran, E.P. Nobi, Mapping the extend of seagrass meadows of gulf of Mannar biosphere reserve, India using IRS ID satellite imagery, Int. J. Biodivers. Conserv. 1 (5) (2009) 187–193; G. Mathews, K. Diraviya Raj, T. Thinesh, J. Patterson, J.K. Patterson Edward, D. Wilhelmsson, Status of seagrass diversity, distribution and abundance in Gulf of Mannar Marine National Park and Palk Bay (Pamban to Thondi), Southeastern India, South Indian Coast, Mar. Bull. 2 (2) (2010) 1–21; T. Thangaradjou, K. Sivakumar, E.P. Nobi, E. Dilipan, Distribution of seagrasses along the Andaman and Nicobar Islands: a post tsunami survey, in: Recent Trends in Biodiversity of Andaman and Nicobar Islands. ZSI, Kolkata, 2010, pp. 157–160; A.F. Newmaster, K.J. Berg, S. Ragupathy, M. Palanisamy, K. Sambandan, S.G. Newmaster, Local knowledge and conservation of seagrasses in the Tamil Nadu state of India, J. Ethnobiol. Ethnomed. 7 (2011) 37; R. Kumar, A.K. Patnaik, "Chilika" – The Newsletter of Chilika Development Authority and Wetlands International – South Asia, vol. 5 (2010) pp. 1–28; P.M. Priyadarsini, N. Lakshman, S.S. Das, S. Jagamohan, B.D. Prasad, Studies on seagrasses in relation to some environmental variables from Chilika lagoon, Odisha, India, Int. Res. J. Environ. Sci. 3 (11) (2014) 92–101; R.D. Kamboj, Biology and status of seagrasses in gulf of Kachchh marine National Park and sanctuary, India, Indian Ocean Turtle Newslett. (2014) 8–11; D. Ganguly, G. Singh, P. Ramachandran, A.P. Selvam, K. Banerjee, R. Ramachandran, Seagrass metabolism and carbon dynamics in a tropical coastal embayment, Ambio. (2017) 1–13.

2010; Priyadarsini et al., 2014). *Halophila beccarii*, the most commonly distributed species reported from all the coastal states except the Islands, acts as a pioneer species in the succession process of mangrove formation (Jagtap et al., 2003). *Cymodocea serrulata* was dominant in the Lakshadweep islands (Nobi et al., 2011). Of the 14 species recorded in India, *Halophila beccarii* has been identified as "vulnerable" under the IUCN Red List (Criterion B2) because of its intertidal habitat, which is under high anthropogenic stress (Patro et al., 2017).

3 IMPORTANCE OF SEAGRASS ECOSYSTEMS

Ecosystem services are the benefits people obtain from ecosystems. These include *provisioning* services such as food, shelter, and medical applications; *regulating* services that affect climate change, carbon sequestration, wave modification, and water quality; *cultural* services that provide recreational and aesthetic benefits; and *supporting* services such as habitats for fishery, turtles, and dugongs (MEA, 2005). The importance of seagrass ecosystems can be seen to derive from the various ecosystem services they provide (Fig. 2).

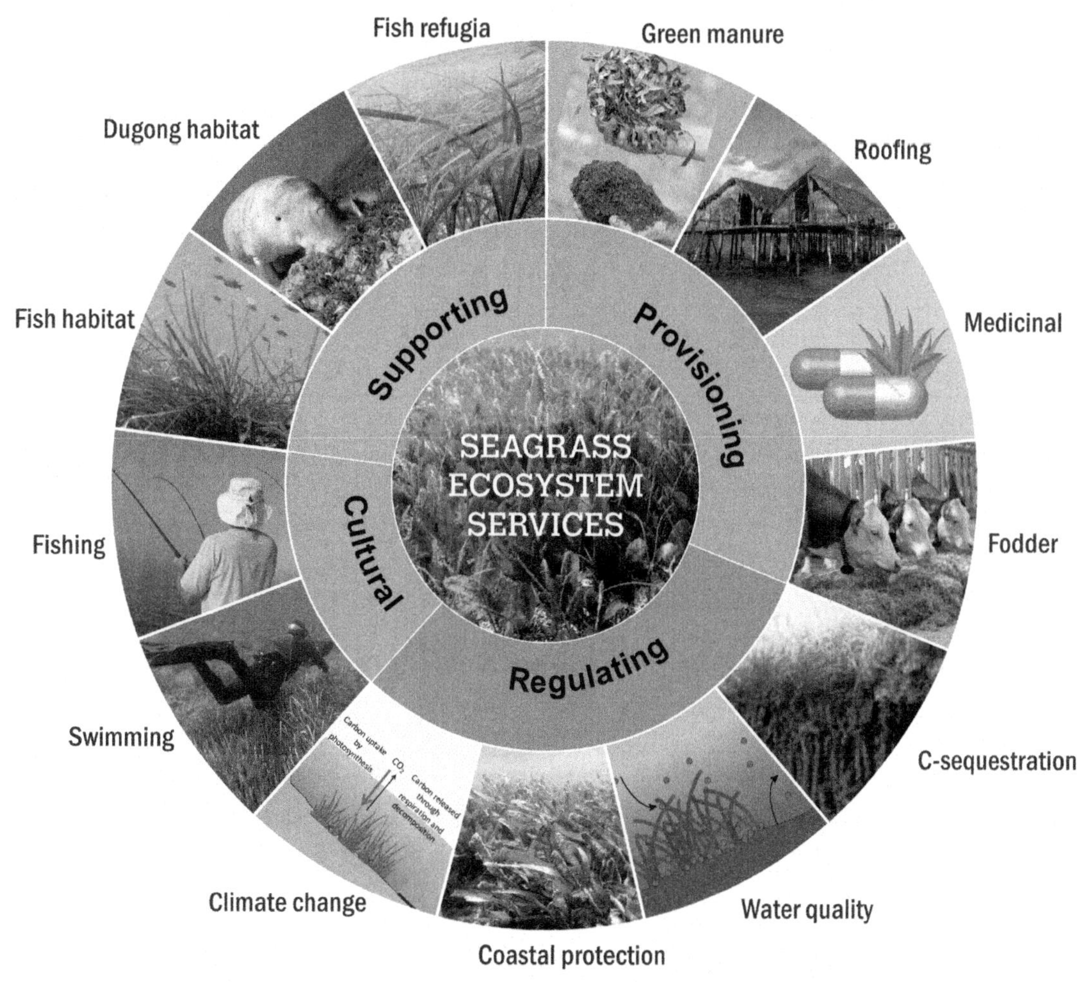

FIG. 2 Ecosystem services from seagrasses.

Provisioning services largely relate to goods such as food and fodder from an ecosystem. Apart from these, different species of seagrasses are reportedly used for treating a variety of diseases (NISCAIR, 2013). The NISCAIR report also provides the results of the chemometric analysis of each seagrass species, indicating the potential for pharmaceutical and nutraceutical products in the future. The various traditional uses of different seagrass species are given in Table 2.

In India, Newmaster et al. (2011) observed that local knowledge systems consisted of a complex classification of seagrass diversity that considered the role of seagrass in the marine ecosystem, including the use of seagrass for medicine (e.g., treatment of heart conditions, seasickness, etc.), food (nutritious seeds), fertilizer (nutrient rich biomass), and livestock feed (goats and sheep). Recently, a study has shown the presence of various biological metabolites in some Indian seagrass that can be used effectively in the food and pharmacological industries. Specifically, *Halodule pinifolia* and *Cymodocea rotundata* exhibited predominant growth-inhibitory activity against all the urinary tract infection (UTI) bacteria (Kannan et al., 2012).

With respect to *supporting services*, as shallow coastal habitats, seagrasses provide key fishing grounds as they offer a complex habitat for a variety of fish and other marine organisms. Seagrass-based fisheries are globally important and are present wherever seagrass exists, supporting subsistence, commercial, and recreational activity (Nordlund et al., 2017). Their high rates of primary production result in well-oxygenated waters that support complex food webs. During photosynthesis, they release oxygen into the water and also pump oxygen into the sediments through their roots, thus creating an oxic environment that promotes nutrient

TABLE 2 Traditional Uses of Various Seagrass Species by Coastal Communities

Species	Parts Used	Traditional Uses
Thalassiahemprichii	Leaf, rhizome, whole plant	Fertilizer, fever, malaria, skin diseases, blood pressure, substrate for bait
Cymodocearotundata	Leaf, whole plant	Fertilizer, tranquillizer for babies, cough, malaria, wounds, fodder
Cymodoceaserrulata	Leaf, whole plant	Fertilizer, tranquillizer for babies, cough, malaria, wounds, fodder
Enhalusacoroides	Root, rhizome, seed	Fertilizer, handicrafts, stings of fishes, seasickness, skin diseases
Syringodiumisoetifolium	Leaf, branches	Fodder, green manure
Thalassodendronciliatum	Leaf, whole plant	Green manure, fever, malaria, smallpox
Halophilaovalis	Leaf	Green manure, skin ailments, burns, boils
Halodulepinifolia	Leaf, branches	Fodder
Haloduleuninervis	Leaf, branches	Fodder

Based on A.F. Newmaster, K.J. Berg, S. Ragupathy, M. Palanisamy, K. Sambandan, S.G. Newmaster, Local knowledge and conservation of seagrasses in the Tamil Nadu state of India, J. Ethnobiol. Ethnomed. 7 (2011) 37.

uptake. While fresh seagrasses are the direct food source for animals such as turtles, dugongs, ducks, fish, sea urchins, and fish, insect larvae and amphipods feed on their decomposed fragments.

Among the *regulating services*, the role played by seagrass ecosystems in reducing the energy of waves and thus protecting the seashore as well as their role in carbon cycling are important. Seagrasses act as ecosystem engineers as they alter water flow, stabilize sediments, and regulate nutrient cycling and food web structure (Hemminga and Duarte, 2000; Gutierrez et al., 2011). By reducing flow velocities in their canopies, seagrass beds promote sedimentation and reduction of grain size (Bos et al., 2007). Carbon sequestration potential is a key aspect of climate regulation and seagrass meadows have shown high carbon sequestration potential as they accumulate from both in situ production and sedimentation of particulate carbon from the water column.

So far, studies on the valuation of seagrass ecosystems are sparse in India, covering only a small fraction of different types of wetlands. Most of the studies focus on valuation of provisioning services while regulating services have seldom received attention (Parikh et al., 2012). *Cultural services* provided by seagrass ecosystems require further research (Nordlund et al., 2017).

Seagrass beds along with mangroves and salt marshes account for up to 70% of the organic carbon in the marine realm. Of the various noteworthy ecosystem services, the carbon sequestration potential of seagrasses to combat climate change by mitigating anthropogenic CO_2 emissions is of growing importance (Hejnowicz et al., 2015). A discussion on recent research in India on carbon sequestration by seagrasses is presented in the next section.

4 CARBON SEQUESTRATION POTENTIAL OF SEAGRASSES

Seagrasses capture carbon dioxide through photosynthesis and incorporate it within their biomass, both above ground and below ground. The extent to which plant biomass accumulates as decay-resistant refractory matter depends on the extent of grazing, export, and burial. Most seagrass ecosystems are net autotrophic as their gross primary production exceeds respiration (Duarte et al., 2010). The proportion of seagrass biomass that accumulates as organic carbon (C_{org}) stored in sediment depends on a low decomposition rate (Breithaupt et al., 2012). In addition, seagrass leaves also trap suspended particulate matter and promote their sedimentation, thus adding to the sedimentary storage component. Thus, seagrass meadows have the capacity to accumulate large carbon pools in the sediment and store it over millennial time scales (Kennedy et al., 2010; Fourqurean et al., 2012; Duarte et al., 2013; Greiner et al., 2013; Serrano et al., 2016). Global seagrass ecosystems are believed to store between 4.2 and 8.4 PgC (Fourqurean et al., 2012). Recent studies have indicated that these ecosystems also contribute to marine carbon sequestration by exporting 24 Tg of carbon per year (30% of the 80 $Tg\,y^{-1}$ carbon sequestered annually in meadows) to deeper layers of the sea, that is, the mixing layer, shelf, and deep sea (Duarte and Krause-Jensen, 2017).

Fig. 3 depicts the biomass variation in the root-to-shoot ratio of different seagrass species recorded from Palk Bay (Purvaja et al., 2017). A recent study from the Indian seagrass

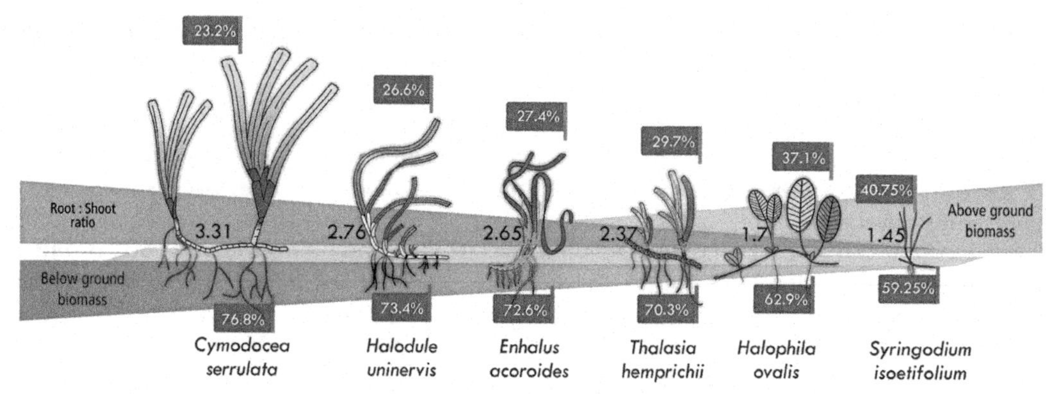

FIG. 3 Biomass variation in root-to-shoot ratio recorded from Palk Bay seagrass species. *R. Purvaja, R.S. Robin, D. Ganguly, G. Hariharan, G. Singh, R. Raghuraman, R. Ramesh, Seagrass meadows as proxy for assessment of ecosystem health, Ocean Coast. Manag. 159 (2017) 34–45. doi:10.1016/j.ocecoaman.2017.11.026.*

ecosystem of Palk Bay indicates that the below ground biomass (rhizome) of seagrass acts as a major compartment in carbon storage, and that the stored carbon is finally sequestered in the sediments (Ganguly et al., 2017). Another species-specific study from Palk Bay and Chilika seagrass meadows from India reported sedimentation rates ranging between 6.2 and 6.9 mm yr^{-1} (Fig. 4) and mass accumulation rates between 0.84 and 1.12 g $cm^{-2}yr^{-1}$ (Fig. 5), with an organic carbon burial rate between 6.97 and 8.99 mol C $m^{-2}yr^{-1}$ (Banerjee et al., 2015).

Preliminary carbon stock assessment by Ganguly et al. (2017) indicates that 1 km^2 of a healthy seagrass meadow in India's coastal waters can store as much as 13.96 Gg C in the top 1 m of the sediment. Similarly, 1 km^2 of a healthy seagrass meadow can sequester 0.44 Gg C yr^{-1}. The comparison of net community production (NCP) between Indian (Palk Bay) seagrass meadows (Ganguly et al., 2017) and the global average (Duarte et al., 2010) shows that production in Indian systems is at least three times higher. Higher growth rates in seagrass ecosystems are translated into an increased accumulation of carbon into both the above-ground biomass (leaves and stem) and the below-ground biomass (root systems and rhizomes) (Russell et al., 2013; Duarte and Krause-Jensen, 2017). Recent estimates indicate that mean above-ground biomass (AGB) from the near-shore region of the Indian Palk Bay (Ganguly et al., 2017) is comparable with the global mean seagrass biomass (224 ± 18 g dwt m^{-2}) (Duarte and Chiscano, 1999).

It is evident from the results presented above that seagrass ecosystems are storehouses of carbon and are capable of capturing atmospheric CO_2 through very efficient sequestration mechanisms. Carbon (C_{org}) is preserved for millennia because seagrass sediment is largely anaerobic. Although extensive literature is available on seagrass carbon sequestration and storage around the globe, research is in its infancy in India. It would be particularly important to estimate the sequestration capacity of dominant seagrass species in India such that conservation efforts of such important seagrass species can be enhanced.

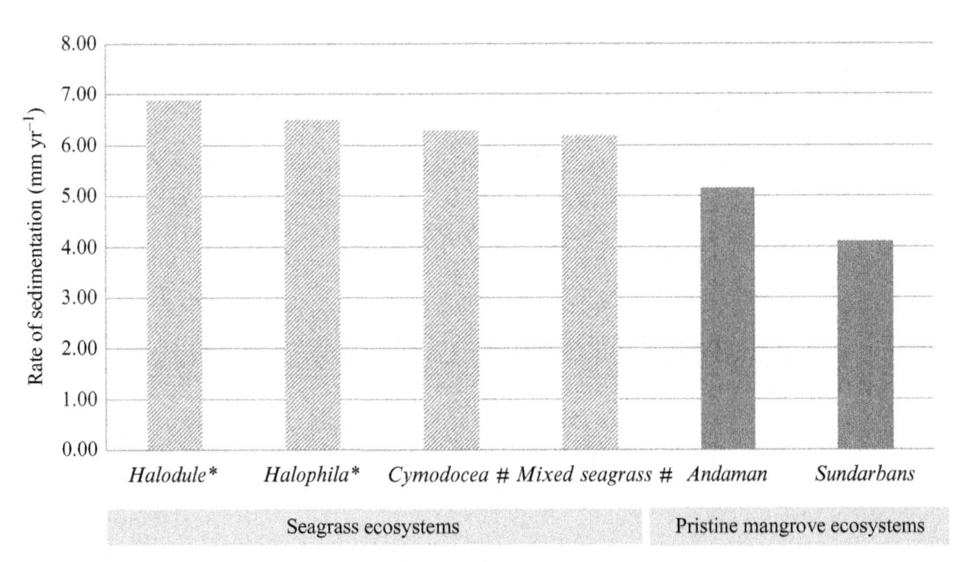

FIG. 4 Comparison of rate of sedimentation between different seagrass species from Chilika Lagoon and Palk Bay and pristine mangrove ecosystems (Banerjee et al., 2012, 2015).

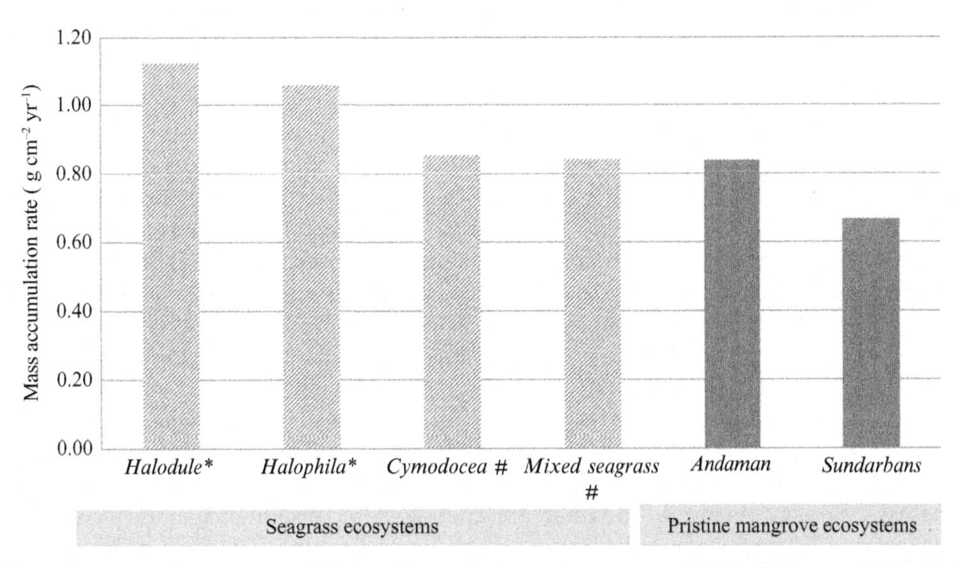

FIG. 5 Comparison of mass accumulation rate between seagrass ecosystems (Chilika Lagoon and Palk Bay) and pristine mangrove ecosystems. *K. Banerjee, B. Senthilkumar, R. Purvaja, R. Ramesh, Sedimentation and trace metal distribution in selected locations of Sundarbans mangroves and Hooghly estuary, northeast coast of India, Environ. Geochem. Health 34 (1) (2012) 27–42; K. Banerjee, A. Paneer Selvam, K. Arumugam, R. Purvaja, R. Ramesh, Seagrass ecosystems as effective sediment stabilizers. In: Presentation on Past and Present Geochemical Process – Impacts on Climate change at JNU, New Delhi, India (2015).*

All this clearly highlights the important role played by seagrass ecosystems in climate change and therefore the importance that needs to be given to the preservation and conservation of seagrass ecosystems, especially in terms of blue carbon. Recently, Ganguly et al. (2017) estimated that the economic values ranged between $1.02 million and $3.65 million per year, based on the regulatory ecosystem services provided from the total seagrass cover in India. Additionally, the monetary values of the stored carbon in the top 1 m sediment of seagrass meadows ranged between $109 million and $146 million.

5 THREATS TO SEAGRASS ECOSYSTEMS

While it is well accepted that ecosystem services provided by seagrass meadows are important, a number of threats to these meadows have been reported. Globally, seagrass habitats have declined in area and several species are threatened due to multiple natural and anthropogenic stressors (Waycott et al., 2009; Short et al., 2011). Natural stressors of seagrass habitat include cyclones, heavy rainfall, coastal uplift and subsidence, grazing herbivores, and diseases, whereas anthropogenic stressors include physical damages due to boating activities, oil spills, and turbidity due to urban, agricultural, and aquacultural runoff (Short and Wyllie-Echeverria, 1996).

The leading stressors of Indian seagrass meadows are mainly human-related activities, both in coastal waters and watershed areas, which lead to subsequent loss in both seagrass area and diversity (Prabhakaran, 2006; Thangaradjou et al., 2007; Singh et al., 2015). Key anthropogenic activities that threaten Indian seagrasses that were identified from the literature include: (i) commercial fishing and trawling activities, (ii) boat activities for recreational purposes, (iii) runoff from coastal aquaculture and agriculture, and (iv) shell harvesting/seaweed cultivation. The ranking among the above activities varies between the systems; however, fishing using trawl/gill nets and boating activities are seen as the most important threats to seagrass in India (Thangaradjou and Nobi, 2009; Mathews et al., 2010). Other prolonged disturbances such as continuous loading of nutrients/fresh water often result in the phase shift of seagrass meadows to macroalgae beds, as seen in the Palk Bay region (Thangaradjou et al., 2013). Alteration in the salinity conditions produced a shift in seagrass species in Chilika Lagoon, with colonization by new species (Kumar and Patnaik, 2010).

Natural disturbances that are responsible for seagrass loss in these regions include cyclones, storms, coastal uplift, grazing by herbivores, and diseases (Jagtap et al., 2003; Mathews et al., 2010; Ragavan et al., 2013). Thangaradjou and Nobi (2009) reported the decrease in seagrass area in the Andaman archipelago as a consequence of the tectonic movement/coastal uplift and sediment dumping on seagrass after the 2004 Indian Ocean tsunami. Increase in rainfall intensity and cyclones have damaged the seagrass meadows in various locations by uprooting the plants and affecting the seawater clarity through high suspended matter inputs. Recently, the leaf-reddening disease was found in species such as *Halophila ovalis* and *Thalassia hemprichi*, which might have an impact on their productivity and distribution (Ragavan et al., 2013). Thangaradjou and Nobi (2009) reported localized effect on reduced leaf cover and biomass distribution in the seagrass meadows of Lakshadweep and Andaman Islands by the grazing of sea turtles. Major threats to different Indian seagrass meadows are summarized in Table 3.

TABLE 3 Threats to Indian Major Seagrass Ecosystems

Location	Major Threats	Impacts	Reference
Palk Bay	High precipitation during northeast monsoon, cyclones	Physical damage to seagrass and reduced light penetration due to high turbidity	Thangaradjou and Nobi (2009)
	Anchoring of boats, propeller damage	Uprooting of seagrass	Thangaradjou and Nobi, (2009) and Mathews et al. (2010)
	Use of push nets, trawl nets, bottom set gill nets	Physical damage to the seagrasses by uprooting the plants and removing the healthy leaves	Thangaradjou and Nobi (2009), Sridhar et al. (2010), Mathews et al. (2010), and D'Souza et al. (2013)
	Nutrient enrichment from aquaculture wastes and proliferation of macroalgae	Eutrophication, growth of algae/ seaweed which competes with seagrass. Diminish light availability and sediment quality	Sridhar et al. (2010) and Thangaradjou et al. (2013)
	Exotic seaweed cultivation	Affects light penetration and seagrass die off	Mathews et al. (2010)
Gulf of Mannar	Southwest monsoonal winds, northeast monsoon, cyclones	Physical damage to seagrass and reduced light penetration due to high turbidity	Thangaradjou and Nobi (2009)
	Shell harvesting of *Tellina angulata*	Physical destruction of seagrass rhizomes and roots	Thangaradjou et al. (2007) and Thangaradjou and Nobi (2009)
	Anchoring of boats, propeller damage	Uprooting of seagrass	Thangaradjou and Nobi (2009) and Mathews et al. (2010)
	Use of push nets, trawl nets, bottom set gill nets	Physical damage to the seagrasses by uprooting the plants and removing the healthy leaves	Thangaradjou and Nobi (2009), Mathews et al. (2010) and D'Souza et al. (2013)
Chilika Lagoon	Natural hazards such as storms, floods	Physical damage to seagrass and reduced light penetration due to high turbidity	Priyadarsini et al. (2014)
	Dredging, inappropriate fishing, anchoring, coastal constructions	Uprooting of seagrass	Priyadarsini et al. (2014)
	Coastal aquaculture wastes	Eutrophication, growth of algae which competes with seagrass. Diminished light availability and sediment quality	Priyadarsini et al. (2014)

(Continued)

TABLE 3 Threats to Indian Major SeagrassEcosystems—cont'd

Location	Major Threats	Impacts	Reference
Gulf of Kachchh	Industrial and domestic pollution	Eutrophication and formation of algal blooms	Kamboj (2014)
	Development of ports and harbors	Increase in sedimentation, solid waste and marine pollution	Kamboj (2014)
	Fishing and boat activities	Physical damage to the seagrass leaves and rhizomes	Kamboj (2014)
Andaman and Nicobar Islands	Intense boating and tourism activities	Physical damage to the seagrasses roots/rhizomes and leaves	Thangaradjou and Nobi (2009)
	Diseases – Leaf Redding	Decoloring of leaves leading to seagrass mortality	Ragavan et al. (2013)
	Tsunami, cyclones, storms	Physical damage, sediment dumping on seagrass and increase turbidity	Thangaradjou and Nobi (2009) and Danielsen et al. (2005)
Lakshadweep	Grazing by green turtles (protected species)	Grazing pressure can modify the species composition	Lal et al. (2010) and Kaladharan et al. (2013)
	Boating and tourism activities	Physical damage to the seagrasses roots/rhizomes and leaves	Thangaradjou and Nobi (2009) and Nobi et al. (2013)
	Sea erosion and siltation	Affects light penetration and seagrass die off	Nobi et al. (2013)
	Disposal of fish waste and untreated solid waste	Localized eutrophication and seagrass damage	Thangaradjou and Nobi (2009)

6 MANAGEMENT OF SEAGRASS ECOSYSTEMS

Sea level rise and an increase in the intensity of cyclones are potential consequences of climate change. In this context, the ability of seagrasses to protect coastlines through mitigation of wave energy assumes importance. However, their other important contribution—the ability to sequester carbon—can be considered to be of higher importance because it is also known that considerable amounts of carbon (that were buried) are released when such ecosystems are degraded. Hence, it is obvious that existing areas under seagrass cover have to be protected and overall, areas under seagrass have to be increased.

There are a number of legislative and policy options that can be used to protect seagrass ecosystems (Ramesh et al., 2018) in India. Of these, the Coastal Regulation Zone Notification 2011 (CRZ, 2011), issued under the Environmental (Protection) Act, 1986, which classifies seagrass meadows as CRZ-Ia (Ecologically Sensitive Areas), is the only explicit legislation that protects seagrasses by prohibiting development activity in their vicinity. Efforts are also being made in marine spatial planning as part of the development of integrated coastal management plans whereby the locations of these important ecosystems are mapped so that they become areas for conservation and protection.

A decline in seagrass beds has prompted the implementation of numerous restoration programs in different parts of the world. In addition, blue carbon sequestration has been found to be enhanced with seagrass restoration (Greiner et al., 2013) which has prompted restoration attempts globally. However, the success of these attempts has been hampered by the difficulties in reducing the causes of disturbance (Paling et al., 2009). Techniques for restoration include natural restoration, transplants, and seeding. The cost of restoration may vary depending on the selection of sites for restoration, techniques, degradation intensity, follow-up measures, and surrounding environmental conditions. According to Bergstrom (2006), the lower end of the global seagrass restoration range is between $83,363 and $244,530 per hectare whereas the higher end is between $1,900,000 (McNeese et al., 2006) and $3,387,000 per hectare (Lewis III. et al., 2006; Paling et al., 2009). The only study on seagrass restoration in the Palk Bay and adjacent Gulf of Mannar with a total cost of INR1,426,750 ($\sim$$22,000 ha^{-1}) reported from India showed an average survival rate (after nine months of transplantation) of 81.5% (*Thalassia hemprichii*), 85.7% (*Cymodocea serrulata*), and 78.6% (*Syringodium isoetifolium*), for three major seagrass species, respectively. The activity carried out in pilot mode covered 200 m^2 in each location (Patterson and D'Souza, 2015). Three techniques—sprigs (Quadrate), plugs, and saplings—were attempted, of which sprigs were the most successful. Nearly one square kilometer was rehabilitated and another area of equal size is under restoration; both attempts are off the coast of Thoothukudi in the Gulf of Mannar. Considering the elevated cost of seagrass restoration, the conservation of seagrass habitats (e.g., $1400 per hectare as per Stowers et al., 2006) is far more cost effective than the restoration of degraded seagrass meadows.

7 CONCLUSIONS AND RECOMMENDATIONS

It is clear that seagrasses are extremely important in the current context of climate change, as these systems are highly efficient in both sequestering and long-term storage of carbon. To increase the area under seagrass meadows, two plans of action may be considered. In the first case, the effective implementation of available legislation may help in the reduction of threats. Ensuring that the provisions of protection accorded to CRZ-I are strictly followed can prevent further degradation and promotion of self-restoration of seagrass meadows. The sequestration capacity of dominant seagrass species in India needs to be estimated so that conservation efforts of such important seagrass species can be enhanced. Research also needs to be undertaken on the use of seagrasses by local communities to determine if protection of seagrass meadows could be enhanced by declaration of such areas as Critically Vulnerable Coastal Areas (CVCA), which are to be managed with the involvement of local communities including fishers (CRZ, 2011). Simultaneously, mapping of areas with potential for growth of seagrasses needs to be accomplished.

Action is necessary to increase the area under seagrass, either by natural expansion or by planting. For the former, areas such as the Chilika lagoon where natural expansion of seagrass meadows has been reported must be studied so as to enable replication in other potential areas. For the latter, trials that have been carried out in Palk Bay and the Gulf of Mannar could help in strengthening this activity. This needs to be supported by research into techniques for seagrass transplantation.

It is also recommended to bring local communities, scientists, resource managers, and government officials together in designing an action plan for seagrass conservation. Such a strategy would be in accordance with Chapter 17 of Agenda 21 of the 1992 Earth Summit at Rio de Janeiro, Brazil, which states that government agencies charged with coastal zone protection must integrate traditional ecological knowledge (TEK) and sociocultural values with management agendas (Wyllie-Echeverria et al., 2002). Ultimately, more awareness needs to be created among all stakeholders about the various ecosystem services provided by seagrasses, especially their role in carbon sequestration.

References

Banerjee, K., Senthilkumar, B., Purvaja, R., Ramesh, R., 2012. Sedimentation and trace metal distribution in selected locations of Sundarbans mangroves and Hooghly estuary, northeast coast of India. Environ. Geochem. Health 34 (1), 27–42.

Banerjee, K., Paneer Selvam, A., Arumugam, K., Purvaja, R., Ramesh, R., 2015. In: Seagrass ecosystems as effective sediment stabilizers.Presentation on Past and Present Geochemical Process – Impacts on Climate change at JNU, New Delhi, India.

Barbier, E.B., Sally, D.H., Kennedy, C., Koch, E.W., Stier, A.C., Silliman, B.R., 2011. The value of estuarine and coastal ecosystem services. Ecol. Monogr 81 (2), 169–193. https://dx.doi.org/10.1890/10-1510.1.

Bergstrom, P., 2006. In: Treat, S.F., Lewis III, R.R. (Eds.), Species selection, success, and costs of multi-year, multi-species submerged aquatic vegetation (SAV) planting in Shallow Creek, Patapsco River, Maryland. Seagrass Restoration: Success, Failure, and the Costs of Both. Selected Papers Presented at a Workshop, Mote Marine Laboratory, Sarasota, Florida, 11–12 March 2003. Lewis Environmental Services, Valrico, Florida, pp. 49–58.

Breithaupt, J.L., Smoak, J.M., Smith, T.J., Sanders, C.J., Hoare, A., 2012. Organic carbon burial rates in mangrove sediments: strengthening the global budget. Glob. Biogeochem. 26(3).

Bos, A.R., Bouma, T.J., de Kort, G.L., van Katwijk, M.M., 2007. Ecosystem engineering by annual intertidal seagrass beds: sediment accretion and modification. Estuar. Coast. Shelf. Sci. 74 (1), 344–348. https://dx.doi.org/10.1016/j.ecss.2007.04.006.

Danielsen, F., Sørensen, M.K., Olwig, M.F., Selvam, V., Parish, F., Burgess, N.D., Hiraishi, T., Karunagaran, V.M., Rasmussen, M.S., Hansen, L.B., Quarto, A., 2005. The Asian tsunami: a protective role for coastal vegetation. Science 310 (5748), 643.

D'Souza, N., Patterson, J.K., Ishwar, N.M., 2013. Survey and assessment of seagrass beds in the Gulf of Mannar and Palk Bay to support strategy to conserve and manage seagrass habitats. SDMRI, Tamil Nadu, pp. 1–29.

Duarte, C.M., Marba, N., Gacia, E., Fourqurean, J.W., Beggins, J., Barron, C., Apostolaki, E.T., 2010. Seagrass community metabolism: assessing the carbon sink capacity of seagrass meadows. Glob. Biogeochem. Cycles 24, 1–8.

Duarte, C.M., Chiscano, C.L., 1999. Seagrass biomass and production: a reassessment. Aquat. Bot. 65 (1), 159–174. https://dx.doi.org/10.1016/S0304-3770(99)00038-8.

Duarte, C.M., Krause-Jensen, D., 2017. Export from seagrass meadows contributes to marine carbon sequestration. Front. Mar. Sci. 4. https://dx.doi.org/10.3389/fmars.2017.00013.

Duarte, C.M., Sintes, T., Marba, N., 2013. Assessing the CO_2 capture potential of seagrass restoration projects. J. Appl. Ecol. 50 (6), 1341–1349.

Duffy, J.E., 2006. Biodiversity and the functioning of seagrass ecosystems. Mar. Ecol. Prog. Ser. 311, 233–250.

Fourqurean, J.W., Duarte, C.M., Kennedy, H., Marba, N., Holmer, M., 2012. Seagrass ecosystems as a globally significant carbon stock. Nat. Geosci. 5, 505–509. https://dx.doi.org/10.1038/ngeo1477.

Ganguly, D., Singh, G., Ramachandran, P., Selvam, A.P., Banerjee, K., Ramachandran, R., 2017. Seagrass metabolism and carbon dynamics in a tropical coastal embayment. Ambio, 1–13.

Geevarghese, G.A., Akhil, B., Magesh, G., Krishnan, P., Purvaja, R., Ramesh, R., 2017. A comprehensive geospatial assessment of seagrass distribution in India. Ocean Coast. Manag. https://dx.doi.org/10.1016/j.ocecoaman.2017.10.032.

Greiner, J.T., McGlathery, K.J., Gunnell, J., McKee, B.A., 2013. Seagrass restoration enhances "blue carbon" sequestration in coastal waters. PLoS ONE 8 (8), e72469. https://dx.doi.org/10.1371/journal.pone.0072469.

Gutierrez, J.L., Jones, C.G., Byers, J.E., Arkema, K.K., Berkenbusch, K., Commito, J.A., Duarte, C.M., Hacker, S.D., Lambrinos, J.G., Hendriks, I.E., Hogarth, P.J., Palomo, M.G., Wild, C., 2011. Physical ecosystem engineers and the functioning of estuaries and coasts. In: Treatise on Estuarine and Coastal Science. Academic Press, Waltham, pp. 53–81. https://dx.doi.org/10.1016/B978-0-12-374711-2.00705-1.

Heck, K.L., Carruthers, T.J., Duarte, C.M., Hughes, A.R., Kendrick, G., Orth, R.J., Williams, S.W., 2008. Trophic transfers from seagrass meadows subsidize diverse marine and terrestrial consumers. Ecosystems 11 (7), 1198–1210.

Hejnowicz, A.P., Kennedy, H., Rudd, M.A., Huxham, M.R., 2015. Harnessing the climate mitigation, conservation and poverty alleviation potential of seagrasses: prospects for developing blue carbon initiatives and payment for ecosystem service programmes. Front. Mar. Sci. 2, 32.

Hemminga, M.A., Duarte, C.M., 2000. Seagrass Ecology. Cambridge University Press, Cambridge.

Jagtap, T.G., Komarpant, D.S., Rodrigues, R.S., 2003. Status of a seagrass ecosystem: an ecologically sensitive wetland habitat from India. Wetlands 23 (1), 161–170. https://dx.doi.org/10.1672/0277-5212(2003)023[0161,SOASEA]2.0.CO;2.

Kaladharan, P., Koya, K.P., Kunhikoya, V.A., AnasuKoya, A., 2013. Turtle herbivory of seagrass ecosystems in the Lakshadweep atolls: concerns and need for conservation measures. J. Mar. Biol. Assoc. 55 (1), 25–29.

Kamboj, R.D., 2014. Biology and status of seagrasses in gulf of Kachchh marine National Park and sanctuary, India. Indian Ocean Turtle Newslett. 8–11.

Kannan, R.R.R., Arumugam, R., Anantharaman, P., 2012. Chemical composition and antibacterial activity of Indian seagrasses against urinary tract pathogens. Food Chem. 135 (4), 2470–2473.

Kennedy, H., Beggins, J., Duarte, C.M., Fourqurean, J.W., Holmer, M., Marbà, N., Middelburg, J.J., 2010. Seagrass sediments as a global carbon sink: isotopic constraints. Glob. Biogeochem. Cycl. 24, GB4026. https://dx.doi.org/10.1029/2010GB003848.

Kumar, R., Patnaik, A.K., 2010. "Chilika" – The Newsletter of Chilika Development Authority and Wetlands International – South Asia. vol. 5, pp. 1–28.

Lal, A., Arthur, R., Marba, N., Lill, A.W.T., Alcoverro, T., 2010. Implications of conserving an ecosystem modifier: increasing green turtle (Cheloniamydas) densities substantially alters seagrass meadows. Biol. Conserv. 143, 2730–2738.

Lewis III., R.R., Marshall, M.J., Bloom, S.A., Hodgson, A.B., Flynn, L.L., 2006. Evaluation of the success of seagrass mitigation at Port Manatee, Tampa Bay, Florida. In: Treat, S.F., Lewis III, R.R. (Eds.), Seagrass Restoration: Success, Failure, and the Costs of Both. Selected Papers Presented at a Workshop, Mote Marine Laboratory, Sarasota, Florida, March 11–12, 2003. Lewis Environmental Services, Valrico, Florida, pp. 19–40.

Mathews, G., Diraviya Raj, K., Thinesh, T., Patterson, J., PattersonEdward, J.K., Wilhelmsson, D., 2010. Status of seagrass diversity, distribution and abundance in Gulf of Mannar Marine National Park and Palk Bay (Pamban to Thondi), Southeastern India. South Indian Coast. Mar. Bull. 2 (2), 1–21.

Mcleod, E., Chmura, G.L., Bouillon, S., Salm, R., Björk, M., Duarte, C.M., Lovelock, C.E., Schlesinger, W.H., Silliman, B.R., 2011. A blueprint for blue carbon: toward an improved understanding of the role of vegetated coastal habitats in sequestering CO_2. Front. Ecol. Environ. 9, 552–556.

McNeese, P.L., Kruer, C.R., Kenworthy, W.J., Schwarzschild, A.C., Wells, P., Hobbs, J., 2006. In: Treat, S.F., Lewis III, R.R. (Eds.), Topographic restoration of boat grounding damage at the Lignumvitae Submerged Land Management area. Seagrass Restoration: Success, Failure, and the Costs of Both. Selected Papers Presented at a Workshop, Mote Marine Laboratory, Sarasota, Florida, 11–12 March 2003. Lewis Environmental Services, Valrico, FL, pp. 131–146.

MEA, 2005. Hassan, R., Scholes, R., Ash, N. (Eds.), Ecosystems and Human Well-being: Current State and Trends. In: vol. 1. Millennium Ecosystem Assessment, Island Press, Washington.

MoEF, 2013. National Mission for a Green India (Under the National Action for Climate Change). http://www.moef.gov.in/sites/default/files/GIM_Mission%20Document-1.pdf. [(Accessed 12 February 2018)].

Nayak, S., Bahuguna, A., 2001. Application of remote sensing data to monitor mangroves and other coastal vegetation of India. Ind. Jour. of Mar. Sci. 30, 195–213.

Newmaster, A.F., Berg, K.J., Ragupathy, S., Palanisamy, M., Sambandan, K., Newmaster, S.G., 2011. Local knowledge and conservation of seagrasses in the Tamil Nadu state of India. J. Ethnobiol. Ethnomed. 7, 37.

NISCAIR, 2013. Seagrasses: The Oxygen Pumps in the Sea. The Wealth of India, Raw Materials series. CSIR-National Institute of Science Communication and Information Resources, New Delhi.

Nobi, E.P., Dilipan, E., Sivakumar, K., Thangaradjou, T., 2011. Distribution and biology of seagrass resources of Lakshadweep group of Islands, India. Ind. J. GeoMar. Sci. 40 (5), 624–634.

Nobi, E.P., Dilipan, E., Thangaradjou, T., Dinesh Kumar, P.K., 2013. Restoration scaling of seagrass habitats in the Oceanic Islands of Lakshadweep, India using geospatial technology. Appl. Geomat. 5 (2), 167–175.

Nordlund, L.M., Unsworth, R.K.F., Gullstrom, M., Cullen-Unsworth, L.C., 2017. Global significance of seagrass fishery activity. Fish Fish., 1–14. https://dx.doi.org/10.1111/faf.12259.

Orth, R.J., Carruthers, T.J., Dennison, W.C., Duarte, C.M., Fourqurean, J.W., Heck, J.K.L., Hughes, A.R., Kendrick, G.A., Kenworthy, W.J., Olyarnik, S., Short, F.T., 2006. A global crisis for seagrass ecosystems. Bioscience 56 (12), 987–996.

Paling, E.I., Fonseca, M., van Katwijk, M.M., van Keulen, M., 2009. Seagrass restoration. In: Coastal Wetlands: An Integrated Ecosystem Approach., pp. 687–713.

Parikh, K.S., Ravindranath, N.H., Murthy, I.K., Mehra, S., Kumar, R., James, E.J., Vivekanandan, E., Mukhopadhyay, P., 2012. The Economics of Ecosystems and Biodiversity-India: Initial Assessment and Scoping Report. Working Document, p. 157.

Patnaik, A.K., 2003. Phyto-diversity of Chilika Lake, Orissa, India. PhD, ThesisUtkal University, pp. 1–105.

Patro, S., Krishnan, P., Samuel, V.D., Purvaja, R., Ramesh, R., 2017. Seagrass and salt marsh ecosystems in south asia: an overview of diversity, distribution, threats and conservation status. In: Wetland Science. Springer, India, pp. 87–104.

Patterson, J.K., D'Souza, N., 2015. Rehabilitation of degraded seagrass area in Tuticorin coast of Gulf of Mannar, Tamil Nadu, to support long term conservation of seagrass habitats. SDMRI, Tuticorin, Tamil Nadu, pp. 1–12. http://www.sdmri.in/index.php/seagrass-rehabilitation.

Prabhakaran, M.P., 2006. Community shift in seagrass ecosystem Minicoy atoll (India). Seagrass Watch News 27, 15.

Priyadarsini, P.M., Lakshman, N., Das, S.S., Jagamohan, S., Prasad, B.D., 2014. Studies on seagrasses in relation to some environmental variables from Chilika lagoon, Odisha, India. Int. Res. J. Environ. Sci. 3 (11), 92–101.

Purvaja, R., Robin, R.S., Ganguly, D., Hariharan, G., Singh, G., Raghuraman, R., Ramesh, R., 2017. Seagrass meadows as proxy for assessment of ecosystem health. Ocean Coast. Manag. 159, 34–45. https://dx.doi.org/10.1016/j.ocecoaman.2017.11.026.

Ragavan, P., Saxena, A., Mohan, P.M., Coomar, T., Ragavan, A., 2013. Leaf reddening in seagrasses of Andaman and Nicobar Islands. Trop. Ecol. 54 (2), 269–273.

Ramesh, R., Banerjee, K., Paneerselvam, A., Lakshmi, A., Krishnan, P., Purvaja, R., 2018. Legislation and policy options for conservation and management of seagrass ecosystems in India. Ocean Coast. Manag. https://dx.doi.org/10.1016/j.ocecoaman.2017.12.025.

Russell, B.D., Connell, S.D., Uthicke, S., Muehllehner, N., Fabricius, K.E., Hall-Spencer, J.M., 2013. Future seagrass beds: Can increased productivity lead to increased carbon storage? Mar. Pollut. Bull. 73 (2), 463–469. https://dx.doi.org/10.1016/j.marpolbul.2013.01.031.

Samal, R.N., 2014. In: Applications of RS and GIS in Wetland management: case study of Chilika lagoon.User Interaction Meet Proceedings, US4, 20–21 January NRSC, India.

Serrano, O., Lavery, P.S., Lopez-Merino, L., Ballesteros, E., Mateo, M.A., 2016. Location and associated carbon storage of erosional escarpments of seagrass Posidonia mats. Front. Mar. Sci. 3, 42.

Short, F., Carruthers, T., Dennison, W., Waycott, M., 2007. Global seagrass distribution and diversity: a bioregional model. J. Exp. Mar. Biol. Ecol. 350 (1–2), 3–20.

Short, F.T., Polidoro, B., Livingstone, S.R., Carpenter, K.E., Bandeira, S., Bujang, J.S., Calumpong, H.P., Carruthers, T.J., Coles, R.G., Dennison, W.C., Erftemeijer, P.L., 2011. Extinction risk assessment of the world's seagrass species. Biol. Conserv. 144 (7), 1961–1971. https://dx.doi.org/10.1016/j.biocon.2011.04.010.

Short, F.T., Wyllie-Echeverria, S., 1996. Natural and human-induced disturbances of seagrass. Environ. Conserv. 23, 17–27.

Singh, G., Ganguly, D., Paneerselvam, A., Kakolee, B., Purvaja, R., Ramesh, R., 2015. Seagrass ecosystem and climate change: an Indian perspective. J. Clim. Change 1 (1, 2), 67–74. https://dx.doi.org/10.3233/JCC-150005.

Sridhar, R., Thangaradjou, T., Kannan, L., Astalakshmi, S., 2010. Assessment of coastal bio-resources of the Palk Bay, India, using IRS-LISS-III data. J. Indian Soc. Remote Sens. 38, 565–575.

Stowers, J.F., Fehrmann, E., Squires, A., 2006. In: Treat, S.F., Lewis III, R.R. (Eds.), Seagrass scarring in Tampa Bay: impact analysis and management options. Seagrass Restoration: Success, Failure, and the Costs of Both. Selected Papers presented at a workshop, Mote Marine Laboratory, Sarasota, Florida, March 11–12, 2003. Lewis Environmental Services, Valrico, FL, pp. 69–78.

Thangaradjou, T., Kannan, L., 2005. Marine sediment texture and distribution of seagrasses in the Gulf of Mannar biosphere reserve. Seaweed Res. Util. 27, 145–154.

Thangaradjou, T., Kannan, L., 2007. Nutrient characteristics and sediment texture of the seagrass beds of the Gulf of Mannar. J. Environ. Biol. 28 (1), 29.

Thangaradjou, T., Nobi, E.P., 2009. Seagrass – Watch, Threats to the Seagrasses of India. vol. 39, pp. 20–21.

Thangaradjou, T., Sridhar, R., Senthilkumar, S., Kannan, S., 2007. Seagrass resource assessment in the Mandapam coast of the Gulf of Mannar biosphere reserve, India. Appl. Ecol. Environ. Res. 6 (1), 139–146.

Thangaradjou, T., Subhashini, P., Raja, S., 2013. Macroalgae competition-challenging seagrass survival. Seagrass – Watch Threats Human Impacts and Mitigation 47, 50–51.

Umamaheswari, R., Ramachandran, S., Nobi, E.P., 2009. Mapping the extend of seagrass meadows of gulf of Mannar biosphere reserve, India using IRS ID satellite imagery. Int. J. Biodivers. Conserv. 1 (5), 187–193.

UNEP-WCMC, Short, F.T., 2016. Global distribution of seagrasses (version 4.0). In: Fourth update to the data layer used in Green and Short (2003). UNEP World Conservation Monitoring Centre, Cambridge, UK. http://data.unepwcmc.org/datasets/7.

Waycott, M., Duarte, C.M., Carruthers, T.J., Orth, R.J., Dennison, W.C., Olyarnik, S., Calladine, A., Fourqurean, J.W., Heck, K.L., Hughes, A.R., Kendrick, G.A., 2009. Accelerating loss of seagrasses across the globe threatens coastal ecosystems. Proc. Natl. Acad. Sci. 106 (30), 12377–12381.

Wyllie-Echeverria, S., Gunnarsson, K., Mateo, M.A., Borg, J.A., Renom, P., Kuo, J., Schanz, A., Hellblom, F., Jackson, E., Pergent, G., Pergent-Martini, C., Johnson, M., Sanchez-Lizaso, J., Boudouresque, C.F., Aioi, K., 2002. Protecting the seagrass biome: Report from the traditional seagrass knowledge working group. Bull. Mar. Sci. 71 (3), 1415–1417.

Further Reading

Jagtap, T.G., 1991. Distribution of seagrasses along the Indian coast. Aqua. Bot. 40, 379–386.

Kannan, L., Thangaradjou, T., Anantharaman, P., 1999. Status of seagrasses of India. Seaweed Res. Util. 21 (1&2), 25–33.

Nair, V., 2002. Status of Flora and Fauna of Gulf of Kachchh. National Institute of Oceanography, Goa.

Ravikumar, K., Ganesan, R., 1990. A new subspecies of halophilaovalis (R. Br.) J.D. Hook. (Hydrocharitaceae) from the eastern coast of peninsular India. Aquat. Bot. 36 (4), 351–358.

Thangaradjou, T., Sivakumar, K., Nobi, E.P., Dilipan, E., 2010. Distribution of seagrasses along the Andaman and Nicobar Islands: a post tsunami survey. In: Recent Trends in Biodiversity of Andaman and Nicobar Islands. ZSI, Kolkata, pp. 157–160.

CHAPTER

15

Managing Mangrove Forests Using Open Source-Based WebGIS

K. Jayakumar

Center for Remote Sensing and Geoinformatics, Sathyabama Institute of Science and Technology, Chennai, India

1 INTRODUCTION

India has a coastline of 7517 km, of which 5423 km belongs to Peninsular India and 2094 km to the Andaman, Nicobar, and Lakshadweep Islands. The mainland coast consists of the following: 43% sandy beaches, 11% rocky coasts, including cliffs, and 46% mud flats or marshy coast (Government of India, 2009). It is endowed with a variety of natural and man-made ecosystems such as mangroves, coral reefs, lagoons, creeks, sands, salt marshes, intertidal mudflats, saltpans, aquaculture ponds, etc. (Sahu et al., 2012). A recent estimate showed that coastal wetlands account for 27.13% (4.01 m ha) (Panigrahy et al., 2012). Among the different flora and fauna associated with the coastal wetlands, mangroves occupy an integral part. The mangroves in India account for about 5% of the world's mangrove vegetation and are spread over an area of about 4662.56 km^2, which is 0.14% of the country's total geographic area along the coastal states/UTs of the country. The very dense mangrove comprises 1403 km^2 (30.10% of mangrove cover), the moderately dense mangrove is 1658.12 km^2 (35.57%), and open mangrove covers an area of 1601.44 km^2 (34.33%). In India, West Bengal has the maximum mangrove cover, followed by Gujarat, the Andaman and Nicobar Islands, and Andhra Pradesh (Forest Survey of India (FSI), 2015).

1.1 Mangrove Distribution in India

The distribution of mangroves in India is given in Table 1. Of the country's total area under mangrove cover, about 59.4% is found on the east coast, 27.4% on the west coast and the remaining 13.2% in the Andaman and Nicobar Islands. Some of the most spectacular

Coastal Management
https://doi.org/10.1016/B978-0-12-810473-6.00016-9

© 2019 Elsevier Inc. All rights reserved.

TABLE 1 Area-Wise Distribution of Mangrove Forests in India (FSI, 2015)

State	Districts	Area (sq. km.)													
		1987	1989	1991	1993	1995	1997	1999	2001	2003	2005	2009	2011	2013	2015
West Bengal	Medinipur, North 24 Pargana, and South 24 Pargana	2076	2109	2119	2119	2119	2123	2125	2081	2120	2136	2152	2155	2097	2106
Orissa	Baleshwar, Bhadrak, Jagatsinghpur, Kendrapara, and Puri	199	192	195	195	195	211	215	219	203	217	221	222	213	231
Andhra Pradesh	Godavari, Krishna, Guntur, Nellore, and Prakasam	495	405	399	378	383	383	397	333	329	329	353	352	352	367
Tamil Nadu	Cuddalore, Nagapattinam, Ramnathapurm, Thanjavur, and Tuticorin	23	47	47	21	21	21	21	23	35	36	39	39	39	47
Gujarat	Ahmedabad, Amerli, Anand, Bharuch, Bhavnagar, Jamnagar, Junagarh, Kuchchh, Navsari, Porbandar, Rajkot, Surat, Vadodara, and Valsad	427	412	397	419	689	901	1031	911	916	991	1046	1058	1103	1107

Maharashtra	Mumbai city, Mumbai Suburb, Raigarh, Ratnagiri, Sindhudurg and Thane	140	114	113	155	155	124	108	118	158	186	186	186	186	222
Goa	North Goa and South Goa	0	3	3	3	3	5	5	5	16	16	17	22	22	26
Karnataka	Uttar Kannada and Udipi	0	0	0	0	2	3	3	2	3	3	3	3	3	3
Kerala	Kannur and Kasaragod	0	0	0	0	0	0	0	0	8	5	5	6	6	9
Andaman and Nicobar islands	Andaman and Nicobar	686	973	971	966	966	966	966	789	568	635	615	617	604	617
Puducherry	Yanam	0	0	0	0	0	0	0	0	1	1	1	1	1.63	2
Daman and Diu	Diu	0	0	0	0	0	0	0	0	1	1	1	1.56	1	3
	Total	4046	4255	4244	4256	4533	4737	4871	4482	4448	4445	4639	4662.56	4628.63	4740

mangrove forests are seen in the Andaman and Nicobar Islands and in the Sundarbans of West Bengal. In India, mangrove habitat is found between 69°.00′ and 89.05° East longitude and 7°.00′–23°.00′ North latitude, which comprises three distinct zones: deltaic, backwater-estuarine, and insular. The deltaic mangroves are luxuriantly present on the east coast (Bay of Bengal) where the gigantic rivers make mighty deltas such as the Gangetic, the Mahanadi, the Godavari, the Krishna, and the Cauvery. The backwater-estuarine types of mangroves exist along the West coast (Arabian Sea), characterized by the typical funnel-shaped estuaries of major rivers (Indus, Narmada, Tapti, etc.), or they occur in the backwater creeks and neritic inlets. The insular mangroves are present in the Andaman and Nicobar Islands, where tidal estuaries, small rivers, neritic islets, and lagoons can support the rich mangrove flora (Mandal and Naskar, 2008).

1.2 Species Distribution in India

Globally, there are 73 true mangrove species in 27 genera and 20 families. In India, there are 34 species of true mangroves on the East and West coasts and the Andaman and Nicobar Islands. Out of which, the mangroves of Bhitrakanika (Orissa) have 31 species, which is the highest number of species found on the East coast of India, followed by Sundarbans (27) and the Andaman and Nicobar Islands (24 species) (Sengupta et al., 2010). The most dominant species in all the mangrove wetlands of India are *R. mucronata, R. apiculata, A. officinalis, A. marina, Ceriops, Excoecaria* sp., and *Acristucyn* sp.; these are uniformly distributed along both coasts of India (Qasim, 1998). The most dominant mangrove species on the East coast are *E. agallocha, C. decandra, S. apetala, A. officinalis,* and *A. marina*. On the West coast, *A. marina* predominates whereas in the Andaman and Nicobar Islands, *R. apiculata, R. mucronata,* and *C. tagal* are uniformly distributed along both coasts of India.

1.3 Benefits of Mangroves

The mangrove wetlands provide numerous benefits to the coastal community, including wood products such as minor timber, poles and posts as well as firewood and nonwood products such as fodder, honey, wax, tannin, dye, and plant materials for thatching. They also provide different aquatic foods for fish, prawn, crabs, mussels, clams, and oysters; mangrove fishery resources can ensure livelihood security to a large section of poor fishing families in the tropical areas (Hossain, 2009; Chan et al., 2012; Jayakumar, 2014). The mangrove areas have high biological productivity associated with heavy leaf production, leaf fall, and rapid decomposition to form detritus. This detritus is consumed by the juveniles of a variety of bivalves, shrimps, and fish, which migrate into the mangrove environment for better feeding and protection. Mangrove trees provide nesting sites for many shore birds while also serving as a home for crab-eating monkeys, proboscis monkeys, fishing cats, lizards, sea turtles, bats, and many more animals, thereby supporting the tourism industry. Tourists can now readily access mangroves through board walks or by using boats. Tourism provides an excellent opportunity for people to learn about the ecological and economic importance of mangroves (Lee, 2003; Sonjai, 2007; Jayakumar, 2014).

Mangroves also protect coastlines from erosion and damage by tidal surges, currents, rising sea levels, and storm energy in the form of waves, storm surges, and wind (Coastal Zone of India, 2012; Chan et al., 2012). Mangroves are less expensive than seawalls but are similar erosion-control structures. It is estimated that mangroves sequester approximately 22.8 million tons of carbon every year (Giri et al., 2011) while also providing provide more than 10% of the essential dissolved organic carbon delivered to the global oceans from land (Dittmar et al., 2006). Mangrove flora act as donors of salt-tolerant genes and are utilized to develop the salinity-resistant crop varieties through recombinant DNA technology (Selvam, 2003). Recognizing these multiple benefits, the conservation and augmentation of mangrove ecosystems has gained much significance.

1.4 Pressure on Mangroves

The mangrove forests are highly vulnerable due to both anthropogenic and extreme environmental factors such as high temperature, high salinity, extreme tides, and strong wind conditions, but they are well adapted and thrive in those conditions. Anthropogenic activities include timber extraction for domestic purpose, conversion of mangroves to other uses such as aquaculture, saltpans, agriculture, tourism, overfishing, solid waste dumping, etc. Climatic factors such as elevated CO_2 levels highly influence the growth and health of mangrove ecosystems. Many of these impacts are focused on the world's coastlines that include a mosaic of mangrove forests, seagrass beds, sandy shores, and coral reef ecosystems. In earlier days, the mapping and monitoring of mangrove forests was difficult due to the prevailing conditions of the mangrove forest and the nonavailability of advanced technology. Globally, during the past three decades, advancements in remote-sensing technology facilitated the decision-makers to carry out mapping, monitoring, conservation, and restoration activities of mangrove ecosystems (Ramasubramanian et al., 2006; Selvam, 2003). Recent technical advancements have helped to integrate geospatial data onto the web through the Internet, which is of utmost importance in mangrove research and management.

1.5 WebGIS and Its Application

The term "WebGIS" refers to applications that distribute spatial data to users through a web browser. Depending on software capabilities, users can display query and analysis as well as geographic data remotely through a web browser. WebGIS is essential for coastal zone management that deals with geographic information for decision support on a distributed and highly dynamic coastal environment. It becomes imperative when the data for any specific location are dynamic and decisions have to be made on a real-time basis (Mujabar and Chandrasekar, 2010). The open source geospatial tools provide tremendous scope for scaling up the outreach on the wetland ecosystems of Kerala. In addition, it is also possible to provide local language versions of the WebGIS. Therefore, it augurs well for implementing the WebGIS applications to allow for an effective people's participatory process in natural resource management (Gaikwad and Prasad, 2010). A study of WebGIS by Mujabar and Chandrasekar (2010) on the southern coastal Tamil Nadu provides several capabilities that can greatly help the geoscientists, coastal planners, engineers, researchers, and policy-makers. It allows users to have visual interaction with the geographic data (shorelines landforms). Jayakumar (2014) used open source and developed WebGIS, then integrated the raster and vector datasets from a period of 74 years (1930–2012) for the Godavari mangrove wetlands, Andhra Pradesh, India.

Rinaudo et al. (2007) made a comparative study between commercial and open source in order to choose the best one considering license costs, maintenance costs, development of the basic and advanced utilities, customization, flexibility, updating, and web migration. They concluded that the open source approach offers more advantages than the commercial approach, such as the increased local expertise and the possibility of sharing the developed solutions without any costs. A study by Jesus et al. (2006) suggested that the WebGIS based on the MapServer-Linux-Apache-PHP-PostGIS connection could be flexible enough to store all the data necessary for analysis of the sandstorm phenomena, which is more economical. Xie et al. (2011) came out with a three-dimensional visualization of China's forest landscape through the Internet using MapServer, an open source WebGIS package. Owusu-Banahene et al. (2011) described the WebGIS of Portege (open source) for the conservation and management of wetlands in Ghana and incorporated four user-friendly interactions within the system: (i) provide a national distributed geodatabase for wetlands; (ii) integrate data, metadata, and information available on wetlands in Ghana in a unified Online Geo-Information System (GIS) and provide systems for updating the wetland database; (iii) web-based GIS for delivery of data, mapping capabilities, and Geo-Information and data delivery services; and iv) host a Ghana Wetland Information System (GWIS) as a platform for developing a National Natural Resources Information System (NNRIS). Laosuwan (2012) used open source-based, web-based design and development for natural resources and environmental management of Mahasarakham Province, Thailand.

2 MANGROVES OF ANDHRA PRADESH

The coastline of Andhra Pradesh is 972 km, making it the second longest coastline in India. Andhra Pradesh (AP) lies between 12°41′–22°N latitudes and 77°–84°40′E longitudes. The major geographic region of the state is a coastal plain (along the east coast, a low-lying area from Nellore to Srikakulam), which covers agricultural land and mangroves (Sudhakar Reddy, 2010). The coastal plain of the Andhra Pradesh coast is comprised of 38% sand, 52% mud, 7% marshy coast and 3% rock. The coastal zone of Andhra Pradesh is potentially a rich terrain from the point of view of agriculture, fisheries, commerce, and transportation. The coastline is smooth with inundations only in the extreme south and between the Godavari and Krishna deltas. AP has an advantage as most of the east-flowing rivers pass through the heart of the state, which brings in copious sediments from the Western and Eastern Ghats and also from the Deccan Plateau.

There are two major rivers that run across the state: the Krishna and the Godavari. The extensive mangroves are found in the estuaries of these rivers and some patches of mangroves are also presented over the coastal districts of Nellore, Prakasam, West Godavari, and Visakhapatnam (Fig. 1). The Godavari delta region is located in the East Godavari district and the Krishna delta region is located between the Krishna and Guntur districts. The total

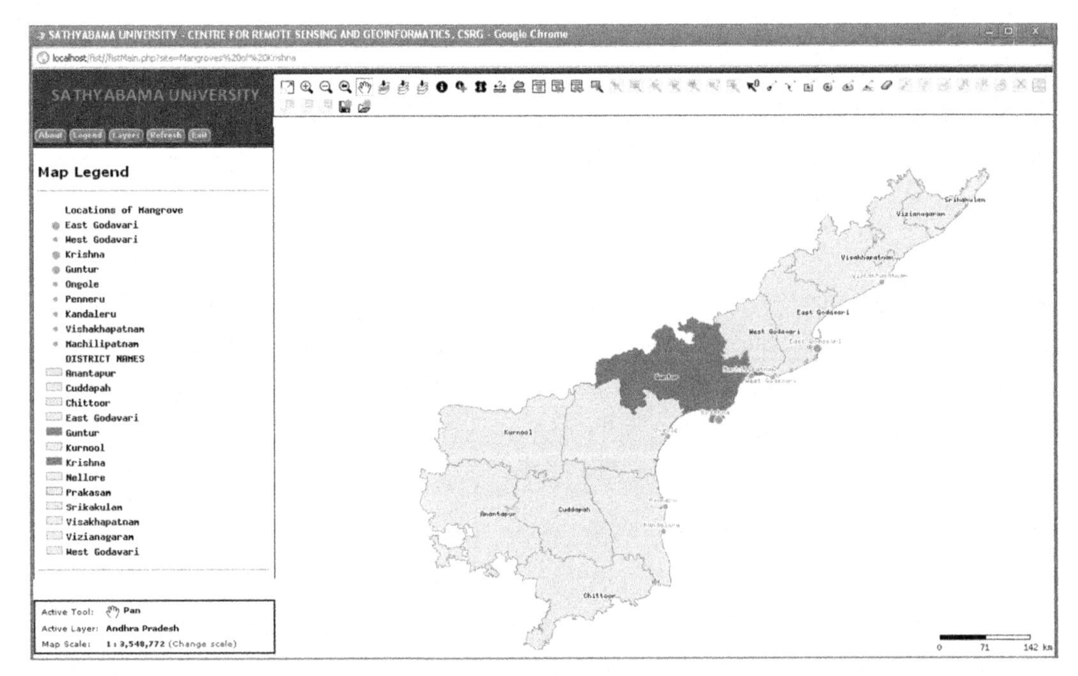

FIG. 1 Locations of mangroves in Andhra Pradesh Coast, India.

area of mangroves in AP is about 35,200 ha (FSI, 2013). The climatic condition of Andhra Pradesh is hot and humid. The mean annual rainfall in this region is 1094 mm and the annual mean temperature is 23°C. The mean daily evapotranspiration ranges from about 3–10 mm per day and the annual average is 1713 mm. The major problems on the AP coast are cyclones, tsunamis, erosion, storm surges, and coastal pollution.

2.1 Mangroves of Krishna

The present study was conducted in the Krishna delta, which falls between the Krishna and Guntur districts of Andhra Pradesh, India (Fig. 1). Geographically, the Krishna mangrove wetland is located between 15°2′N and 15°55′N latitudes and 80°42′and 81°01′E longitudes. The study area is bounded by the Bay of Bengal in the East, West Godavari and the Khammam district of Telangana in the North, the Nalgonda and Mahbubnagar districts of Telangana in the West, and the Prakasam district in the South (Fig. 2). The main occupations of the coastal communities are agriculture and fisheries followed by trade and commerce.

The Krishna River is the fourth-longest river in India and the second-longest river in Andhra Pradesh. This river originates in the Western Ghats near Mahabaleswar of Wai Taluk, Satara district of Maharashtra in the Western part of the state and merges into the Bay of Bengal at Hamasalladeevi in Andhra Pradesh. The main tributaries of the Krishna include the Koyna, Bhima, Mallaprabha, Ghataprabha, Yerla, Warna, Dindi, Musi, Tungabhadra, and Dudhganga rivers. The Krishna River bifurcates into Nadimeru and Gollamattapaya

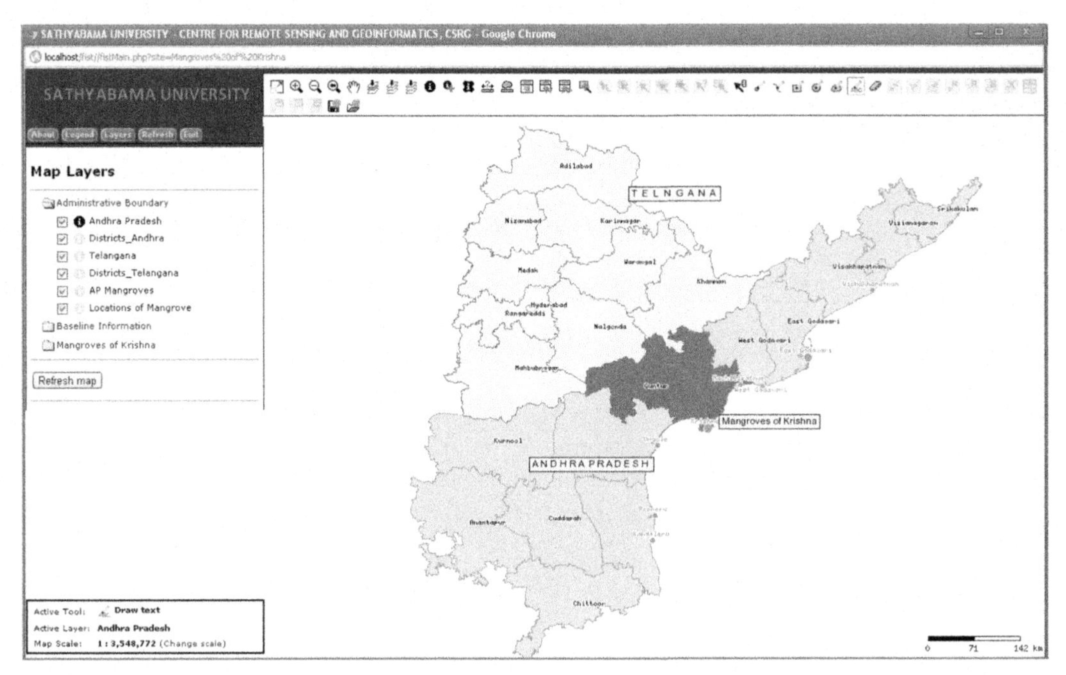

FIG. 2 Study area map of mangroves of Krishna, Andhra Pradesh.

and the total catchment area is 258,948 sq. km; its total length is about 1400 km. According to the Andhra Pradesh Forest Department, the total area of the Krishna mangrove wetlands is about 15,263 ha, which holds rich flora and fauna. In order to protect the last surviving mangrove forests in the Krishna estuary, it has been declared the Krishna Wildlife Sanctuary (KWS). The mangrove forest of Krishna comprises eight Reserve Forests, which include Sorlagondi RF, Nachugunta RF, Yelichetla Dibba RF, Lankevani Dibba RF, Molagunta RF, Kottapalem RF, Kottapalem RF, and Aduvuladivi RF, as shown in Fig. 3. There are 12 true mangroves and 17 mangrove associates identified and the same are listed in Table 2.

3 TECHNIQUES OF WebGIS

In this study, two types of data were used. The first one was remote-sensing data and the second one was ancillary data such as topographical maps, ground-truth data using GPS, and the open source software of MS4W (MapServer for Windows), FIST (Flexible Internet Spatial Template), and PostgreSQL (ORDMS) (Object-Relational Database Management System). The satellite data of Landsat-5 TM (Thematic Mapper) data of 1990, Landsat-7 ETM + (Enhanced Thematic Mapper Plus) data of 2000, and Resourcesat-1 LISS-III (Linear Imaging Self Scanning Sensor) of 2011 (cloud-free data of 2010 not available) were used (Table 3). The satellite data from 1990 and 2000 and the data from 2011 covering the study area were downloaded from the United States Geological Survey (USGS) website (http://glovis.usgs.gov) and the Bhuvan website (http://bhuvan.nrsc.gov.in), respectively.

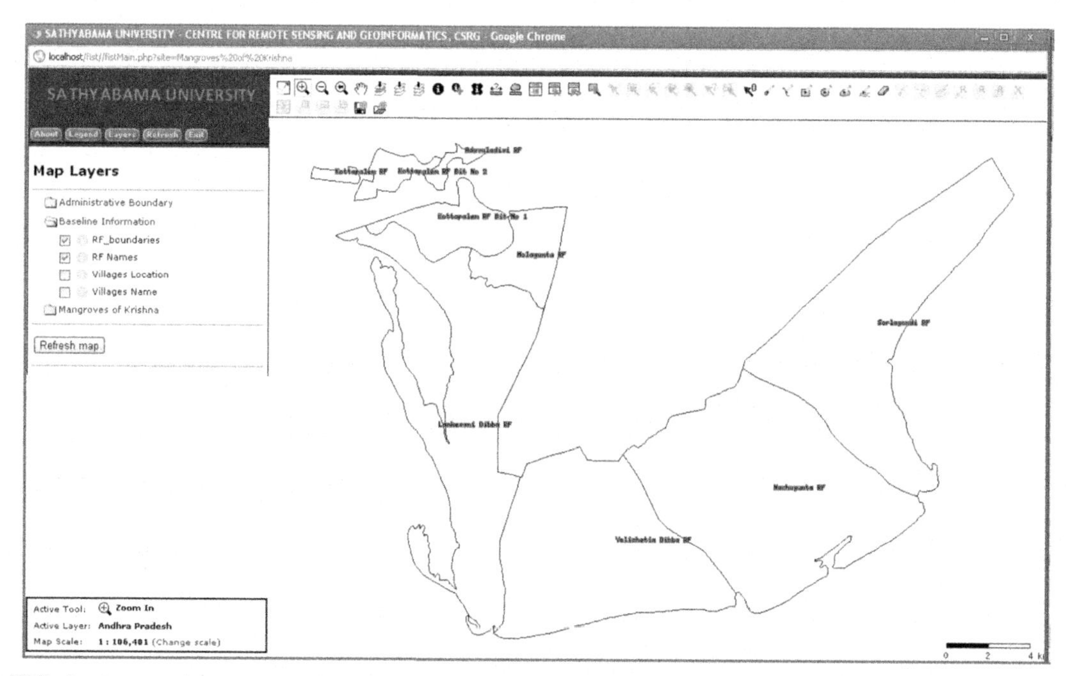

FIG. 3 Reserve forest boundaries of mangroves of Krishna.

TABLE 2 List of True Mangroves and Their Associates in Krishna Wetland

Species	Name of the Mangrove species	Sl. No.	Name of the mangrove associates
1	*Avicennia marina*	1	*Acanthus ilicifolius*
2	*Avicennia officinalis*	2	*Aeluroplls lagopoides*
3	*Aegiceras corniculatum*	3	*Caesalpinia msta*
4	*Bruguiera cylindrica*	4	*ClerodendmllJ inerm*
5	*Bruguiera gymnorrhiza*	5	*Dalbergia spinosa*
6	*Ceriops decandra*	6	*Derns trifoliata*
7	*Excoecaria agallocha*	7	*Fimbnstyhs jermginea*
8	*Lumnitzera racemosa*	8	*Hibisms tifiamls*
9	*Rhizophora apiculata*	9	*Ipomoea pes-caprae*
10	*Rhizophora mucronata*	10	*Ipomoea tuba*
11	*Sonneratia apetala*	11	*Myriostacbya Wighana*
12	*Xylocarpus granatum*	12	*Porteresia coarctata*
		13	*Salicornia brachiata*
		14	*Sesuvium Portulacastrum*
		15	*Suaeda Nudiflora*
		16	*Suaeda maritima*
		17	*Suaeda monoica*

TABLE 3 Satellite Images and Their Characteristic

Year	Satellites	Sensors	Bands	Wavelength (micrometers)	Resolution (meters)	Radiometric (bit)
1990	Landsat-5	TM	Band 1—Green	0.45–0.52	30	8
			Band 2—Red	0.52–0.60	30	
			Band 2—Red	0.52–0.60	30	
			Band 3—Blue	0.63–0.69	30	
			Band 4—NIR	0.79–0.90	30	
			Band 5—SWIR 1	1.55–1.75	30	
			Band 6—TIR	10.40–12.50	120 (30)	
			Band 7—Mid-NIR	2.08–2.35	30	
2000	Landsat-7	ETM +	Band 1—Blue	0.45–0.52	30	8
			Band 2—Green	0.53–0.61	30	
			Band 3—Red	0.63–0.69	30	
			Band 4—NIR	0.77–0.90	30	
			Band 5—SWIR	1.55–1.75	30	
			Band 6—TIR	10.4–12.5	60 × 30	
			Band 7—SWIR	2.09–2.35	30	
			Band 8—PAN	0.52–0.90	15	
2012	IRS R1	LISS 3	Band 2—Green	0.52–0.59	23.5	7
			Band 3—Red	0.62–0.68	23.5	
			Band 4—NIR	0.77–0.86	23.5	
			Band 5—SWIR	1.55–1.70		

The topographical maps at 1:50,000 scales of 66 A/13&14 and 66 E/1&2 were collected from the Survey of India (SOI). The topographical maps were scanned, georeferenced, and digitized baseline information and used as reference map. The three satellite datasets were geocoded with geographical projection WGS 84 (World Geodetic System) datum parameters. Processed satellite images were visually interpreted using onscreen digitization methods; nine coastal wetland classes were delineated using ERDAS IMAGINE 2011 and ARCGIS 10 software. The following wetland classes were delineated: dense mangroves, degraded mangroves, sparse mangroves, sand, waterbodies, aquaculture, agriculture, mud flat, and casuarinas. The classification of wetland classes was verified with a ground-truth survey using GPS and Google Earth maps and then finalized mangrove wetland maps were derived from satellite data. The data covering three decades of Krishna mangroves were used to assess the changes in the region. The main aim of the present research work was to integrate the spatial information on the study area, such as locations of mangroves in AP, mangroves of the Krishna wetland (1990, 2000, and 2011), administrative boundaries, Reserve Forest boundaries (RFs), and the locations of villages in and around the Krishna mangrove forest. To fulfill this task, the administrative boundaries were downloaded from the NIC website (http://krishna.nic.in/district-map.aspx) and georeferenced and digitized using ARCGIS software. The open source software of MS4W, (http://www.maptools.org/ms4w/) FIST http://190.136.181.39/fist/fistMain.php?site), and PostgreSQL (http://www.postgresql.org/download/) were downloaded from the Internet and installed. To integrate the spatial datasets of the Krishna mangroves, the popular three-tier architecture was used. This architecture was divided the application tier, the middle tier, and a data management tier, as shown in Fig. 4. The main aim of this architecture was to solve a number of recurring design and development problems, thereby making the application development work more efficiently. The application tier consists of client-side components, which are used to send requests to the server and to view the results (input/output). The middle tier is the heart of any solution and consists of the server side components, including the web server and the application server. Finally, the data management tier accounts for the organization of both spatial and attribute data in the application. In some cases, one server is used as a data management tier and a middle tier. In other cases, each tier can be on a separate server.

4 RESULTS

The Krishna mangrove wetland maps were prepared for 1990, 2000, and 2011 over the period of 31 years using decadal remote-sensing data. Table 4 shows the spatial extent of mangroves and their surrounding land use and land cover (LULC) classes. The changes in LULC are presented in Fig. 5. The bar diagram shows that the dense mangroves have rapidly increased during the study period, whereas the other land use classes are abnormal. In the present study, remote sensing and GIS techniques were used to map and monitor the Krishna mangrove wetland over the three decades of data from satellite images. A WebGIS was developed using open source software. The WebGIS is a hybrid of the Internet and software and hardware technology to integrate and disseminate the spatial datasets of the Krishna mangrove wetland in a single platform. The major benefit of WebGIS include saving time and money, standardizing the data, avoiding data duplication, avoiding multiple ownership and user-friendly interface, etc. ownership problem and users friendly. WebGIS is multidisciplinary and provides services to academic and scientific communities as well as the public in several disciplines to interact, view, and query the spatial data for the decision-making process at the grass-roots level. Users can use these spatial datasets for their research purposes, which can lead to new research in the future. Apart from this, it may also act as a fast remedial measure during emergency periods such as tsunamis, storms, surges, cyclones, etc.

4.1 Home Page of WebGIS

The home page of the website provides directions to get the information from WebGIS (Fig. 6). Anyone can access the spatial datasets of the study area by clicking on "Mangroves of Krishna" under the public map services. The main section is mapped, with

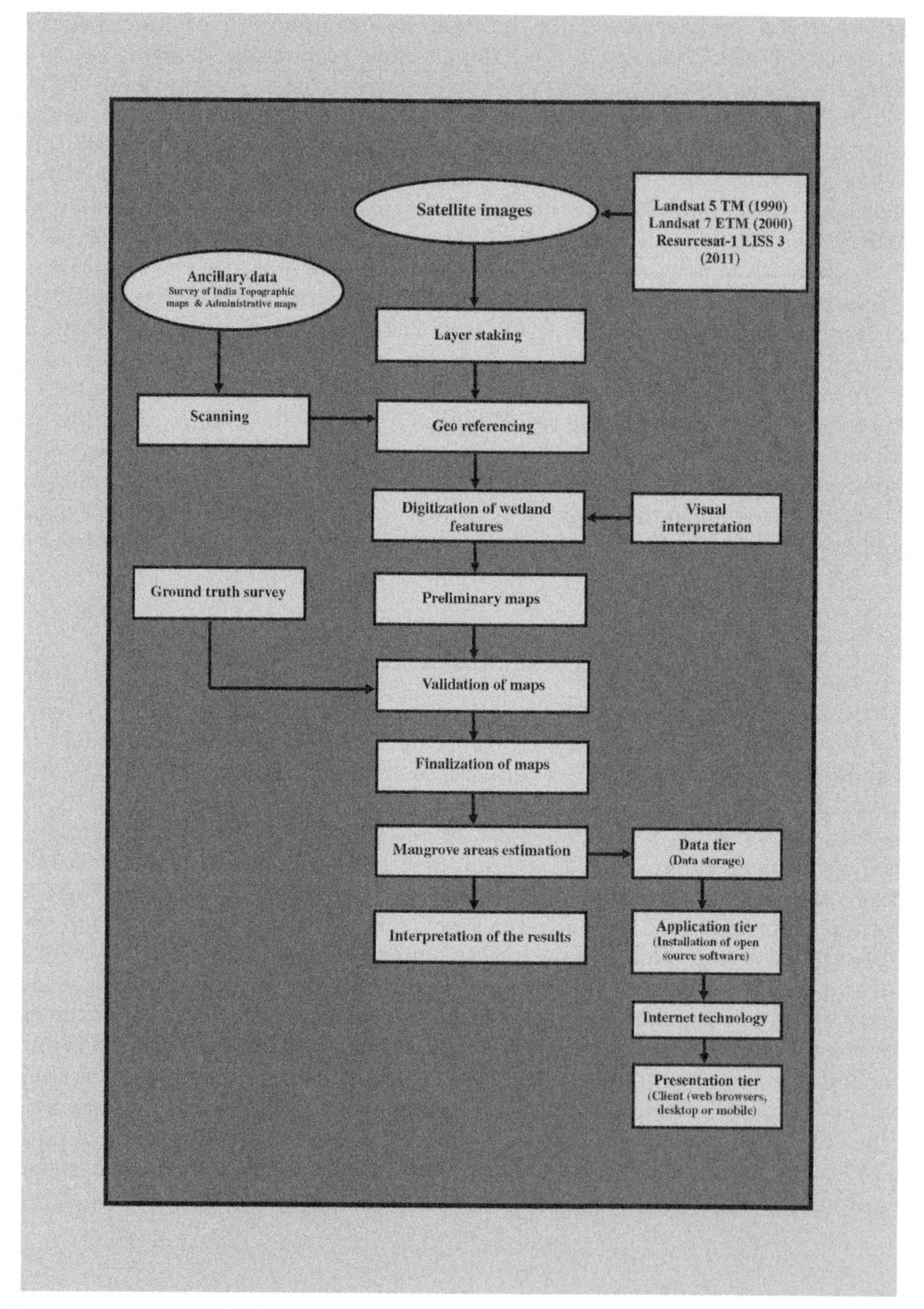

FIG. 4 A flow chart of methodology.

TABLE 4 Wetland Features, Changes Areas in Ha for the Years of 1990, 2000, and 2011

Sl.No.	Land use /land cover classes	1990	2000	2010	Change during different		
					1990–2000	2000–2010	1990–2010
1	Dense Mangroves	7948	8850	9150	−902	−300	−1202
2	Sparse Mangroves	1667	1443	3373	224	−1930	−1706
Total Mangroves		**9615**	**10,293**	**12,523**	678	2230	2908
3	Degraded Mangroves	2454	1363	1339	1091	24	1115
4	Aquaculture	863	20,758	14,838	−19,895	5920	−13,975
5	Water Bodies	5765	5878	6319	−113	−441	−554
6	Agriculture	26,535	12,111	15,448	14,593	−3456	11137
7	Sand	1128	2200	2686	1072	−486	−1558
8	Mud Flat	14,656	10,638	8999	4018	1639	5657
9	Casuarina	3184	959	2048	2225	−1089	−1864
Total		**64,200**	**64,200**	**64,200**			

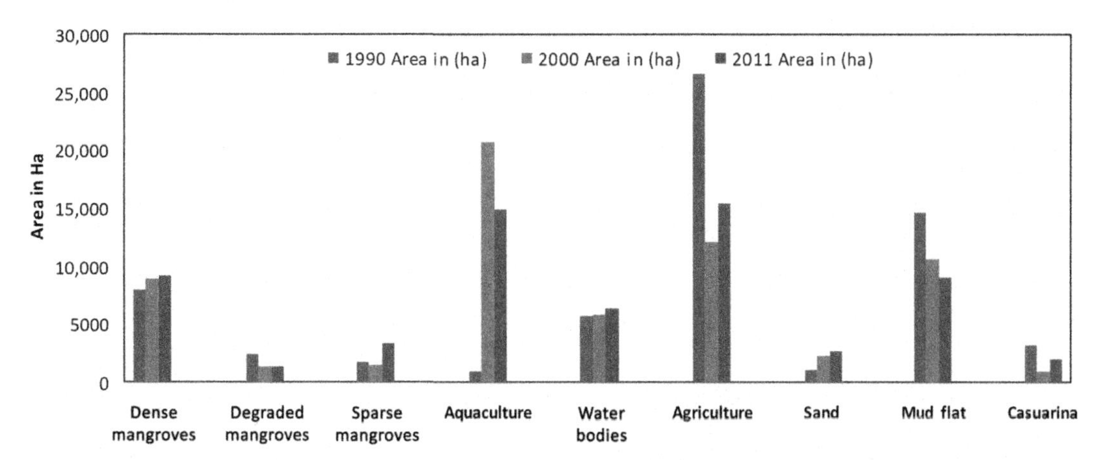

FIG. 5 Landuse/landcover changes over 31 years.

three folders on the left side containing administrative boundaries, baseline information, and mangrove wetland maps for 1990, 2000, and 2011 (Fig. 7). The first folder contains the administrative boundaries of the Andhra Pradesh and Telangana states and mangroves locations in Andhra Pradesh. The second folder contains baseline information such as location of villages, Reserve Forest boundaries, waterbodies, etc. The third folder contains mangroves of Krishna wetland maps from 1990, 2000, and 2011. The user can directly access the spatial datasets of

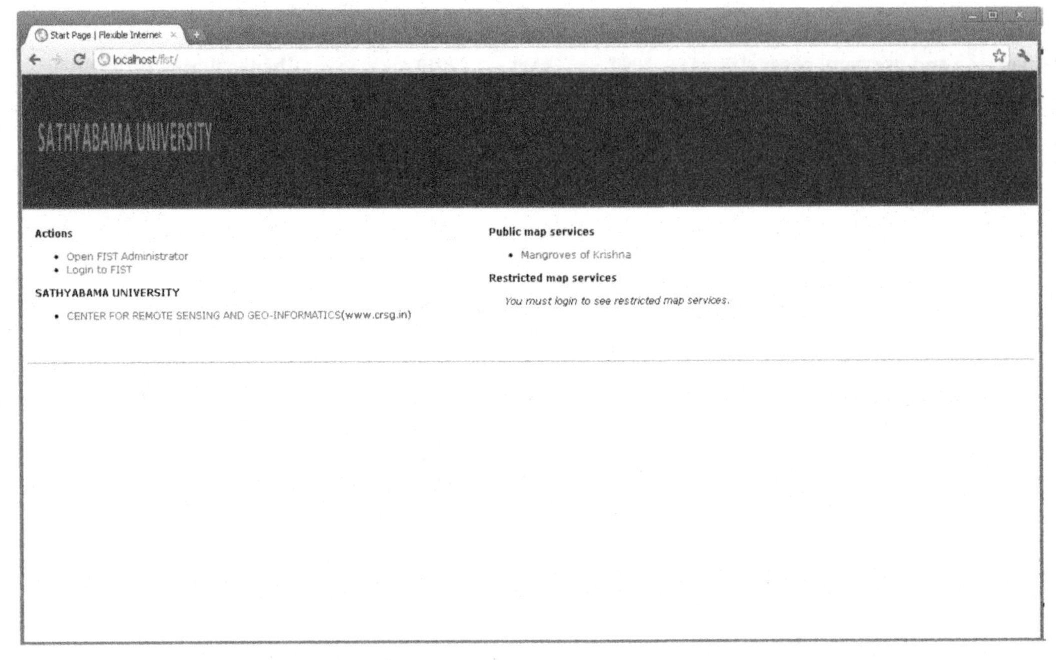

FIG. 6 Home page of WebGIS.

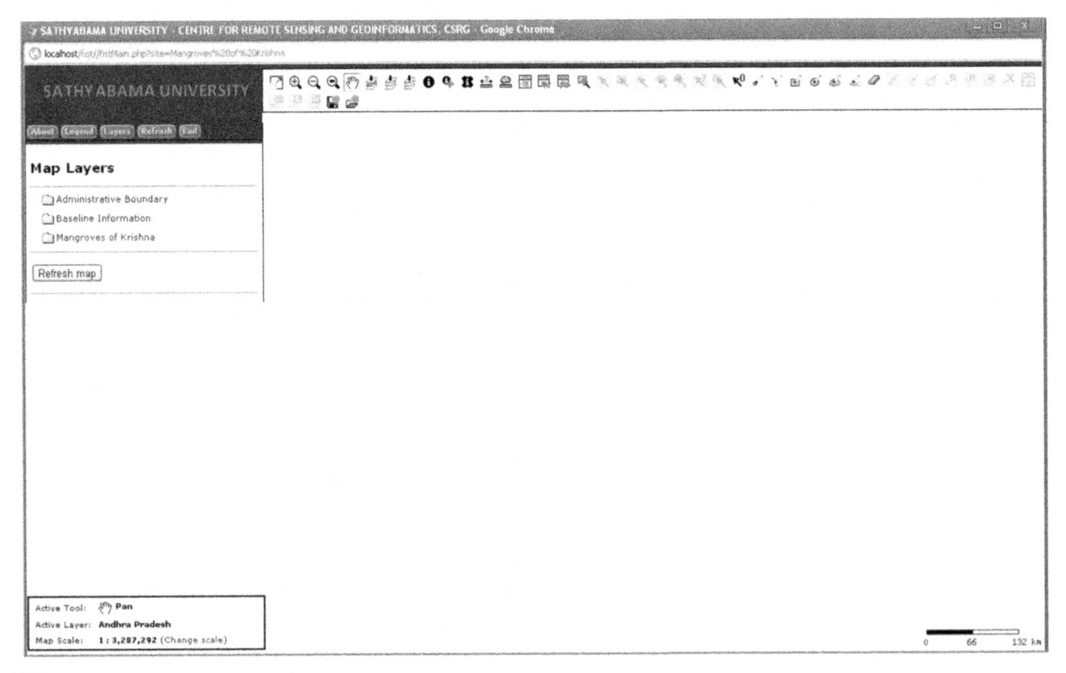

FIG. 7 Map section of WebGIS.

the mangroves of Krishna by clicking on any of the folders, then selecting the desired file by clicking on the check box. Further, the toolbar is also available in the map section to navigate within the maps whenever the users require more details.

4.2 Mangrove Wetland Map of 1990

The Landsat-5 TM satellite image was used to derive the mangrove wetland map of Krishna. The coastal wetland features that include dense mangroves, degraded mangroves, sparse mangroves, aquaculture, agriculture, mud flats, casuarinas, sand, and waterbodies were mapped and added to the WebGIS. This spatial map of the Krishna mangrove wetlands in WebGIS is illustrated in Fig. 8. During the period of study, aquaculture emerged. In and around these mangrove areas, degraded mangroves, mud flats, and sand have been enclosed. Apart from this, most of the land use was under agriculture, followed by waterbodies. The map shows the extent of mangroves, including dense mangroves, sparse mangroves, and degraded mangroves, which are spread over the western, southern, and eastern parts of the Krishna delta. However, sparse and degraded mangroves are observed in and around the RF boundaries.

4.3 Mangrove Wetland Map of 2000

The Landsat-7 ETM+ satellite data were used to delineate the mangrove wetland map of 2000. This spatial information was then integrated into the WebGIS platform. The spatial map

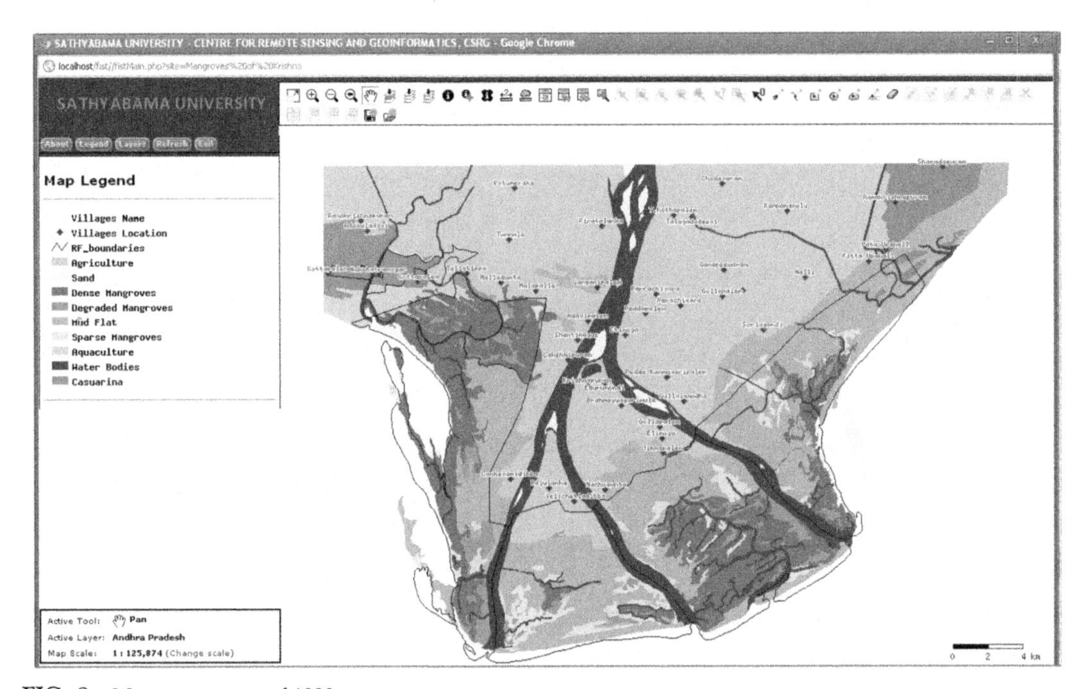

FIG. 8 Mangroves map of 1990.

in WebGIS clearly illustrates that the LULC of the region comprises 10,293 ha of mangroves, which includes dense and sparse mangroves. This was then followed by 1363 ha of degraded mangroves, 12,111 ha of mud flats, 20,758 ha of aquaculture, 3159 ha of sand (beach and sand dunes), 3184 ha of casuarina, 10,941 ha of agriculture including crop land, fallow, and settlements while waterbodies comprised the remainder of the 13,757 (Fig. 9). It is noticed that the spatial extent of the aquaculture followed by mud flats and casuarinas is high during the study period in the wetland.

4.4 Mangrove Wetland Map of 2011

The mangrove wetland map of Krishna for the year 2011 was prepared by using Resourcesat-1 data. The mangrove wetland map of Krishna was incorporated into the WebGIS platform (Fig. 10). The spatial map of the Krishna mangrove wetland in the WebGIS platform shows that the area under mangroves increased to 7948 ha in 1990, 8850 ha in 2000, and 9150 ha in 2011. Also, the area under degraded mangroves changes to 2454 ha in 1990, 1363 ha in 2000, and 1339 ha 2011 while the area under sparse mangroves changes to 1667 ha in 1990, 1443 ha in 2000, and 1373 ha in 2011. In the case of aquaculture, the change in area is noticed as 863 ha in 1990, 20,758 ha in 2000, and 14,838 ha in 2011. For mud flats, the change in the area was found to be 14,656 ha in 1990, 10,638 ha in 2000, and 8999 ha in 2011. Apart from this, agriculture, casuarina, waterbodies, and sand demonstrate irregular changes. It is noticed that about 5920 ha of agriculture and mud flat areas were converted into aquaculture. The spatial extent of the mangroves indicates that 23% expansion and growth

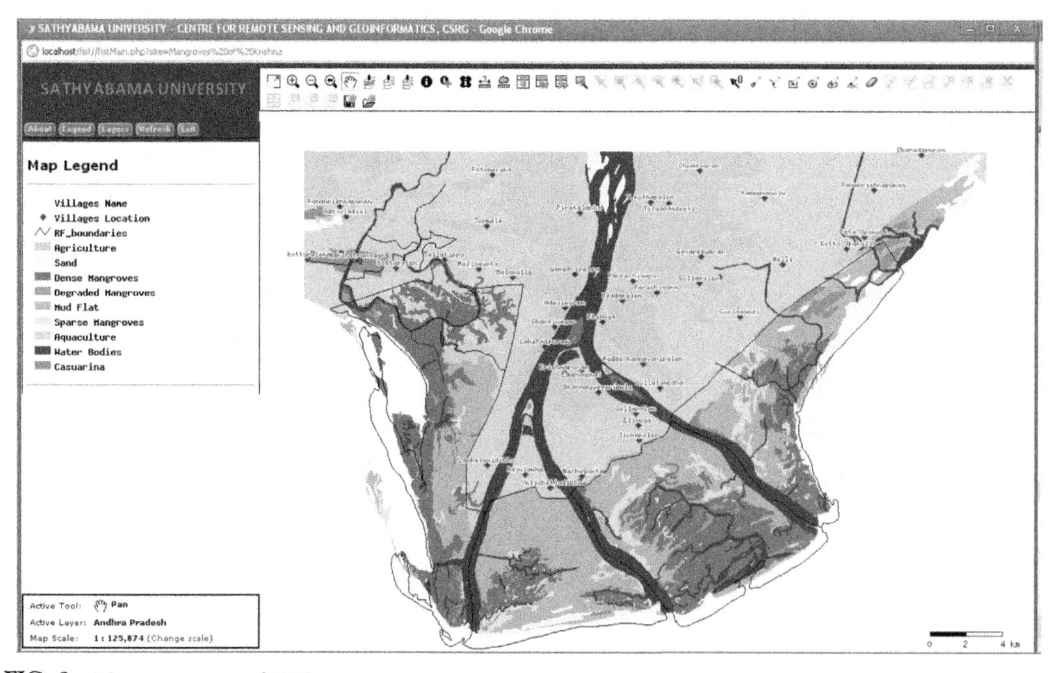

FIG. 9 Mangroves map of 2000.

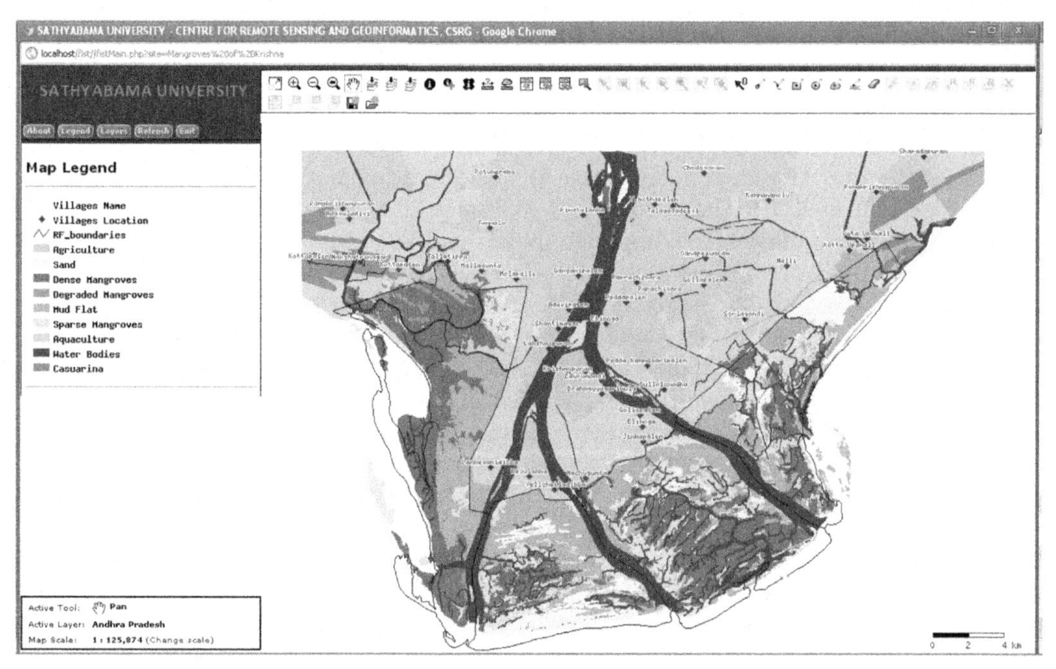

FIG. 10 Mangroves map of 2011.

of the mangroves occurred in all directions, but this was found to be very high in the first to third RFs due to the mangrove restoration progam from MSSRF, the Andhra Pradesh Forest Department, and local NGOs in the early 20th century. After the tsumani, people from the Guntur and Krishna districts realized the importance of mangroves and started the restoration/afforestation of mangrove forests in and around the RF boundaries. It may be concluded that the sparse mangroves found in the fringes of dense mangroves are in the growing stage.

4.5 Functions of WebGIS

In the present study, an open source-based WebGIS was developed to regularly monitor the mangroves of the Krishna wetland. In this study, three components—data servers, maps and application servers, and web servers—were used. These components allow easy access to spatial data and attribute data to various stakeholders while also providing reliable information on the Krishna mangroves. Spatial information of study area is easy to download and read the database of each layer through PostgresSQL by converting the shapefiles into tables using the command prompt. These functions in WebGIS may help everyone to access the spatial information on the Krishna mangroves in an interactive mode using a web browser via the Internet. In the devised WebGIS, 45 tools are available, including zoom in, zoom out, pan, measurement tools, identification features, select features, draw features, and export maps. With the help of these tools, the user can effectively perform various operations.

4.6 Wetland Changes of Krishna Mangroves During 1990 and 2000 and 2011

For the present study, the total area covered is 64,200 ha, and out of this, the total RF area is 24,953 ha. According to the Forest Department, the total areas of mangrove extent were 15,263 ha and the mangrove area occupies only about 9615 ha, 10,293 ha, and 12,523 ha during 1990, 2000, and 2011, respectively (Table 4). The degraded mangrove area was decreased to about 2454 ha in 1990, 1363 ha in 2000, and 1339 ha in 2011. It is clearly noticed that within the reserve forest boundaries, the degraded mangroves were found to be decreasing. At the same time, mangrove areas were gradually increased, which indicates the positive end result of afforestation measures taken by the Andhra Pradesh Forest Department and nongovernmental organizations such as the M. S. Swaminathan Research Foundation (MSSRF) and others. The average rate increase in forest cover per year is about 94 ha. The trend observed from the study clearly illustrates that, if there is no disturbance from natural/man-made incidents, then this mangrove wetland would be fully covered in its complete mangrove wetland area of 15,263 ha in another 29 years. This is may be possibly when people from local community involve in restoration work with the help of the forest department, wherein remote sensing coupled with WebGIS techniques can play vital role in the planning and implementation of restoration programs. Apart from the changes in the mangrove forest cover, the other major land use and land cover changes that occurred in the study area were dynamically in saltpans and aquaculture ponds in the mid 1990s. The aquaculture ponds/saltpans increased from 863 ha to 20,758 ha from 1990 to 2000 and decreased from 20,758 ha to 14,838 ha in about 11 years (2000 – 2011). For the last four decades, shrimp farming has also contributed significantly in the LULC of the region. The impact of aquaculture ponds/saltpans on mangrove forests was high in India. The rapid growth of shrimp farms developed in the study area during the early 1990s (Fig. 7), contributed to the land use conversion of agriculture fields, mangrove vegetation, and mud flats into aquaculture fields. The casuarina plantation occupied 3184 ha, 959 ha, and 2048 ha during 1990, 2000, and 2011, respectively (Fig. 11D). The variation in casuarina growth may be due to the interference of the locals. After the tsunami, the local community with the help of NGOs and the Andhra Pradesh Forest Department started a plantation program in the sandy area, which may have caused the increase in casuarina (Fig. 11H). The mud flat area decreased from 14,656 ha in 1990 to 10,638 ha in 2000, then further decreased to 8999 ha in 2011. It showed that the decrease in mud flats may be due to the restoration program of the government and the NGOs (Fig. 11B and C). The sandy area comprised about 1128 ha in 1990, 2200 ha in 2000, and 2686 in 2011. Agriculture covered an area of about 25,534 ha in 1990, 10,941 ha in 2000, and 14,397 ha in 2011 while the area under waterbodies was found to be 5765 ha in 1990, 5878 ha in 2000, and 6319 ha in 2011. These variations clearly indicate that the conversion of LULC changes near the mangrove region was high in agriculture, mud flats, and adjacent waterbodies.

4.7 Findings

(a) The WebGIS-based study illustrates that the Krishna mangrove wetland area is on the increasing trend, which may be attributed to the mangrove restoration techniques developed and demonstrated by MSSRF and the Andhra Pradesh Forest Department. (b) The Andhra

FIG. 11 (A) Mangrove associates occupying mud flat area near mangrove forest, (B) The fishbone canal method constructed by MSSRF near Sorlagondi village, (C) Mangrove grows healthily in the area restored by MSSRF and Forest Department, (D) Aquaculture forms, (E) local community cleared mangroves for domestic purpose, (F) Cattle grazing, (G) shows sandy area near Krishna river mouth, and (H) represent the casuarinas plantation on the sandy area.

Pradesh Forest Department still adopts the model of MSSRF, planting the mangroves within the RF areas. (c) The awareness program conducted by the MSSRF and FD made the local people realize the importance of mangroves and their ability to protect human lives during natural disasters such as tsunamis, storm surges, cyclones, etc. (d) The major problem in the study area is the conversion of mangroves, agriculture, and mud flat areas for aquaculture ponds. (e) Increasing aquaculture ponds doubles the soil salinity, which will lead to a hypersalinity condition that also halts the natural regeneration of mangroves while making the region unsuitable for activities such as agriculture. (f) The WebGIS-based LULC information may help planners and decision-makers in planting or restoring mangroves in mud flat areas and nonmangroves in sandy areas. (g) Apart from this, cattle grazing, fuel wood collection, pollutions, garbage dumps, etc., threaten mangrove health (Fig. 11E–G).

5 CONCLUSION

The mangrove wetland of Krishna is dynamic, undergoing significant changes due to natural and anthropogenic causes. The changes in the mangroves are either from external forces or are due to pressures from within the system. The present study used RS, GIS, and GPS techniques coupled with the WebGIS technique, which is a powerful tool for mapping and monitoring mangroves in the Krishna delta. It is an effective and easy way to measure the changes in the mangroves of the Krishna delta. Apart from this, the current study also demonstrated an open source-based WebGIS on the mangrove environment. The major advantages of WebGIS are (i) reduced cost and time, (ii) support in avoiding data duplication, and (iii) data standardization. The technique is also user friendly and provides sufficient data for fast remedial measures as well as for future planning and management. The results reveal that the areas under mangrove (dense and sparse) were increased from 9,615ha in 1990 to 10,293ha in 2000, then to 12,523ha in 2011. Other land use features, especially agriculture and mud flats, were converted into aquaculture ponds that increased significantly from 863ha in 1990 to 20,758ha in 2000, then to 14,838ha in 2011. The measurement of the change in mangrove wetland areas will be very useful data for future planning at the micro level. From the present study, it can be concluded that WebGIS techniques offer spatial mapping and effective monitoring of mangrove wetlands, which also supports interactive public participation via the Internet at any location. The technique will be of immense use for planning and management of mangrove wetlands for sustained development.

References

Chan, H.T., Spalding, M., Baba, S., Kainuma, M., Sarre, A., Johnson, S., Sato, K., 2012. Tropical forest update. Int. Trop. Timber Organ. 21 (2), 1–24.

Coastal Zones of India (CZI), 2012. Space Applications Centre–ISRO. Ministry of Environment and Forests, Government of India, Ahmedabad, p. 609.

Dittmar, T., Hertkorn, N., Kattner, G., Lara, R.J., 2006. Mangroves, a major source of dissolved organic carbon to the oceans. Glob. Biogeochem. Cycles 20, 1–7.

Forest Survey of India, 2013. India State of Forest Report. Ministry of Environment and Forests. Government of India, pp. 11–32.

Forest Survey of India, 2015. India State of Forest Report. Ministry of Environment and Forests, Government of India, New Delhi, pp. 35–39.

Gaikwad, S., Prasad, S.N., 2010. In: A Web GIS for Wetlands of Kerala using Open Source Geospatial Software.Proceedings of Symposium on Lake 2010: Wetlands, Biodiversity and Climate Change. 22-24 December, 2010, Indian Institute of Science Campus. Available at http://www.ces.iisc.ernet.in/energy/lake2010/prg_schedule23.htm.

Giri, C., Ochieng, E., Tieszen, L.L., Zhu, Z., Singh, A., Loveland, T., Masek, J., Duke, N., 2011. Status and distribution of mangrove forests of the world using earth observation satellite data. Glob. Ecol. Biogeogr. 20 (1), 154–159.

Government of India, 2009. State of Environment Report India. Ministry of Environment and Forests, Government of India, New Delhi, p. 179.

Hossain, M.S., 2009. Coastal community resilience assessment: using analytical hierarchy process. In: Hossain, M.S. (Ed.), Climate Change Resilience by Mangrove Ecosystem. PRDI, Dhaka, Bangladesh, pp. 1–11.

Jayakumar, K., 2014. Remote Sensing and GIS Application in the Management of Godavari Mangroves Wetlands Andhra Pradesh, South India. Ph.D. thesis, University of Madras, p. 150.

Jesus, J., Panagopoulos, T., Blumberg, D., Orlovsky, L., Ben Asher, J., 2006. Monitoring dust storms in Central Asia with open-source WebGIS assistance. WSEAS Trans. Environ. Dev. 6 (2), 895–898.

Laosuwan, T., 2012. A web-based GIS development for natural resources and environmental management. J. Appl. Technol. Environ. Sanit. 2 (2), 103–108.

Lee, G.P., 2003. Mangroves in the Northern Territory. Department of Infrastructure, Planning and Environment, Darwin, p. 44.

Mandal, R.N., Naskar, K.R., 2008. Diversity and classification of Indian mangroves: a review. Trop. Ecol. 49 (2), 131–146.

Mujabar, P.S., Chandrasekar, N., 2010. Web based coastal GIS for southern coastal Tamilnadu by using ArcIMS server technology. Int. J. Geomat. Geosci. 1 (3), 649–662.

Owusu-Banahene, W., Nti, I.K., Sallis, P.J., 2011. In: Integrating geo-spatial information infrastructure into conservation and management of wetlands in Ghana. 2011 Second International Conference on Intelligent Systems, Modelling and Simulation, 25–27 January, 2011. IEEE-Computer Society, Kuala Lumpur, pp. 91–94.

Panigrahy, S., Murthy, T.V.R., Patel, J.G., Singh, T.S., 2012. Wetlands of India: inventory and assessment at 1:50,000 scale using geospatial techniques. Curr. Sci. 106 (6), 852–856.

Qasim, S.Z., 1998. Glimpses of the Indian Ocean. Universities Press Limited, India, p. 206.

Ramasubramanian, R., Gnanappazham, L., Ravishankar, T., Navamuniyammal, M., 2006. Mangroves of Godavari—analysis through remote sensing approach. Wetl. Ecol. Manag. 14, 29–37.

Rinaudo, F., Agosto, E., Ardissone, P., 2007. In: GIS and Web-GIS, Commercial and Open Source Platforms: General Rules for Cultural Heritage Documentation.XXI International CIPA Symposium, 01–06 October, 2007, Athens, Greece.

Sahu, K.C., Baliarsingh, S.K., Srichandan, S., Lotliker, A., Kumar, T.S., 2012. Validation of PFZ advisories - a case study along Ganjam coast of Orissa, east coast of India. Indian Stream Res. J. 1 (12), 1–6.

Selvam, V., 2003. Environmental classifications of mangrove wetlands of India. Curr. Sci. 84 (6), 757–765.

Sengupta, R., Sirohi, M.S., Ahmad, A., Chowdhury, P.R., Gautam, S., Garg, Y., Alam, M., Radhakrishnan, T., 2010. Mangroves: soldiers of our coasts. In: Mangroves for the Future (MFF). TERI, New Delhi, India, p. 32.

Sonjai, H., 2007. Pernetta, J.C. (Ed.), National Report on Mangroves in South China Sea–Thailand. Reversing Environmental Degradation Trends in the South China Sea and Gulf of Thailand. UNEP/GEF/SCS. Technical Publication No. 7.

Sudhakar Reddy, C., 2010. Gap analysis for protected areas of Andhra Pradesh, India for conserving biodiversity. J. Am. Sci. 6 (11), 472–484.

Xie, X., Wang, Q., Dai, L., Su, D., Wang, X., Qi, G., Ye, Y., 2011. Application of China's National Forest continuous inventory database. Environ. Manag. 48, 1095–1106.

CHAPTER

16

Ecological Studies in the Coastal Waters of Kalpakkam, East Coast of India, Bay of Bengal

K.K. Satpathy, A.K. Mohanty*, G. Sahu*, S. Biswas*, M.S. Achary*, Bharat Kumar*, R.K. Padhi*, N.P.I. Das*, S.N. Panigrahi*, M.K. Samantara*, S.K. Sarkar[†], R.C. Panigrahy[‡]*

***Environment and Safety Division, Indira Gandhi Centre for Atomic Research, Kalpakkam, India**
[†]Department of Marine Sciences, University of Calcutta, Kolkata, India
[‡]Department of Marine Sciences, Berhampur University, Berhampur, India

1 INTRODUCTION

The interrelationship between biotic and abiotic factors and understanding their behavior in variable environmental circumstances are salients part of ecology. The biotic factors or the organisms are mainly regulated by the various external circumstances, namely the physical and chemical characteristics of the ecosystem, the population density in various trophic levels, climatic conditions, and various natural processes. Hence the changes brought to the ecosystem and their inhabitants are crucial to record and understand in a larger time frame. Hydrographic and chemical features such as temperature, pH, salinity, dissolved oxygen, carbon dioxide content, suspended matter, nutrients, chlorophyll, and water transparency play a significant role in the productivity of a coastal ecosystem (Strauss, 1989). Phytoplankton growth, metabolism, and abundance are basically governed by three dissolved inorganic macronutrients such as nitrogen, phosphorous, and silicon that are present in the aquatic ecosystem (Raymont, 1980; Grant and Gross, 1996). Thus, studies on the origin, abundance, and utilization of nutrients have gained significance in coastal research in the last few decades. Although a large amount

Coastal Management
https://doi.org/10.1016/B978-0-12-810473-6.00017-0

© 2019 Elsevier Inc. All rights reserved.

of information with respect to the general hydrography and biology of Kalpakkam on the southeast coast of India is available (Nair and Ganapathy, 1983; Satpathy et al., 1987; Satpathy and Nair, 1990, 1996; Satpathy, 1996), the data on nutrients have been scarce. Moreover, a continuous and systematic study on nutrients has not been reported so far. In view of this, coastal waters at Kalpakkam are being monitored continuously for nutrient (nitrite, nitrate, ammonia, total nitrogen, phosphate, total phosphorous, and silicate) content with the following objectives: (I) to study the seasonal variations in nutrient content, (II) to find out any major changes over the years due to anthropogenic impacts, and III) to create baseline data for future impact studies. In this chapter, a detailed account of the hydrographic features, including nutrient distribution in the coastal waters of Kalpakkam, is discussed.

As the basis of the trophic chain, phytoplankton constitutes the most important biological community in any aquatic system (Monbet, 1992; Sin et al., 1999). About 95% of the total production in a marine ecosystem is contributed by the phytoplankters. For assessing fishery yields, knowledge on the rate of gross primary productivity is important. Hence, the abundance of phytoplankton can be considered the best index for the quantitative assessment of fishery potential of an area. In addition, the phytoplankton assemblages are often used as the indicators of water quality, including pollution. As a matter of fact, predicting changes in the dominance and diversity of phytoplankton as the indicators of water quality has promoted analysis of this community using different strategies, such as long-term monitoring of dominant species and their relationship with seasonal changes of environmental conditions. During the past, extensive work has been carried out pertaining to the qualitative and quantitative ecology of phytoplankton from both the east and west coasts of India (Mani et al., 1986; Devassy and Goes, 1988). However, such studies in the southeast coast of India in general and in the Kalpakkam coastal waters in particular are very scant. Sparse observation on phytoplankton distribution in the coastal waters of Kalpakkam (Sargunam, 1994; Poornima et al., 2005) has motivated this investigation. Keeping this view in the backdrop, the recent results of studies on (i) the qualitative and quantitative abundance of phytoplankton, (ii) seasonal variations in phytoplankton community organization, and (iii) the influence of environmental variables on phytoplankton species assemblage in the Kalpakkam coastal waters are also discussed in this chapter.

Zooplankton, the free floating animal component of the water column in every aquatic ecosystem, is the most important secondary producer of the food chain, playing pivotal roles in the energy and matter transport processes. Both as primary and secondary consumers, the zooplankton community covers about 10% of the total marine biomass on which the whole class of fisheries depend. Many zooplankton species are used as water quality indicators. Rao (1958) used the chaetognaths to locate the current pattern in the Indian seas. Later, the abundance of polychaetes was used as the indicator of water quality with special reference to the detection of the organic pollution load in coastal ecosystems (Ganapati, 1975). Owing to such multidimensional economic and ecological utility, zooplankton constitutes a core subject of research in all marine biological investigations. Zooplankton studies in this coast are limited to the observations made by Sreekumaran Nair et al. (1981) and Santhakumari and Saraswathy (1981) from the southern part of the western Bay of Bengal between Cape Comorin and Tuticorin. Moreover, zooplankton studies from the Kalpakkam coast are very scant. Only a single account of zooplankton distribution in the coastal waters of Kalpakkam during a monsoon transition period has been reported so far (Saravanane et al., 1999). Hence, in order to understand the valuable functions and variations in the zooplankton community structure in the coastal waters of Kalpakkam, a comprehensive study has been undertaken.

In tropical regions, the seasonal variations in climatic conditions influence the hydrological variables while influencing fish behavior and species composition. Various abiotic factors are known to influence species composition, abundance, and distribution of fish in coastal regions (Ansari et al., 2003). Coastal biodiversity in India gets affected by the hydrographic condition that depends on the monsoonal rainfall. The southeast coast of India (Tamil Nadu and its neighborhood), with a coastline of about 1000 km, is influenced by the northeast monsoon (Dhar and Rakhecha, 1983), giving the Tamil Nadu coast different hydrographic features compared to other Indian coasts (Satpathy et al., 2010a). Studies on ichthyofaunal diversity along the Indian coast can be traced back to Day (1878), who reported nearly 800 species. Subsequent sporadic works on marine fish of India were mostly confined to the west coast or the northern part of the east coast. Very little has been represented from the large stretch of coast between Chennai and the Gulf of Mannar (GoM) except at Chennai and GoM. A few studies (Prabhu and Dhawan, 1974; Radhakrishnan, 1974; Kasim and Khan, 1986) on the seasonal fish catch along the Indian coast are reported in the literature. However, there are no such data available from the southeast coast. The Kalpakkam coast, situated between Chennai and GoM, is an ideal study location to bridge the gap. In the present endeavour, the fish assemblage of the Kalpakkam coast was studied for the first time.

Crustaceans are important members from the benthic communities, which include crabs, lobsters, crayfish, shrimp, krill, and barnacles. Many of the edible species present in crustaceans form a larger chunk of human food while, at the same time, the small species of crustaceans play an important role in the ecosystems. Marine crabs are also considered an important indicator of coastal pollution. Crabs are the most diverse group of crustaceans, ranking third after shrimp and lobster by virtue of importance as an esteemed seafood delicacy and also by the value of the fisheries they support (Varadharajan and Soundarapandian, 2012). Brachyura crab fisheries comprise up to 20% of all crustaceans caught and farmed worldwide, with an annual consumption of 1.4 million tons. More crab species are found in tropical and subtropical regions compared to temperate and cold regions (Sakthivel and Fernando, 2012). Brachyurans are the most significant because of their great diversity, which consists of 93 families, 1271 genera, and 6793 species recorded worldwide (Ng et al., 2008). In India, 705 brachyuran crab species belonging to 28 families and 270 genera have been recorded so far, of which only 15 species are edible (Sruthi et al., 2014). The present study on crab diversity forms the first report from this location.

Bottom-dwelling organisms, generally filter feeders in nature, play a vital role in the biogeochemical cycling of elements in coastal ecosystems by consuming organic matter from the water column. They inhabit the submerged surfaces and colonization takes place due to intra- and interspecific competition and the seasonal breeding pattern of the organisms, which depends upon the physicochemical and biological properties of the region. The present study has been undertaken to observe the variations in colonization patterns of macrobenthos after a long gap, especially after the megatsunami event of December 2004 (Sahu et al., 2011). Therefore, the monthly and seasonal status of the benthic community on the southeast coast of India is discussed here as part of the ecological studies.

A growing need for energy production by power plants makes it imperative to establish the real impact of discharge from power stations on the marine environment (Crema and

Bonvicini, 1981). Considerable research has been carried out on benthic communities, which are considered biofouling organisms in cooling water systems. However, less attention has been paid to nonfouling benthic species that are found to be a suitable group of organisms that can be used for environmental indicators indicating different kinds of pollution (Saravanane et al., 1998; Satpathy et al., 2008a,b). Bivalves are widely used as bioindicators of all types of environmental pollution in coastal areas, as they provide a time-integrated indication of environmental contamination. The impact of heated effluents is more evident in these organisms because most of them are sessile, sedentary, and immobile. Bivalves of the family Donacidae inhabit the exposed intertidal sandy beaches and form, by far, the largest group worldwide living in such highly dynamic environments (Ansell, 1983). Members of the genus *Donax* are commonly the main primary consumers in soft-bottom communities. They are subjected to predation by a wide variety of invertebrates, fish, birds, and mammals (Luzzatto and Penchaszadeh, 2001; Peterson et al., 2000; Salas et al., 2001). Hence, this is an important trophic link in surf zone food webs (McLachlan et al., 1996). In India, *Donax cuneatus* is collected as a source of nutritious food (Azariah, 1999; Shanmugam et al., 2007). On the east coast of southern India, *D. cuneatus* is abundant and members of this genus are sensitive to thermal and chlorine stresses (Bharathi et al., 2001; Stenton-Dozey and Brown, 1994). Hence, this species was chosen for the present study.

The Bay of Bengal sustains a strong potential for impacts from riverine nutrient loads due to the very high nutrient yields in its catchment basins, for example, via the Ganges/Brahmaputra, Godavari, and Mahanadi rivers. The subsurface stratification is dominated by salinity and the upwelling lead to a coastward increase in salinity. The surface circulation in the Bay of Bengal also undergoes large seasonal changes, but the consequence and extent of these changes on the composition of subsurface waters is not visibly identified. The residence time of the intermediate waters, or the OMZ (100–1000 m) of the bay, is 12 years. Southern and northern regions of the bay have distinct mixed layer depths (MLD) along the western boundary that shoaled from 25 m at 12°N to ~2 m at 19°N. The doming of the isotherm was around 17°N, where a cold-core eddy was evident. The east coast of India, in its southern region (off Tamil Nadu), has a wide continental shelf area that spreads 22,411 km^2 with a depth of 50 m and 11,205 km^2 with a depth of 51–200 m. Here we report the vertical expansion of hypoxia within the OMZ along the outer shelf and slope region and the emergence of an inner shelf hypoxia that was not apparent in the southwestern Bay of Bengal or the Indian Ocean.

2 STUDY AREA

The Kalpakkam coast (12° 33′ N Lat. and 80° 11′ E Long.), *an important nuclear hub of India*, is situated about 80 km south of Chennai (Fig. 1). It harbors a functional nuclear power plant (Madras Atomic Power Station, MAPS) and a desalination plant. A prototype fast breeder reactor (PFBR) of 500 MW(e), another two MGD desalination plant, and a fast reactor fuel cycle facility (FRFCF) are under construction at this site. In addition, a nuclear R&D center (Indira Gandhi Centre for Atomic Research), a centralized water management facility (CWMF), a reprocessing plant, and a waste immobilization plant have been operating since the mid-1980s. MAPS uses 35 m^3/s of seawater for condenser cooling purposes, which is

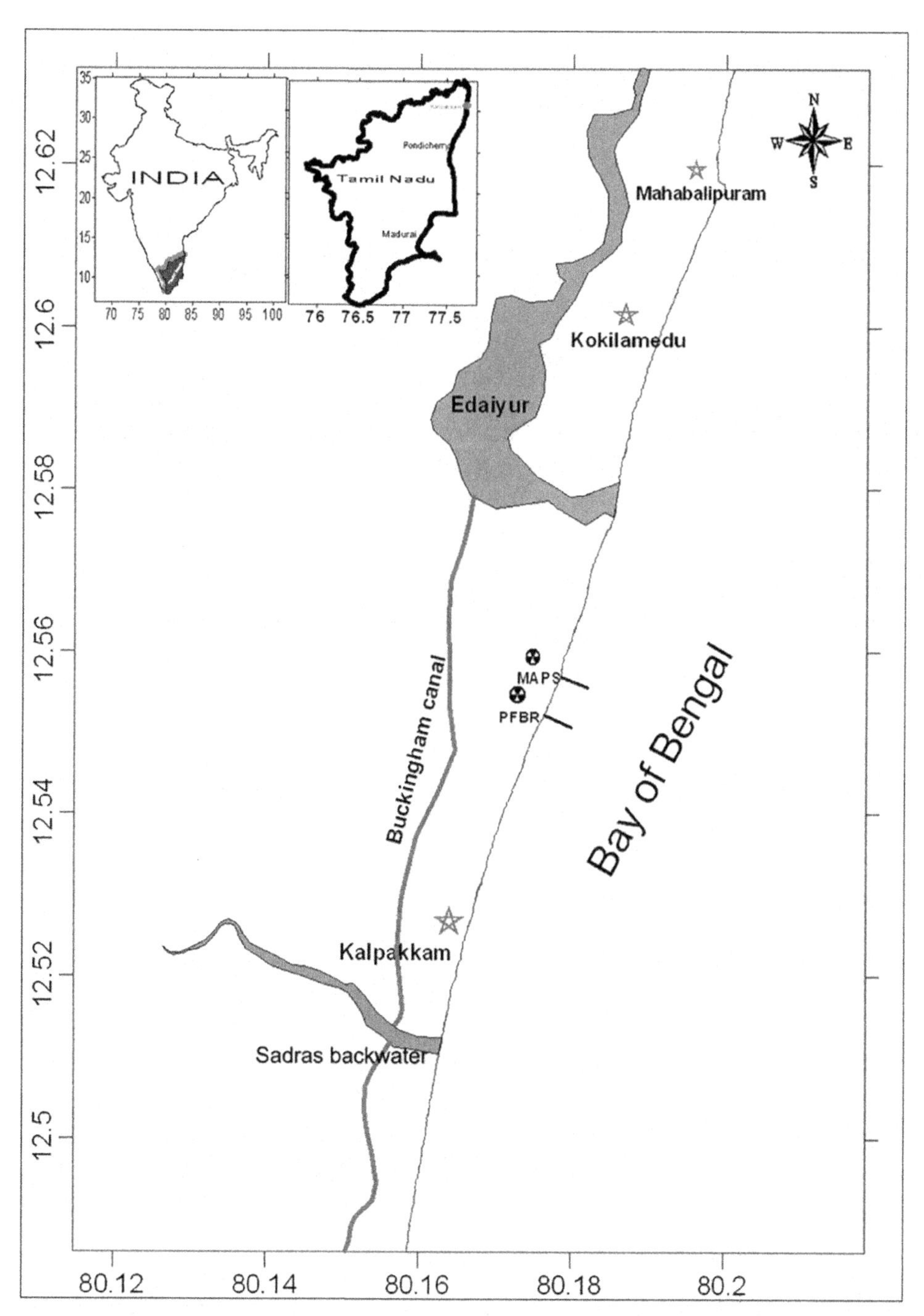

FIG. 1 Kalpakkam coast, east coast of India.

drawn through an intake structure located about 468 m inside the sea. The heated seawater is released into the coastal waters after the heat is extracted. Edaiyur and the Sadras backwater systems are important features that are connected to the Buckingham canal that runs parallel to the coast. The whole year has been divided into three seasons: (1) summer (February-June), (2) Southwest (SW) monsoon (July-September), and (3) Northeast (NE) monsoon (October-January) (Nair and Ganapathy, 1983), based on the pattern of rainfall and associated changes in hydrographic characteristics at the Kalpakkam coast. Seasonal monsoon wind reversal is a unique feature of the Indian Ocean that results in subsequent changes in the coastal circulation pattern (La Fond, 1957; Wyrtki, 1973), which are also felt at this location. The reversal of wind occurs during the transition period between the southwest (SW) and northeast (NE) monsoons. That transition takes place during September/October and the NE to SW transition occurs during February/March. The coastal current, which is poleward during the SW monsoon, changes to equator-ward during the SW to NE monsoon transition, whereas a reverse current pattern is observed during the NE to SW monsoon transition period (Varkey et al., 1996; Haugen et al., 2003; Satpathy et al., 2009, 2010b).

The two backwaters open onto the coast, discharging copius amounts of freshwater to the coastal milieu for a period of 2–3 months during the NE monsoon period. The bulk of the rainfall (~60%) in this region of peninsular India is received from the NE monsoon. The average annual rainfall at Kalpakkam is about 1250 mm. However, the littoral drift during the postmonsoon period causes the formation of a sand bar between the backwaters and the sea, leading to a stoppage of low saline water inflow from the backwaters to the sea. The Sadras backwater is relatively polluted due to domestic waste discharge as well as the discharge from the Buckingham canal, as compared to the Edaiyur backwater, which receives Buckingham canal discharge and adjoining agricultural land drainage. A large number of aquaculture farms have come up along the Buckingham canal, affecting the quality of this brackish water canal significantly. Moreover, a large number of small industries have come up in the entire belt and thus associated anthropogenic input into this water body has also increased. In addition to about 50,000 inhabitants of the township, two fishermen villages located at both sides of the township have sizable populations. The Kalpakkam coast was severely affected by the 2004 tsunami, which devastated the entire east coast of India and had maximum impact at this location.

3 PHYSICOCHEMICAL PROPERTIES OF COASTAL WATERS

3.1 Material and Methods

Samples were collected monthly at the MAPS jetty (about 450 m away from the shore) during the study. Precleaned polythene bottles were used for sample collection. Dissolved oxygen (DO) was estimated by Winkler's titrimetric method (Grasshoff et al., 1983). Knudsen's method (Grasshoff et al., 1983) was followed for salinity estimation. The turbidity meter CyberScan IR TB 100 (±0.1 NTU accuracy) was used for measuring the turbidity of the water sample shaving. pH measurement was carried out by a pH meter (CyberScan PCD 5500) with ±0.01 accuracy. Samples for nutrient analysis were filtered through 0.45μ millipore filter paper and preserved at −20°C; analyses were carried out within 2–3 days of collection.

Dissolved micronutrients such as nitrite, nitrate, ammonia, silicate, and phosphate along with TN and TP were estimated by the spectrophotometric method following standard procedures (Parsons et al., 1984; Grasshoff et al., 1983). Chlorophyll-*a* (chl-*a*) was analyzed by spectrophotometry following the method of Parsons et al. (1984). For all the spectrophotometric analyses, a double beam UV-visible spectrophotometer (Chemito Spectrascan UV 2600) was used. Correlation analysis was carried out by using an XLStat Pro 2008.

3.2 Results and Discussions

3.2.1 Hydrographic Parameters

The results obtained were pooled into seasonal averages for individual parameters. Water temperatures ranged from 25.4°C to 31.5°C with relatively high values measured during postmonsoon months (Fig. 2). Coastal water pH ranged from 7.9 to 8.4. It did not show any particular trend during different seasons and seasonal average values ranged from 7.95 to 8.15 (Fig. 3). Salinity values ranged from 25.45 to 35.97 psu. It showed a significant seasonal trend with the lowest and the highest salinity being observed during the premonsoon and monsoon periods, respectively (Fig. 4). Dilution of coastal water by the addition of fresh water from the two backwaters during the NE monsoon period could be the reason for the lower salinity values observed during monsoon. The DO contents varied between 4.5 and 7.6 mg/L. Relatively high DO content was observed during premonsoon and monsoon periods as compared to the postmonsoon period (Fig. 5). The noticeable increase in DO observed during the premonsoon and monsoon periods could be attributed to the phytoplankton production during premonsoon/summer and the input of DO-rich freshwater into the coastal waters during the monsoon, respectively (Mohanty et al., 2010; Satpathy et al., 2011). However, a freshwater influx was found to be the most important factor that controlled the level of DO in coastal waters, as was evident from the strong negative correlation ($P < .0001$) between DO and salinity (Table 1). The negative correlation of DO with chl-*a* showed that the photosynthetic release was not the primary source of DO in these coastal waters, which further

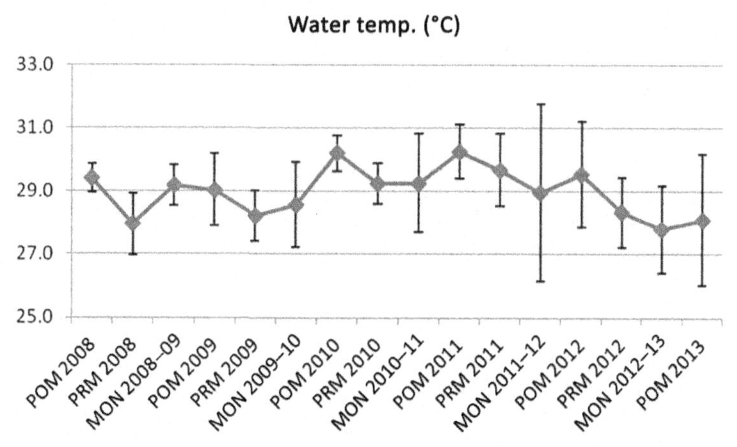

FIG. 2 Variations in seawater temperature at Kalpakkam coast, east coast of India.

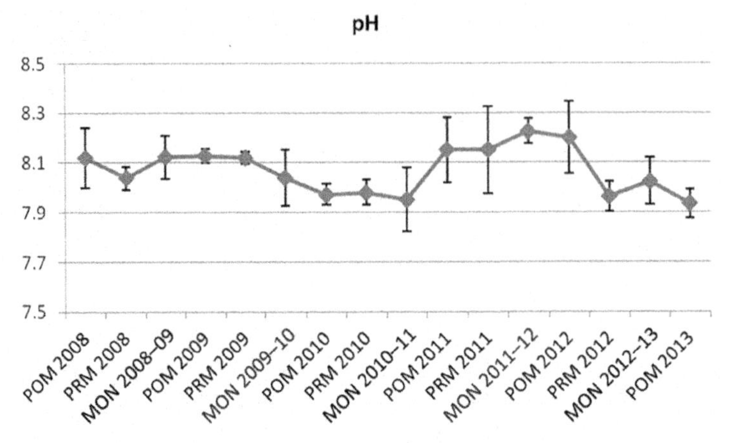

FIG. 3 Variations in seawater pH at Kalpakkam coast, east coast of India.

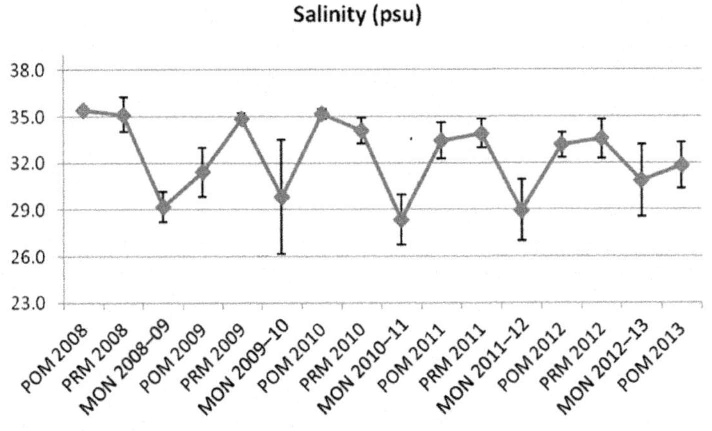

FIG. 4 Variations in seawater salinity at Kalpakkam coast, east coast of India.

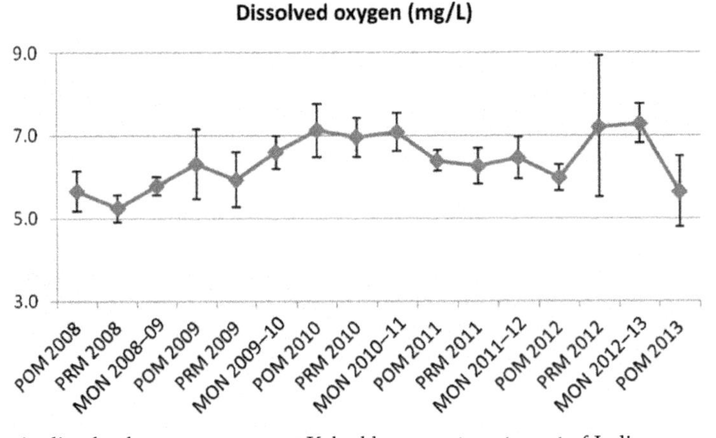

FIG. 5 Variations in dissolved oxygen content at Kalpakkam coast, east coast of India.

TABLE 1 Correlation Matrix (Pearson) of Physico-Chemical Parameters

Variables	Temp	pH	Salinity	Turbidity	DO	Nitrite	Nitrate	Ammonia	TN	Silicate	Phosphate	TP	Chl-a
Correlation Matrix (Correlation Coefficient)													
Temp													
pH													
Salinity													
Turbidity	−0.093		0.202										
DO	−0.188		−0.376	−0.230									
Nitrite		−0.114	−0.080										
Nitrate	0.082	−0.266	−0.231	0.113	−0.092	0.224							
Ammonia													
TN		−0.265	−0.306	0.085		0.166	0.836						
Silicate			−0.414		0.183		−0.082						
Phosphate			−0.079	0.099		0.320		0.246					
TP						0.169		0.264			0.814		
Chl-a			0.543	0.335	−0.129		−0.182		−0.186	−0.275			
***P*-values (Significance Level)**													
Temp													
pH													
Salinity													
Turbidity	0.043		<0.0001										
DO	<0.0001		<0.0001	<0.0001									
Nitrite		0.013	0.081										
Nitrate	0.073	<0.0001	<0.0001	0.014	0.044	<0.0001							
Ammonia													
TN		<0.0001	<0.0001	0.062		0.000	<0.0001						
Silicate			<0.0001		<0.0001		0.075						
Phosphate			0.087	0.031		<0.0001		<0.0001					
TP						0.000		<0.0001			<0.0001		
Chl-a			<0.0001	<0.0001	0.005		<0.0001		<0.0001	<0.0001			

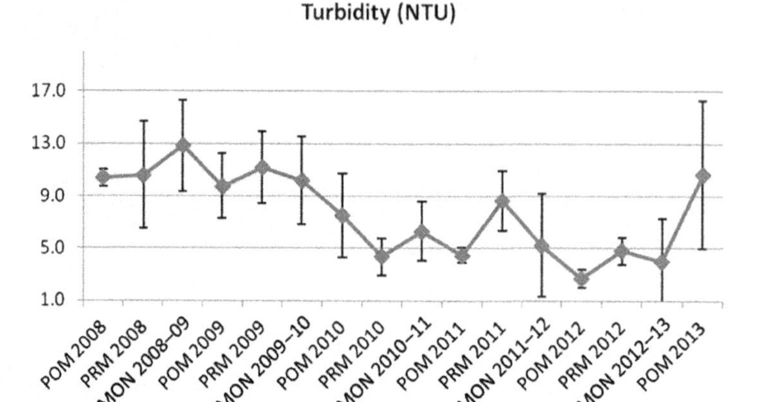

FIG. 6 Variations in seawater turbidity at Kalpakkam coast, east coast of India.

denotes that in aquatic systems, oxygenation is the result of an imbalance between the process of photosynthesis, the degradation of organic matter, and reaeration (Garnier et al., 2000). Water turbidity ranged from 0.34 to 24.59 NTU (Fig. 6). Phytoplankters played an important role in determining turbidity values (Kalimurthy, 1973) as a strong positive correlation ($P < .0001$) was observed between turbidity and chl-*a*.

3.2.2 *Nutrients*

In shallow coastal ecosystems, the major sources of nutrients are from rivers, groundwater, atmospheric transport, and benthic fluxes (Malone et al., 1988; Paerl, 1997; Conley, 2000). Being the most unstable form of dissolved inorganic nitrogen species present in seawater, the nitrite level showed wide fluctuations during the present investigation (Santschi et al., 1990; Chandran and Ramamoorthi, 1984). The concentration of nitrite ranged from BDL-5.88. Though it did not show any particular seasonal trend, postmonsoon values remained relatively low as compared to the other seasons (Fig. 7). Nitrate concentration ranged from below the detection limit (BDL) to 9.18 μmol/L. It also did not show any particular seasonal trend (Fig. 8). Quick assimilation by phytoplankton and enhancement by surface runoff resulted in large-scale spatio-temporal variations of nitrate in the coastal milieu (Qasim, 1977; De Souza, 1983; Prasanna Kumar et al., 2004). Nitrate showed strong a negative correlation with salinity, denoting the freshwater influx as the main source of this nutrient in coastal waters, as has been reported for other coastal waters (Choudhury and Panigrahy, 1991; Satpathy, 1996). The ammonia concentration did not show any typical trend and BDL values were observed on many occasions. It ranged from BDL-11.74 μmol/L. Excretory release by marine invertebrates and utilization by phytoplankton as a nutrient significantly affects spatiotemporal variations of ammonia in the marine environment (Olson, 1980; Gilbert et al., 1982; Sankaranrayanan and Qasim, 1969). Total nitrogen values during the present study did not show any seasonal trend (Fig. 9). TN concentration ranged from 3.74 to 47.68 μmol/L (Fig. 10). Observation of a strong negative correlation between TN and chl-*a* in the present study indicated that nitrogenous nutrients were rapidly utilized by the phytoplankton community at this location, as has been observed in the case of nitrate too.

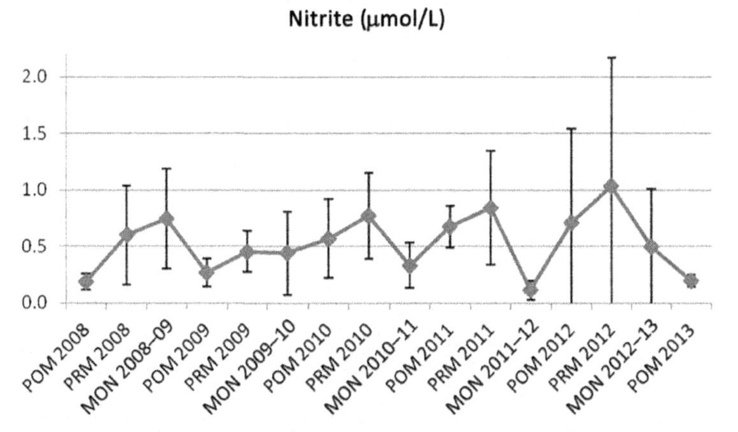

FIG. 7 Variations in nitrite concentration at Kalpakkam coast, east coast of India.

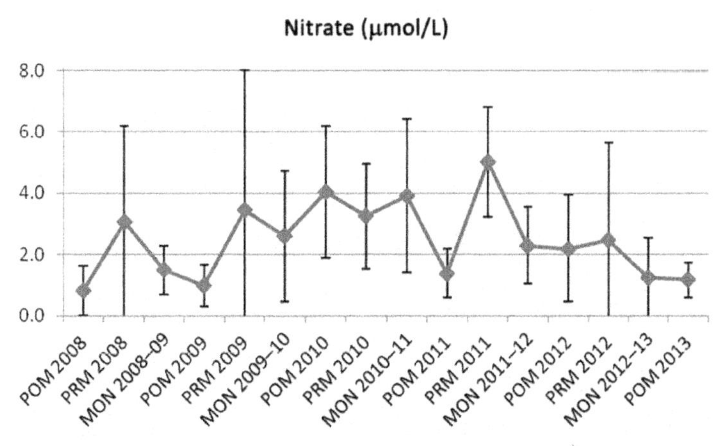

FIG. 8 Variations in nitrate content at Kalpakkam coast, east coast of India.

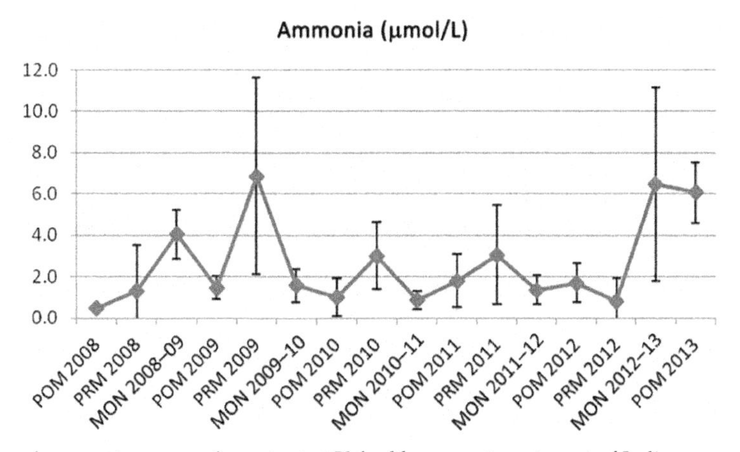

FIG. 9 Variations in seawater ammonia content at Kalpakkam coast, east coast of India.

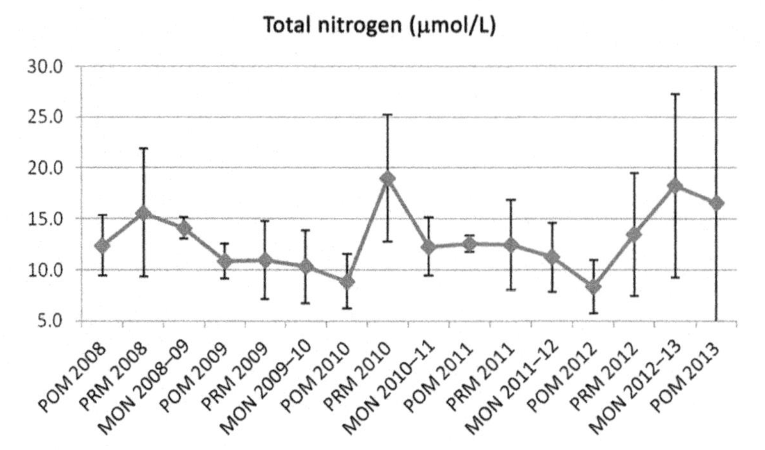

FIG. 10 Variations in total nitrogen concentration at Kalpakkam coast, east coast of India.

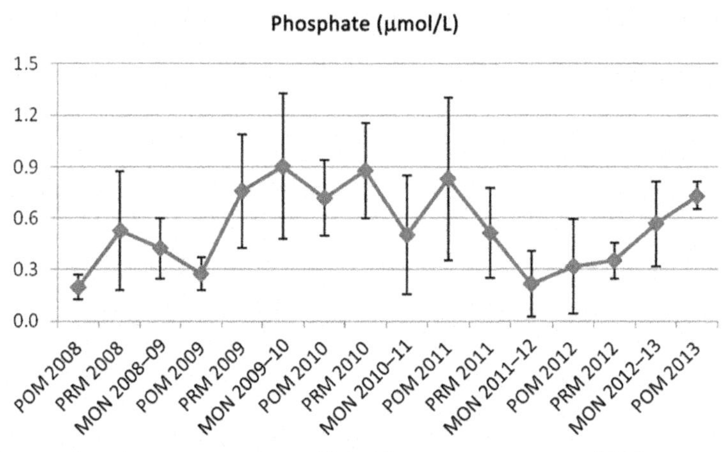

FIG. 11 Variations in phosphate concentration at Kalpakkam coast, east coast of India.

Phosphate concentration ranged from BDL-2.29. Like the nitrogenous nutrients, phosphate also did not show any seasonal trend (Fig. 11). TP concentrations ranged from 0.14 to 3.24 μmol/L. It showed a similar trend as that of the phosphate (Fig. 12). Concentration of phosphate in seawater depends upon the fresh water input, phytoplankton uptake, addition through localized upwelling, and regeneration through microbial decomposition of organic matter. It constitutes the most important inorganic nutrient that can limit phytoplankton production in tropical coastal marine ecosystems (Cole and Sanford, 1989). Phosphate and TP showed a similar trend seasonally at this location, which is supported by the strong positive correlation between them. Silicate concentration ranged from 2.33 to 25.40 μmol/L. The temporal variation in silicate content in coastal waters is influenced by several factors, more

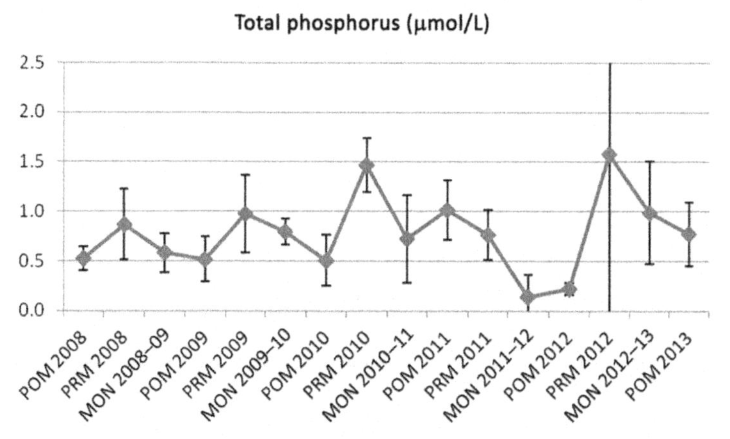

FIG. 12 Variations in total phosphorus contents at Kalpakkam coast, east coast of India.

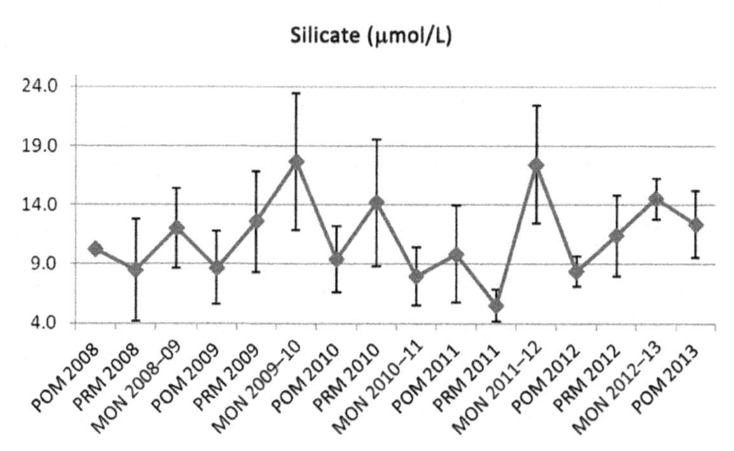

FIG. 13 Variations in silicate concentration at Kalpakkam coast, east coast of India.

importantly the proportional physical mixing of seawater with freshwater, adsorption into suspended sediment (Lal, 1978), chemical interaction with clay minerals (Aston, 1980; Gouda and Panigrahy, 1992), coprecipitation with humic compounds and iron (Stephens and Oppenheime, 1972), and phytoplankton uptake (Liss and Spencer, 1970). Freshwater input from the backwaters into the coastal water and surface runoff could be the reason for the observed higher values during monsoon periods (Fig. 13) in the present study. Strong negative correlation of silicate with salinity and strong positive correlation with DO indicated that the freshwater that is rich in DO could be an important source of silicate at this location as enrichment of silicate in coastal waters generally takes place through land drainage enriched with weathered silicate material (Lal, 1978).

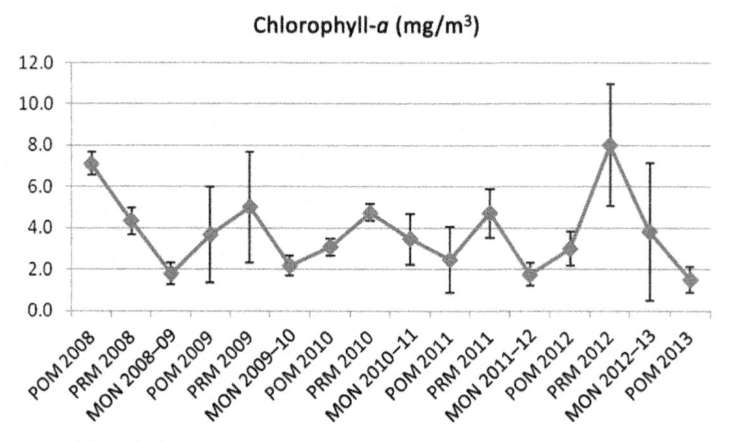

FIG. 14 Variations in chlorophyll-a concentration at Kalpakkam coast, east coast of India.

3.2.3 *Chlorophyll-a*

The chl-*a* content of coastal waters ranged from 0.28 to 14.46 mg/m^3. It showed a clear seasonal trend with relatively high values during premonsoon and low values during the monsoon and postmonsoon periods (Fig. 14). Chl-*a* values observed during the present study were in the following order: monsoon <postmonsoon <premonsoon. The typical marine conditions prevailed during the premonsoon period, supporting high phytoplankton production at this location (Satpathy et al., 2010a). Similar observations have been reported from other coastal waters (Prasannakumar et al., 2000; Madhupratap et al., 2001; Prasanna Kumar et al., 2002; Gilbert et al., 1982). Relatively high phytoplankton growth during the premonsoon period could be attributed to the upwelling, which is a regular phenomenon in this region (Murty and Varadachari, 1968; La Fond, 1957). Chl-*a* concentration decreased significantly during the NE monsoon period, which could be due to unfavorable conditions for phytoplankton growth due to a reduction in salinity and an elevation in turbidity as a result of precipitation and land drainage. This is further supported by the fact that chl-*a* showed strong positive correlation ($P \geq 0.001$) with salinity.

4 PHYTOPLANKTON DIVERSITY

4.1 Methodology

In order to assess the phytoplankton dynamics, samples were collected monthly from two transects between March 2008 and February 2009, covering a seasonal cycle (Fig. 15). The transects were parallel to the shoreline at 0.5 km (T1) and 5 km (T2) away from the coast. Each transect comprised six sampling locations and was chosen as per the point sources and/or anthropogenic disturbances. The point sources are explained as per their characteristics and degree of discharge. Opposite to the opening of the Sadras and Edaiyur backwaters, stations B1 and B2 and E1 and E2 were respectively located. Stations A1 and A2 were taken as control whereas C1 and C2 were in the vicinity of the fishing villages. Station D1 and D2

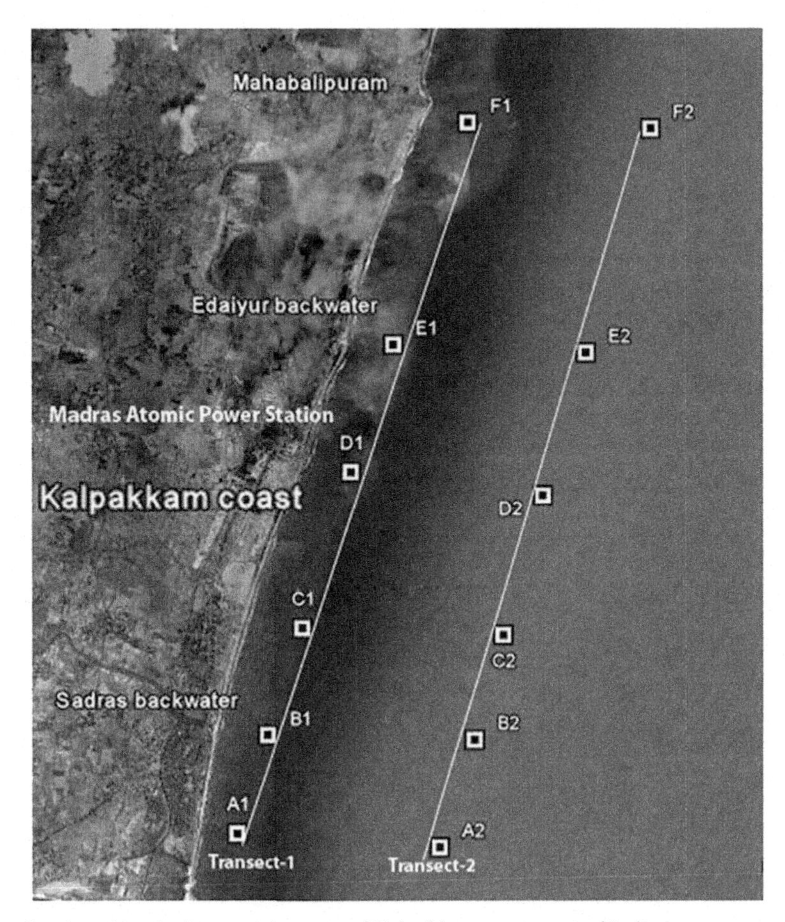

FIG. 15 Sampling locations in the coastal waters of Kalpakkam, east coast of India.

were located at the seawater intake and outfall of the MAPS condenser cooling unit. Stations F1 and F2 were located at Mahabalipuram, where year-round tourist activities prevail. Samples for the identification of phytoplankton species composition and abundance were preserved onboard with Lugol's solution. Phytoplankton species were identified following standard manuals (Desikachary, 1987; Carmelo, 1996) and counted by microscopic examination on an inverted microscope (Zeiss) in accordance with the Utermöhl method (Vollenweider, 1974).

4.2 Results

4.2.1 Phytoplankton Composition

During the study, the phytoplankton community in the coastal waters of Kalpakkam comprised 219 species from three different groups: Bacillariophyta, Cyanophyta, and Dinophyta.

Diatoms represented the most diverse and abundant group (78.08%) with 171 species, whereas dinoflagellates were recorded (19.63%) with 43 species during most of the year. Among 219 species recorded, 198 were encountered in T1 and 193 were from T2, representing 90.41% and 88.12%, respectively, of the total taxa. *Skeletonema costatum* was abundant throughout the year in both the transects and was varied between 0.58% and 11.69% of total density. At T1, *Melosira* sp. was relatively plenty during the intermonsoon period; *Skeletonema costatum, Chaetoceros curvisetus,* and *Coscinodiscus radiatus* during the SWM season; and species such as *Pseudonitzschia pungen, Thalassionema nitzschioides, Thalassiothrix frauenfeldii,* and *Chaetoceros curvisetus* were abundant during the NEM period. Similarly, at T2, dominance of species such as Biddulphia longicuris, Chaetoceros curvisetus, Coscinodiscus apiculatus, and Leptocylindrus danicus during intermonsoon period; Thalassionema nitzschioides, Thalassiothrix frauenfeldii, and Chaetoceros curvisetus during SWM; and Trichodesmium erythraeum, Thalassionema nitzschioides, Chaetoceros curvisetus, and Skeletonema costatum during NEM season were observed.

Phytoplankton density indicated a relatively higher population at T1 when compared to T2 (Fig. 16A). T1 has an annual mean population of 1.28×10^5 cells/L whereas T2 had a population of 0.64×10^5 cells/L. They ranged between 0.15×10^5–4.48×10^5 cell/L and

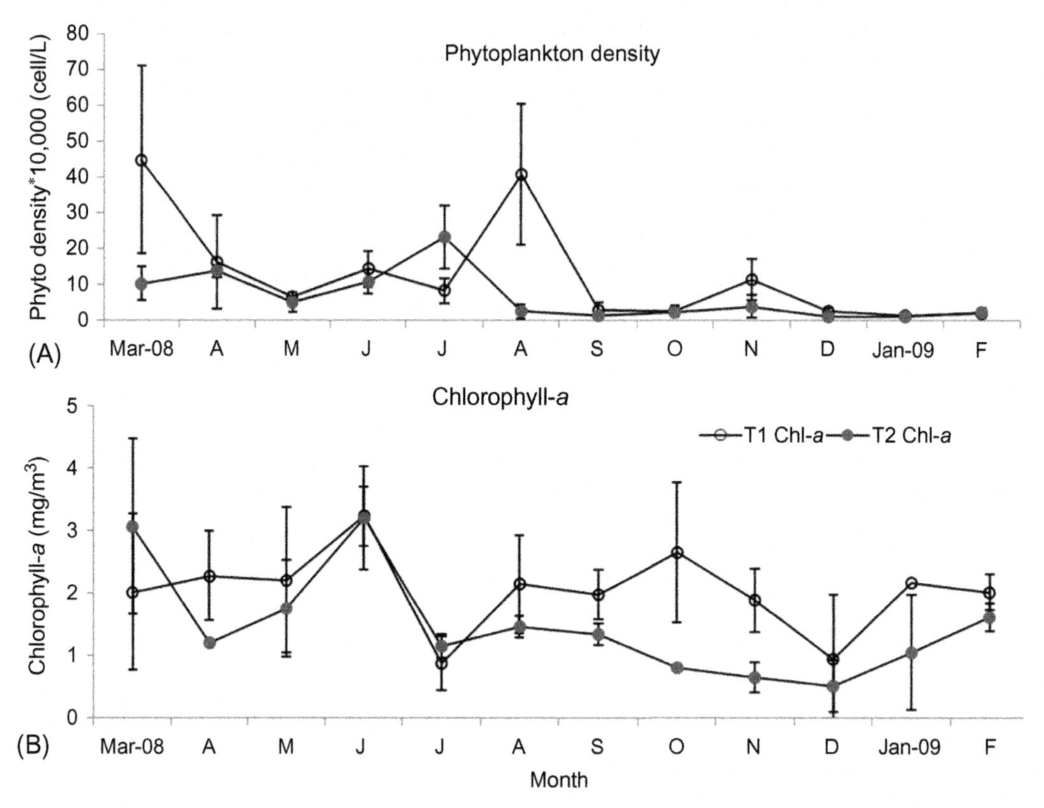

FIG. 16 Monthly variation of phytoplankton density (A) and Chl-a (B) concentration in transects T1 and T2 during 2008–09 at Kalpakkam Coast. The vertical bars.

0.11×10^5–2.3×10^5 cells/L at T1 and T2, respectively. The abundance oscillated frequently but showed a regular annual pattern similar in both transects, with low abundance during September-February and peak during March-August. Chl-*a* concentration ranged between 0.52 and 3.23 mg/m^3 in both transects (Fig. 16B). It was moderately high during the SWM and the early part of the intermonsoon period, in contrast to the later part of the year. The near-shore transect had higher pigment concentrations throughout the study period.

4.2.2 Principal Component Analysis

Principal component analysis on water quality characteristics, nutrients, and phytoplankton density and biomass (Chl-*a*) was performed by considering a six-dimensional space for transects T1 and T2. Inspection of the loads revealed that the first component explains about 58.6% of the total variability, whereas the second component (PC2 or Factor 2) accounts for 34.8% of the total variance. The projections of the variables onto the PC1 versus PC2 plot for T1 (Fig. 17A) produces very prominent clusters resulting from differences in the ecological status of the ecosystem. In one association, it groups turbidity, SiO_4, PO_4, DO, and TP, which could be an indication of proliferation of diatoms and other planktonic species that grow in a better phosphate and silicate ambience and ultimately contribute to lesser transparency. That ultimately helped to increase the dissolved oxygen status. The above condition could be ascribed to the intermonsoon and SWM season, where there is a proliferation of plankton communities. This is also supported by another cluster formed in the same factor plot that associates temperature, salinity, phytoplankton density, and chlorophyll-*a*. The other cluster includes the contribution of nitrogen toward the N: P ratio which was high during NEM seasons. Similarly in the factor plots (PC1 versus PC2) for T2 (Fig. 17B), Chl-a and phytoplankton density are associated with NO_3, indicative of proliferation of algal communities, most likely the cyanobacteria component that enhances the nitrogen status of the environment. The other group formed by the association of turbidity, temperature, NO_2, NO_2+NO_3, and TN acknowledges the heavy influx of fresh water to the marine environment that enhances the nitrogenous nutrients and turbidity of the system. Among the other combinations N/P, NH_3, and SiO_4, PO_4, and TP formed groups.

4.3 Discussion

4.3.1 Nutrient Ratios and Limitations

Ambient nutrient ratios (molar ratios) are used as an indicator of "potential" nutrient limitation. The atomic ratio of Si:N:P in marine diatoms is about 16:16:1 (Redfield, 1958; Brzezinski, 1985). Any deviation in the Redfield ratio indicates the potential for N, P, or Si limitation of the ecosystem. Two ambient nutrient ratios were calculated for each nutrient in this study and the Redfield ratio was applied to predict the following limitations: (1) N limitation when N:P <16 and N:Si <1; (2) P limitation when N:P > 16 and Si:P > 16; and (3) Si limitation when N:Si > 1 and Si:P <16 (Dubravko et al., 1995). Scatter plots (Fig. 18) of the atomic Si:P against N:P ratios in the surface waters of Kalpakkam indicate potential nutrient limitations on phytoplankton biomass growth. The data points (Fig. 18) in the upper left quadrant (N:P $<16:1$) are indicative of N limitation, in the upper right quadrant are indicative of P limitation, and

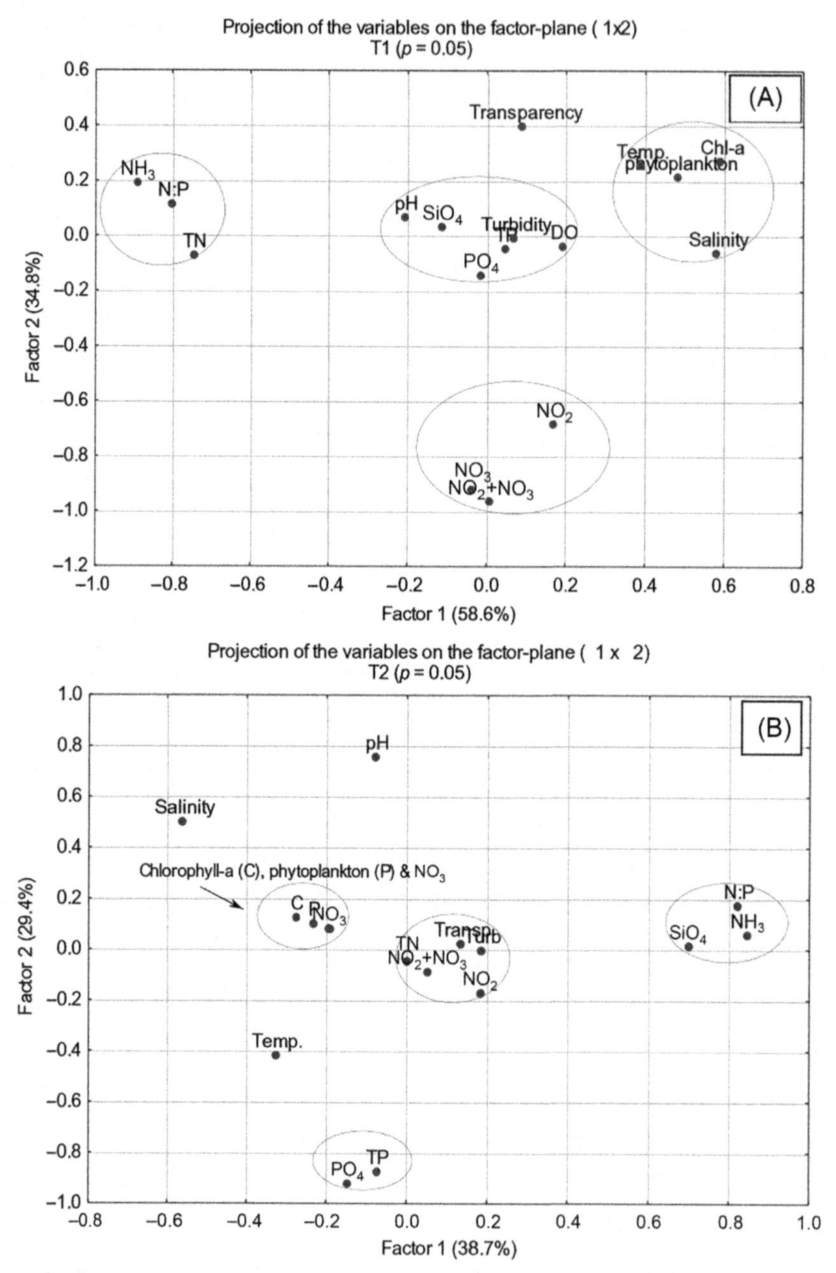

FIG. 17 Projections of first and second factor scores obtained from PCA for transect T1 (A) and T2 (B). The proposed alternative classifications are indicated by the circles.

of Si limitation in the lower left quadrant. T1 and T2 exhibited a similar fashion for nutrient limitation in three different seasons. At T1 during the NEM season, 67% of cases illustrated acute P limitation along with 21% Si limitation and 12% N limitation. Similarly, during the SWM and intermonsoon seasons, the majority of cases at T1 exhibited Si (63%) and

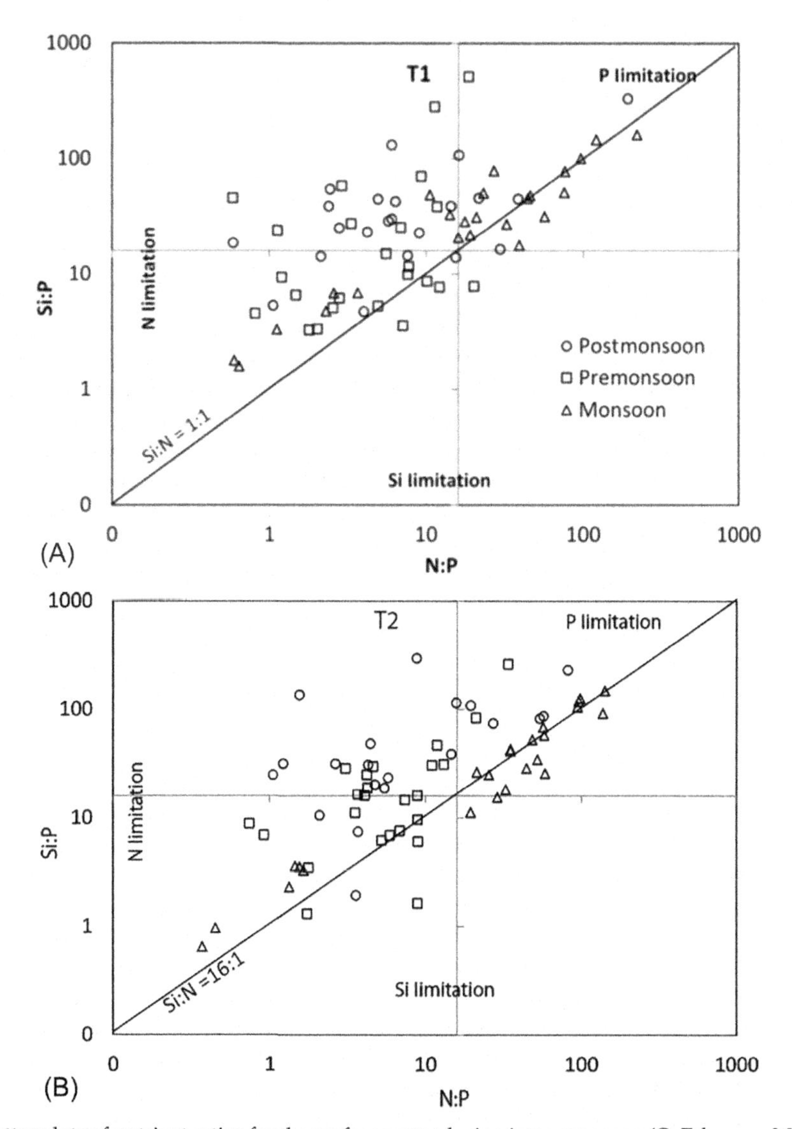

FIG. 18 Scatter plots of nutrient ratios for the surface water during inter-monsoon (O, February–May), Southwest monsoon (□, June–September) and Northeast monsoon (Δ, October–January) seasons at transects T1 (A) and T2 (B). N:P and Si represents dissolved inorganic forms. Stoichiometric (potential) limitation for N, P, and Si is indicated by the number of data points in various quadrants.

N limitations (76%), respectively (Fig. 18A). T2 demonstrated a P limitation (75%) during NEM, potential N limitation (50%) during intermonsoon, and Si limitation during the SWM monsoon season (Fig. 18B). The phytoplankton population has revealed a relatively higher density between March and May (intermonsoon), which had a dip during October and January (NEM). During the P limiting NEM season, the phytoplankton population at both T1 and T2 was dominated by the nitrogen-fixing cyanobacteria and a very low population of the diatoms that usually dominated during the intermonsoon and SWM periods.

We found that the filamentous Cyanobacteria, *Trichodesmium erythraeum,* was plenty and at its peak during NEM, contributing 37.52% of total phytoplankton density. Such increased cyanobacteria concentrations in surface layers has also been previously reported from other areas (Bustillos-Guzmán et al., 1995; Bel Hassen et al., 2009) and could be due to their ability to emerge as photosynthetically competent species in high light intensities (Kana and Gilbert, 1987) or to their ability to fix atmospheric nitrogen to meet their nitrogenous nutrient requirements (Haselkorn and Buikema, 1992).

4.3.2 Relationship Between Phytoplankton Species and Physicochemical Parameters

A group of 16 major species has been chosen from the taxonomic composition of the phytoplankton group that was encountered during the study. The species that occurred in all the stations at least 10 months of the year and/or constituted at least 20% of the total population has been chosen for establishing a relationship with the physicochemical characteristics of this particular coastal ecosystem. The relationship was established by the use of factor loading generated during the extraction of PCA and varimax normalized rotation as the medium of rotation (Table 2). There were three significant factors encountered with an Eigen value of $\sim$1. Factor 1 constituted the taxonomic group Chaetoceros eibenii, Chaetoceros decipiens, Chaetoceros lorenzianus, Chaetoceros sp, Proboscia alata, and Pseudonitzschia delicatissima along with salinity, chlorophyll and DO, which indicated their habitat preference of saline water with better oxygen conditions. This group is negatively related to another group that associates *Chaetoceros diversus* and *Thalassionema nitzschoides* with pH, turbidity, NH_3, SiO_4, and N:P. In Factor 2 *Chaetoceros curvisetus, Melosira* sp., *Skeletonema costatum,* salinity, NO_2+NO_3 and PO_4, and phytoplankton density were reversely related to *Chaetoceros diversus, Proboscia alata, Pseudonitzschia pungens, Pseudonitzschia australis,* turbidity, SiO_4, DO, and N:P. Factor 3 has a combination of *Asterionellopsis glacialis, Pseudonitzschia pungens, Thalassiothrix frauenfeldii, Thalassiothrix longissima,* TP, and Chl-*a* that negatively related to the group *Chaetoceros diversus,* temperature, transparency NO_2, and TN.

4.3.3 Phytoplankton Blooms at Kalpakkam

A prominent discoloration of coastal waters by the blue-green alga *Trichodesmium erythraeum* was encountered in the Bay of Bengal near Kalpakkam on March 16, 2007. The bloom persisted for only 1 day exhibiting visible alteration in physicochemical properties and phytoplankton community structure. The *Trichodesmium* cell density was 4.14×10^6 cells/L, sharing 74.19% of the total cell count (5.57×10^6 cells/L). Only 24 species of phytoplankton were encountered on the day of bloom as compared to the highest number of 44 species in a single observation during pre- and postbloom periods. Concentration of chlorophyll-*a* and phaeopigments increased about 20 times on the day of bloom compared to the prebloom values (Mohanty et al., 2007). An abrupt increase in ammonia, total nitrogen, and phosphate was noticed on the day of bloom. Similarly, an intense bloom of *Trichodesmium erythraeum* was observed in the coastal waters of Kalpakkam during the postnortheast monsoon period of 2008. A significant reduction in nitrate concentration was noticed during the bloom period, whereas a relatively high concentration of phosphate and total phosphorous was observed. An abrupt increase in ammonia concentration to the tune of 284.36 μmol/L was observed, which coincided with the highest *Trichodesmium* density (2.88×10^7 cells/L). Contribution of *Trichodesmium* to the total phytoplankton density ranged from 7.79% to

TABLE 2 Principal Component Analysis: Varimax Normalized Rotated Component Matrix for Different Physicochemical Characteristics and 16 Major Phytoplankton Species Encountered During the Study, Establishing Their Relationship

Variables	Factor 1	Factor 2	Factor 3
Temperature	–	–	0.91
Salinity	−0.80	−0.53	–
pH	0.98	–	–
Turbidity	0.62	0.60	–
NO_2+NO_3	–	−0.94	–
NH_3	1.00	–	–
TN	–	–	0.86
PO_4	–	−0.98	–
TP	–	–	−0.92
SiO_4	0.72	0.68	–
Phytoplankton	–	−0.81	–
Chl-*a*	−0.80	–	−0.54
DO	−0.56	0.75	–
N:P	0.67	0.74	–
Asterionellopsis glacialis	–	–	−0.96
Chaetoceros curvisetus	–	−0.93	–
Chaetoceros decipiens	−1.00	–	–
Chaetoceros diversus	0.56	0.66	0.51
Chaetoceros eibenii	−0.93	–	–
Chaetoceros lorenzianus	−0.98	–	–
Chaetoceros sp.	−1.00	–	–
Melosira sp.	–	−1.00	–
Proboscia alata	−0.52	0.79	–
Pseudonitzschia pungens	–	0.66	−0.75
Pseudonitzschia australis	–	0.82	–
Pseudonitzschia delicatissima	−0.99	–	–
Skeletonema costatum	–	−0.98	–
Thalassionema nitzschoides	0.93	–	–
Thalassiothrix frauenfeldii	–	–	−0.93
Thalassiothrix longissima	–	–	−0.96

97.01%. A distinct variation in phytoplankton species number and phytoplankton diversity indices was noticed. The lowest diversity indices coincided with the observed highest *Trichodesmium* density. Concentrations of chlorophyll-*a* (maximum 42.15 mg/m^3) and phaeophytin (maximum 46.23 mg/m^3) increased abnormally during the bloom.

A monospecies bloom of diatom *Asterionellopsis glacialis* was observed in the coastal waters of Kalpakkam during the post NE monsoon of 2015. The highest cell density (5.63×10^7 cells/L) of *A. glacialis* was observed on Jan. 17, 2015, accompanied with brownish discoloration. The contribution of other phytoplankters during bloom was relatively low. Contribution of *A. glacialis* during the peak bloom period was 97.12% of the total phytoplankton density. During the bloom event, 115 species of phytoplankton consisting of 64 centrales, 36 pennates, 12 dinoflagellates, and three cyanobacteria were identified. Physicochemical properties did not show any significant variation during the study. However, a significantly high concentration of total nitrogen (TN) and total phosphorus (TP) was observed on the day of peak bloom. Similarly, the highest concentration of chl-*a* (15.99 mg/m^3) coincided with the highest cell density.

5 ZOOPLANKTON DIVERSITY

5.1 Methodology

5.1.1 Sampling Strategy and Biomass Estimation of Zooplankton

Zooplankton samples were collected twice a month from the an MAPS Jetty by a conical plankton net with a mouth area of 1.25 m^2 (mesh size 200 μm) fitted with a flow meter (Hydrobios). The samples were preserved with 5% formalin and brought to the laboratory for further analysis. The zooplankton biomass was determined by the dry weight (g. 10/m^3) method. In addition to the above, the physicochemical parameters and phytoplankton density were estimated.

5.1.2 Qualitative and Quantitative Analysis

One sample of each collection was used exclusively for qualitative analysis. As far as practicable, individual animals were thoroughly examined under a binocular research microscope for their morphological attributes and were identified up to the genera/species level. For the purpose of identification of different species or genera, standard literature such as those of Kasturirangan (1963) for copepods, Rao (1958) for Chaetognaths, and Conway et al. (2003) and Newell and Newell (1967) for other components were followed. Besides these, the invertebrates were used to ascertain the taxonomic status of some planktonic animals, including larval forms.

The residual zooplankton organism, retained on the filtering screen of the volume determination apparatus, was resuspended in 5% formaldehyde solution. The larger organisms such as the copepods of families Pontellideae, Eucalanidae, and Euchaetaidae; holoplanktons such as Mysids, Lucifers, Siphonophores, Chaetognaths, Ctenophores, Dolioloids, and Salps; and meroplankton such as fish eggs and larvae, larvae of decapods (zoea, megalopa, alima, phyllosoma, mysis), polychaete larvae (Trochophores) were sorted out. The residual mixture was then diluted to exactly 100 mL with 5% formalin. One ml of this aliquot was taken in a Sedgewick-rafter cell using a Stempel's pipette. A stereo binocular research microscope (Nikon Eclipse-50*i*) was used for identification and quantification.

5.1.3 Relative Abundance

For the purpose of computing relative abundance, the entire bulk of holoplankton was divided into nine groups: Copepods, Lucifers, and other Crustaceans, Siphonophores, Chaetognaths, Dolioloids and Salps, Appendicularians, Ctenophores, and others. Similarly, the meroplankton fraction was divided into three groups: larvae of crustaceans, fish eggs and larvae, and others. From the bulk density of each sample and group-wise density, the percentage contribution of each group was computed and presented in a pie chart.

5.2 Results and Discussion

Results of qualitative aspects of the study period showed that, in total, 116 species of mesozooplankton (size 0.2–20 mm) were found, out of which 11 are larval plankton, 99 are holoplankton (Copepod-58 and noncopepod holoplankton-41), and the rest are unidentified species. The zooplankton community cis omprised of 10 phylum of holoplankton and five phylum of meroplankton. The holoplankton constitutes 9 classes and 15 orders. A comparative account of the qualitative and quantitative results of the present study with other coastal waters is given in Table 3.

Cthe copepod population was found to predominate over the other groups in the zooplankton community. The copepod population is comprised of five orderss: Calanoida (12 families), Harpacticoida (four families), Cyclopoida (one family), Poecilostomatoida (four families), and Siphonostomatoida (one family). Calanoids dominated the copepod community, followed by cyclopoids, poicilostomatoids, and harpacticoids. Calanoids are arguably the most common and abundant planktonic copepods anywhere in the world's oceans (Madhupratap, 1999). During the study period, the lowest (18) and highest (38) zooplankton species were observed in the monsoon and postmonsoon seasons, respectively (Fig. 19). A visible difference in seasonal variation between the years 2008–10 and 2011–14 could be

TABLE 3 Comparison of Qualitative and Quantitative Zooplankton Results with Other Coastal Waters

Authors	Study Area	No. of Species	Population Density (Ind. 10/m^3)
Kumar and Perumal (2011)	Ayyapattinam coast, East coast of India	45	3.3×10^5–2.01×10^8 Ind. 10/m^3
Perumal et al. (2009)	Nagapattinam Coast, South East Coast of India	110	4.8×10^7–1.5×10^8 Ind. 10/m^3
Present study	Kalpakkam coastal water	116	2.8×10^5–8.5×10^5 Ind. 10/m^3
Goswami and Shrivastava (1996)	Northern Arabian Sea	116	2.4×10^3–5.1×10^3Ind. 10/m^3
Robin et al. (2009)	Southern Kerala Coast, Southwest coast of India	120	600–366010/m^3

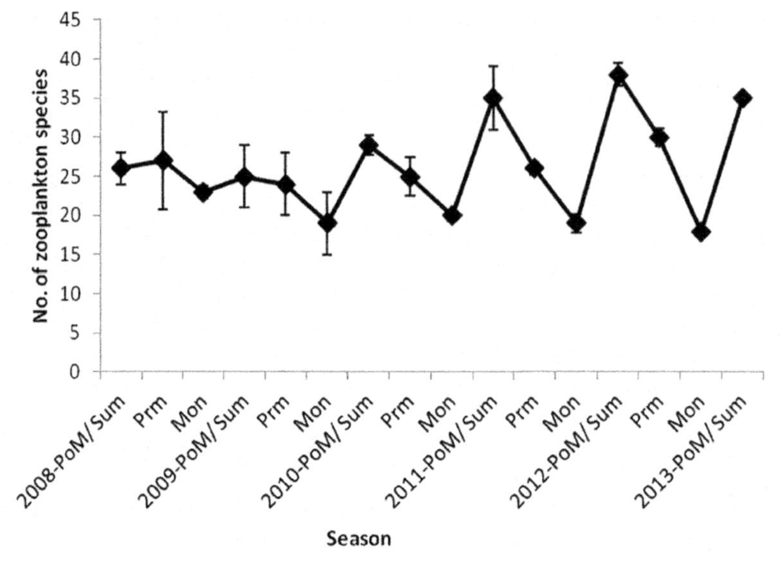

FIG. 19 Seasonal variations in number of zooplankton species.

due to the sampling strategy adopted during respective periods. During 2008–10, sampling was carried out on a weekly basis and the results were pooled to get seasonal figures, whereas samples were collected monthly during 2011–14 and pooled to seasonal figures.

The relative abundance of different groups of copepods is given in Fig. 20. Among copepods, *Canthocalanus pauper, Acrocalanus gibber, Acartia erythraea, Paracalanus parvus,* and *Oithona rigida* were found dominant and contributed about 50%–70% to the total population throughout the study period. *Acartia centrura, Centropages furcatus, P. aculeatus, Labidocera*

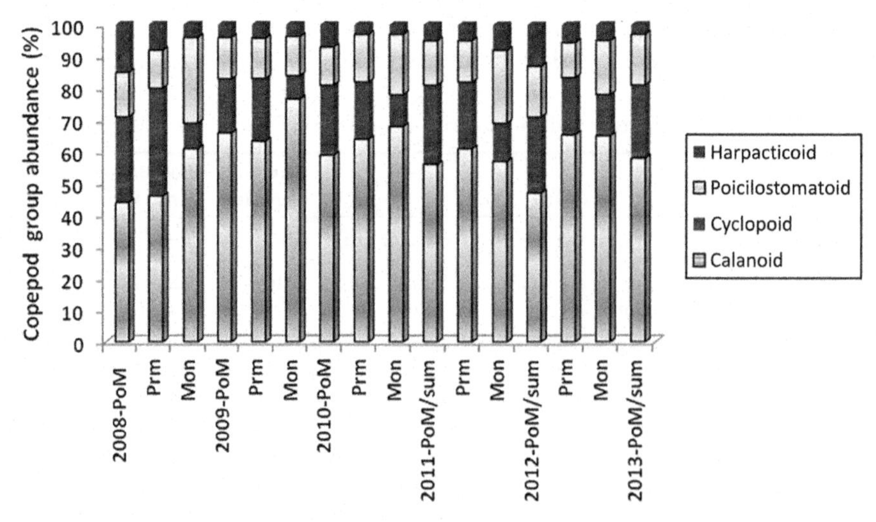

FIG. 20 Seasonal variations in relative abundance of different groups of copeods.

acutifrons, Euterpina acutifrons, and *Corycaeus danae* were also found throughout the year, but contributed ~10%–15% to the total population. In the copepod community, the herbivorous forms such as *Paracalanus parvus, Acrocalanus gibber, A. gracilis,* and some other herbivorous species were dominant during the postmonsoon/summer and premonsoon periods, which could be due to the abundant food supply (phytoplankton) prevailing then. This community was subsequently succeeded by a few numbers of carnivore species (*Pontella andersonii, Pontellopsis scotti, Undinula vulagaris,* and *Labidocera acuta*) for a short period, but in good numbers (contributed ~10%–20%) to feed upon the rich herbivore population. A similar observation was also reported by Rakhesh et al. (2006), Sreekumaran Nair et al. (1981), and Sahu et al. (2010). Some Cyclopoids such as *Oithona rigida, Oithona similis, Corycaeus longistylis,* and *C. catus* (appeared in other periods also) thrived well during the monsoon period, although their relative abundance was comparatively less. It is believed that cyclopoids are euryhaline in nature and adapted to variable salinity (Hwang et al., 2010). Among copepods, *Oithona* sp. and *Euterpina acutifrons* females showed their ovigerous stage abundantly during the monsoon period. Similarly, the highest meroplankton abundance was found during the monsoon period (19%) and the lowest during the postmonsoon period (8%) (Fig. 21). Among noncopepod holoplankton, *Evadne tergestina, Penilia avirostris,* and *Lucifer* sp. were the most dominant forms.

A pronounced variation in zooplankton density was observed during the study period. The minimum (2.2×10^5 Ind. $10/m^3$) and maximum (8.7×10^5 Ind. $10/m^3$) density were found

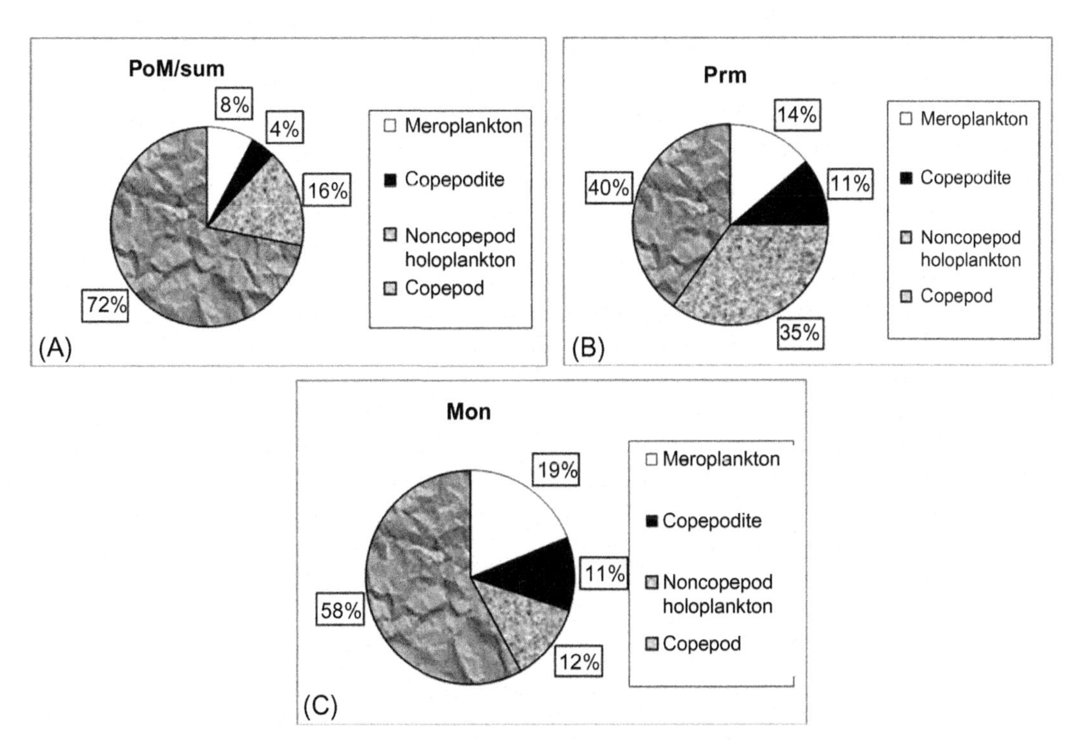

FIG. 21 Seasonal variations in relative abundance of different groups of zooplankton.

during the monsoon and premonsoon periods, respectively (Fig. 22). The rich abundance as well as number of species during postmonsoon/summer and premonsoon could be attributed to the stable as well as optimal salinity, temperature, light, nutrient levels, etc., prevailing during these periods (Goswami and Shrivastava, 1996; Perumal et al., 2009). This was further confirmed by the positive correlation of zooplankton density with salinity ($P \leq .005$), temperature ($P \leq .05$), and phytoplankton ($P \leq .005$) during the present study (Table 4). Goswami and Padmavati (1996) also reported the abundance of phytoplankton during postmonsoon months, which supported a higher population of herbivores. Zooplankton biomass values ranged between 0.20 and 0.42 g dry wt. 10/m^3 (Fig. 23) and this was in tune with population density.

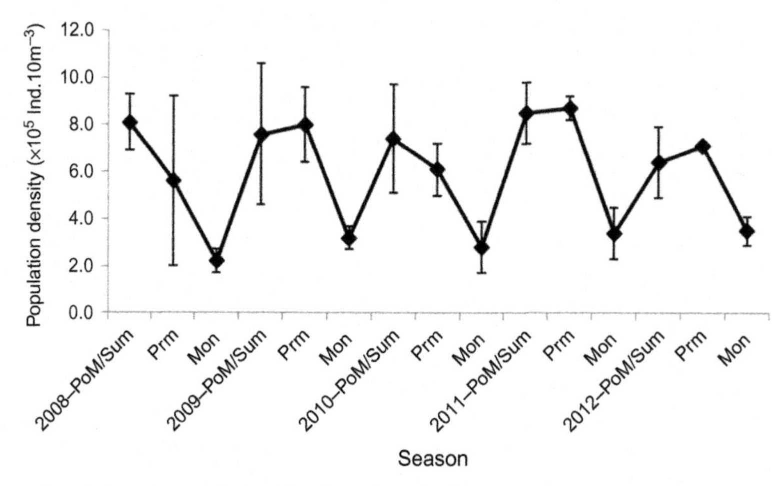

FIG. 22 Seasonal variations in population density of zooplankton.

TABLE 4 Correlation Between Zooplankton Population and Different Physico-Chemical Parameters

Parameters	Correlation Coefficient	*P*-Values
Salinity	0.604[a]	0.002
Temperature	0.5[b]	0.013
DO	0.451[b]	0.027
Turbidity	0.373[c]	0.073
NO_3	0.408[b]	0.048
NH_3	0.381[c]	0.062
PO_4	0.346[c]	0.097
Phytoplankton	0.583[a]	0.003

[a] $P \leq .005$
[b] $P \leq .05$
[c] $P \leq .1$

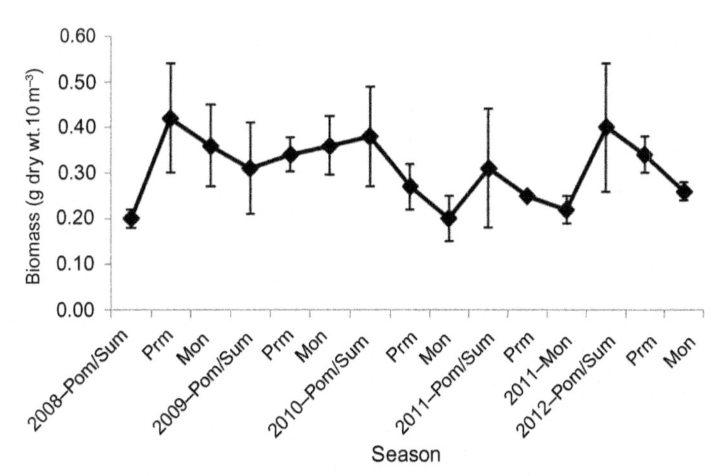

FIG. 23 Seasonal variations in biomass of zooplankton.

During the course of this study, a blooming condition of *Trichodesmium erythraeum*, a marine nitrogen-fixing Cyanobacteria, was encountered. During the appearance of the bloom (contributed to the extent of 80%–90% of total phytoplankton density) in this coastal water, a steep reduction in the zooplankton density (prebloom—4.2×10^5; bloom—3.4×10^5; postbloom—2.1×10^5) and biomass was observed. Copepods contributed ~64.2% of the total zooplankton abundance during the prebloom period in which calanoid copepods were dominant (41%). Carnivore copepods (25%) (Cyclopoids—*Oithona rigida* and *O. brevicornis*, and Poicilostomatoids—*Oncaea venusta*) dominated over the herbivore (23%) population (members of the Paracalanidae family such as *Acrocalanus gibber*, *Paracalanus aculeatus*, *P. parvus*, *P. crassirostris*, etc.) during the peak bloom period. Noncopepod holoplankters were dominant during the prebloom period (prebloom—13%, bloom—11%, and postbloom—6%). The cladoceran *Evadne tergestina* (35% of the total noncopepod crustacean holoplankton) was found to be the most important solitary species during the bloom period, but strikingly, it was totally absent during pre- and postbloom periods. Meroplankters were found to be abundant during the prebloom condition (16% of the total zooplankton abundance) comprising polychaete larvae, veliger larvae of bivalves, cirriped nauplii, etc. Among the meroplankters, the cirripede nauplii (59% of the total meroplankton) were the most dominant form during the bloom period.

6 FISH DIVERSITY

6.1 Materials and Methods

Samples of fishes were collected at weekly intervals over a period of three years (May 2007 to May 2010) from landing centers at Pudupattinam (12° 29′ N; 80° 09′ E) and Sadras (12° 31′ N; 80° 09′ E) as well as from the MAPS pumphouse (12° 33′ N; 80° 10′ E). Details of materials and methods for fish diversity studies are given elsewhere (Biswas et al., 2014). The fish specimens were identified as per taxonomic keys provided in the studies of Day (1878),

Munro (1982), and Talwar and Kacker (1984). A species accumulation curve was plotted to understand the accumulation of species through time. Jaccard's coefficient (*JC*) of similarity was expressed in a dendrogram to cluster the monthly occurrence of fish species to determine seasonal patterns. Water samples were collected weekly and analyzed following standard methods. Canonical component analysis (CCA) was carried out by means of the CANOCO 4.5 software (ter Braak and Šmilauer, 2002) to understand how species respond to the above environmental variables. A Monte-Carlo simulation (199 permutations) was performed to test for the significance of the overall ordination and the first canonical axis.

6.2 Results

6.2.1 Community Composition

The present study recorded 244 coastal fish species, which represented 163 genera, 87 families, and 21 orders. The Carangidae family represented the most number of species (22), followed by Clupeidae, Apogonidae, Leiognathidae, and Tetraodontidae with nine species each and Sciaenidae with eight species. Out of the 163 genera recorded, 19 genera collectively contributed 30% of the total species encountered. The species accumulation curve (Fig. 24) showed that *c.* 30% of the species were recorded in July 2008 following severe cyclonic weather, due to which many offshore and demersal species were observed. Even at the end of three years, the graph shows an increasing trend that clearly indicates the possibility of getting more species with more sampling efforts. The habitats of the recorded species showed that 44% were typically reef-associated (RA), closely followed by demersal (DL) species (37%), with the rest being bentho-pelagic (BP) (7%), pelagic-neritic (PN) (11%), and pelagic-oceanic (PO) (1%).

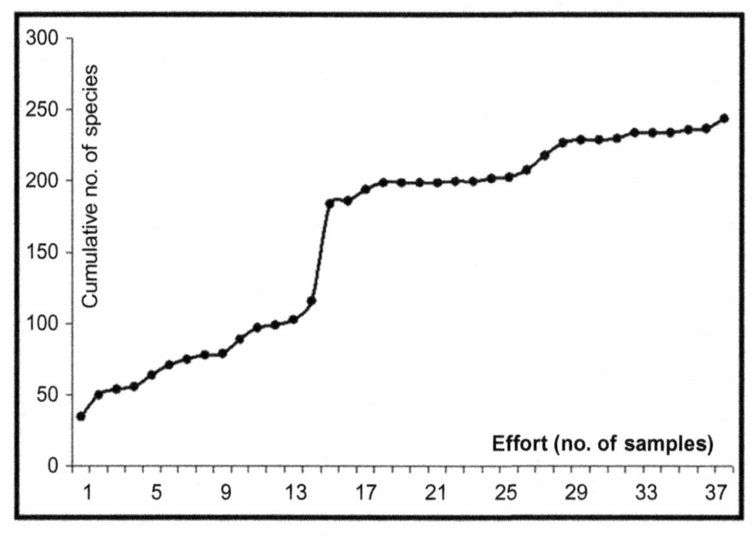

FIG. 24 Species accumulation curve.

6.2.2 Statistical Analysis

Species occurrence on a time scale using JC showed three major clusters of months coinciding with three seasons observed at Kalpakkam: premonsoon (PrM), monsoon (M), and postmonsoon (PoM). Seasonal variability data indicate the prominence of some families. The PrM period was dominated by the families Megalopidae, Nemipteridae, Sciaenidae, and Cynoglossidae whereas the PoM period was dominated by Carangidae and the M period was dominated by Chirocentridae. During the PrM and PoM period, the Dasyatidae and Engraulidae families were common whereas the families of Leiognathidae, Terapontidae, and Scombridae were observed throughout the year.

A specific pattern in seasonal distribution was observed in the case of some species in the species occurrence analysis. Species such as the Indian oil sardine *Sardinella longiceps* Valenciennes 1847 (PrM and M), Dussumier's ponyfish *Leiognathus dussumieri* Valenciennes 1835 (M), pugnose ponyfish *Secutor insidiator* Bloch 1787, and deep pugnose ponyfish *Secutor ruconius* Hamilton 1822 (M and PoM) were dominant during particular seasons. Additionally, 16 species found in the PrM period, nine in the M period, and 15 in the PoM season were categorized as very common (VC). During the study period, eight species were VC and mostly dominant throughout the year. The observed changes in fish diversity during different seasons could be attributed to temporal variations in environmental parameters.

Results of CCA on the fish communities of the Kalpakkam coastal waters are given in the biplot (Fig. 25). The first two canonical axes explained 13.1% of cumulative variance in the species dataset. The vectors of the species scores and the environmental variables collectively explained 21.4% of the variance in the species-environment relationship on the first axis and 15.8% along the second axis. This is confirmed by the species-environmental correlation coefficient, 0.902 for the first and 0.954 for the second axis, which showed their capacity to explain variations in fish community composition (Table 5). Total inertia for all four axes was 3.314 and the sum of all canonical eigenvalues was 1.167. Monte-Carlo permutation test results showed significance for the first canonical axis ($F = 2.113$, $P < .01$) and for the overall canonical axis ($F = 1.413$, $P < .01$) (Table 5). Species located near the centroid were mostly common, based on occurrence patterns, whereas the rare species were located at the extreme end of the plot (Fig. 25). A high negative correlation (−0.553) was observed between salinity and rainfall (Table 6).

6.3 Discussion

This study provides baseline data on fish diversity for the Kalpakkam coast in particular, and adds to previous records of fish from the southeast coast of India. Out of the 244 species recorded here, 11 are new to the Tamil Nadu coast (Biswas et al., 2012a), four are new to the east coast of India (Biswas et al., 2012b), and two species, the stargazer snake eel *Brachysomophis cirrocheilos* and the *Torquigener brevipinnis* Regan 1903 are new to the east and west Indian coasts (Biswas et al., 2010a,b). Two species such as *Heteroconger tomberua* Castle and Randall 1999 and *Opistognathus macrolepis* Peters 1866 form the first report from the Indian Ocean region (Biswas et al., 2012c, 2013). An unusual abundance of reef fish diversity (44%) was observed along the Kalpakkam coast. Although coral reefs are absent at the Kalpakkam coast, the rocky patches located to the north of the study area could be the reason

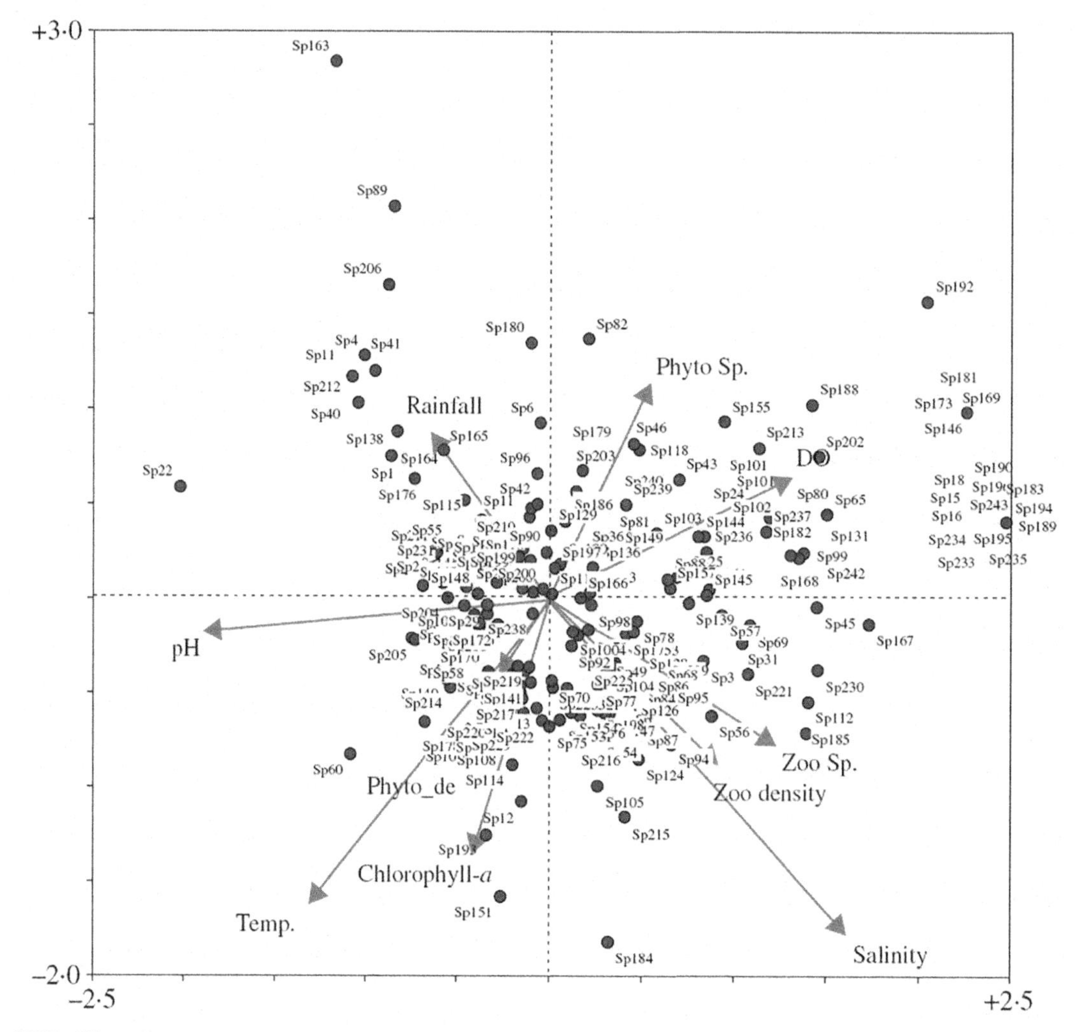

FIG. 25 CCA biplot ordination for fish community composition of Kalpakkam coastal waters along with the environmental gradients.

for the presence of reef fish. The large number of DL species present at this location throughout the year also suggestsed that these assemblages have restricted movement and are residential in nature. Longhurst and Pauly (1987) and Blaber et al. (1990) showed that nearshore DL fish assemblages are generally dominated by the families such as Leiognathidae, Sciaenidae, and Tetraodontidae and the order Pleuronectiformes. The present study also confirmed the presence of such a fish community in the coastal waters of Kalpakkam.

Temporal analysis using JC based on monthly pooled data shows three distinctive clusters of fish assemblage that broadly correspond to the three seasons observed at Kalpakkam: PrM, M, and PoM. A little displacement in PrM months in the cluster, where October 2008 and 2009

TABLE 5 CCA Summary of Fish Diversity Studies at Kalpakkam Coast

	Axes				Total Inertia
	1	2	3	4	
Eigenvalues	0.249	0.185	0.163	0.129	3.314
Species-environment correlations	0.902	0.954	0.931	0.888	
Cumulative percentage variance of species data	7.5	13.1	18	21.9	
Cumulative percentage variance of species-environment relation	21.4	37.2	51.2	62.3	
Sum of all unconstrained eigenvalues					3.314
Sum of all canonical eigenvalues					1.167

Monte Carlo Test (199 Permutations)			
Significance of first canonical axis	*Eigenvalue*	*F-ratio*	*P-value*
	0.249	2.113	0.0050
Significance of all canonical axes	*Trace*	*F-ratio*	*P-value*
	1.167	1.413	0.0050

TABLE 6 Canonical Correlation Matrix of Environmental and Habitat Parameters with Their Variance Inflation Factor

	Species Axis1	Species Axis2	Temp.	pH	Salinity	DO	Chl-a	Psp	PD	Zsp	ZD	Rainfall
Environmental Axis1	0.902*											
Environmental Axis2		0.954*										
Temp.	−0.409*	−0.512*										
pH	−0.584*											
Salinity	0.500*	−0.563*										
DO	0.405*											
Chl-a												
Psp												
PD												
Zsp				−0.459*								
ZD												
Rainfall					−0.553*							
Variance Inflation factor			1.5018	3.4628	3.3443	1.4181	2.6251	1.5633	1.367	1.6661	1.4907	2.3518

$*P < 0.05$.

Temp., temperature; *DO*, dissolved oxygen; *Chl-a*, chlorophyll a; *Psp*, phytoplankton species; *PD*, phytoplankton density; *Zsp*, zooplankton species; *ZD*, zooplankton density.

(monsoon months) was observed, could be due to the late arrival of the monsoon for 2009 and 2010. Some minor inconsistencies in clusters such as the overlapping of PoM and M (October 2007, November 2007, December 2007, January 2008, January 2009, and November 2009) were observed, which could be due to similar species assemblage and species composition at the end of one season resembling that of the beginning of the next season. The above poor species richness observed during the M season was the result of increased freshwater discharge into the sea from the nearby backwaters during the same period, leading to a lowering of salinity, a main deciding factor for species abundance (Ansari et al., 2003). A similar decline in fish species richness has been reported by Claridge et al. (1986) and Ansari et al. (1995). A substantial increase in species richness during the PoM, reaching a peak in the PrM, was observed (Fig. 26).

Studies on fish assemblage variation with time along the Indian coast are poorly known (Biswas et al., 2014). Studies on fish diversity have suggested that the occurrence of fish follows seasonal patterns, which has also been observed here. Some of the examples reported earlier in this regard are: the white-spotted spinefoot *Siganus canaliculatus* Park 1797 was observed in the GoM during the M and PoM seasons, *T. lepturus* during the M season in the Arabian Sea Coast (Al-Nahdi et al., 2009), tardoore *Opisthopterus tardoore* Cuvier 1829 during the M and PoM in the Goa coast (Prabhu and Dhawan, 1974), the Chinese silver pomfret *Pampus chinensis* Euphrasen 1788 during the PrM in Veraval (Kasim and Khan, 1986),

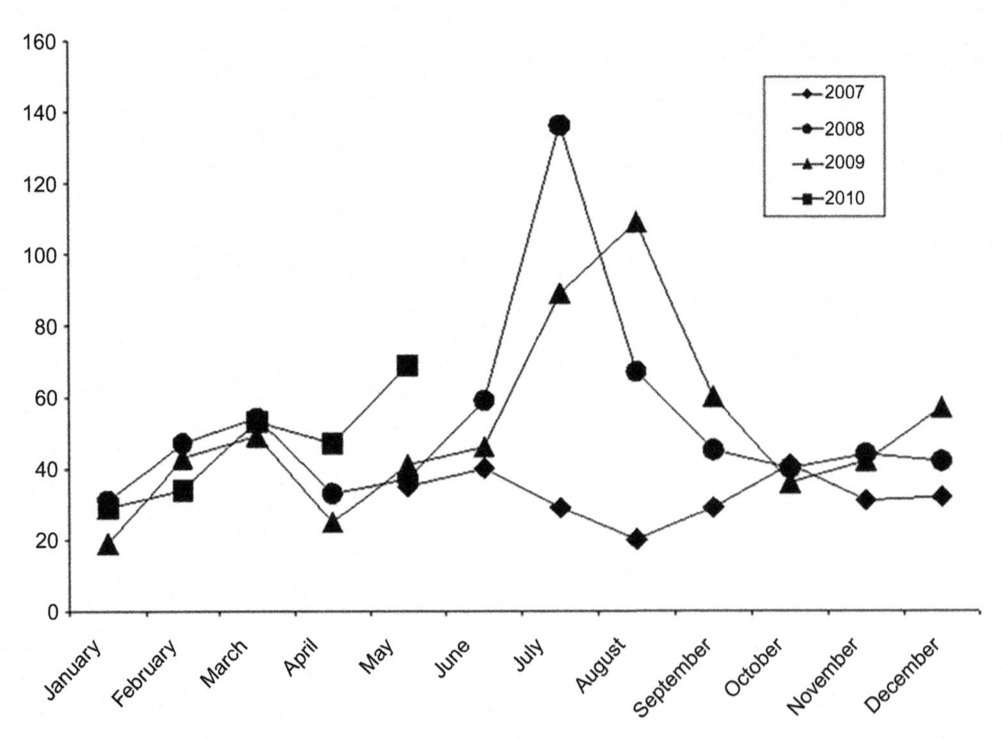

FIG. 26 Monthly variation of number of species.

Lutjanus species during the PrM and PoM, and goatfish (*Upeneus* and *Parupeneus* species) during the PrM and M in Vizhinjam (Radhakrishnan, 1974). Despite differences in the southwest and northeast monsoon climates, similar seasonal occurrences were observed for the above-mentioned fish species at the Kalpakkam coast. Although no effort has been made to corelate between assemblage structure and environmental parameters, substantial information on commercial fish catches and their seasonality is available (Kasim and Khan, 1986; Lalithambikadevi, 1993; Naomi et al., 2011). CCA results revealed that rainfall and pH and the covariates of rainfall, such as salinity, decided most of the fish assemblage structure variation on the Kalpakkam coast. Rainfall variations influenced the salinity and temperature regimes of this region (Katsaros and Buettner, 1969) and salinity, the primary factor, appeared to strongly influence the community structure of fish (Marais, 1988) along the coast. Relatively high values of salinity were observed during PrM followed by PoM, which coincided with an influx of species. Our observations corroborated the findings of other authors (Thorman, 1986; Lazzari et al., 1999; Akin et al., 2003). A number of species such as bluelined hind *Cephalopholis formosa* Shaw 1812, yellowtail scad *Atule mate* Cuvier 1833, tuberculated flathead *Sorsogona tuberculata* Cuvier 1829, three spotted flounder *Pseudorhombus triocellatus* Bloch Schneider 1801, and fringefin zebra sole *Zebrias quagga* Kaup 1858 were found to be associated with temperature, phytoplankton density, and chlorophyll-*a* gradient. Secondary productivity variables, namely zooplankton species and zooplankton density, were found to be associated with salinity. The mangrove red snapper *Lutianus argentimaculatus* Forsskål 1775, blackfin scad *Alepes melanoptera* Swainson 1839, blacktip soldierfish *Myripristis botche* Cuvier 1829, Gon's cardinalfish *Archamia bleekeri* Gunther 1859, herring scad *Alepes vari* Cuvier 1833, and spotted scat *Scatophagus argus* L. 1766 are a few of the mentioned species that were associated with these environmental variables.

7 CRAB DIVERSITY

The sampling was done on a weekly basis at the water cooling system of the (MAPS), the fish landing centers (Pudupattinam, Sadras and Kokilamedu), and the artificially established mangroves at Sadras. The common species of crabs were identified on the spot and photographed. All the unidentified crabs were photographed, noted for date, and preserved in 10% formalin for further identification in the laboratory.

Owing to the ecological role of crabs in the coastal milieu, they are regarded as one of the best indicators to validate the environmental health of a coastal industrial setup. Considering this, a coastal biodiversity study program on the brachyuran crabs of Kalpakkam and their ecological role was initiated. A total of 60 species of Brachyuran crabs belonging to 41 genera and 17 families were recorded from the Kalpakkam coast. The most diverse family is Portunidae (18 species). The Portunidae is represented by five genera (*Charybdis, Thalamita, Portunus, Scylla,* and *Podophthalmus*) consisting of nine *Charybdis* species, five species of *Portunus,* two species of *Thalamita,* and one species of *Podophthalmus.* Among these, six species—*Charybdis lucifera, Portunus pelagicus, Portunus reticulates, Portunus sanguinolentus, Scylla serrata,* and *Podophthalmus vigil*—are edible. The family Leucosiidae is represented by six genera consisting of seven species: *Ixa inermis, Ixa cylindrus, Lyphira perplexa, Arcania undecimspinos, Coleusia urania, Myra subgranulata,* and *Philyra globus*. The family Grapsidae is represented by six genera (*Plagusia, Grapsus, Ocypode, Epiresarma, Precon,* and *Neosarmatium*) consisting of six species. Biofouling associated crabs were also recorded from the MAPS cooling water screen, namely *Sphaerozius nudus* and *Actumnus setifer*. The rare family Corystidae has been recorded from these study sites with a single representative from the genus *Jonas*.

Brachyuran crabs are represented by 33 species, 19 genera, and nine families from the fish landing center (Pudupattinam, Sadras, and Kokilamedu). Seven species (*Scylla serrata, Neosarmatium meinerti, Episesarma mederi, Muradium tetragonum, Uca triangularis, Uca lacteal,* and *Metaplax distinct*) belonging to sixgenera and five families of mangrove crabs were recorded from the artificially developed mangroves at Sadras. Five species such as *Charybdis lucifera, Charybdis hellerii, Thalamita crenata, Portunus sanguinolentus, Portunus reticulates,* and *Menippe rumphii* that were recorded from fish landing centers as well as from the MAPS cooling water system were absent in the Sadras artificially developed mangrove ecosystem. Only one species, *Scylla serrata,* was observed to be common in the fish landing centers and the Sadras artificially developed mangroves (Table. 7). The species found exclusively in the artificial mangroves ecosystem at Sadras showed that it provides a suitable breeding ground for these species. Kalpakkam is found to be represented by a high number of crab species. The rich diversity of brachyuran crabs from the Kalpakkam coast shows the benign nature of the coastal water and sediment. At the same time, frequent observation of commercial crab species at this coast implies that the presence of industrial establishments since three decades has not affected the local fisheries.

TABLE 7 List of Brachyuran Crabs Recorded at Three Stations from Kalpakkam Coast

No.	Family	Species Name	Fish Landing Center	MAPS Cooling Water System	Sadras Artificially Developed Mangroves
1	Portunidae	*Charybdis lucifera*	X	X	–
2		*Charybdis feriata*	X	–	–
3		*Charybdis annulata*	X	–	–
4		*Charybdis hellerii*	X	X	–
5		*Charybdis orientalis*	–	X	–
6		*Charybdis natator*	X	X	–
7		*Charybdis callianassa*	–	X	–
8		*Charybdis hoplites*	–	X	–
9		*Charybdis granulate*	X	–	–
10		*Portunus sanguinolentus*	X	X	–

TABLE 7 List of Brachyuran Crabs Recorded at Three Stations from Kalpakkam Coast—cont'd

No.	Family	Species Name	Fish Landing Center	MAPS Cooling Water System	Sadras Artificially Developed Mangroves
11		*Portunus reticulates*	X	X	–
12		*Portunus pelagicus*	X	–	–
13		*Portunus gladiator*	X	–	–
14		*Portunus hastoide*	X	–	–
15		*Scylla serrata*	X	–	X
16		*Thalamita crenata*	X	X	–
17		*Thalamita picta*	–	X	–
18		*Podophthalmus vigil*	X	–	–
19	Grapsidae	*Grapsus albolineatus*	–	X	–
20		*Ocypode platytarsis*	–	–	–
21		*Plagusia depressa tuberculata*	–	X	–
22		*Episesarma mederi*	–	–	X
23		*Percnon planissimum*	–	X	–
24		*Neosarmatium meinerti*	–	–	X
25	Leucosiidae	*Ixa inermis*	X	–	–
26		*Ixa cylindrus*	X	–	–
27		*Lyphira perplexa*	X	–	–
28		*Arcania undecimspinos*	X	–	–
29		*Coleusia urania*	X	–	–
30		*Myra subgranulata*	X	–	–
31		*Philyra globus*	X	–	–
32	Xanthidae	*Atergatis integerrimus*	–	X	–
33		*Quingue tentatus*	–	X	–

(Continued)

TABLE 7 List of Brachyuran Crabs Recorded at Three Stations from Kalpakkam Coast—cont'd

No.	Family	Species Name	Fish Landing Center	MAPS Cooling Water System	Sadras Artificially Developed Mangroves
34		*Leptodius sanguineus*	–	X	–
35		*Neoxanthius michaelae*	–	X	–
36		*Neoxanthius* sp.	–	X	–
37	Calappidae	*Calappa lophos*	X	–	–
38		*Calappa capellonis*	X	–	–
39		*Calappa clypeata*	X	–	–
40	Matutidae	*Matuta victor*	X	–	–
41		*Matuta planipes*	X	–	–
42		*Ashtoret miersii*	X	–	–
43	Menippidae	*Menippe rumphii*	X	X	–
44		*Eriphia smithi*	–	X	–
45		*Sphaerozius nitidus*	–	X	–
46	Epialtidae	*Naxioides robillardi*	–	X	–
47		*Doclea muricata*	–	X	–
48		*Doclea canalifera*	–	X	–
49		*Doclea armata*	–	X	–
50	Galenidae	*Galene bispinosa*	X	–	–
51		*Halimede ochtodes*	–	X	–
52	Ocypodidae	*Uca triangularis*	–	–	X
53		*Uca lactea*	–	–	X
54	Varunidae	*Metaplax distincta*	–	–	X
55	Corystidae	*Jonas* sp.	X	–	–
56	Parthenopoidea	*Parthenope longimanus*	X	–	–
57	Pilumninae	*Actumnus setifer*	–	X	–
58	Dorippidae	*Dorippoides facchino*	X	–	–
59	Dromiidae	*Lauridromia dehaani*	X	–	–
60	Sesarmidae	*Muradium tetragonum*	–	–	X

X, present; –, absent.

8 MARINE MACROBENTHOS

8.1 Material and Methods

The present study has been undertaken in the coastal waters of Kalpakkam in the vicinity of MAPS. Teak wood (each $12 \times 9 \times 0.3$ cm) and metal (each $12 \times 9 \times 0.1$ cm) panels were suspended on epoxy-coated mild steel frames from the MAPS jetty where the water depth is ~8 m. In total, nine types of metals [copper based—admiralty brass, aluminum brass, copper, monel and cupro-nickel; noncopper based—SS-316, SS-304, mild steel (MS), and titanium] were used for this study. The panels were suspended at 1 m below the lowest low water mark, ~450 m away from the shoreline. Three series of observations (weekly, monthly, and cumulative at 30 d intervals) were made. Weekly and monthly observations were considered under short-term observations and cumulative was considered under long-term observation. Different parameters, namely species composition, population density, growth rate, both percentage of number and percentage of area coverage, and settlement biomass (g/100 cm^2) were used to study the settlement pattern of benthos. The fouling density was assessed by counting the organisms settled on the panels. The growth rate was observed by measuring the size of macrofoulers at different time intervals. In addition, seawater samples were collected for the estimation of hydrographical parameters such as temperature, pH, salinity, dissolved oxygen (DO), and turbidity and biological parameters such as chlorophyll-*a*. A detailed description of the methods is given elsewhere (Satpathy et al., 2010a). The data were grouped into three seasons.

8.2 Results and Discussion

8.2.1 Community Structure of Macrobenthos

A substantial temporal variation in settlement patterns was observed with respect to the composition of organisms, the number of organisms settled on the test panels, the growth rate of organisms, percentage of total number, percentage of area coverage, and biomass. The total number of taxa observed in Kalpakkam coastal waters was found to be about 115 during the present investigation. It is worthwhile to mention here that wide variations in the number of species (Balaji, 1988) have been reported from the east and west coasts of India. For example, 121 taxa from Visakhapatnam harbor, 37 taxa from Kakinada (Rao and Balaji, 1988), 85 taxa from Bombay waters (Venugopalan, 1987), and 42 and 65 taxa from Goa (Anil and Wagh, 1988) and Cochin harbor (Nair and Nair, 1987) respectively, have been reported. It is interesting to mention here that some earlier reports from this location studied during the same time of year showed a substantially different number of taxa (Sasikumar et al., 1989; Rajagopal et al., 1997). Direct comparison of above data is not justified for plausible reasons, such as differences in the exposure methodology, substratum, and level of competency in systematic identification.

Barnacles: Among the different groups, barnacles were found to be the most common and dominant community; their accumulation on the test panels was observed throughout the study period (Sahu et al., 2011). A similar kind of year-round barnacle settlement has also been reported previously from this location (Nair et al., 1988; Sasikumar et al., 1990). During the study period, barnacles were represented by four species—*Balanus amphitrite*, *B. tintinabulum*, *B. reticulatus*, and *B. variegates*—that were found to be the most dominant on weekly (12.4%–99%) as well as monthly (5.9%–85.2%) panels. Barnacle settlement was continuous with peaks during June-July and November-April. A similar pattern of barnacle settlement has also been reported from Kalpakkam coastal waters (Rajagopal et al., 1997; Nair et al., 1988). To assess the growth rate, rostro-carinal diameter measurements were made and the maximum size reached was 10–1 mm during our observations.

Hydroids: Hydroids were second only to barnacles in abundance as well as in seasonal occurrence and were dominated by *Obelia* sp. They started appearing on the panels after 5 d of immersion. A maximum length of 17 mm was observed. Their percentage contribution to total population varied between 0.64% and 81.62% during the study period. The peak settlement of this species was during July-August (premonsoon) and January-March (postmonsoon). The present observation is found to be similar with that of the earlier findings (Nair et al., 1988; Sasikumar et al., 1989). It is important to mention here that the scyphozoan jellyfish bloom and its ingress to the MAPS cooling water system is an occasional problem (Masilamani et al., 2000) at Kalpakkam. The jellyfish abundance depends on various factors such as availability of copepods and fish larvae, oxygen depletion, nutrient release, etc. (Brodeur et al., 2002; Hirose et al., 2009; Møller and Riisgård, 2007). A previous study at this locality has attributed jellyfish bloom to increased copepod population, abundance of fish larvae, and stable oceanic salinity conditions (Masilamani et al., 2000). Jellyfish as predators of zooplanktons, fish eggs, and larvae occasionally dominate the aquatic ecosystem, which increases the probability of high impingement (Lynam et al., 2006). In addition, the feeble swimming capacity and drifting with water current makes jellyfish vulnerable to impingement in power plant intake systems that draw huge quantities of water.

Ascidians: Ascidians are a very important group of organisms with worldwide geographical distribution (Whoi, 1952). *Didemnum psammathodes* and *Lissoclinum fragile* are the major ascidian species encountered during the present observation. The occurrence of ascidians was generally restricted to March-April and June-August, with peak settlement during March-April. Such dominance of ascidians during a certain period could be ascribed to increased larval density. The ascidians have the tendency to form a dormancy bag and when favorable conditions set in, the cells rebuild the tissues and develop into an adult ascidian. Such interaction of the breeding period of organisms in the development of benthic communities was reported elsewhere (Chalmer, 1982). A total absence of ascidians was encountered during September-December. A complete disappearance of ascidians during the monsoon period (June-September) was also reported from New Mangalore Port (west coast of India) (Khandeparker et al., 1995). The lack of settlement during the monsoon period was attributed to low salinity and increased suspended load (turbidity). Peak colonization of ascidians during February-June (Rajagopal et al., 1997) and April-July (Nair et al., 1988) is more or less comparable with the present findings.

Sea anemones: Sea anemones also form a prominent fraction of the benthic assemblages and were represented by *Sertularia* sp. and *Aiptasia* sp. in the present study. Their presence was recorded during weekly as well as monthly observations. Their settlement started from September-October onward and they were particularly abundant during the NE monsoon period (Masilamani et al., 2000). Their growth was 1.5–8 mm diameter during 7–30 d observation period. Settlement was less during the premonsoon period. The present observation with regard to the sea anemone settlement agrees with the earlier reports (Nair et al., 1988; Sasikumar et al., 1989; Rajagopal et al., 1997).

Green mussels: Green mussels (*Perna viridis*) are the most important constituent of the macrobenthic community in tropical waters (Masilamani et al., 2001). In the present study, green mussels were found as the climax community. This could be due to the fast growing and competitively superior nature of green mussels, which established dominance on panel surfaces so that most other fouling organisms are left with little space to settle (Masilamani et al., 2002a,b). A surprising finding of the present study is that, despite being the climax community and the most dominant species, green mussels completely disappeared during 2010. In the present study, excluding 2010, their percentage composition varied from 11.0% to 62.2% of the total macrobenthic population. Also, their colonization was generally observed during May-September with peak settlement during May-June and August-September, which continued until November. The first peak coincided with the seasonal temperature and salinity maxima of the present study. A maximum *P. viridis* settlement during relatively high temperature and salinity conditions was reported previously from this location (Rajagopal et al., 1997) and also from Kovalam and Ennore, a little north of Kalpakkam (Selvaraj, 1984). The second peak settlement of green mussels coincided with the phytoplankton density maxima and relatively high saline condition in August-September, which indicated the influence of food availability and salinity on mussel larvae abundance and settlement (Pieters et al., 1980; Newell et al., 1982; Paul, 1942).

Other organisms: Other groups of organisms included bryozoans (Ectoprocta), oysters, polychaete worms, flat worms, and some other crustaceans such as crabs (both larvae and juveniles), amphipods, and juvenile lobsters. The settlement pattern of bryozoans (Ectoprocta) did not show any definite trend in their temporal variation on short-term panels. The appearance of juvenile oysters (*Crassostrea madrasensis, Ostrea edulis*) was observed in almost all the months, with peak settlement during August. During the present study, a considerable contribution (~7%–20%) of oysters to the benthic community was noticed. However, in previous reports, the settlement of oysters was found to be very negligible from this locality (Rajagopal et al., 1997; Sasikumar et al., 1989; Nair et al., 1988). Though the availability of polychaete worms (*Serpula vermicularis, Hydroides norvegica*) (0.05%–2.1% monthly and 2%–56% cumulative) was observed during most of the study period, it showed the peak settlement in January. During cumulative observation (28 d), polychaete density was found to be relatively high, which gradually disappeared during subsequent observations. The settlement of flat worms was relatively less compared to the other organisms during short-term observations.

8.2.2 Population Density, Area Coverage, and Biomass of the Macrobenthic Community

Low-fouling density was observed during the northeast monsoon period (Fig. 27), whereas during the same period, maximum area coverage on the same panels was observed (Fig. 28). High area coverage during these months could be ascribed to a relatively high growth rate of organisms, which may be due to low fouling intensity and a fewer number of species leading to less competition for survival (Satpathy et al., 2010c). Therefore, a particular individual that colonizes during this period could grow very fast, as there was no crowding by others. The biomass of macrobenthos results showed relatively high values during premonsoon seasons

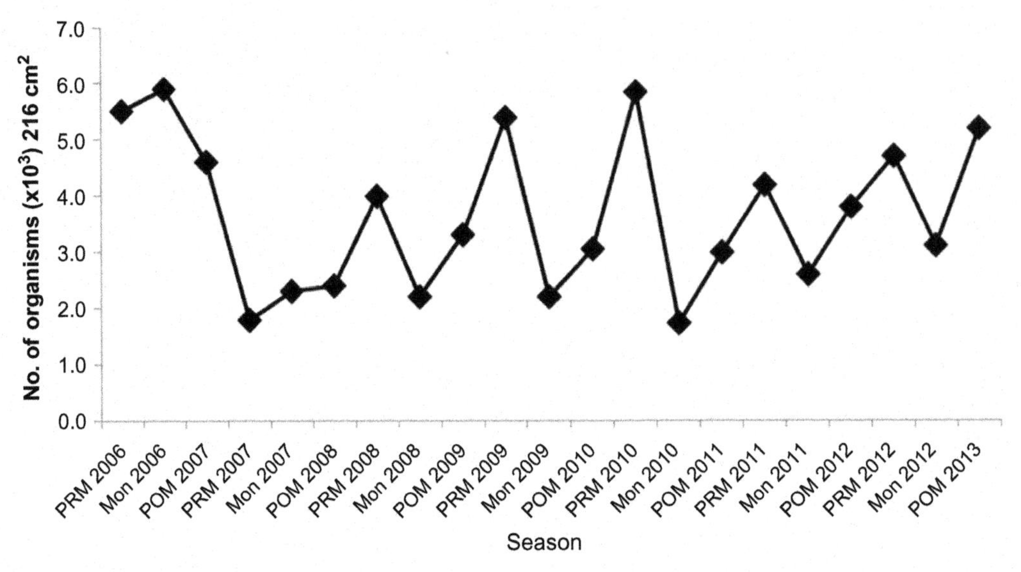

FIG. 27 Seasonal variations in macrobenthic population density.

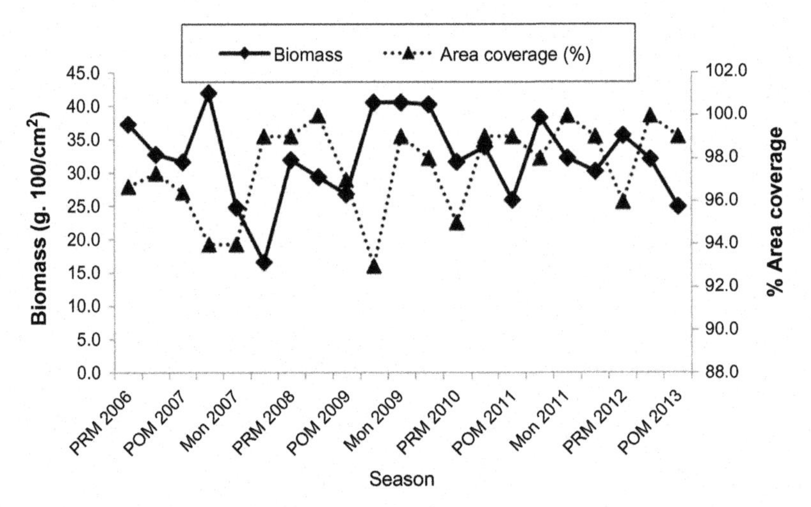

FIG. 28 Seasonal variations in macrobenthic biomass and area coverage.

and low values during postmonsoon seasons (Fig. 28). The premonsoon period (June-September) was found favorable for the recruitment of green mussels, which was reflected from the abundance of mussel larvae in the zooplankton population during the postmonsoon/summer period (April/May); they were the major contributors of macrobenthos biomass. In our study, we found that the bivalve (mostly mussel larvae) larvae were positively correlated with salinity and phytoplankton abundance. This may be the possible cause of mussel recruitment during the premonsoon period.

The percentage composition of barnacles showed a visible reduction during the long-term observation (Fig. 29A–D), which was due to the dominance of polychaete worms (56%) in 28 d and later on by the huge bulk of green mussels (41% in 56 d, 84% in 112 d, and 90% in 150 d). After 28 d, the settlement of juvenile green mussels occurred along with the preexisting community consisting of barnacles, hydroids, oysters, polychaete worms, flat worms, and sea anemones. On the 56 d panel, mussels attained 0.5–1 cm in size, whereas on 112 and 150 d, the panels were fully covered with adult green mussels of size 3–5 cm. (Fig. 29A–C). The relative abundance of the fouling community for 28, 56, 112, and 150 d observations are given in Fig. 29A–D.

8.2.3 Findings From Metal Panel Analyses

Results of metal panel analyses revealed that density and percentage of area coverage were found to be negligible on copper-based panels, whereas noncopper-based panels showed 100% area coverage and pronounced density; however, this was comparatively lower than wooden ones. Among noncopper-based panels, MS showed relatively less biomass (Fig. 30). This could be due to its corrosive nature, which leads to metal loss from the initial weight of the panel. Surprisingly, during the 2011 monsoon, the biomass of panels was found

FIG. 29 (A) View of weekly panel, (B) monthly, and (C) 112 days old panel exposed to coastal waters of Kalpakkam.

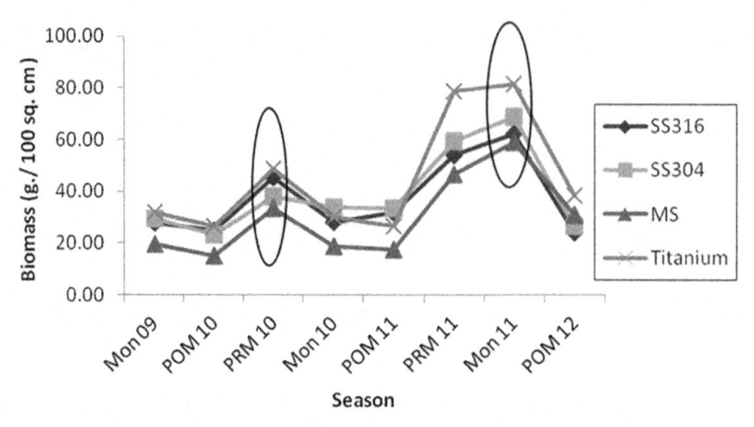

FIG. 30 Variations in biomass of fouling organisms on metal panels.

almost double that of premonsoon 2010 (Fig. 30), although the premonsoon period is considered as the principal recruitment period for most of the benthos (Balaji, 1988). This could be attributed to the heavy rainfall during premonsoon 2010 causing a lowering of salinity in the coastal waters, which led to a fall of the benthic population and total elimination of green mussels. As green mussels are the major contributors of panel biomass, its negligible appearance during premonsoon 2010 lead to a reduction in total biomass.

8.2.4 Larval Abundance in Relation to Macro-Benthic Community

Apart from the above studies, a study on larval abundance and their relation to macrobenthos settlements was also carried out in the coastal waters of Kalpakkam. Results showed that salinity, temperature, and food availability were the essential factors controlling the larval abundance as well as the settlement of its adults. The premonsoon period was found to be the most suitable period for larval growth, development, and survival to adult stages for most of the organisms. Among all the major groups, bivalves were found to establish a good relationship between larvae and adults (Table 8). Cluster analyses indicated a good association of bivalve larvae with polychaete larvae, whereas in the macrobenthic assemblage, the biotic relation of bivalves was established with barnacles. Biotic interactions

TABLE 8 Regression Analysis Between the Larval Supply and Their Recruitment Observed on Panels

Organism	Regression Coefficient (R^2)
Barnacle	0.032
Bivalve	0.453
Polychaete worm	0.065
Ascidian	0.054
Bryozoan	0.078

between ascidians and bryozoans were found throughout the study period. Principal component analysis further supported the above observations of associations between different larval and adult groups (Fig. 31). Results of the present study, the first of its kind from this locality, indicate that variations in larval abundance are likely to play a significant role in the formation and development of benthic communities. Studies including breeding patterns, larval life cycles, transportation, and succession capability of the fouling community have much importance for this location, as seawater is being used by various installations and thus needs investigation to understand the above phenomena.

9 IMPACT OF THERMAL DISCHARGE ON ABUNDANCE OF WEDGE CLAM *DONAX CUNEATUS* IN THE VICINITY OF MAPS

9.1 Sampling Design

The sampling plan was devised so as to make a comparison of *Donax* populations at different spatial intervals along the coast from a heated effluent outlet and to determine the impact boundary. Twenty stations were selected both on the south and north sides of the mixing zone, 12 locations (S0, S10, S20, S40, S80, S100, S 200, S400, S600, S800, S1000, and S2000) were selected on south side at a distance of 0 (near mixing point), 10, 20, 40, 80, 100, 200, 400, 800, 1000, and 2000 m, respectively, and similarly eight locations (N10, N20, N40, N80, N100, C3-N500, C2-N1000, and C1-N5500) were selected on the north from the effluent outlet (Fig. 32). The present study was conducted during January 2008. More locations were chosen on the south side of the discharge point, as during October-February the heated water plume shows southerly movement (Satpathy et al., 1986). Locations C1-C3 were considered as control stations because they are located in the opposite direction to plume flow and spatially far away from the outfall area. Underwood (1994) and Lardicci et al. (1999) have indicated the importance of multiple control stations while assessing the impact of power plant effluents on benthic macroinvertebrates. Hence, three control sites were selected in the present study. Moreover, the random comparison between control and impacted location may not provide the zone of influence. Hence, the study was designed to assess the impact of the costal power plant effluent on the population size of the intertidal bivalve (*D. cuneatus*) at different spatial scales. A preliminary investigation by Suresh et al. (1992) has indicated that the nature of impact was local; hence, more locations on smaller spatial scales were adopted in this study. The most common approaches to assess the impact of human activities in an ecological process in experimental fashion are related to the family of general linear models such as analysis of variance (ANOVA), which are flexible, robust, and powerful hypothesis-testing procedures (Downes et al., 2002).

9.2 Methods

Donax cuneatus was collected quantitatively from the mid-water mark of the intertidal area at different stations. At each location, three replicates of 1 m^2 sand samples were excavated up to 30 cm depth, and the sand was sieved on a 1 mm screen. All wedge clams were transferred to the laboratory immediately for further investigation. Water temperature was measured by

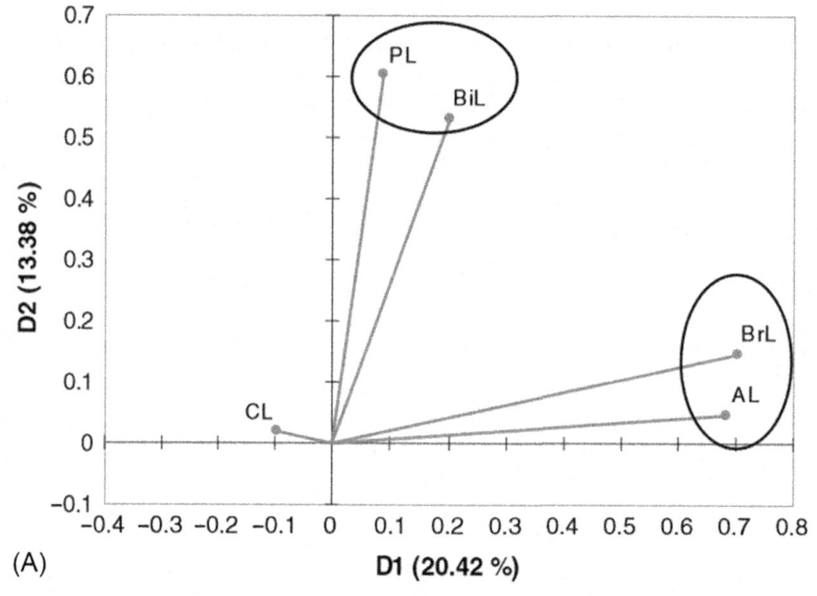

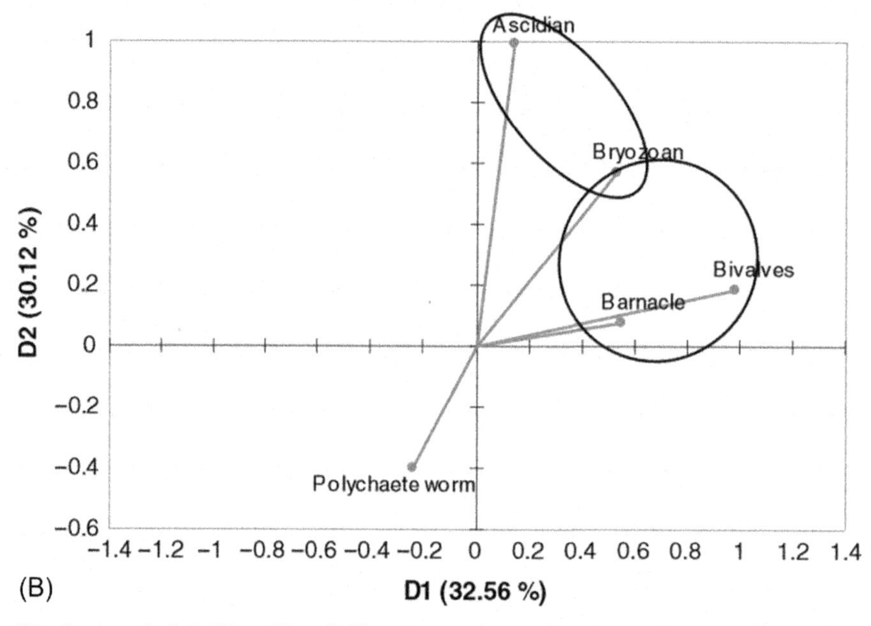

FIG. 31 Distribution of adult (A) and larval (B) groups on factor plane (*PL*, Polychaete larva; *BiL*, Bivalve larva; *BrL*, Bryozoan larva; *AL*, Ascidian Larva; *CL*, Cirripede larva).

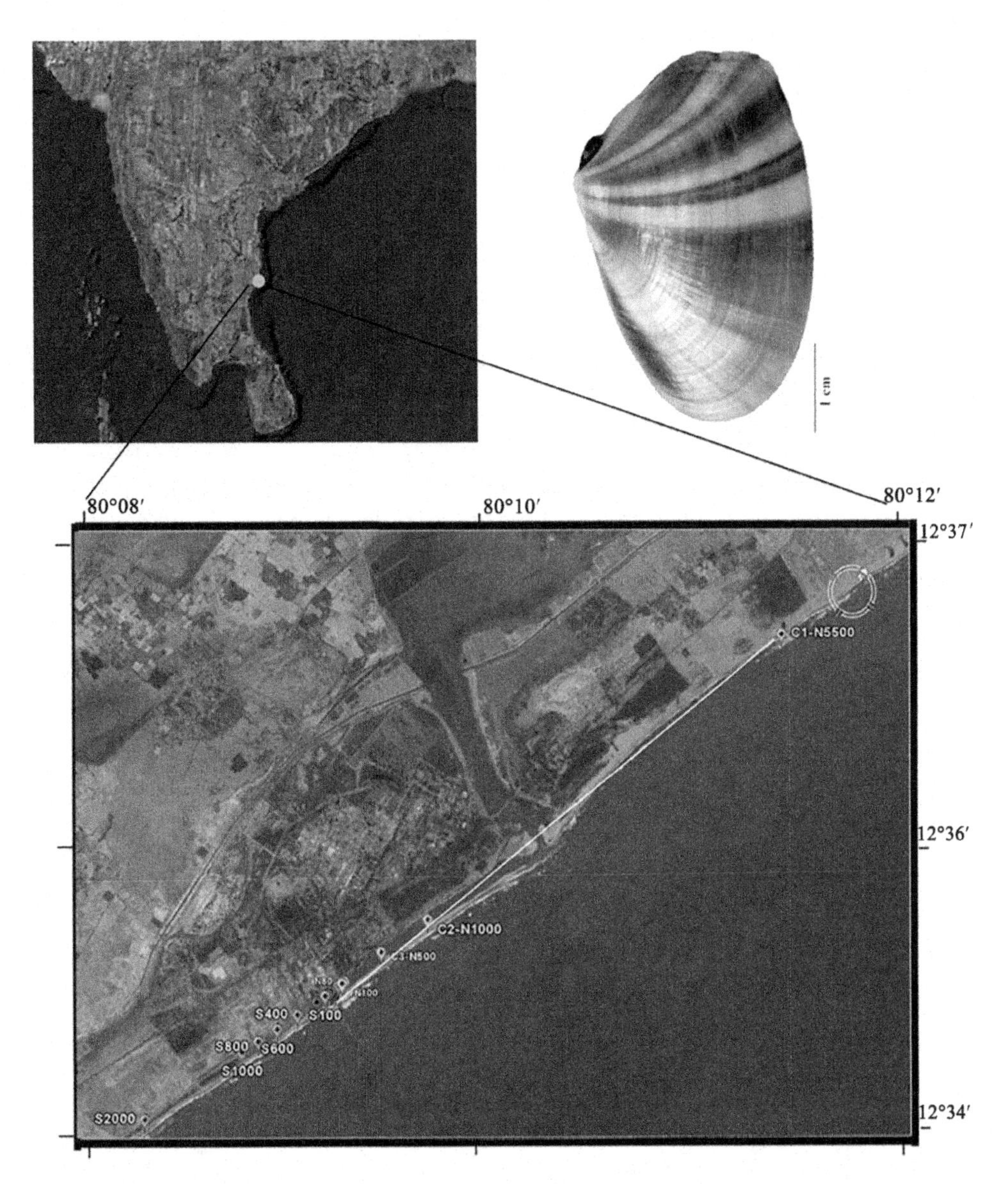

FIG. 32 Study area map showing the sampling locations and the organisms (*Donax cuneatus*) selected for the present study.

a mercury thermometer with an accuracy of ±0.1°C. Sediment samples were collected in triplicate in the form of 30cm cores at the same points. Sediment was sorted into seven grades using a series of mesh sieves (2, 1, 0.5, 0.250, 0.125, 0.0625mm), following the Wentworth scale (Buchanan, 1984). Sediments were categorized based on grain size following the method of Gray (1981). The total organic carbon (TOC) of sediments was analyzed in a computer-controlled TOC solid sample analyzer (Shimadzu-TOC-VCPH/CPN, Japan). Differences in the abundance pattern among 20 locations were compared with the analysis of variance, employing Tukey's pair-wise comparison. To increase the power of analysis, a square root transformation of all data was performed. Normality and homogeneity of variance were tested before executing the analysis. Similarly, multivariate clustering was carried out among stations for their similarities in habitat variability. Parameters such as water temperature, substrate characteristics, beach slope, and total organic content were taken into account.

9.3 Results and Discussion

Variations in shore water temperature at different sampling locations are given in Fig. 33. Near the effluent mixing zone, the beach was submerged in water and the swash zone was totally absent due to its steepness. Therefore, *D. cuneatus* sampling was not possible at this location but all other abiotic parameters were taken into account to know the spatial change in the physicochemical environment. The *D. cuneatus* population on the swash zone ranged from 1.3±1.5 to 88.3±9.6. On the south side of the mixing zone, the abundance of clams increased with increasing distance (Fig. 34). A meager population of wedge clams was observed up to 100m (S100) south from the mixing point and the abundance gradually increased at different spatial scales, which stabilized at S400 and reached a maximum at S1000 (64.0±3.6). The abundance pattern was same on north side too but less abundance was observed up to 80m (N80). Maximum abundance was observed at control location C3-N500 (88.3±9.6) while the other two control locations C1-N5500 (51±3.6) and C2-N1000 (52±6.5) showed relatively fewer numbers.

ANOVA revealed that a significant difference ($P < .01$) existed among stations located near the mixing zone (S10 to S200 south and N10 to N80 north) and stations away from the mixing zone (S400 to S2000 south and N100 to C1-N5500) with reference to the population size of

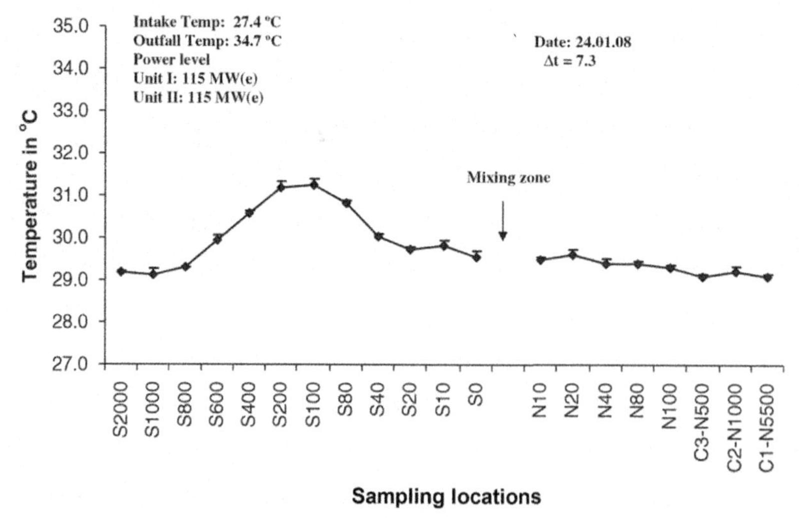

FIG. 33 Variations in shore water temperature from mixing zone with southward current direction (S0–S2000: Sampling locations towards south of mixing point at 0–2000 m, respectively; N10–N5500: Sampling locations towards north of mixing point at 0–5500 m, respectively).

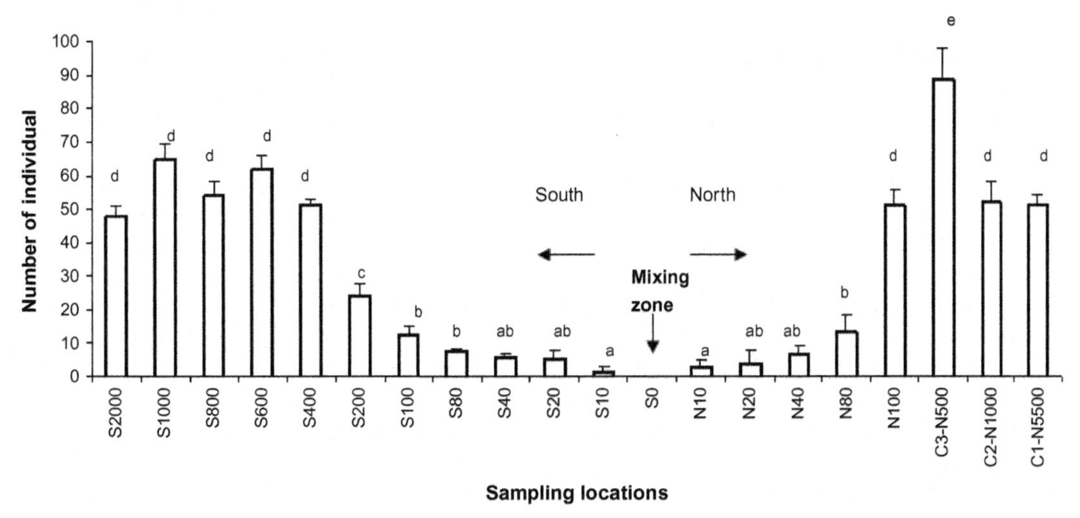

FIG. 34 Abundance of *D. cuneatus* at different spatial scale away from mixing zone (Bars followed by same letter(s) are not statistically significant at $P < 0.05$).

D. cuneatus. Interestingly, stations located after S400 and N100 were not statistically significant with the two control stations (C1-N5500 and C2-N1000). Moreover, control station C1-N5500 was distinct from all other stations with reference to abundance. The above statistics showed that the primary impact was not more than 400m south and 100m north from the mixing zone. Tukey's pairwise comparison not only revealed the difference among the control and the impacted site, but it also indicated the impact boundary. From the mixing zone, 40m on either side was highly impacted, 80–100m toward the plume flow (south) was moderately impacted, and 80m north of the mixing point also witnessed moderate impact. Past 100m (N100) north was not affected by effluents whereas, the area between 100 and 400m south was influenced marginally. A preliminary investigation made by Suresh et al. (1992) in the vicinity of MAPS observed low densities (15–27no./m^2) of *D. cuneatus* at 0.5 and 1km north of the mixing zone when the plume flow was toward the north. Even though the general pattern of increased density from the discharge point is in agreement with the present study, the impact boundary observed presently was very much less compared to the previous findings. This could be due to characteristics of the costal current, as during the NE monsoon (southerly current) period, the current is not strong as compared to that of the SW monsoon (northerly current) (Satpathy et al., 1986). Strong currents may carry heated plumes to longer distances and extend the impact boundary. Moreover, the ambient temperature was also comparatively high during the study of Suresh et al. (1992). Similarly, Hameed and coworkers studied the Kalpakkam coast during 2000–02 and they have reported that the shore attached the movement of the thermal plume. According to them, the effluent impact is only on intertidal benthic populations and has no bearing on the sublittoral macrobenthic density (Hameed, 2005; Hameed et al., 2007). They have reported that the intertidal benthic animals are absent in the mixing zone and their density decreased up to 500m on either side of the mixing zone, but they have not explained the influence of environmental variables on density and macrobenthic assemblages occurring during their study. Moreover, each taxon has a different tolerance to pollution and hence studies on species level might help in identifying the indicator species prevailing in the particular region.

In the present study, the minimum population of *D. cuneatus* was witnessed between 0 and 200 m south and 80 m north from the mixing zone and the population recovered at 400 m south of the mixing zone, which is a fairly short distance when compared to previous findings. In this context it, is interesting to mention that Hameed has reported the impact of heated effluent up to 9.0 km north of the mixing point when the plume flow was toward the north (SW monsoon). This appears to be quite unlikely unless supported by logical evidence along with all physical and chemical parameters because macrobenthos density is governed by synergistic interactions of the environment. Because population density of the organism shows a complex pattern to synergistic effects, it is better to consider other environmental variables apart from thermal stress alone. The statistical difference among control stations was produced by the variability in microclimate and the absence of stress. Station C3-N500 was the most abundant station, which is statistically different from all others. It could be due to favorable environmental conditions such as easily penetrable sand and a favorable swash climate. Moreover, station C3-N500 was totally free from human disturbance because it is a prohibited area falling under the Department of Atomic Energy (DAE) campus. On the other hand, abundance at station S2000 far away from the impact zone was marginally less because of the *Donax* harvest by coastal fishers living near this area. A comparatively high population of *D. cuneatus* was encountered at stations C3-N500, S600, S800, and S1000 where 0.12–0.25 mm sand size was dominant, which might facilitate better burrowing and thereby reduce the predation pressure, as has been reported by Nel et al. (2001) for another species of *Donax*. Moreover, medium and fine sands usually have an abundant meiofauna and macrofauna because of more organic matter per unit area (Gray, 1981). In agreement with the above statement, the *Donax* population was high at stations S600 to S1000 where medium and fine sand were dominant and TOC content was comparatively high. In contrast, abundance was relatively less at sites S10, S20, and S100 in spite of high TOC content. This disparity could be due to the influence of other parameters such as sediment characteristics, slope, and temperature. The abundance was low at stations S40 to S200 where the mean water temperature was comparatively high. However, when all the observations were taken into consideration, no clear correlation could be observed between population density and temperature. A similar pattern in population distribution was also observed by Hameed et al. (2007) for macrobenthos where they correlated the density with temperature. In general, multivariate clustering analysis based on the environmental characteristics and population trends of *D. cuneatus* showed similarity. This showed that the microclimate prevailing at those particular locations is structuring the *D. cuneatus* population. Hence, the abundance pattern of *D. cuneatus* on the sandy beaches of the Kalpakkam coast is governed by multiple interacting factors.

10 OBSERVATION OF OXYGEN MINIMUM ZONE IN THE SHALLOW REGIONS OFF THE TAMIL NADU COAST

Oxygen depletion in the subsurface waters and the formation of hypoxic/anoxic systems have expanded significantly in coastal systems around the world over the last few decades (Diaz and Rosenberg, 2008). Depletion in dissolved oxygen concentration in the water column leads to hypoxic ($O_2 \leq 2$ mg/L) and/or anoxic ($O_2 \leq 0$ mg/L) conditions and the further formation of oxygen minimum zones (OMZ) in the coastal areas. Oxygen deficiency in the shelf region could be a critical determinant for fisheries and ecological and biogeochemical processes (Childress and Seibel, 1998), along with the economic condition of the region. The rise or expansion of hypoxia and anoxia at par with urbanization and industrialization indicates a major perturbation to the structure and functioning of coastal marine ecosystems. Here we report the vertical expansion of hypoxia within the OMZ along the outer shelf and slope region and the emergence of an inner-shelf hypoxia that was not apparent in the southwestern Bay of Bengal (BoB).

Off BoB, hypoxic conditions are persistent on the outer shelf beyond 100 m and the OMZ ($O_2 < 0.7$ mg/L) is present from 150 m to about 600 m (Helly and Levin, 2004; Rao et al., 1994). Minimum oxygen concentrations within the Indian ocean OMZ are generally deeper (~800 m) than the Atlantic and Pacific Oceans (Stramma et al., 2008). Long-term data (1960–90) on the Indian Ocean, the Arabian Sea (AS), and BoB shows constancy in the OMZ over the past few decades (Stramma et al., 2008; Sarma, 2002) (Table 9). However, shelf water column hypoxia and anoxia were not prevalent in BoB.

BoB sustains a strong potential for impacts from riverine nutrient loads due to the very high nutrient yields in its catchment basins, for example, via the Ganges/Brahmaputra, Godavari, and Mahanadi rivers (Rabalais et al., 2010). It is the most open of all the systems receiving high nutrient inputs, with no physical barriers separating its coastal zone from the open ocean. During the southwest monsoon (July-September) period, isopycnals from depths up to about 70 m surfaced due to upwelling forced by local winds and the geostrophic velocity in the upwelling band is in the direction of the winds. The residence time of the intermediate

TABLE 9 Summary of Previously Reported Hypoxia and Anoxia and Their Proximity Within the Southern Bay of Bengal Spreading Central and Western Parts

Area in BoB	Hypoxic Depth	Anoxic Depth	Year	Observed OMZ[a]	References
8–20°N, 80–100°E (15°N, 90°E)	90–200 m	200–490 m	1893–2004	180–490 m	Paulmier and Ruiz-Pino (2009)
West and Central	91–104 m	104–389 m	1906–1990	91–582 m	Helly and Levin (2004)
6–12°N, 80–98°E	100 m	–	1994–2001	100–1000 m	Sarma (2002)
Central (20°N) Western (12–20°N; 81–85°E)	– 150–300 m	150–700 m	2001	150–500 m	Fernandes and Ramaiah (2009)
Central (11–15°N)	100–120 m	120–500 m	2001–2003	120–600 m	Sardessai et al. (2007)
11–13°N, 79–80°E	Beyond 24 m (inner shelf region) 59 m (outer shelf and slope region)	– 98 m (slope)	Sept. 2010 July 2010	55–215 m and beyond	Present investigation

The oxygen minimum zones encountered in the previous observations are presented for a comparison with the present oxygen scenario within the southwestern Bay of Bengal.
[a]Limit of OMZ: $O_2 \leq 0.7$ mg/l (0.5ml/l or 22 μmol/l)[3]; Hypoxic depth ($0.14 < O_2 \leq 2$ mg/l); Anoxic depth ($O_2 = 0$ mg/l).

waters or the OMZ (100–1000 m) of the bay is 12 years (Sarma, 2002). The southern and northern regions of the bay have distinct mixed layer depths (MLD) in their western boundaries that shoaled from 25 m at 12°N to ~2 m at 19°N. The doming of isotherm has been recorded around 17°N, where a cold-core eddy was evident. The Tamil Nadu coast has a wide continental shelf area that spreads over 22,411 km^2 with a depth of 50 m and 11,205 km^2 with a depth of 51–200 m.

During a recent expedition (September 2010) when the southwest monsoon wind was apparently stronger, we measured a vertical expansion of hypoxia ($0.14 < O_2 \leq 2$ mg/L) at a minimum depth of 59 m and further anoxia ($O_2 = 0$ mg/L) at a depth of ~98 m (Fig. 35) within 30 km off the Tamil Nadu coast. The onset of shallow water (59 m) hypoxia within the OMZ was associated with a sporadic event of inner-shelf hypoxia (24 m) in a broad section of the southern Indian east coast. However, earlier reports only explained the hypoxia and further OMZs at the lowest depth of 91 m in the continental slope and ridge (Helly and Levin, 2004; Sarma, 2002). In the present investigation, hypoxia prevailed in several places across the area, extending between 11–13°N and 79–80°E, and spreading from the shelf break to the inner shelf encompassing ~8000 km^2 (Satpathy et al., 2013). We presumed the phenomenon of hypoxia was not persistent and was spreading vertically within the OMZ of BoB due to the local wind-driven upwelling and other physical forcings. In order to confirm further the upwelling driven hypoxia, which used to be a seasonal occurrence, data from the year-round observations of the research cruises conducted by the National Institute of Ocean Technology, the Ministry of Earth Sciences, India, were analyzed. It is detected that a shoreward pushing of the upwelled low oxygen bottom water caused a severe hypoxic condition in the inner shelf

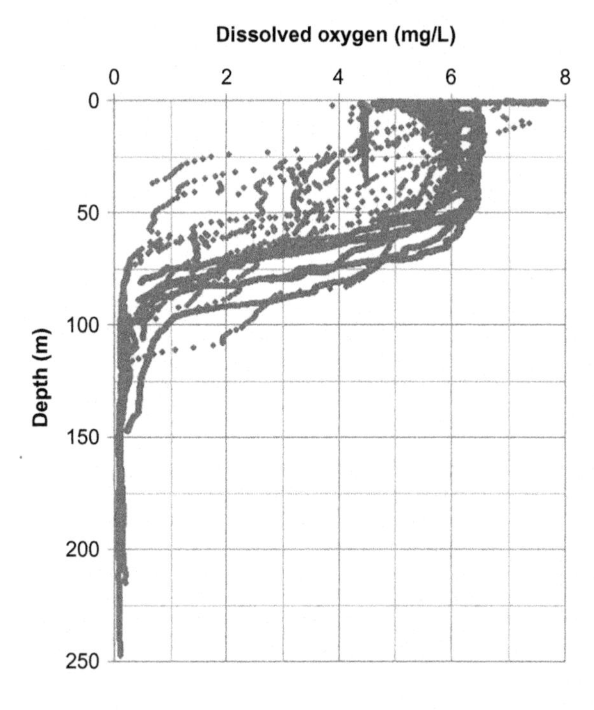

FIG. 35 Vertical distribution of oxygen profile in south western Bay of Bengal, (11–13° N; 79–80° E) up to 250 m depth during 2010.

region. Interestingly, the observations indicated a year-round hypoxic event in the inner shelf region, which was evident even in the absence of a strong local wind that causes upwelling. The year-round observations during 2010 revealed that the phenomenon was exceptional in its vertical and temporal extent, persisting over the shelf area during February, July, and September. Even though intense hypoxia is a permanent feature of the OMZ that intersects the continental slope (>800 m in this system), possibly there are no prior records of hypoxia over the continental shelf (February, July, and September 2010) and severe anoxia in the OMZ for this marine ecosystem.

11 CONCLUSION

A significant difference in the physicochemical characteristics of coastal waters was observed during the monsoon season as compared to the pre- and postmonsoon periods. A relatively high concentration of all the nutrients and DO observed during the NE monsoon season showed that coastal water at this location is influenced by freshwater input for a brief period. The above observation was supported by statistical analysis, which showed two temporally distinct water masses: one during the monsoon period and the other during the pre- and postmonsoon periods. Distinct environmental states of coastal waters such as nitrogen enrichment, phytoplankton growth, phosphate enrichment, low dissolved oxygen, and high temperature condition and nitrite enrichment were observed from the PCA, which might have existed at this location on a temporal scale. Cluster analysis also depicted the existence of a well-marked spatial heterogeneity in the study area.

The phytoplankton community structure and density at the Kalpakkam coastal ecosystem was characterized by high species richness and abundance. The community was dominated by diatoms during the SWM and intermonsoon season while the NEM season was dominated by the cyanobacteria population. Nutrient stoichiometry played an important role in the distribution of planktonic populations spatially and temporally. Zooplankton studies from this part of SW-BOB, that is, from Kalpakkam costal waters, revealed that salinity and phytoplankton were the chief controlling parameters for zooplankton growth and proliferation, whereas temperature had a very negligible role for the same. Zooplankton taxa showed seasonal specificity for their prevalence in the present milieu where they were found as the main regulator of food, either as grazers or as prey.

The study also reinforces the concept that interannual climate and hydrodynamic variations might have a strong influence on coastal fish communities and, consequently, on the recruitment of coastal fish populations. The study showed that coastal fish assemblage on the southeast coast of India is influenced by the monsoonal pattern. Though the present study would form a baseline for ecological management purposes, future studies in this regard with respect to different monsoon periods will yield a clear picture on fish diversity patterns at this location. Kalpakkam is found to be represented by a high number of crab species. The rich diversity of brachyuran crabs from the Kalpakkam coast shows the benign nature of the coastal water and sediment. At the same time, frequent observation of commercial crab species at the coast implies that the presence of industrial establishments for more than 30 years in the region has not affected the local fisheries.

Studies on the settlement pattern of macrobenthos in the coastal waters of Kalpakkam revealed a wide variation in their colonization. A shift in the peak settlement period of green mussels (*Perna viridis*) compared to previous studies from this location was also observed. Interactions among the macrobenthic community showed association of bivalves with barnacles. Similarly, in a larval pool, the cooccurrence of bivalves and polychaetes was observed. The study also showed that bivalves are the only group that established a successful relationship between its larvae and adults. Other than bivalves, no other larval dispersal was found to contribute to their adult settlement. Hence, to get more realistic estimates, better measurements of settlement and larval supply are essential, making in-depth knowledge about recruitment dynamics necessary.

The present study showed that the effluent from the Madras Atomic Power Plant has a local effect that covers ~100 m both south and north of the discharge point. The area after 100 m north was not affected by effluents, whereas there was a little impact in the zone between 100 and 400 m south. Even though *D. cuneatus* showed a clear pattern of increase in abundance with an increase in the distance from the mixing zone, a direct correlation between temperature and population density could not be established. Hence it can be concluded that the abundance pattern of *D. cuneatus* on the sandy beaches of Kalpakkam is not governed by a single major factor but is due to the results of multiple interacting factors. Further in-depth investigation is required to estimate the impact zone at either side during different seasons.

It appears that the development of the present coastal hypoxia in BoB is a result of enhanced terrestrial nutrient loading as reflected in the sharp increase in their concentration, which in turn could have promoted microbial activity resulting in elevated respiration. This resulted in greater demand for oxygen leading to a depletion of dissolved oxygen, which further gets adversely affected as the water column stratifies. The rate of exchange between coastal and open water masses and the addition of highly nutrient enriched river discharge from catchment areas results in persistent hypoxia. However, in open water circulation systems (BoB), with increasing eutrophication, hypoxia might become a common phenomenon in coastal areas. BoB is highly susceptible because it receives large river runoff and is located close to regions of high population density as well as intense agriculture. Irrespective of a seasonal prototype emergence of hypoxia and anoxia in the AS, BoB witnessed year-round hypoxic and anoxic zones in the present study area that were not registered in the earlier expeditions.

References

Akin, S., Winemiller, K.O., Gelwick, F.P., 2003. Seasonal and spatial variations in fish and macrocrustacean assemblage structure in Mad Island Marsh Estuary, Texas. Estuar. Coast. Shelf Sci. 57 (1–2), 269–282.

Al-Nahdi, A., Al-Marzouqi, A., Al-Rasadi, E., Groeneveld, J.C., 2009. The size composition, reproductive biology, age and growth of largehead cutlassfish *Trichiurus lepturus* Linnaeus from the Arabian Sea coast of Oman. Indian J. Fish. 56 (2), 73–79.

Anil, A.C., Wagh, A.B., 1988. Aspects of biofouling community development in the Zuary estuary, Goa, India. In: Thompson, M.P., Sarojini, R., Nagabhushanam, R. (Eds.), Marine Biodeterioration: Advanced Technique Applicable to the Indian Ocean. Oxford & IBH publishing (P) Ltd, New Delhi, India, pp. 529–537.

Ansari, Z.A., Chatterji, A., Ingole, B.S., Sreepada, R.A., Rivonkar, C.U., Parulekar, A.H., 1995. Community structure and seasonal variation of an inshore demersal fish community at Goa, West Coast of India. Estuar. Coast. Shelf Sci. 41, 593–610.

Ansari, Z.A., Sreepada, R.A., Dalal, S.G., Ingole, B.S., Chatterji, A., 2003. Environmental influences on the trawl catches in a bay-estuarine system of Goa, west coast of India. Estuar. Coast. Shelf Sci. 56, 503–515.

Ansell, A.D., 1983. The biology of the genus *Donax*. In: McLachlan, A., Erasmus, T., Junk, W. (Eds.), Developments in Hydrobiology, vol. 19. Sandy Beaches as Ecosystems. Dr. W. Junk Publ., The Hague, The Netherlands, pp. 607–635.

Aston, S.R., 1980. Nutrients dissolved gasses and general biochemistry in estuaries. In: Olausson, E., Cato, I. (Eds.), Chemistry and Biogeochemistry of Estuaries. John Wiley & Sons, New York, pp. 233–262.

Azariah, J., 1999. Biopiracy, Environment and Culture.Paper presented at international conference on Genetics, Law and Society Conference, Saint Paul, Minnesota, Oct. 11–14.

Balaji, M., 1988. Investigations on Biofouling at Two Ports in Andhra Pradesh, India and Some Aspects of Toxicity of Copper to the Fouling Bivalve *Mytilopsis sallei* (Recluz) (Ph.D. thesis). Andhra University, India, pp. 1–209.

Bel Hassen, M., Drira, Z., Hamza, A., Ayadi, H., Akrout, F., Messaoudi, S., Issaoui, H., Aleya, L., Bouain, A., 2009. Phytoplankton dynamics related to water mass properties in the Gulf of Gabes: ecological implications. J. Mar. Syst. 75, 216–226.

Bharathi, C.H., Sandeep, B.V., Rao, S., 2001. The effect of mercuric chloride on the respiration of marine intertidal bivalve *Donax cuneata*. J. Pollut. Res. 20, 5–7.

Biswas, S., Mishra, S.S., Satpathy, K.K., Selvanayagam, M., 2010a. First record of stargazer snake eel *Brachysomophis cirrocheilos* (Osteichthyes: Ophichthidae) from India. Mar. Biodivers. Rec. 3 (e85), 1–3.

Biswas, S., Mishra, S.S., Satpathy, K.K., Selvanayagam, M., 2010b. Discovery of *Torquigener brevipinnis* (Osteichthyes: Tetraodontidae) from the Indian coast. Mar. Biodivers. Rec. 3 (e123), 1–4.

Biswas, S., Mishra, S.S., Das, N.P.I., Nayak, L., Selvanayagam, M., Satpathy, K.K., 2012a. First record of eleven reef inhabiting fishes from Tamil Nadu coast of India, Bay of Bengal. Proc. Zool. Soc. 65 (2), 105–113.

Biswas, S., Mishra, S.S., Das, N.P.I., Selvanayagam, M., Nayak, L., Satpathy, K.K., 2012b. New records of four reef-associated fishes from East coast of India. Acta Ichthyol. Piscat. 42 (3), 253–258.

Biswas, S., Mishra, S.S., Satpathy, K.K., Das, N.P.I., Selvanayagam, M., Nayak, L., 2012c. A new record of a garden eel, *Heteroconger tomberua* (Actinopterygii: Anguilliformes: Congridae), from the Indian Ocean. Acta Ichthyol. Piscat. 42 (1), 65–68.

Biswas, S., Mishra, S.S., Das, N.P.I., Satpathy, K.K., Nayak, L., Selvanayagam, M., 2013. First record and range-extension of Bigscale Jawfish, *Opistognathus macrolepis* (Perciformes: Opistognathidae), from Indian Ocean. Mar. Biodivers. Rec. 6 (e8), 1–4.

Biswas, S., Jahir Hussain, K., Das, N.P.I., Russell, B.C., Satpathy, K.K., Mishra, S.S., 2014. Imprint of monsoonal patterns on the fish assemblage in coastal waters of south-east India: a case study. J. Fish Biol. 85(3). https://dx.doi.org/10.1111/jfb.12461.

Blaber, S.J.M., Brewer, D.T., Salini, J.I., Kerr, J., 1990. Biomasses, catch rates and abundances of demersal fishes, particularly predators of prawns, in a tropical bay in the Gulf of Carpentaria, Australia. Mar. Biol. 107, 397–408.

Brzezinski, M.A., 1985. The Si:C:N ratio of marine diatoms: interspecific variability and the effect of some environmental variables. J. Phycol. 21, 347–357.

Brodeur, R.D., Sugisaki, H., Hunt Jr., G.L., 2002. Increases in jellyfish biomass in the Bering Sea: implications for the ecosystem. Mar. Ecol. Prog. Ser. 233, 89–103.

Buchanan, J.B., 1984. Sediment analysis. In: Holme, N.A., McIntyre, A.D. (Eds.), Methods for the Study of Marine Benthos. Blackwell, Oxford, pp. 41–63.

Bustillos-Guzmán, J., Claustre, H., Marty, J., 1995. Specific phytoplankton signatures and their relationship to hydrographic conditions in the coastal northwestern Mediterranean Sea. Mar. Ecol. Progress Ser. 124, 247–258.

Carmelo, R., 1996. Identifying Marine Diatoms and Dinoflagellates. Academic Press Inc., New York, 598 pp.

Chalmer, P.N., 1982. Settlement patterns of species in a marine fouling community and some mechanisms of succession. J. Exp. Mar. Biol. Ecol. 58, 73–86.

Chandran, R., Ramamoorthi, K., 1984. Hydrobiological studies in the gradient zone of Vellar Estuary 1-Physico-chemical parameters. Mahasagar. Bull. Natl. Inst. Oceanogr. 17, 69–77.

Childress, J.J., Seibel, B.A., 1998. Life at stable low oxygen: adaptations of animals to oceanic oxygen minimum layers. J. Exp. Biol. 201, 1223–1232.

Choudhury, S.B., Panigrahy, R.C., 1991. Seasonal distribution and behavior of nutrients in the creek and Coastal waters of Gopalpur, East Coast of India. Mahasagar. Bull. Natl. Inst. Oceanogr. 24, 81–88.

Claridge, I.N., Potter, I.C., Hardisty, M.W., 1986. Seasonal changes in movements, abundance, size, composition and diversity of the fish fauna of the Severn Estuary. J. Mar. Biol. Assoc. U. K. 66, 229–258.

Cole, C.V., Sanford, R.L., 1989. Biological aspects of the phosphorus cycle. In: Proceedings of a Symposium on Phosphorous Requirements for Sustainable Agriculture in Asia and Oceania, 6–10 March, SCOPE/UNEP.

Conley, D.J., 2000. Biogeochemical nutrient cycles and nutrient management strategies. Hydrobiologia 410, 87–96.

Conway, D.V.P., White, R.G., Hugues-Dit-Ciles, J., Gallienne, C.P., Robins, D.B., 2003. Guide to the Coastal and Surface Zooplankton of the South-Western Indian Ocean. Occasional Publication No. 15. Marine Biological Association, UK, 354 pp.

Crema, R., Bonvicini, A.M., 1981. The structure of benthic communities in an area of thermal discharge from a coastal power station. Mar. Pollut. Bull. 11, 221–224.

Day, F., 1878. The Fishes of India, Being a Natural History of the Fishes Known to Inhabit the Seas and Freshwaters of India, Burma and Ceylon, vol. I & II. Bernard Quaritch, London.

De Souza, S.N., 1983. Study on the behaviour of nutrients in the Mandovi estuary during premonsoon. Estuar. Coast. Shelf Sci. 16, 299–308.

Desikachary, T.V., 1987. Atlas of Diatoms, Fascicles—III & IV. Madras Science Foundation, Madras, 222–400 Plates.

Devassy, V.P., Goes, J.I., 1988. Phytoplankton community structure and succession in tropical estuarine complex. Estuar. Coast. Shelf Sci. 27, 671–685.

Dhar, O.N., Rakhecha, P.R., 1983. Fore shadowing northeast monsoon rainfall over Tamil Nadu, India. Mon. Weather Rev. 111 (1), 109–112.

Diaz, R.J., Rosenberg, R., 2008. Spreading dead zones and consequences for marine ecosystems. Science 321, 926–929.

Downes, B.J., Barmuta, L.A., Fairweather, P.G., Faith, D.P., Keough, M.J., Lake, P.S., Mapstone, B.D., Quinn, G.P., 2002. Monitoring Ecological Impacts: Concepts and Practice in Flowing Waters. Cambridge University Press, Cambridge, UK.

Dubravko, J., Rabalais, N.N., Turner, R.E., Dortch, Q., 1995. Changes in nutrient structure of river-dominated coastal waters: stoichiometric nutrient balance and its consequences. Estuar. Coast. Shelf Sci. 40, 339–356.

Fernandes, V., Ramaiah, N., 2009. Mesozooplankton community in the Bay of Bengal (India): spatial variability during the summer monsoon. Aquat. Ecol. 43 (4), 951–963.

Ganapati, P.N., 1975. Estuarine Pollution. Bulletin of the Department of Marine Sciences, University of Cochin, VII-1-9.

Garnier, J., Billen, G., Palfner, L., 2000. Understanding the oxygen budget and related ecological processes in the river Mosel: the RIVERSTRAHLER approach. Hydrobiologia 410, 151–166.

Gilbert, P.M., Biggs, D.C., McCarthy, J.J., 1982. Utilization of ammonium and nitrate during austral summer in the Scotia Sea. Deep Sea Res. 29, 837–850.

Goswami, S.C., Padmavati, G., 1996. Zooplankton production, composition and diversity in the coastal waters of Goa. Indian J. Mar. Sci. 25, 91–97.

Goswami, S.C., Shrivastava, Y., 1996. Zooplankton standing stock, community structure and diversity in the northern Arabian sea. In: Proceedings of the Second Workshop on Scientific Results of FORV Sagar Sampada, pp. 127–137.

Gouda, R., Panigrahy, R.C., 1992. Seasonal distribution and behavior of silicate in the Rushikulya estuary, East coast of India. Indian J. Mar. Sci. 24, 111–115.

Grant, G.M., Gross, E., 1996. Oceanography—A View of the Earth. Prentice-Hall, USA.

Grasshoff, K., Ehrhardt, M., Kremling, K., 1983. Methods of Seawater Analysis. Wiley-VCH, New York.

Gray, J., 1981. The Ecology of Marine Sediments: An Introduction to the Structure and Function of Benthic Communities. Cambridge University Press, London.

Hameed, S.P., 2005. A study on the littoral benthos for biological impact assessment in the vicinity of Madras Atomic Power Station thermal outfall (Kalpakkam, India). In: Proceedings of Annual Conference of Indian Nuclear Society, Mumbai, India, pp. 1–11.

Hameed, S.P., Syed Mohamed, H.E., Krishnamoorthy, R., 2007. Biological impact assessment of thermal discharge in the vicinity of Madras Atomic Power Station, Kalpakkam, India. In: Proceedings of 15th National Symposium on Environment, Coimbatore, India, pp. 39–45.

Haselkorn, R., Buikema, W.J., 1992. Nitrogen fixation in cyanobacteria. In: Stacey, G., Burris, R.H., Evans, H.J. (Eds.), Biological Nitrogen Fixation. Chapman & Hall, New York, pp. 166–190.

Haugen, V.E., Vinayachandran, P.N., Yamagata, T., 2003. Comment on "Indian Ocean: validation of the Miami Isopycnic coordinate Ocean Model and ENSO events during 1958–1998". J. Geophys. Res. 108 (C6), 3179. https://dx.doi.org/10.1029/2002JC001624.

Hirose, M., Mukai, T., Hwang, D., Iida, K., 2009. The acoustic characteristics of three jellyfish species: *Nemopilema nomurai, Cyanea nozakii*, and *Aurelia aurita*. ICES J. Mar. Sci. 66, 1233–1237.

Helly, J.J., Levin, L.A., 2004. Global distribution of naturally occurring marine hypoxia on continental margins. Deep Sea Res. I 51, 1159–1168.

Hwang, J.S., Kumar, R., Dahms, H.U., Tseng, L.C., Chen, Q.C., 2010. Interannual, seasonal, and diurnal variation in vertical and horizontal distribution patterns of 6 *Oithona* spp. (Copepoda: Cyclopoida) in the South China Sea. Zool. Stud. 49, 220–229.

Kalimurthy, M., 1973. Observations on the transparency of the waters of the Pulicat Lake with particular reference to plankton production. Hydrobiologia 41, 3–11.

Kana, T., Gilbert, P.M., 1987. Effect on irradiance up to 2000 μE m^{-2} s^{-2} on marine Synechococcus WH 7803-I. Growth: pigmentation and cell composition. Deep Sea Res. 34, 479–516.

Kasim, H.M., Khan, M.Z., 1986. A preliminary account of the gillnet fishery off Veraval during 1979–82. Indian J. Fish. 33 (2), 155–162.

Kasturirangan, L.R., 1963. A Key to the Identification of the More Common Planktonic Copepoda of India Coastal Waters. Publication No. 2, Indian National Committee on Oceanic Research, Council of Scientific and Industrial Research, New Delhi 87 pp.

Katsaros, K., Buettner, K.J.K., 1969. Influence of rainfall on temperature and salinity of the ocean surface. J. Appl. Meteorol. 8, 15–18.

Khandeparker, D.C., Anil, A.C., Venkat, K., 1995. Larvae of fouling organisms and macrofouling at New Mangalore port, west coast of India. Indian J. Mar. Sci. 24, 37–40.

Kumar, C.S., Perumal, P., 2011. Hydrobiological investigations in Ayyampattinam coast (Southeast coast of India) with special reference to zooplankton. Asian J. Biol. Sci. 4, 25–34.

La Fond, E.C., 1957. Oceanographic studies in the Bay of Bengal. Proc. Indian Acad. Sci. 46, 1–46.

Lal, D., 1978. Transfer of Chemical species through estuaries to oceans. In: Proceeding of UNESCO/SCOR Workshop, Melreux, Belgium, pp. 166–170.

Lalithambikadevi, C.B., 1993. Seasonal fluctuation in the distribution of eggs and larvae of flat fishes (Pleuronectiformes - Pisces) in the Cochin backwater. J. Indian Fish. Assoc. 23, 21–34.

Lardicci, C., Rossi, F., Maltagliati, F., 1999. Detection of thermal pollution: variability of benthic communities at two different spatial scales in an area influenced by a coastal power station. Mar. Pollut. Bull. 38, 296–303.

Lazzari, M.A., Sherman, S., Brown, C.S., King, J., Joule, B.J., Chenoweth, S.B., Langton, R.W., 1999. Seasonal and annual variations in abundance and species composition of two near-shore fish communities in Maine. Estuaries 22 (3A), 636–647.

Liss, P.S., Spencer, C.P., 1970. A biological process in the removal of silicate from seawater. Geochim. Cosmochim. Acta 34, 1073–1088.

Longhurst, A.R., Pauly, D., 1987. Ecology of Tropical Oceans. Academic Press, California.

Luzzatto, D.C., Penchaszadeh, P.E., 2001. Regeneration of the inhalant siphon of *Donax hanleyanus* (Philippi, 1847) (Bivalvia, Donacidae) from Argentina. J. Shellfish Res. 20, 149–153.

Lynam, C.P., Gibbons, M.J., Axelsen, B.E., Sparks, C.A.J., Coetzee, J., Heywood, B.G., et al., 2006. Jellyfish overtake fish in a heavily fished ecosystem. Curr. Biol. 16, R492–R493. https://dx.doi.org/10.1016/j.cub.2006.09.012.

Madhupratap, M., Nair, K.N.V., Gopalakrishnan, T.C., Haridas, P., Nair, K.K.C., Venugopal, P., Gauns, M., 2001. Arabian Sea oceanography and fisheries off the west coast of India. Curr. Sci. 81, 355–361.

Madhupratap, 1999. Free-living copepods of the Arabian Sea: distributions and research perspectives. Indian J. Mar. Sci. 28, 146–149.

Malone, T.C., Crocker, L.H., Pike, S.E., Wendler, B.W., 1988. Influences of river flow on the dynamics of phytoplankton production in a partially stratified estuary. Mar. Ecol. Progr. Ser. 48, 235–249.

Mani, P., Krishnamurthy, K., Palaniappan, P., 1986. Ecology of phytoplankton blooms in the Vellar estuary, east coast of India. Indian J. Mar. Sci. 15, 24–28.

Marais, J.F.K., 1988. Some factors that influence fish abundance in South African estuaries. S. Afr. J. Mar. Sci. 6 (1), 67–77.

Masilamani, J.G., Jesudoss, K.S., Nandakumar, K., Satpathy, K.K., Nair, K.V.K., Azariah, J., 2000. Jellyfish ingress: a threat to the smooth operation of coastal power plants. Curr. Sci. 79 (5), 567–569.

Masilamani, J.G., Nandakumar, K., Satpathy, K.K., Jesudoss, K.S., Nair, K.V.K., Azariah, J., 2002a. Lethal and sublethal effects of chlorination on green mussel *Perna viridis* in the context of biofouling control in power plant cooling water system. Mar. Environ. Res. 53 (1), 65–76.

Masilamani, J.G., Satpathy, K.K., Jesudoss, K.S., Nandakumar, K., Nair, K.V.K., Azariah, J., 2002b. Influence of temperature on the physiological responses of the bivalve *Brachidontes striatulus* and its significance in fouling control. Mar. Environ. Res 53 (1), 51–63.

Masilamani, J.G., Satpathy, K.K., Jesudoss, K.S., Nandakumar, K., Nair, K.V.K., Azariah, J., 2001. Excretory products of green mussel *Perna viridis* (L.) and their implication on power plant operation. Turk. J. Zool. 25, 117–125.

McLachlan, A., Dugan, J.E., Defeo, O., Ansell, A.D., Hubbard, D.M., Jaramillo, E., Penchaszadeh, P., 1996. Beach clam fisheries. In: Ansell, A.D., Gibson, R.N., Barnes, M. (Eds.), Oceanography and Marine Biology: An Annual Review. UCL Press, London, pp. 163–232.

Mohanty, A.K., Sahu, G., Biswas, S., Natesan, U., Prasad, M.V.R., Satpathy, K.K., 2010. Spatio-temporal variation in physico-chemical properties of coastal waters of Kalpakkam, southeast coast of India. J. Mar. Biol. Assoc. India 52 (1), 75–84.

Mohanty, A.K., Satpathy, K.K., Sahu, G., Sasmal, S.K., Sahu, B.K., Panigrahi, R.C., 2007. Red tide of *Noctiluca scintillans* & its impact on the coastal water quality of the near shore waters, off the Rushikulya River, Bay of Bengal. Curr. Sci. 93 (5), 616–618.

Møller, L.F., Riisgård, H.U., 2007. Feeding, bioenergetics and growth in the common jellyfish *Aurelia aurita* and two hydromedusae, *Sarsia tubulosa* and *Aequorea vitrina*. Mar. Ecol. Prog. Ser. 346, 167–177. https://doi.org/10.3354/meps06959.

Monbet, Y., 1992. Control of phytoplankton biomass in estuaries: a comparative analysis of microtidal and macrotidal estuaries. Estuaries 15, 563–571.

Munro, I.S.R., 1982. The Marine and Freshwater Fishes of Ceylon. Narendra Publishing House, Delhi.

Murty, C.S., Varadachari, V.V.R., 1968. Upwelling along the east coast of India. Bull. Natl. Inst. Sci. India 36, 80–86.

Nair, K.V.K., Ganapathy, S., 1983. Baseline ecology of Edaiyur-Sadras estuarine system at Kalpakkam. I: General hydrographic and chemical feature. Mahasagar 16, 143–151.

Nair, N.U., Nair, N.B., 1987. Marine biofouling dynamics in and around the Cochin harbor. In: Proceedings of the National Seminar on Estuarine Management, Trivandrum, India.

Nair, K.V.K., Murugan, P., Eswaran, M.S., 1988. Macrofoulants in Kalpakkam coastal waters, east coast of India. Indian J. Mar. Sci. 17, 341–343.

Naomi, T.S., George, R.M., Sreeram, M.P., Sanil, N.K., Balachandran, K., Thomas, V.J., Geetha, P.M., 2011. Finfish diversity in the trawl fisheries of southern Kerala. Mar. Fish. Inform. Ser. T&E Ser. (207), 11–21.

Nel, R., McLachlan, A., Winter, P.E., 2001. The effect of grain size on the burrowing of two Donax species. J. Exp. Mar. Biol. Ecol. 265, 219–238.

Newell, G.E., Newell, R.C., 1967. Marine Plankton: A Practical Guide. Hutchinson Educational, London, 221 pp.

Newell, R.I.E., Hilbish, T.J., Koehn, R.K., Newell, C.J., 1982. Temporal variation in the reproductive cycle of *Mytilus edulis* L. (Bivalvia, Mytilidae) from localities in the east coast of the United States. Biol. Bull. 162, 299–310.

Ng, P.K.L., Guinot, D., Davie, P.J.F., 2008. Systema Brachyurorum: Part I. An annotated checklist of extant brachyuran crabs of the world. Raffles Bull. Zool. (Suppl. 17), 1–286.

Olson, R.J., 1980. Nitrate and ammonium uptake in Antarctic waters. Limnol. Oceanogr. 26, 1064–1074.

Paerl, H.W., 1997. Coastal eutrophication and harmful algal blooms: importance of atmospheric deposition and groundwater as "new" nitrogen and other nutrient sources. Limnol. Oceanogr. 42, 1154–1165.

Parsons, T.R., Maita, Y., Lalli, C.M., 1984. A Manual of Chemical and Biological Methods for Seawater Analysis. Pergamon Press, Oxford.

Paul, M.D., 1942. Studies on the growth and breeding of certain sedentary organisms in the madras harbor. Proc. Indian Acad. Sci. 150, 1–42.

Paulmier, A., Ruiz-Pino, D., 2009. Oxygen minimum zones (OMZs) in the modern ocean. Prog. Oceanogr. 80, 113–128.

Perumal, N.V., Rajkumar, M., Perumal, P., Rajasekar, K.T., 2009. Seasonal variations of plankton diversity in the Kaduviyar estuary, Nagapattinam, southeast coast of India. J. Environ. Biol. 30, 1035–1046.

Peterson, C., Hickerson, H., Johnson, G.G., 2000. Short-term consequences of nourishment and bulldozing on the dominant large invertebrates of a Sandy Beach. J. Coast. Res. 16, 368–378.

Pieters, H., Kluytmans, J.H., Zandee, D.I., Cadee, G.C., 1980. Tissue composition and reproduction of *Mytilu edulis* in relation to food availability. Neth. J. Sea Res. 14, 349–361.

Poornima, E.H., Rajadurai, M., Rao, T.S., Anupkumar, B., Rajmohan, R., Narasimhan, S.V., Rao, V.N.R., Venugopalan, V.P., 2005. Impact of thermal discharge from a tropical coastal power plant on phytoplankton. J. Therm. Biol. 30, 307–316.

Prabhu, M.S., Dhawan, R.M., 1974. Marine fisheries resources in the 20 and 40 meter regions off the Goa coast. Indian J. Fish. 21 (1), 40–53.

Prasanna Kumar, S., Narvekar, J., Kumar, A., Shaji, C., Anand, P., Sabu, P., Rijomon, G., Josia, J., Jayaraj, K.A., Radhika, A., Nair, K.K.C., 2004. Intrusion of the Bay of Bengal water into the Arabian Sea during winter monsoon and associated chemical biological response. Geophys. Res. Lett. 31, L15304. https://dx.doi.org/10.1029/2004GL020247.

Prasannakumar, S., Madhupratap, M., Dileep Kumar, M., Gauns, M., Muraleedharan, P.M., Sarma, V.V., De Souza, S.N., 2000. Physical control of primary productivity on a seasonal scale in central and eastern Arabian Sea. Proc. Indian Acad. Sci. 109, 433–441.

Prasanna Kumar, S., Muraleedharan, P.M., Prasad, T.G., Gauns, M., Ramaiah, N., de Souza, S.N., Sardesai, S., Madhupratap, M., 2002. Why is the Bay of Bengal less productive during summer monsoon compared to the Arabian Sea? Geophys. Res. Lett. 29 (24), 2235. https://dx.doi.org/10.1029/2002GL016013.

Qasim, S.Z., 1977. Biological productivity of the Indian Ocean. Indian J. Mar. Sci. 6, 122–137.

Rabalais, N.N., Diaz, R.J., Levin, L.A., Turner, R.E., Gilbert, D., Zhang, J., 2010. Dynamics and distribution of natural and human-caused hypoxia. Biogeosciences 7, 585–619.

Radhakrishnan, N., 1974. Demersal fisheries of Vizhinjam. Indian J. Fish. 21 (1), 29–39.

Rajagopal, S., Nair, K.V.K., Van Der Velde, G., Jenner, H.A., 1997. Seasonal settlement and succession of fouling communities in Kalpakkam, east coast of India. Neth. J. Aquat. Ecol. 30, 1–17.

Rakhesh, M., Raman, A.V., Sudarshan, D., 2006. Discriminating zooplankton assemblages in neritic and oceanic waters: a case study for the Northeast coast of India, Bay of Bengal. Mar. Environ. Res. 61, 93–109.

Rao, T.S.S., 1958. Studies on chaetognatha in the Indian seas. II. Distribution in relation to currents. Andhra Univ. Mem. Oceanogr. Search 62, 162–167.

Rao, S.K., Balaji, M., 1988. Biological Fouling at Kakinada, Godavari Estuary: Marine Biodeterioration. Oxford IBH Publisher, New Delhi.

Rao, C.K., Naqvi, S.W.A., Kumar, M.D., Varaprasad, D.A., Jayakumar, D.A., George, M.D., Singbal, S.Y.S., 1994. Hydrochemistry of the Bay of Bengal: possible reason for a different water column cycling of carbon and nitrogen from the Arabian Sea. Mar. Chem. 47, 279–290.

Raymont, J.E.G., 1980. Plankton and Productivity in the Oceans, vol. I—Phytoplankton. Pergamon Press, Oxford.

Redfield, A., 1958. The biological control of chemical factors in the environment. Am. Sci. 46, 205–221.

Robin, R.S., Srinivasan, M., Chandrasekar, K., 2009. Distribution of zooplankton from Arabian Sea, along Southern Kerala (southwest coast of India) during the cruise. Curr. Res. J. Biol. Sci. 155–159.

Sahu, G., Achary, M.S., Satpathy, K.K., Mohanty, A.K., Biswas, S., Prasad, M.V.R., 2011. Studies on the settlement and succession of macrofouling organisms in the Kalpakkam coastal waters, southeast coast of India. Indian J. Geo-Mar. Sci. 40 (6), 747–761.

Sahu, G., Mohanty, A.K., Singhsamanta, B., Mahapatra, D., Panigrahy, R.C., Satpathy, K.K., Sahu, B.K., 2010. Zooplankton diversity in the near shore waters of Bay of Bengal, off the Rushikulya estuary. ICFAI Univ. J. Environ. Sci. IV (2), 61–85.

Sakthivel, K., Fernando, A., 2012. Brachyuran crabs diversity in Mudasal Odai and Nagapattinam coast of south east India. Arthropod 1 (4), 136–143.

Salas, C., Tirado, C., Manjon-Cabeza, M.E., 2001. Sublethal foot-predation on Donacidae (Mollusca: Bivalvia). J. Sea Res. 46, 43–56.

Sankaranrayanan, V.N., Qasim, S.Z., 1969. Nutrients of the Cochin Backwaters in relation to environmental characteristics. Mar. Biol. 2, 236–247.

Santhakumari, V., Saraswathy, M., 1981. Zooplankton along the Tamil Nadu coast. Mahasarar. Bull. Natl. Inst. Oceanogr. 14 (4), 289–302.

Santschi, P., Honener, P., Benoit, G., Brink, M.B., 1990. Chemical process at the sediment–water interface. Mar. Chem. 30, 269–315.

Saravanane, N., Nandakumar, K., Durairaj, G., Nair, K.V.K., 1999. Plankton as indicators of coastal water bodies during southwest to northeast monsoon transition at Kalpakkam. Curr. Sci. 78, 173–176.

Saravanane, N., Satpathy, K.K., Nair, K.V.K., Durairaj, G., 1998. Preliminary observations on the recovery of tropical phytoplankton after entrainment. J. Thermal Biol. 23 (2), 91–97.

Sardessai, S., Ramaiah, N., Prasanna Kumar, S., de Sousa, S.N., 2007. Influence of environmental forcings on the seasonality of dissolved oxygen and nutrients in the Bay of Bengal. J. Mar. Res. 65, 301–316.

Sargunam, C.A., 1994. Studies on the Microfouling in the Coastal Waters of Kalpakkam, West Coast of India, with Reference to Diatoms (Bacillariophyceae) (Ph.D. thesis). University of Madras, India.

Sarma, V.V.S.S., 2002. An evaluation of physical and biogeochemical processes regulating the oxygen minimum zone in the water column of the Bay of Bengal. Global Biogeochem. Cycles 16 (4), 1099–1108.

Sasikumar, N., Nair, K.V.K., Azariah, J., 1990. Colonization of marine foulants at a power plant site. Proc. Indian Acad. Sci. 99, 525–531.

Sasikumar, N., Rajagopal, S., Nair, K.V.K., 1989. Seasonal and vertical distribution of macrofoulants in Kalpakkam coastal waters. Indian J. Mar. Sci. 18, 270–275.

Satpathy, K.K., 1996. Seasonal distribution of nutrients in the coastal waters of Kalpakkam, East Coast of India. Indian J. Mar. Sci. 25, 221–224.

Satpathy, K.K., Nair, K.V.K., 1990. Impact of power plant discharge on the physico-chemical characteristics of Kalpakkam coastal waters. Mahasagar 23, 117–125.

Satpathy, K.K., Nair, K.V.K., 1996. Occurrence of phytoplankton bloom and its effect on coastal water quality. Indian J. Mar. Sci. 25, 145–147.

Satpathy, K.K., Eswaran, M.S., Nair, K.V.K., 1986. Distribution of temperature in the vicinity of a condenser outfall in Kalpakkam coastal waters. J. Mar. Biol. Assoc. India 28, 151–158.

Satpathy, K.K., Mathur, P.K., Nair, K.V.K., 1987. Contribution of Edaiyur-Sadras estuarine systems to the hydrographic characteristics of Kalpakkam coastal waters. J. Mar. Biol. Assoc. India 29, 344–350.

Satpathy, K.K., Mohanty, A.K., Prasad, M.V.R., Natesan, U., Sarkar, S.K., 2008a. Post-Tsunami changes in water quality of Kalpakkam coastal waters, east coast of India with special references to nutrients. Asian J. Water Environ. Pollut. 5, 15–30.

Satpathy, K.K., Jebakumar, K.E., Bhaskar, S., 2008b. A novel technique for the measurement of temperature in outflow water from a coastal power plant, with notes on chlorination and phytoplankton determination. J. Thermal Biol. 33 (4), 209–212.

Satpathy, K.K., Sahu, G., Mohanty, A.K., Prasad, M.V.R., Panigrahy, R.C., 2009. Phytoplankton community structure and its variability during southwest to northeast monsoon transition in the coastal waters of Kalpakkam, east coast of India. Int. J. Ocean Oceanogr. 3, 43–74.

Satpathy, K.K., Mohanty, A.K., Natesan, U., Prasad, M.V.R., Sarkar, S.K., 2010a. Seasonal variation in physicochemical properties of coastal waters of Kalpakkam, east coast of India with special emphasis on nutrients. Environ. Monit. Assess. 164, 153–171.

Satpathy, K.K., Mohanty, A.K., Sahu, G., Sarkar, S.K., Natesan, U., Venkatesan, R., Prasad, M.V.R., 2010b. Variations of physicochemical properties in Kalpakkam coastal waters, east coast of India, during southwest to northeast monsoon transition period. Environ. Monit. Assess. 171, 411–424.

Satpathy, K.K., Mohanty, A.K., Sahu, G., Biswas, S., Selvanyagam, M., 2010c. Biofouling and its control in seawater cooled power plant cooling water system—a review. In: Tsvetkov, P. (Ed.), Nuclear Power. Sciyo Publications, Croatia, pp. 191–242.

Satpathy, K.K., Mohanty, A.K., Sahu, G., Sarguru, S., Sarkar, S.K., Natesan, U., 2011. Spatio-temporal variation in physico-chemical properties of coastal waters off Kalpakkam, southeast coast of India, during summer, pre-monsoon and post-monsoon period. Environ. Monit. Assess. 180, 41–62.

Satpathy, K.K., Panigrahi, S., Mohanty, A.K., Sahu, G., Achary, M.S., Bramha, S.N., Padhi, R.K., Samantara, M.K., Selvanayagam, M., Sarkar, S.K., 2013. Severe oxygen depletion in the shallow regions of Bay of Bengal off Tamil Nadu coast. Curr. Sci. 104 (11), 1467.

Selvaraj, V., 1984. Community structure in the coastal ecosystem with special reference to the green mussel, *Perna viridis* (L.) (Ph.D. thesis). University of Madras, India.

Shanmugam, A., Palpandi, C., Sambasivam, S., 2007. Some valuable fatty acids exposed from wedge clam *Donax cuneatus* (Linnaeus). Afr. J. Biochem. Res. 1, 14–18.

Sin, Y., Wetzel, R.L., Anderson, I.C., 1999. Spatial and temporal characteristics of nutrient and phytoplankton dynamics in the York River estuary, Virginia: analysis of long-term data. Estuaries 22, 260–275.

Sreekumaran Nair, S.R., Nair, V.R., Achuthankutty, C.T., Madhupratap, M., 1981. Zooplankton composition and diversity in western Bay of Bengal. J. Plankton Res. 3, 493–508.

Sruthi, S., Nansimole, A., Gayathri Devi, T.V., Tresa, Radhakrishnan, 2014. Diversity of Brachyuran Crabs of Kanyakumari Area, South West Coasts of India. Int. J. Sci. Res. 3(12).

Stenton-Dozey, J.M.E., Brown, A.C., 1994. Exposure of the sandy-beach bivalve Donax serra Roding to a heated and chlorinated effluent: 1. Effects of temperature on burrowing and survival. J. Shellfish Res. 13, 443–449.

Stephens, C., Oppenheime, C.H., 1972. Silica contents in the Northwestern Florida Gulf coast. Contrib. Mar. Sci. 16, 99–108.

Stramma, L., Johnson, G.C., Sprintall, J., Mohrholz, V., 2008. Expanding oxygen-minimum zones in the tropical oceans. Science 320, 655–658.

Strauss, S.D., 1989. New methods, chemical improve control of biological fouling. Power, 51–52.

Suresh, K., Shafiq Ahamed, M., Durairaj, G., 1992. The impact of heated effluents from a power plant on the macrofauna and meiofauna of a sandy shore. In: Proceedings of National Symposium on Environment, Bombay, India, pp. 21–24.
Talwar, P.K., Kacker, R.K., 1984. Commercial Sea-Fishes of India. Zoological Survey of India, Calcutta.
ter Braak, C.J.F., Šmilauer, P., 2002. CANOCO Reference Manual and CanoDraw for Windows User's Guide: Software for Canonical Community Ordination (version 4.5). Microcomputer Power, Ithaca, NY, USA, www.canoco.com.
Thorman, S., 1986. Seasonal colonisation and effects of salinity and temperature on species richness and abundance of fish of some brackish and estuarine shallow waters in Sweden. Ecography 9 (2), 126–132.
Underwood, A.J., 1994. On beyond BACI: sampling designs that might reliably detect environmental disturbances. Ecol. Appl. 4, 3–15.
Varadharajan, D., Soundarapandian, P., 2012. Commercially important crab fishery from Arukkattuthurai to Pasipattinam, South East Coast of India. J. Mar. Sci. Res. Dev. 2, 110.
Varkey, M.J., Murty, V.S.N., Suryanaryan, A., 1996. Physical oceanography of the Bay of Bengal and Andaman Sea. In: Ansell, A.D., Gibson, R.N., Barnes, M. (Eds.), Oceanography and Marine Biology. UCL Press, 34, 1–70.
Venugopalan, V.P., 1987. Studies on Biofouling in the Offshore Waters of Arabian Sea (Ph.D. Thesis). University of Bombay, India, 214 pp.
Vollenweider, R.A., 1974. A Manual on Methods for Measuring Primary Production in Aquatic Environments, 2nd ed. IBP Handbook No. 12. Blackwell Scientific, Oxford, UK.
Woods Hole Oceanographic Institution, 1952. Marine Fouling and Its Prevention. United States Naval Institute, Annapolis, Maryland, 388 pp.
Wyrtki, K., 1973. Physical oceanography of the Indian Ocean. In: Zeitzshel, B. (Ed.), Biology of the Indian Ocean. Springer Verlag, Berlin, New York.

Further Reading

Chandramohan, P., Sreenivas, N., 1998. Diel variations in zooplankton populations in mangrove ecosystem at Gadesu canal, southeast cost of India. Indian J. Mar. Sci. 27, 486–488.
Chidambaram, K., Menon, M.D., 1945. The correlation of the west coast fisheries with plankton and certain oceanographical factors. Proc. Indian Acad. Sci. 27, 355–367.
Delsman, H.C., 1939. Preliminary plankton investigation in the Java Sea. Treubia 17, 139–181.
Elliott, M., Nedwell. S., Jones, N.V., Read, S.J., Cutts, N.D., Hemingway, K.L., 1998. Intertidal Sand and Mudflats & Subtidal Mobile Sandbanks: An Overview of Dynamic and Sensitivity Characteristics for Conservation Management of Marine SACs. Scottish Association for Marine Science (UK Marine SACs Project).
George, P.C., 1953. The marine plankton of the coastal waters of Calicut with observations on the hydrological conditions. J. Zool. Soc. India 5, 76–103.
Gouda, R., Panigrahy, R.C., 1996. Ecology of phytoplankton in coastal waters of Gopalpur, east coast of India. Indian J. Mar. Sci. 25, 146–150.
Govindasamy, C., Kannan, L., Azariah, J., 2000. Seasonal variation in physic-chemical parameters properties and primary production in the coastal water biotopes of Coromandel Coast, India. J. Environ. Biol. 21, 1–7.
Hanninen, J., Vourinen, I., Helminen, H., Kirkkala, I., Lehtila, 2000. Trends and gradients in nutrient concentrations and loading in the Archipelago Sea, Northern Baltic. Estuar. Coast. Shelf Sci. 50, 153–171.
Jayaraman, R., 1951. Observation on the chemistry of the waters off the Bay of Bengal off the Madras City during 1948–1949. Proc. Indian Acad. Sci. 33, 92–99.
Jayaraman, R., 1954. Seasonal variations in salinity, dissolved oxygen and nutrient salts in the inshore waters of the Gulf of Mannar and Palk Bay near Mandapam (South India). Indian J. Fish. 1, 345–364.
Kow, T.Ah., 1953. The feed and feeding relationships of the fishes of Singapore. Straits Fish. Publ. 1, 123.
Nydahl, A., Panigrahi, S., Wikner, J., 2013. Increased microbial activity in a warmer and wetter climate enhances the risk of coastal hypoxia. FEMS Microbiol. Ecol. https://dx.doi.org/10.1111/1574-6941.12123.
Panigrahy, R.C., Sahu, J.P., Mishra, P.M., 1984. Studies on some hydrographic features in the surface waters of Bay of Bengal at Gopalpur, Orissa coast. Bull. Environ. Sci. 1, 10–14.
Panigrahy, R.C., Mishra, S., Sahu, G., Mohanty, A.K., 2006. Seasonal distribution of phytoplankton in the surf waters off Gopalpur, east coast of India. J. Mar. Biol. Assoc. India 48, 156–160.
Phillips, D.J.H., 1980. Quantitative Aquatic Biological Indicators: Their Use to Monitor Trace Metal and Organochlorine Pollution. Lond. Application Science Ltd., U.K

Phillips, D.J.H., 1990. Use of microalgae and invertebrates as monitors of metal levels in estuaries and coastal waters. In: Furness, R.W., Rainbow, P.S. (Eds.), Heavy Metals in the Marine Environment. CRC Press, Florida, pp. 81–99.

Rajasegar, M., 2003. Physico-chemical characteristics of the Vellar estuary in relation to shrimp farming. J. Environ. Biol. 24, 95–101.

Ramamurthy, S., 1953. Hydrobiological studies in Madras coastal waters. J. Madras Univ. 23, 143–163.

Ramaraju, V.S., Sarma, V.V., Rao, P.B., Rao, V.S., 1992. Water masses of Visakhapatnam shelf. In: Physical Processes in Indian seas (Proceedings of First Convention, ISPSO, 1992), pp. 75–78.

Sasmal, S.K., Sahu, B.K., Panigrahy, R.C., 1986. Monthly variations in some chemical characteristics of near shore waters along the south Orissa coast. Indian J. Mar. Sci. 15, 199–200.

Satpathy, K.K., Venugopalan, V.P., Nair, K.V.K., Mathur, P.K., 1992. Proc. Nat. Symp. Environ. 8–10.

Seshappa, G., 1953. Phosphate content of mudbanks along the Malabar Coast. Nature 171, 526–527.

Somayajulu, Y.K., Ramanamurthy, T.V., Prasannakumar, S., Sastry, J.S., 1987. Hydrographic characteristics of central Bay of Bengal waters during southwest monsoon of 1983. Indian J. Mar. Sci. 16, 207–217.

Suryanarayan, A., Rao, D.P., 1992. Coastal circulation and upwelling index along the east coast of India. In: Physical Processes in the Indian Seas (Proceedings of First Convention, ISPSO, 1992), pp. 125–129.

Verlecar, X.N., Pereira, N., Desai, S.R., Jena, K.B., Snigdha, 2006. Marine pollution detection through biomarkers in marine bivalves. Curr. Sci. India 91, 1153–1157.

CHAPTER

17

Management of Coastal Groundwater Resources

Manivannan Vengadesan, Elango Lakshmanan
Department of Geology, Anna University, Chennai, India

1 INTRODUCTION

Groundwater, the largest freshwater resource, is used for domestic, agricultural, and industrial purposes in most of the regions of the world. Coastal regions are not an exception, as groundwater is extensively pumped and used to meet various requirements. Proper and efficient management of coastal groundwater resources is very essential and is an integral part of coastal zone management. Coastal groundwater resources are threatened due to overexploitation. As the groundwater-bearing formations are hydraulically connected to the sea, overpumping results into the entry of seawater into freshwater aquifers. Thus, seawater intrusion is the landward movement of seawater into the aquifer, which is usually caused by overextraction of groundwater. Further land-use changes due to urbanization lead to a reduction in rainfall recharge. Also, a sea level rise due to climate change results in increase of the extent of seawater intrusion. Seawater intrusion results in degradation of groundwater quality, where even 1% of seawater (250 mg/L chloride) can make freshwater unfit for drinking (Werner et al., 2013). Seawater intrusion has been assessed by various researchers in different parts of the world, such as Malaysia (Abdullah et al., 2010); Australia (Werner, 2010); North Oman (El-Kaliouby and Abdalla, 2015); Israel (Shalev et al., 2009); Morocco (Najib et al., 2017); the Nile Delta (Nofal et al., 2015); Turkey (Karahanoglu and Doyuran, 2003); Croatia (Terzic et al., 2010); Korea (Park et al., 2012); Germany (Yang et al., 2015); Northern Italy (Colombani et al., 2016); Argentina (Carretero et al., 2013); Tunisia (Paniconi et al., 2001); Taiwan (Chen et al., 2015); United Arab Emirates (Sherif et al., 2012); Libya (Sadeg and Karahanolu, 2001); Northern China (Zhao et al., 2016); China (Zhou et al., 2000); Spain (Vallejos et al., 2015); and Ireland (Perriquet et al., 2014). Further, many researchers have also suggested methods to mitigate seawater intrusion such as treated waste water recharge (Sales et al., 2017), removal of the saltwater wedge (Liu et al., 2017), an artificial

https://doi.org/10.1016/B978-0-12-810473-6.00018-2
© 2019 Elsevier Inc. All rights reserved.

intelligence approach (Triki et al., 2017), and also suggested different management measures (Barazzuoli et al., 2008; Chami et al., 2009; Das and Datta, 1999; Kourakos and Mantoglou, 2011; Reichard et al., 1999; Tsao, 1982; Zacharias and Koussouris, 2000).

In India, more than 85% of the population depends on groundwater (Parimala and Elango, 2013). Nearly 7516km of the country is bounded by sea whereas 25% of the country's population lives in the coastal zone. The high population density along the coastal areas is attributed to the easy availability of water. Three metropolitan cities out of four such as Chennai, Mumbai, and Kolkata as well as several urban centers such as Puri, Visakhapatnam, Nellore, Puducherry, Cuddalore, Nagapattinam, Tuticorin, Trivandrum, Cochin, Mangaluru, Goa, Surat, and Diu are located on the coast. The coastal zone is the most industrialized area in the country. The geology of the coastal region comprises Deccan basalts and crystalline rocks on the western coast whereas unconsolidated quaternary and tertiary sedimentary formations are found on the eastern coast. The lithological units of the Indian coasts range from recent fluvial and marine deposits to Archaean crystallines (CGWB, 2014). About 14 major, 44 medium, and 55 minor rivers/streams discharge into the sea. The salinity of the Arabian Sea is higher than the Bay of Bengal due to high evaporation in the Arabian Sea and less river discharge (National Institute of Oceanography (NIO), 2006). Many researchers have carried out seawater intrusion studies throughout the Indian coast (Das et al., 2016; Kumar and Divya, 2012; Kumar et al., 2013; Kumar et al., 2014; Laluraj et al., 2005; Naik et al., 2007; Nair et al., 2015; Rao et al., 2005; Rina et al., 2013; Singaraja et al., 2014; Vetrimurugan and Elango, 2015). This chapter describes the concept of freshwater-seawater interface, methods of identification of seawater intrusion, and methods of mitigation, with two case studies.

2 RELATIONSHIP BETWEEN FRESHWATER AND SEAWATER

Seawater intrusion into coastal aquifers has been the subject of studies for more than a century. Researchers were curious to understand the location of the boundary between freshwater and seawater in the aquifers. The first physical formulations of saltwater intrusion were initially made by Drabbe and Ghyben (1889) and Herzberg (1901) by assuming that they are immiscible. The Ghyben-Herzberg relation (Fig. 1) considering the density difference between the seawater and freshwater is,

$$z = \frac{\rho_f}{\left(\rho_s - \rho_f\right)} h$$

where h is thickness of the freshwater zone above sea level, z is thickness of the freshwater zone below sea level, ρ_f is density of freshwater, and ρ_s is density of saltwater.

Considering the density of freshwater (1.000 g/cm^3 at 20°C) and seawater (1.025 g/cm^3 at 20°C), the equation given by Ghyben-Herzberg simplifies to

$$z = 40h$$

Thus, for every meter of freshwater in an unconfined aquifer above sea level, there will be 40m of freshwater in the aquifer below sea level (Drabbe and Ghyben, 1889; Herzberg, 1901). That is, if the groundwater level declines by a meter, the interface will rise by 40m. Though

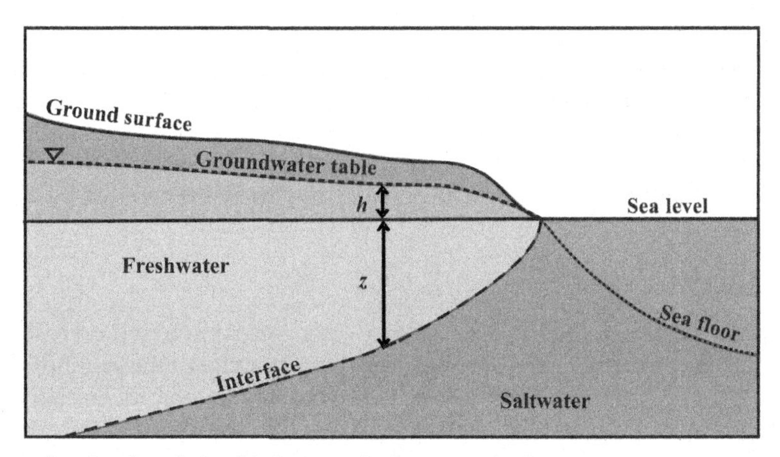

FIG. 1 Diagram showing the relationship between freshwater and saltwater.

the Ghyben-Herzberg relationship is an approximate method of estimating the interface, this relationship emphasises the importance of maintaining the groundwater level above the sea level. In reality, there is no sharp interface between freshwater and seawater. The concentration of water increases in the aquifer toward the sea. Hence, subsequent researchers such as Ataie-Ashtiani et al. (1999), Chen and Hsu (2004), Dausman and Langevin (2004), Eeman et al. (2011), and Michael et al. (2005) carried out several laboratory and field experimental studies about the freshwater-saltwater mixing zone. The thickness of the mixing zone varies widely between a few meters in laboratory experiments and few meters to kilometers in field situations. The mixing zones in coastal aquifers are influenced by geologic structure, tidal fluctuations, temporal variation in groundwater recharge, sea level rise, and pumping activities. Fluctuations in both inland freshwater heads and seawater levels can play an important role in modifying mixing zone thickness. The position and thickness of the mixing zone should be reasonably well characterized for effective management of coastal groundwater resources.

3 IDENTIFICATION OF SEAWATER INTRUSION

The coastal aquifers are subjected to the intrusion of seawater and several regions have been affected due to this in recent decades. Many methods can be used to identify the extent of seawater intrusion, such as groundwater levels, concentration of ions, and resistivity values. It is necessary to identify seawater intrusion for the proper management of groundwater in coastal regions. The common methods used to identify seawater intrusion are discussed in this section.

3.1 Groundwater Head

Measuring the groundwater head with respect to the sea level is the simplest way to identify the seawater intrusion. The Ghyben-Herzberg relation is used to delineate the location of the freshwater-saltwater interface. If the groundwater level is one meter above mean sea level, then the interface will present at 40 m below mean sea level. The interface will present near the mean sea level if the groundwater level above mean sea level is zero. So it is necessary to maintain the groundwater level above mean sea level to avoid seawater intrusion. Inouchi et al. (1985) determined the interface in the estuaries of the Nakka and Kiki Rivers in Japan based on the Ghyben-Herzberg relation. The depth of the freshwater-saline water interface using hydraulic heads in the coastal aquifer near Beihai, Chinam was estimated by Zhou (2011).

3.2 Geophysical Methods

Geophysical methods are used to identify seawater intrusion based on resistivity values. Resistivity values are indirectly proportional to the conductivity of the groundwater. Usually, groundwater will have high resistivity values. But in coastal regions, low resistivity values are observed, indicating the presence of seawater. In case of absence of monitoring wells, geophysical methods find useful application in evaluating the extent of seawater intrusion. The zone of mixing between freshwater and saltwater in Chennai, India, was studied by Sathish et al. (2011) using the high resistivity electrical resistivity tomography technique. The penetration of saltwater in the coastal aquifers of Israel was studied by Melloul and Goldenberg (1997), using the time domain electrical method resistivity values.

3.3 Geochemical Methods

Electrical conductivity and the concentration of ions present in the groundwater are keys to identify seawater intrusion. Electrical conductivity of groundwater greater than 3000 μS/cm is considered a seawater-intruded region (Karahanoglu, 1997). Major ions such as sodium (Na), chloride (Cl), calcium (Ca), and magnesium (Mg), and minor ions such as bromide (Br), fluoride (F), and iodide (I) are commonly used ions in identifying seawater intrusion. An excess of Cl over Na ions with the molar ratio of $Na/Cl = 0.86$ and the excess of Mg with molar ratio of $Mg/Ca = 4.5–5.2$ indicate seawater intrusion (Jones et al., 1999). Seawater intrusion was identified in the fast-growing coastal area of Chennai using EC, Na, and Cl by Kumar et al. (2013). The groundwater quality index for seawater intrusion was developed by Tomaszkiewicz et al. (2014) using geochemical parameters and a pilot study was carried out along the eastern coast of the Mediterranean Sea.

3.4 Isotopic Methods

Oxygen ($\delta^{18}O$) and hydrogen ($\delta^{2}H$) are stable isotopes that have been widely used to study the origin and dynamics of groundwater. The source of various water types and interactions between them was studied by the distribution of these isotopes. Each waterbody has its specific isotopic signatures. The mechanism of seawater intrusion is confirmed by the high contrast in the isotopic signature between groundwater and seawater. Isotopic ratios are generally expressed by stable isotope contents of light elements. Due to the small differences observed in the isotopic ratios (e.g., $^{2}H/^{1}H$), the isotope concentrations are expressed as deviations (δ) between the ratio of the sample and the same ratio in an internationally accepted standard (IAEA, 1994). Vienna-Standard Mean Oceanic Water (VSMOW) is a standard ocean water sample used as a reference in case of stable isotopes of hydrogen and oxygen in water samples. Isotopically depleted samples will have a negative δ value (lower than the standard) of isotopic ratio whereas isotopically enriched samples will have a positive δ value. The extent of seawater intrusion was identified up to a distance of 13 km inland during March 2012 and up to 14.7 km during May 2012 in Chennai by Nair et al. (2015) using isotopic ratios such as oxygen-18 ($\delta^{18}O$) and deuterium ($\delta^{2}H$). The temporal changes in the extent of seawater intrusion in the coastal carbonate aquifer of northeastern China using $\delta^{18}O$ and $\delta^{2}H$ was studied by Han et al. (2015).

4 MITIGATION MEASURES

Mitigation measures should be followed to preserve the freshwater reserves and also to remediate the regions affected by seawater intrusion. Mitigation of seawater intrusion plays a major role in coastal zone management. The different methods for mitigating seawater intrusion are discussed in this section.

4.1 Reduction in Pumping

Overextraction of groundwater in the coastal aquifer leads to a reduction in the availability of freshwater. Additional pumping of groundwater induces further movement of seawater into the aquifer toward the groundwater extraction. The mixing of seawater with groundwater affects the quality and normal usefulness of groundwater. Hence, overpumping of groundwater should be reduced in the coastal aquifer. This method is cost-free and more effective toward the mitigation of seawater intrusion. A decrease in saltwater upconing due to a reduction in groundwater pumping in the coastal aquifer of Cebu City was studied by Scholze et al. (2002). An increase in groundwater level was observed by a reduction in pumping in the coastal aquifer of Chennai (Pandian et al. 2016).

4.2 Rearranging of Pumping Wells

Overextraction of groundwater in the seawater-intruded area will increase the movement of interface landward. The pumping wells that are affected by seawater can be rearranged toward landward, which is away from the interface (Fig. 2). This method is cost-effective and an optimum pumping of the freshwater should be carried out to mitigate seawater intrusion. In Chennai, the groundwater pumping stations located close to the sea were shifted away from the sea to prevent seawater intrusion. Rearranging pumping wells in parts of the Nile Delta aquifer reduced seawater intrusion, as found by Sherif and Al-Rashed (2001).

4.3 Increasing Groundwater Recharge

Recharging the groundwater increases the volume of available freshwater, which leads to the movement of the interface toward the sea. The main source of water for recharging groundwater is precipitation. The rainwater can be collected and stored either in tanks or

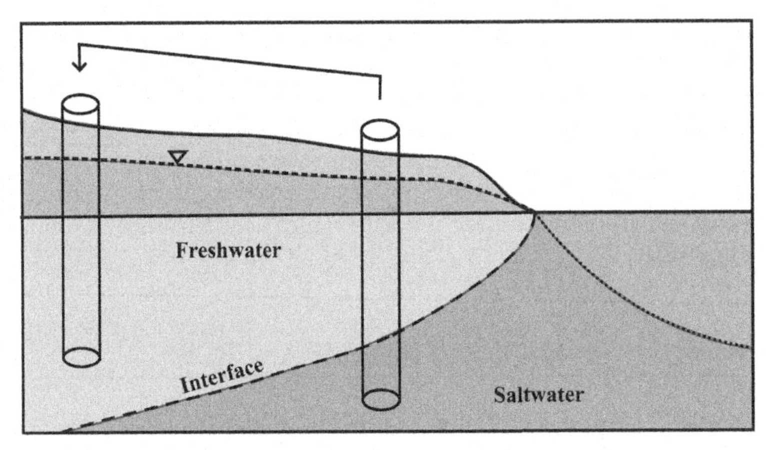

FIG. 2 Conceptual diagram of rearranging pumping wells.

ponds, which will be used to recharge groundwater. A percolation tank or pond (Fig. 3) is commonly used to collect rainwater and infiltrate through the permeable base to enhance freshwater storage in unconfined aquifers. This is a cost-effective method for mitigation that is feasible to construct. Recharge shafts are constructed in the percolation pond to increase the efficiency of the recharge. The groundwater quality was improved by constructing a percolation pond in Chennai (Christy and Elango, 2017).

4.4 Injection Wells

Injection wells are tube wells constructed for the purpose of freshwater recharge, which directly induces the movement of the interface toward the sea (Fig. 4). Freshwater is injected under high pressure through the injection wells, which allows the backward movement of

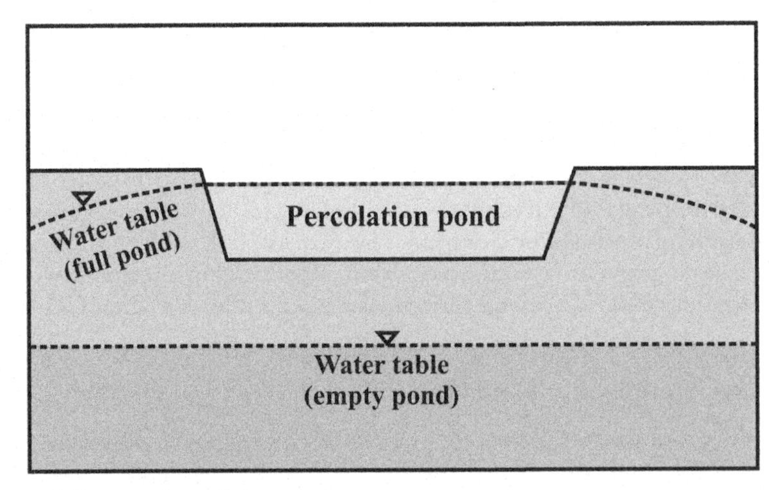

FIG. 3 Conceptual diagram of percolation pond.

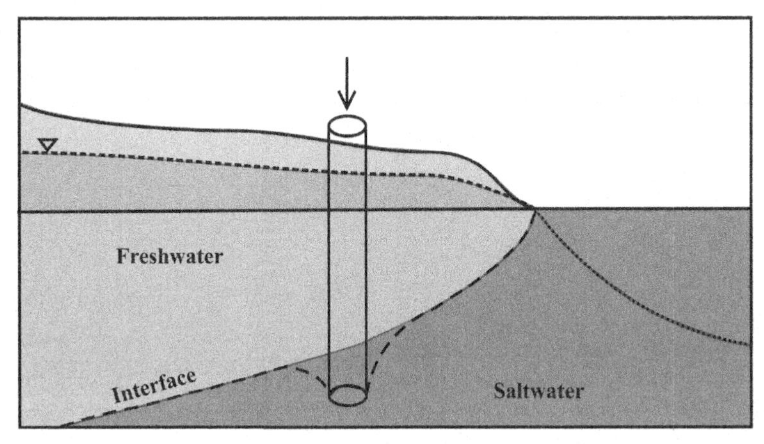

FIG. 4 Conceptual diagram of injection wells.

seawater. This method depends upon the availability of freshwater used to inject and so it is not commonly practiced. Recharging by injection wells and increasing the injection rate can move the interface toward the coast, which has been studied in Crete, Greece, by Papadopoulou et al. (2005). Luyun et al. (2011) also found that more effective saltwater repulsion can be achieved if recharge is applied near the toe of the saltwater wedge.

4.5 Saltwater Pumping Wells

Pumping of saltwater from a seawater-intruded coastal aquifer will reduce the volume of available saltwater and restore the quality of groundwater (Fig. 5). This method is expensive and it is difficult to pump out the saltwater completely. Mitigation of seawater intrusion was carried out by pumping out saltwater in Chennai (Sherif and Hamza, 2001).

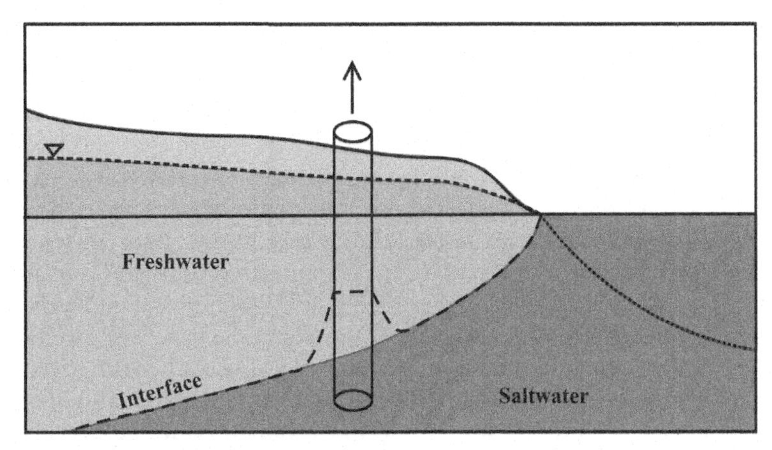

FIG. 5 Conceptual diagram of brackish water pumping.

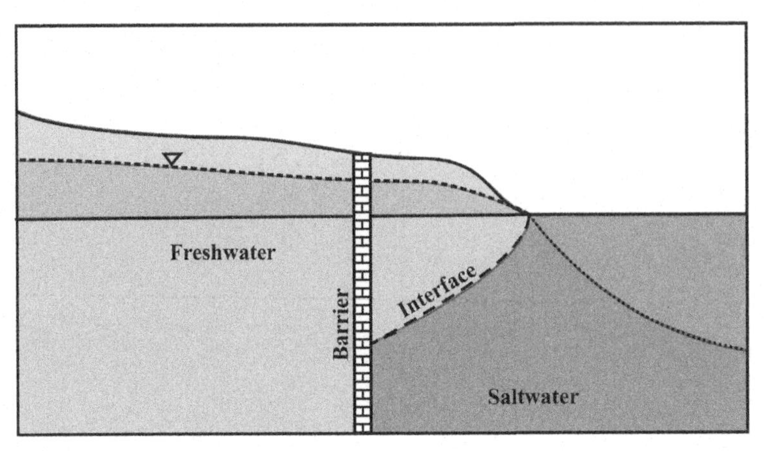

FIG. 6 Conceptual diagram of subsurface barrier.

4.6 Subsurface Barrier

The subsurface barrier is constructed below the surface to arrest the inland movement of seawater (Fig. 6). This barrier will increase the usable storage of groundwater within the aquifer, thus improving the availability of freshwater. The barrier was constructed using the cement grouting method, which rests on an impervious stratum. A subsurface barrier protects the aquifer from seawater intrusion in a limestone coastal aquifer on Okinawa Island in the western Pacific Ocean (Sugio et al., 1987).

5 CASE STUDY I: IDENTIFYING SEAWATER INTRUSION IN THE COASTAL AQUIFER OF PUDUCHERRY, INDIA

Puducherry is an urban region situated on the southeastern coast of India where groundwater is the major source for domestic needs. It covers about 293 sq km in area with a coastal length of 30 km. The population growth is high due to urbanization, which leads to the demand in availability of freshwater. The impact of seawater interaction in the coastal aquifers of the Sankaraparani river basin, Puducherry was studied. The entire region is made of porous sedimentary formations ranging from cretaceous to recent in age. The Tertiary aquifer is overlain by the Alluvium aquifer near the coast and Cretaceous aquifer is present in the western part of the region. Groundwater occurs in both confined and unconfined conditions. Groundwater samples were collected from 42 locations during June and September 2017. Groundwater level, electrical conductivity, and concentration of major ions in groundwater were measured. Seawater intrusion was identified using geochemical methods in this region. The groundwater head in this region varies from 7 to 54 m. The EC of the groundwater ranges from 454 to 4800 μS/cm. Wells located near the sea had electrical conductivity greater than 3000 μS/cm, indicating seawater intrusion (Fig. 7). Based on the ionic ratio of Na/Cl, the seawater intrusion was observed in the southern part of the Alluvium aquifer (Fig. 8) whereas the Tertiary aquifer was not affected by seawater intrusion. Monitoring groundwater

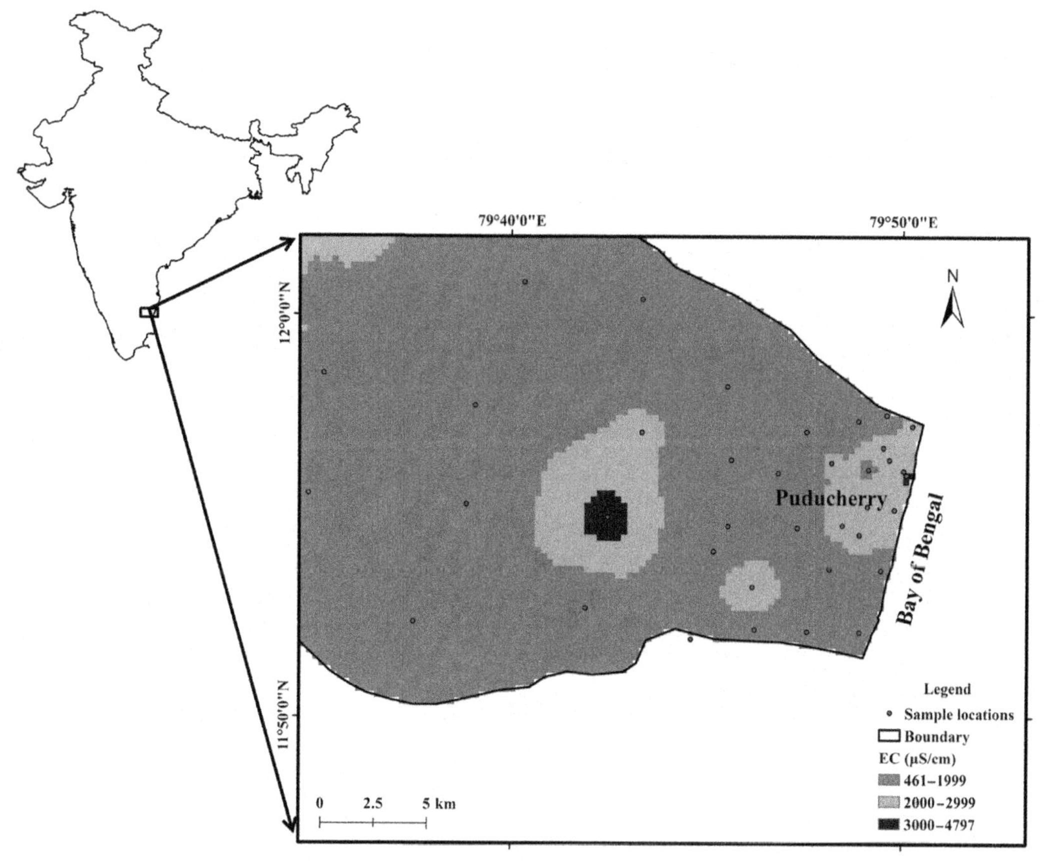

FIG. 7 Spatial variation of EC of groundwater.

pumping and proper groundwater management such as practicing mitigation methods will reduce the seawater intrusion in this region.

6 CASE STUDY II: MANAGEMENT OF GROUNDWATER TO MITIGATE SEAWATER INTRUSION IN CHENNAI, INDIA

Chennai is a metropolitan city with high population density. The need for water resources is increasing due to a drastic growth in urbanization. The response of the aquifer to variations in pumping and rainfall recharge due to projected climate change by using groundwater modeling in a heavily exploited aquifer of Chennai was assessed. This area predominantly consists of fluvial deposits that include clay, silt, sand, gravel, and pebble layers. The underlying clay layers divide the deposit into two layers, with the top layer functioning as an unconfined aquifer and the bottom layer as a semiconfined aquifer. Charnockites of Archaean age are at the base, which is overlain by marine sediments, sand, marine cla,y and gravel

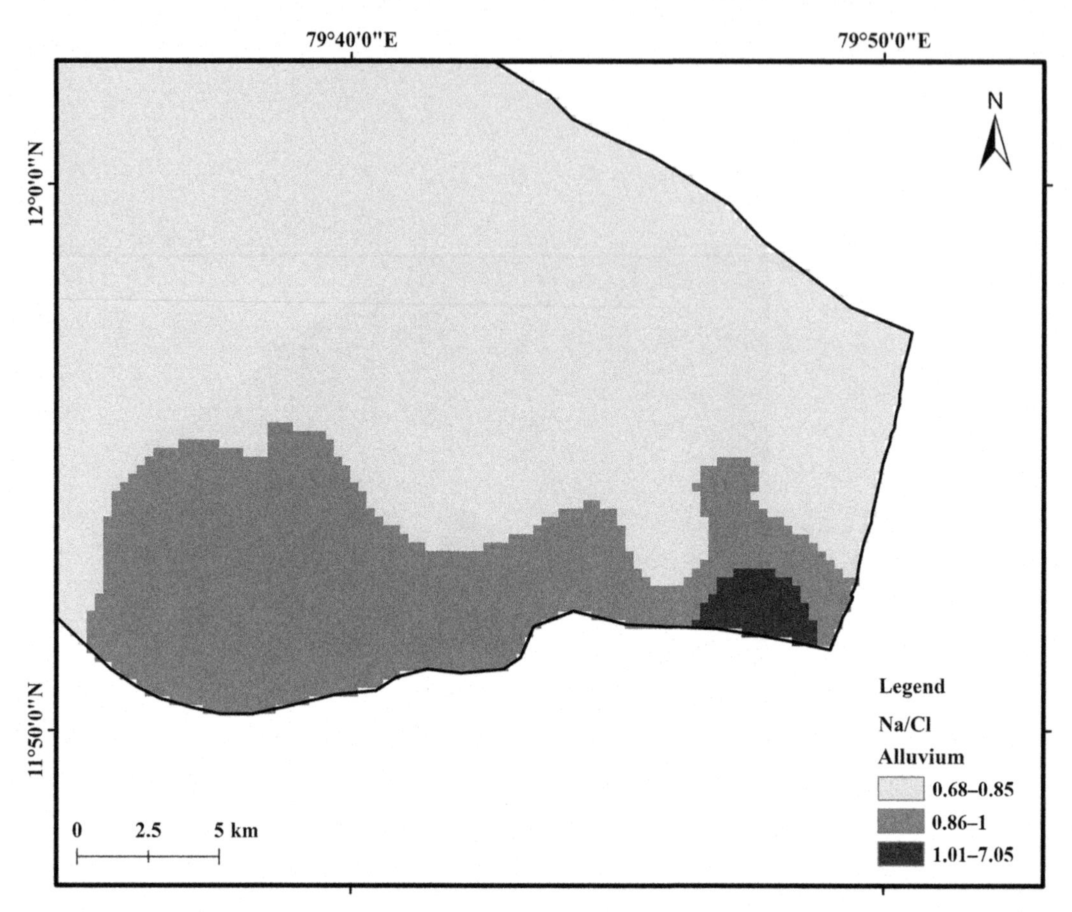

FIG. 8 Spatial variation of Na/Cl ratio in Alluvium aquifer.

formations followed by clay and silt. FEFLOW software was used to carry out finite element groundwater flow modeling from March 1988 to December 2030. Steady-state calibration was done to match observed and simulated groundwater heads by varying aquifer parameters within the allowable range. Transient state calibration was carried out during the period of March 1988 to December 2002. The calibrated model was validated by comparing the simulated and observed groundwater head from January 2003 to December 2012 (Fig. 9).

Until 2030, the groundwater head was predicted under eight different scenarios such as:

1. 10% increase in rainfall recharge and normal pumping
2. 10% decrease in rainfall recharge and normal pumping
3. 10% increase in pumping.
4. 10% increase in rainfall recharge with 10% increase in pumping
5. 10% decrease in rainfall recharge and 10% increase in pumping

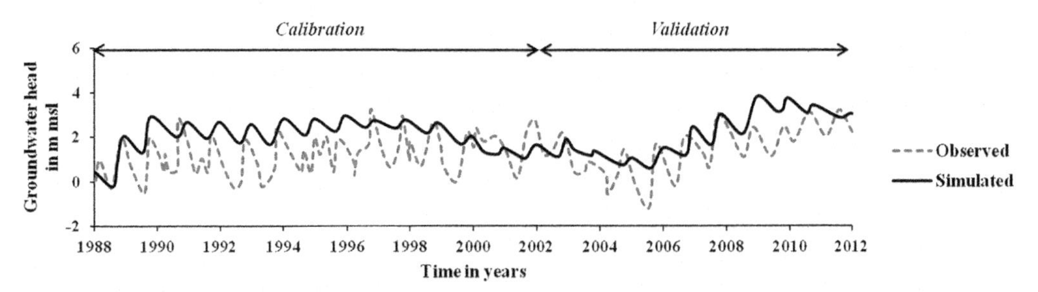

FIG. 9 Temporal variation in observed and simulated groundwater head.

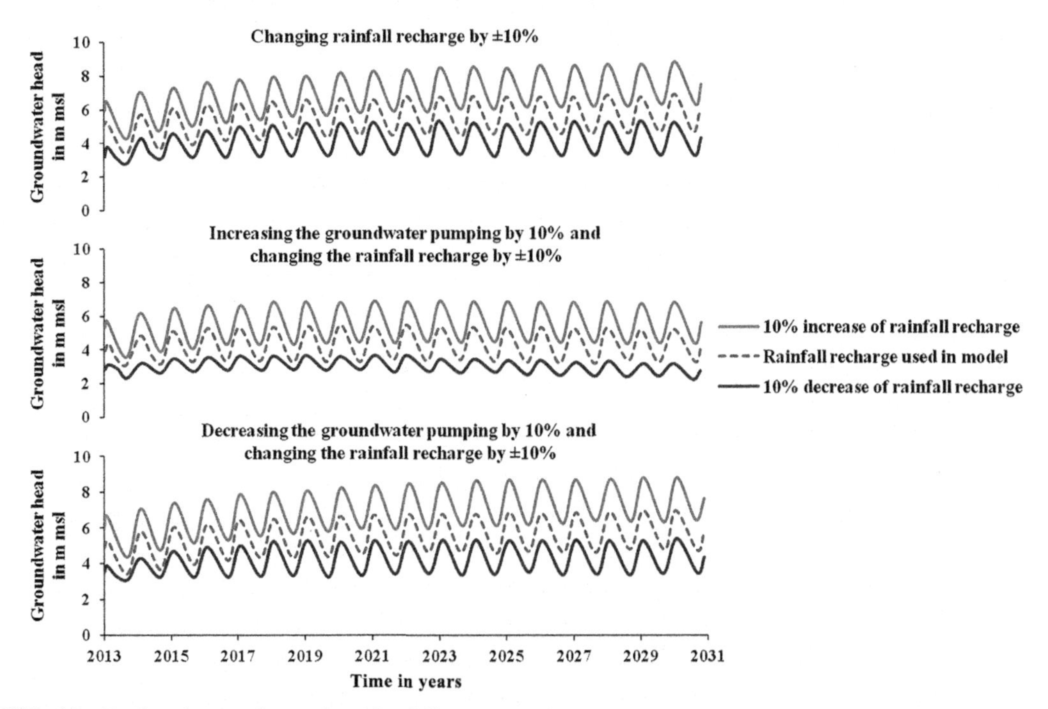

FIG. 10 Predicted groundwater head by different scenarios.

6. 10% decrease in pumping
7. 10% decrease in rainfall recharge and 10% decrease in pumping
8. 10% increase in rainfall recharge and 10% decrease in pumping

This prediction indicated that a 10% increase of recharge and a 10% decrease of pumping will cause a three-meter increase in groundwater head in the upper aquifer, respectively, by the end of 2030 (Pandian et al., 2016) (Fig. 10). Therefore, groundwater management should be followed by increasing the recharge and reducing the pumping to reduce seawater intrusion in Chennai.

7 CONCLUSION

Seawater intrusion has become a major issue in several coastal regions of the world over the last few decades. This chapter discusses the relation between freshwater and saltwater, identifying tools and mitigation methods of seawater intrusion along with two case studies. Measurement of groundwater level, geochemical methods, and geophysical methods are commonly used to identify seawater intrusion. Though there are several methods to mitigate seawater intrusion, the economy plays a major role. A reduction in pumping and a rearranging of pumping wells are cost-effective but not practicable. The abstraction of seawater and a subsurface barrier are expensive methods, hence they are not normally considered. Increasing groundwater recharge by percolation ponds is economically feasible and can be easily adopted. The quality of the groundwater is also improved by recharging through a percolation pond. Groundwater modeling can be used as a tool for sustainable groundwater management by understanding the dynamics of seawater intrusion under different hydraulic stresses as well as planning strategies.

Acknowledgment

The authors sincerely thank Anna University, Chennai, India for funding support through Anna Centenary Research Fellowship (Lr. No. CFR/ACRF/2017/40) and the Department of Science and Technology, New Delhi (Grant No. DST/CCP/NCC&CV/135/2017(G)).

References

Abdullah, M.H., Praveena, S.M., Aris, A.Z., 2010. A numerical modelling of seawater intrusion into an oceanic island aquifer, Sipadan Island, Malaysia. Sains Malays 39, 525–532.

Ataie-Ashtiani, B., Volker, R.E., Lockington, D.A., 1999. Tidal effects on sea water intrusion in unconfined aquifers. J. Hydrol. 216, 17–31.

Barazzuoli, P., Nocchi, M., Rigati, R., Salleolini, M., 2008. A conceptual and numerical model for groundwater management: a case study on a coastal aquifer in southern Tuscany, Italy. Hydrogeol. J. 16 (8), 1557–1576.

Carretero, S., Rapaglia, J., Bokuniewicz, H., Kruse, E., 2013. Impact of sea-level rise on saltwater intrusion length into the coastal aquifer, partido dela costa, Argentina. Cont. Shelf Res. 61–62, 62–70.

CGWB, 2014. Report on status of groundwater quality in coastal aquifers of India. CGWB, Faridabad, Haryana.

Chami, D., Moujabber, M., Scardigno, A., 2009. Regional water balance and economic assessment as tools for water management in coastal Lebanon. Water Resour. Manag. 23 (11), 2361–2378.

Chen, B.F., Hsu, S.M., 2004. Numerical study of tidal effects on seawater intrusion in confined and unconfined aquifers by time-independent finite-difference method. J. Waterw. Port Cost Ocean Eng. 130, 191–206.

Chen, B.W., Liu, C.W., Hsu, H.M., 2015. Modeling assessment of a saltwater intrusion and a transport time scale response to sea-level rise in a tidal estuary. Environ. Fluid Mech. 15, 491–514.

Christy, M.R., Elango, L., 2017. Percolation pond as a method of managed aquifer recharge in a coastal saline aquifer: a case study on the criteria for site selection and its impacts. J. Earth Syst. Sci. https://dx.doi.org/10.1007/s12040-017-0845-8.

Colombani, N., Osti, A., Volta, G., Mastrocicco, M., 2016. Impact of climate change on salinization of coastal water resources. Water Resour. Manage. 30, 2483–2496.

Das, A., Datta, B., 1999. Development of management models for sustainable use of coastal aquifers. J. Irrig. Drain. Eng. 125 (3), 112–121.

Das, P.P., Sahoo, H.K., Mohapatra, P.P., 2016. Hydrogeochemical evolution and potability evaluation of saline contaminated coastal aquifer system of Rajnagar, Odisha, India: a geospatial perspective. Earth Syst. Sci. 125 (6), 1157–1174.

Dausman, A., Langevin, C.D., 2004. Movement of the saltwater interface in the surficial aquifer system in response to hydrologic stresses and water management practices, Broward County, Florida. US Geol. Surv. Sci. Investig. Rep. 5256, 73.

Drabbe, J., Ghyben, B.W., 1889. Nota in verband met de voorgenomen putboring nabij Amsterdam. Tijdschrift van het Koninklijk Instituut van Ingenieurs, 8–22.

Eeman, S., Leinse, A., Raats, P.A.C., van der Zee, S.E.A.T.M., 2011. Analysis of the thickness of a fresh water lens and of the transition zone between this lens and upwelling saline water. Adv. Water Resour. 34, 291–302.

El-Kaliouby, H., Abdalla, O., 2015. Application of time-domain electromagnetic method in mapping saltwater intrusion of a coastal alluvial aquifer, North Oman. J. Appl. Geophys. 115, 59–64.

Han, D., Post, E.A.V., Song, X., 2015. Groundwater salinization processes and reversibility of seawater intrusion in coastal carbonate aquifers Dongmei. J. Hydrol. 531, 1067–1080.

Herzberg, A., 1901. Die Wasserversorgung einger Nordsee bader. J. Gasbeleuchtung Wasserversorgung 44, 815–819. 842–844.

IAEA-TECDOC, 1994. Isotope hydrology investigations in Latin America. p. 835.

Inouchi, K., Kishi, Y., Kakinuma, T., 1985. The regional unsteady interface between fresh water and salt water in a confined coastal aquifer. J. Hydrol. 77, 307–331.

Jones, B.F., Vengosh, A., Rosenthal, E., Yechieli, Y., 1999. Geo- chemical investigation of groundwater quality. In: Seawater Intrusion in Coastal Aquifers: Concepts, Methods and Practices. Springer, Dordrecht, The Netherlands, pp. 51–71.

Karahanoglu, N., 1997. Assessment of sea-water intrusion in a coastal aquifer by using correlation, principal component, and factor analyses. Water Environ. Res. 69 (3), 331–341.

Karahanoglu, N., Doyuran, V., 2003. Finite element simulation of seawater intrusion into a quarry-site coastal aquifer, Kocaeli-Darıca, Turkey. Environ. Geol. 44, 456–466.

Kourakos, G., Mantoglou, A., 2011. Simulation and multi-objective management of coastal aquifers in semi-arid regions. Water Resour. Manage. 25 (4), 1063–1074.

Kumar, B.R.B., Divya, M.P., 2012. Spatial evaluation of groundwater quality in Kazhakuttam block, Thiruvananthapuram District, Kerala. J. Geol. Soc. India 80, 48–56.

Kumar, P.J.S., Elango, L., James, E.J., 2014. Assessment of hydrochemistry and groundwater quality in the coastal area of south Chennai, India. Arab. J. Geosci. 7 (7), 2641–2653.

Kumar, P.J.S., Jose, A., Jamer, E.J., 2013. Spatial and seasonal variation in groundwater quality in parts of Cuddalore District, South India. Natl. Acad. Sci. Lett 36 (2), 167–179.

Laluraj, C.M., Gopinath, G., Dineshkumar, P.K., 2005. Groundwater chemistry of shallow aquifers in the coastal zones of Cochin. Ind. Appl. Ecol. Environ. Res. 3 (1), 133–139.

Liu, H., Yoshikawa, N., Tanaki, S., 2017. Effective method of removing saltwater wedge for preserving agricultural water quality. Paddy Water Environ. 15 (2), 331–341.

Luyun, R., Momii, K., Nakagawa, K., 2011. Effects of recharge wells and flow barriers on seawater intrusion. Ground Water 49, 239–249.

Melloul, A.J., Goldenberg, L.C., 1997. Monitoring of seawater intrusion in coastal aquifers: basic and local concerns. J. Environ. Manage. 51, 73–86.

Michael, H.A., Mulligan, A.E., Harvey, C.F., 2005. Seasonal oscillations in water exchange between aquifer and the coastal ocean. Nature 436, 1145–1148.

Naik, P.K., Dehury, B.N., Tiwary, A.N., 2007. Groundwater pollution around an industrial area in the coastal stretch of Maharastra state, India. Environ. Monitor. Assess. 132, 207–233.

Nair, S.I., Pandian, S.R., Schneider, V., Elango, L., 2015. Geochemical and isotopic signatures for the identification of seawater intrusion in an alluvial aquifer. J. Earth Syst. Sci. https://dx.doi.org/10.1007/s12040-015-0600-y.

Najib, S., Fadili, A., Mehdi, K., Riss, J., Makan, A., 2017. Contribution of hydrochemical and geoelectrical approaches to investigate salinization process and seawater intrusion in the coastal aquifers of Chaouia, Morocco. J. Contam. Hydrol. 198, 24–36.

National Institute of Oceanography (NIO), 2006. SAGAR a pocketbook on the ocean with special reference to the waters around India. National Institute of Oceanography (NIO), Dona Paula, Goa.

Nofal, E.R., Amer, M.A., El-Didy, S.M., Fekry, M.A., 2015. Delineation and modeling of seawater intrusion into the Nile Delta Aquifer: a new perspective. Water Sci. 29, 156–166.

Pandian, S.R., Nair, S.I., Elango, L., 2016. Evaluation of impact of climate change on seawater intrusion in a coastal aquifer by finite element modelling. J. Clim. Change 2 (2), 111–118.

Paniconi, C., Khlaifi, I., Lecca, G., Giacomelli, A., Tarhouni, J., 2001. Modeling and analysis of seawater intrusion in the coastal aquifer of Eastern Cap-Bon, Tunisia. Transport Porous Media 43, 3–28.

Papadopoulou, M.P., Karatzas, G.P., Koukadaki, M.A., Trichakis, Y., 2005. Modelling the saltwater intrusion phenomenon in coastal aquifers – a case study in the industrial zone of Herakleio in Crete. Global NEST J. 7, 197–203.

Parimala, R.S., Elango, L., 2013. Impact of recharge from a check dam on groundwater quality and assessment of suitability for drinking and irrigation purposes. Arab. J. Geosci. https://dx.doi.org/10.1007/s12517-013-0989-z.

Park, Y.H., Jang, K., Ju, W.J., Yeo, W.I., 2012. Hydrogeological characterization of seawater intrusion in tidally-forced coastal fractured bedrock aquifer. J. Hydrol. 446–447, 77–89.

Perriquet, M., Leonardi, V., Henry, T., Jourde, H., 2014. Saltwater wedge variation in a nonanthropogenic coastal karst aquifer influenced by a strong tidal range (Burren, Ireland). J. Hydrol. 519, 2350–2365.

Rao, N.S., Nirmala, I.S., Suryanarayana, K., 2005. Groundwater quality in a coastal area: A case study from Andhra Pradesh, India. Environ. Geol. 48 (4–5), 543–550.

Reichard, E.G., Crawford, S.M., Land, M.T., Paybins, K.S., 1999. Management of groundwater supply and water quality in the Los Angeles Basin, California. 260, IAHS-AISH Publication, Houston, TX, pp. 91–92.

Rina, K., Datta, P.S., Singh, K.C., Mukherjee, S., 2013. Isotopes and ion chemistry to identify salinization of coastal aquifers of Sabarmati River Basin. Curr. Sci. 104, 3.

Sadeg, A.S., Karahanolu, N., 2001. Numerical assessment of seawater intrusion in the Tripoli region, Libya. Environ. Geol. 40, 1151–1168.

Sales, J., Tamoh, K., Lopez-Gonzalez, J.L., Gaaloul, N., Candela, L., 2017. Controlling seawater intrusion by treated wastewater recharge. Numerical modelling and cost-benefit analysis (CBA) at Korba case study (Cap Bon, Tunisia). Desalin. Water Treat. 76, 184–195.

Sathish, S., Elango, L., Rajesh, R., Sarma, V.S., 2011. Application of three dimensional electrical resistivity tomography to identify seawater intrusion. Earth Sci. India 4 (1), 21–28.

Scholze, O., Hillmer, G., Schneider, W., 2002. Protection of the groundwater resources of Metropolis CEBU (Philippines) in consideration of saltwater intrusion into the coastal aquifer.17th Salt Water Intrusion Meeting, Delft, The Netherlands.

Shalev, E., Lazar, A., Wollman, S., Kington, S., Yechieli, Y., Gvirtzman, H., 2009. Biased monitoring of fresh water-salt water mixing zone in coastal aquifers. Ground Water 47, 49–56.

Sherif, M., Al-Rashed, M., 2001. Vertical and horizontal simulation of seawater intrusion in the Nile Delta Aquifer.1st International Conference on Saltwater Intrusion and Coastal Aquifers, Monitoring, Modelling, and Management (Morocco).

Sherif, M., Kacimov, A., Javadi, A., Ebraheem, A.A., 2012. Modeling groundwater flow and seawater intrusion in the coastal aquifer of Wadi Ham, UAE. Water Resour. Manage. 26, 751–774.

Sherif, M.M., Hamza, I.K., 2001. Mitigation of seawater intrusion by pumping brackish water. Transport Porous Media 43, 29–44.

Singaraja, C., Chidambaram, S., Anandhan, P., Prasanna, M.V., Thivya, C., Thilagavathi, R., 2014. A study on the status of saltwater intrusion in the coastal hard rock aquifer of South India. Environ. Dev. Sustain. https://dx.doi.org/10.1007/s10668-014-9554-5.

Sugio, S., Nakada, K., Urish, D.W., 1987. Subsurface seawater intrusion barriers analysis. J. Hydraul. Eng. 113, 767–779.

Terzic, J., Peh, Z., Markovic, T., 2010. Hydrochemical properties of transition zone between fresh groundwater and seawater in karst environment of the Adriatic islands, Croatia. Environ. Earth Sci. 59, 1629–1642.

Tomaszkiewicz, M., Najm, A.M., Fadel, E.M., 2014. Development of a groundwater quality index for Seawater intrusion in coastal aquifers. Environ. Model. Softw. 57, 13–26.

Triki, C., Zekri, S., Al-Maktoumi, A., Fallahnia, M., 2017. An artificial intelligence approach for the stochastic management of coastal aquifers. Water Resou.Manage. 31 (15), 4925–4939.

Tsao, Y.S., 1982. Simulation model for the management of groundwater in the Yun-Lin basin. 136 IAHS-AISH Publication, pp. 333–343.

Vallejos, A., Sola, F., Pulido-Bosh, A., 2015. Processes influencing groundwater level and the freshwater-saltwater interface in a coastal aquifer. Water Resour. Manage. 29, 679–697.

Vetrimurugan, E., Elango, L., 2015. Groundwater chemistry and quality in an intensively cultivated river delta. Water Qual. Expo. Health 7, 125–141.

Werner, A.D., 2010. A review of seawater intrusion and its management in Australia. Hydrogeol. J. 18, 281–285.

Werner, D.A., Bakker, M., Post, E.A.V., Vandenbohede, A., Lu, C., Ashtiani-Ataie, B., Simmons, T.C., Barry, D.A., 2013. Seawater intrusion processes, investigation and management: recent advances and future challenges. Adv. Water Resour. 51, 3–26.

Yang, J., Graf, T., Ptak, T., 2015. Impact of climate change on freshwater resources in a heterogeneous coastal aquifer of Bremerhaven, Germany: a three-dimensional modeling study. J. Contam. Hydrol. 177–178, 107–121.

Zacharias, I., Koussouris, T., 2000. Sustainable water management in the European Islands. Phys. Chem. Earth Part B: Hydrol. Oceans Atmos. 25 (3), 233–236.

Zhao, J., Lin, J., Wu, J., Yang, Y., Wu, J., 2016. Numerical modeling of seawater intrusion in Zhoushuizi district of Dalian City in northern China. Environ. Earth Sci. 75, 805.

Zhou, X., 2011. A method for estimating the fresh wateresalt water interface with hydraulic heads in a coastal aquifer and its application. Geosci. Front. 2 (2), 199–203.

Zhou, X., Chen, M., Ju, X., Ning, X., Wang, J., 2000. Numerical simulation of sea water intrusion near Beihai, China. Environ. Geol. 40, 223–233.

CHAPTER

18

Practices of Tsunami Evacuation Planning in Padang, Indonesia

Faisal Ashar, Dilanthi Amaratunga, Pournima Sridarran, Richard Haigh

Global Disaster Resilience Centre, University of Huddersfield, Huddersfield, United Kingdom

1 INTRODUCTION

Indonesia is considered to be a very vulnerable country that is prone to disasters because it has >18,000 islands located along the Pacific *"ring of fire"* of active volcanoes and tectonic faults. The national population is approximately 224 million inhabitants, comprising a mix of ethnicities, religions, customs, and traditions. Of the 471 districts/cities, 383 are disaster-prone areas with a large and uneven population distribution (Hadi, 2009a).

Irsyam et al. (2010) calculated >14,000 earthquake occurrences with a magnitude of $M > 5.0$ in Indonesia between 1897 and 2009. The largest earthquakes in the last 6 years were the 2004 Aceh earthquakes and tsunami ($Mw = 9.2$), the 2005 Nias earthquake ($Mw = 8.7$), the 2006 Yogyakarta earthquake ($Mw = 6.3$), the 2009 Tasikmalaya earthquake ($Mw = 7.4$), and the 2009 Padang earthquake ($Mw = 7.6$).

Losses from earthquakes are not only measured in the loss of human lives but also by the amount of damage to the housing and infrastructure. In the 2004 Aceh earthquake, 120,000 houses were damaged. Subsequently, 306,234 and 13,577 houses were damaged in the 2006 Yogyakarta and 2007 Bengkulu earthquakes, respectively (Hadi, 2009b). More recently, the 2009 Padang earthquake destroyed 114,797 houses.

Among disaster-prone cities in Indonesia, Padang is identified as the region that is the most likely to be devastated by a massive tsunami in the near future. Padang is the third-biggest city and the capital of the Western Sumatra Province of Indonesia. It is situated on the coast of the Indian Ocean. Many international research communities have studied the potential tsunami hazard in Padang. Among them, some noteworthy studies are (Borrero et al., 2006), (Taubenböck et al., 2009), and (McCloskey et al., 2008). A study conducted by Singh

Coastal Management
https://doi.org/10.1016/B978-0-12-810473-6.00019-4

© 2019 Elsevier Inc. All rights reserved.

et al. (2010) estimates that it takes about 20–32 min for a tsunami to hit Padang after a massive earthquake that generates a tsunami. On the other hand, Padang is located on very flat liquefiable ground, and one has to walk >3 km to reach an altitude of 5 m from the coast.

The impact of a potential tsunami hitting Padang is expected to be high owing to the large population that lives in the coastal region. The key reason for this density is that the primary economic activities of Padang, including critical infrastructure, medical services, schools, public offices, and transportation networks, are constructed parallel to the coastline. This multiplies the exposure of the population to potential tsunami waves. It is estimated that about 50% of residents live in the lowland areas around the coast or in areas that are between 0 and 5 m above sea level. The population of the city is 844,316 (BPS, 2011), and thus, it would be challenging to evacuate about 400,000 people to a tsunami safe zone on short notice. On one hand, the time will be limited for people to reach a safe place or higher. On the other hand, insufficient transportation facilities that are filled with people who are panicking would cause traffic jams.

This frightening threat that could occur in the near future and the high level of vulnerability of the city impose a strong need for the enforcement of tsunami hazard preparedness in the city.

2 TSUNAMIS IN INDONESIA AND PADANG

The source event that could create a tsunami is an earthquake in Sumatra in the Sunda Arc subduction zones and submarine features (Hamilton, 1979). When an earthquake occurs, the sea bed over the fracture zone is catastrophically displaced, causing collapse or upheaval of the overlying water mass. The Sunda Arc is one of the most active plate tectonic margins in the world. Studies on the Sunda Arc show that massive megathrust earthquakes have not been evenly distributed over time. Historical records with intervals of about 250 years show that specific sections have been active while others have been relatively quiet (in front of Padang). Most parts of the Sunda Arc are considered capable of generating significant megathrust earthquakes and potentially producing significant tsunamis (UNESCO-IOC, 2009). Natawidjaja (2007) and UNESCO-IOC (2009) state that research about earthquakes in Sumatra have often been conducted, and the studies produced adequate data records from 700 years ago.

Padang is located close to the subduction zone with earthquakes that can cause tsunamis. Tsunami waves travel quickl, reaching the coast in very short periods of time (GITEWS, 2008a). In that way, tsunamis in Padang are called “local tsunamis” or “near-field tsunamis.”

According to Singh et al. (2010), as stated in Section 1, the time interval between the first strong earthquake and the first tsunami wave that hits the coast of Padang is about 20–30 min. Thus, it is expected that the residents can save themselves as soon as possible within 10 min. In contrast to the expectation, the reality is that the residents have to walk 3–5 km to reach a safe area. It clearly shows that the time for tsunami evacuation in Padang is very short.

Several studies have been conducted to develop an evacuation and preparedness plan for Padang. Notably, the Padang Consensus on Official Tsunami Hazard Map-Protocol Nov 2008 (GITEWS, 2008b) has been determined to be an evacuation map for Padang.

3 TSUNAMI EVACUATION PLAN

The primary problem is evacuating the potentially vulnerable population on time to safe places in the event of a tsunami. Usually, in densely populated areas, detailed planning is required so that evacuation can be executed as efficiently as possible. Thus, the evacuation plan should be established, implemented, and monitored by local decision-makers (Scheer et al., 2011). Generally, a tsunami evacuation plan is prepared for evacuation before and during a tsunami (Spahn et al., 2010).

A tsunami evacuation plan (TEP) is a plan that will be invoked and undertaken if a tsunami warning has been activated. Thus, the TEP will affect various preparedness measures to be activated in the event of a tsunami warning. The primary aim of a TEP should, therefore, be to guide all affected persons along the evacuation routes toward safe places, and on time (Scheer et al., 2012).

Because a TEP will activate a variety of preparedness measures in the case of a tsunami alert, usually, an audiovisual signal or a telecommunication-based message is suggested. However, in some cases, especially in the context of a near-field tsunami or a local tsunami, the warning may happen only a few minutes before the arrival of the first tsunami wave. Therefore, the ability to sense and assess the earthquake from the first nonambiguous signal is essential, even though no tsunami will occur (Scheer et al., 2011). The following section explains some approaches to evacuation planning processes.

3.1 Approaches of Evacuation Planning Processes

3.1.1 SCHEMA Project Methodology

The SCHEMA project (Scenarios for Hazard-induced Emergencies Management) develops the TEP in three phases: the first valid evacuation plan, the midterm revision, and long-term revision and integration. In order to maintain the first valid evacuation plan continuously, the midterm revision will be executed. Further, long-term studies will be conducted to run integration with early warning systems and other emergency plans and examine legal obligations. Ideally, the entire evacuation plan should be assessed in conjunction with the affected population to gain maximum acceptance (Scheer et al., 2012). UNESCO-IOC (2015) agrees with this by stating that evacuation planning is a long process that should be included in the standard operation procedure and considered as an ongoing effort in successive iterations.

According to Scheer et al. (2011), the purpose of a TEP is to protect people who are exposed to a tsunami by showing how to get to a safe place effectively and on time. Thus, the SCHEMA project developed a TEP based on two parameters. The first parameter is the number of people affected as well as locations, roads, and distances. The second parameter is an assumption of maximal expected wave height and estimated time of arrival (ETA) of the first wave. The TEP is developed based on the result of local tsunami scenarios, scientific data from the tsunami characteristics of the planned area, the possibility of the extent of damage, the timing of the movement of refugees to a safe or vertical shelter, and the use of iterations for each stage of the plan.

Scheer et al. (2011) suggests a flow chart of the steps of evacuation planning that follows the three-step methodology:

1. Risk and impact analysis for the definition of the plan background and input.
2. Evacuation plan production and implementation.
3. Evacuation plan deployment, with monitoring and updating.

The flow chart shows the three main steps of evacuation planning and the iteration of each step to the previous step. There is an improvement or refinement of the activity before proceeding to the next stage. It is even possible to jump from step 3 to step 1 if there are essential things that must be fixed.

The evacuation map based on the SCHEMA project map recommends that the evacuation areas should be located outside the tsunami hazard areas. The map further indicates some evacuation buildings that are proposed on the way for the displaced population who do not have enough time. Evacuation routes or trails are also clearly illustrated, and evacuation routes forbidden for the displaced population are also illustrated in the map. This map is the basic or simple template of a tsunami evacuation map.

In short, the SCHEMA approach provides a complete methodology that enables communities to prepare for a tsunami disaster. This method is based on scientific insights that are specific to tsunami characteristics, and thus, further calculations are essential to compile it more efficiently.

The SCHEMA also detects some constraints in compiling TEP, including:

- Absence of early warning system (EWS).
- Absence of analysis tools.
- Absence of shelter sites.
- Lack of acceptance by population.
- Lack of acceptance by decision-makers.
- Evacuation of special population.
- Tsunami warning limitations.

Agreeing above, Scheer et al. (2011) says that a TEP will not be successful without EWS installment and integration, dissemination of all information, community preparedness, and the awareness of the people.

3.1.2 *GITEWS Project*

GITEWS is a joint project between Germany and Indonesia to enhance the preparedness of the government and communities who face disasters in several tsunami-threatened cities in Indonesia. According to PROTECTS and GIZ (2009), "Saving yourself from the tsunami is the ability to get out of the reach of the tsunami and (the) inundated water just in time."

Therefore the principles of evacuation planning based on the GITEWS approach are as follows;

- Who should be involved in planning? It is all stakeholders from local government, universities, NGOs, etc.
- Who should know the evacuation plan? Every member of society.
- As Indonesia faces a local tsunami hazard, time is the most important factor.

- The time for tsunami evacuation covers before and during the tsunami and is part of the tsunami contingency plan.
- Evacuation planning requires a good understanding of the tsunami risks facing the community.
- The official evacuation plan gives trusted and binding referral to the institutions under it.
- Evacuation plans should be specially designed for specific area conditions.
- Evacuation plans need to inform citizens when to evacuate.

From the above principles, the evacuation planning process was developed by GITEWS with five steps of the planning process:

1. Prepare for the planning.
2. Understand your community's tsunami risk.
3. Design your evacuation strategy and map.
4. Assess, endorse, and disseminate your evacuation plan.
5. Test, evaluate, and improve your evacuation plan.

A TEP process based on the GITEWS approach provides two key outcomes: tsunami evacuation maps and evacuation strategies. The evacuation map highlights the zones as red and yellow areas. The red zone is considered as the most dangerous area and the one most likely to be hit by a tsunami. The yellow area is regarded as relatively safe unless there is a worst-case scenario (Anggraeni and Artanti, 2013; Gede and Hoppe, 2010; Spahn and Kesper, 2010).

As mentioned above, GITEWS suggests five steps of the planning process, and every step contains topics of discussion and output at each level. These outputs are in the form of a plan. The first step of this approach begins with preparing the plan, then understanding the risks that are faced by the community, and followed by designing the evacuation strategy and maps. Step four is reviewing, legitimating, and spreading the evacuation plan, and finally testing, evaluating, and improving it (PROTECTS and GIZ, 2009; Spahn et al., 2010).

Based on the discussion above, the tsunami evacuation planning process can be compared between the SCHEMA and GITEWS methods. SCHEMA uses more accurate counts while GITEWS is largely based on discussions with all stakeholders, especially affected communities. The output of the SCHEMA is a scalable and well-scaled map. However, if the scientific data or calculations such as bathymetry data or tsunami modeling is unavailable, the GITEWS approach can be used. The tsunami evacuation planning process with the GITEWS approach has included communities or all stakeholders. Thus, GITEWS is more appropriate to be applied in Indonesia. Besides, SCHEMA's approach has advantages in monitoring activities from start to finish because iterations of each stage are included in their methodology approach stage.

3.2 Tsunami Evacuation Planning in Padang

TEP is prepared for evacuation before and during a tsunami (Spahn et al., 2010). Therefore, a TEP should integrate the phases of disaster preparedness with good planning that must be evaluated continuously. Furthermore, the validity of an evacuation plan has to be checked continuously, appropriate updates generated, and the overall maintenance guaranteed

(FEMA, 2003). Further, planning and evaluation are also on a continuum of the cycle of disaster preparedness (Socialresearchmethods, 2006).

With Padang having a population of is approximately 800,000, a large number of vehicles are on the highways and congestion often occurs in the city on a typical day. From the revenue point of view, the city is not economically stable, and thus, it has a limited budget in terms of city operation costs (BPS, 2011). In other words, the city has its unique characteristics and is different from other cities.

However, beyond these issues, there is the frightening threat that could occur in the near future: a tsunami. Due to the high level of vulnerability, this city has to be prepared for a tsunami hazard. Although some parts of the TEP have been already implemented, not all the variables of the TEP activities have been addressed.

In this context, this chapter reports an evaluation of the TEP process in Padang, based on the two models of TEP (Scheer et al., 2011; Spahn et al., 2010; UNESCO-IOC, 2009). The evaluation is conducted by formulating a disaster preparedness measure, based on the study of Seymour et al. (2010) and Simpson (2008). Thus, this study offers an evaluation of the TEP process in Padang using the three models of TEP (Scheer et al., 2011; Spahn et al., 2010; UNESCO-IOC, 2009).

4 CASE STUDY: PADANG

The case study was conducted in Padang. Semistructured interviews and a questionnaire wereconducted to collect data. The survey is conducted on a sample of 257 households, consisting of 126 women and 131 men. A total of 214 respondents live in the red zone or tsunami inundation area, and 43 respondents are in safe zone area. The semistructured interviews were conducted among the experts from the government, NGOs, and KSB. Fourteen in-depth interviews were conducted, and the respondents are referred to as INT01-INT14 within the text as their identities were anonymized.

4.1 Geography, Topography, and Oceanography

The Republic of Indonesia (RI), commonly called Indonesia, is a country in Southeast Asia that is crossed by the equator. It is located between the continents of Asia and Australia and between the Pacific and Indian Oceans. Indonesia is the biggest archipelago in the world, consisting of 13,466 islands. The five largest islands are Sumatra, Java, Kalimantan, Sulawesi, and Irian. Jakarta is the capital of the state and currently consists of 34 provinces, including the province of West Sumatra, which has Padang as its provincial capital (Indonesia, 2016).

Padang, the case study in this research, is located on the west coast of the island of Sumatra. The coordinates are between 0°44′00′-10°8′35′ latitude and 100°05′05″-100°34′09″ longitude (BPS, 2016). Padang is located on the same island (the island of Sumatra) as Aceh. Aceh was severely affected by the earthquake and tsunami on December 26, 2004, and the tsunami waves also hit several other countries such as Thailand and Sri Lanka.

According to Government Regulation No. 17 of 1980, Padang's administrative area is 694.96 km^2, which is equivalent to 1.65% of the area of the West Sumatra Province. The

TABLE 1 Slope Area in Padang

No	Slope	Area (Hectare)
1	0%–2%	16,292.45
2	2%–15%	5481.53
3	15%–40%	13,148.96
4	>40%	34,202.35

Sources: BAPPEDA Padang and ArcGIS analysis.

original Padang consisted of three districts with 15 villages. It was later developed into 11 districts with 193 villages. Law No. 22 in 1999 on Regional Government was followed by Government Regulation No. 25 in 2000, which granted an additional extended administration that consisted of 1414.96 km^2 (720.00 km^2 of which is the region of the sea) and the merger of several villages, thus becoming 104 villages, among which Koto Tangah is the biggest at 232.25 km^2 (BPS, 2016; Padang, 2011; PP No.17/1980, 1980).

Padang has a varied topography with a mixture of gently sloping land and undulating hills, some of which are steep. Most of the topography in Padang has an average slope rate of >40%. The areas with slopes between 0 and 2% are found in the districts of Padang Barat, Padang Timur, Padang Utara, Nanggalo, part of the Kuranji subdistrict, Padang Selatan, Lubuk Begalung, and Koto Tangah. Regions with gradients of 0–2%, which are relatively flat, and most of the coastal areas are prone to tsunamis. Data tables and maps of slopes can be seen below (Table 1 and Fig. 1).

The elevation of the land area of Padang varies between 0 and 1853 m above sea level, and the highest area is the district of Lubuk Kilangan. The areas with heights of 0–5 m or 0%–2% slope along the coastline are considered as potentially a tsunami inundation area, a prone zone, or a red zone.

Five large rivers and 16 small rivers run through Padang. The longest river is Batang Kandis, which is 20 km long. The large rivers will be the biggest obstacles to the process of tsunami evacuation in Padang. This is because, based on the experience of previous large earthquakes, the bridges experience a shift in position (abutment dislocation) due to earthquake vibration, and are sometimes severely damaged. Thus, after a massive earthquake occurs, some of the bridges cannot be used or bypassed by people and vehicles. Moreover, it takes a long time and considerable cost to repair a bridge.

Padang has a coastline of ±84 km, or 68.126 km (excluding small islands), according to PP No.17/1980 (1980). Administratively, there are six districts in direct contact with the coast: Koto Tangah, Utara Padang, Padang Barat, Padang Selatan, Lubuk Begalung, and Bungus Teluk Kabung. Residents who live in these six districts are vulnerable to potential tsunamis. Padang is divided into two portions based on topography. The first part is the north, and the second is the South. The northern part includes the districts of Koto Tangah, Utara Padang, and Padang Barat. The southern part contains the districts of Padang Selatan, Lubuk Kilangan, and Bungus Teluk Kabung.

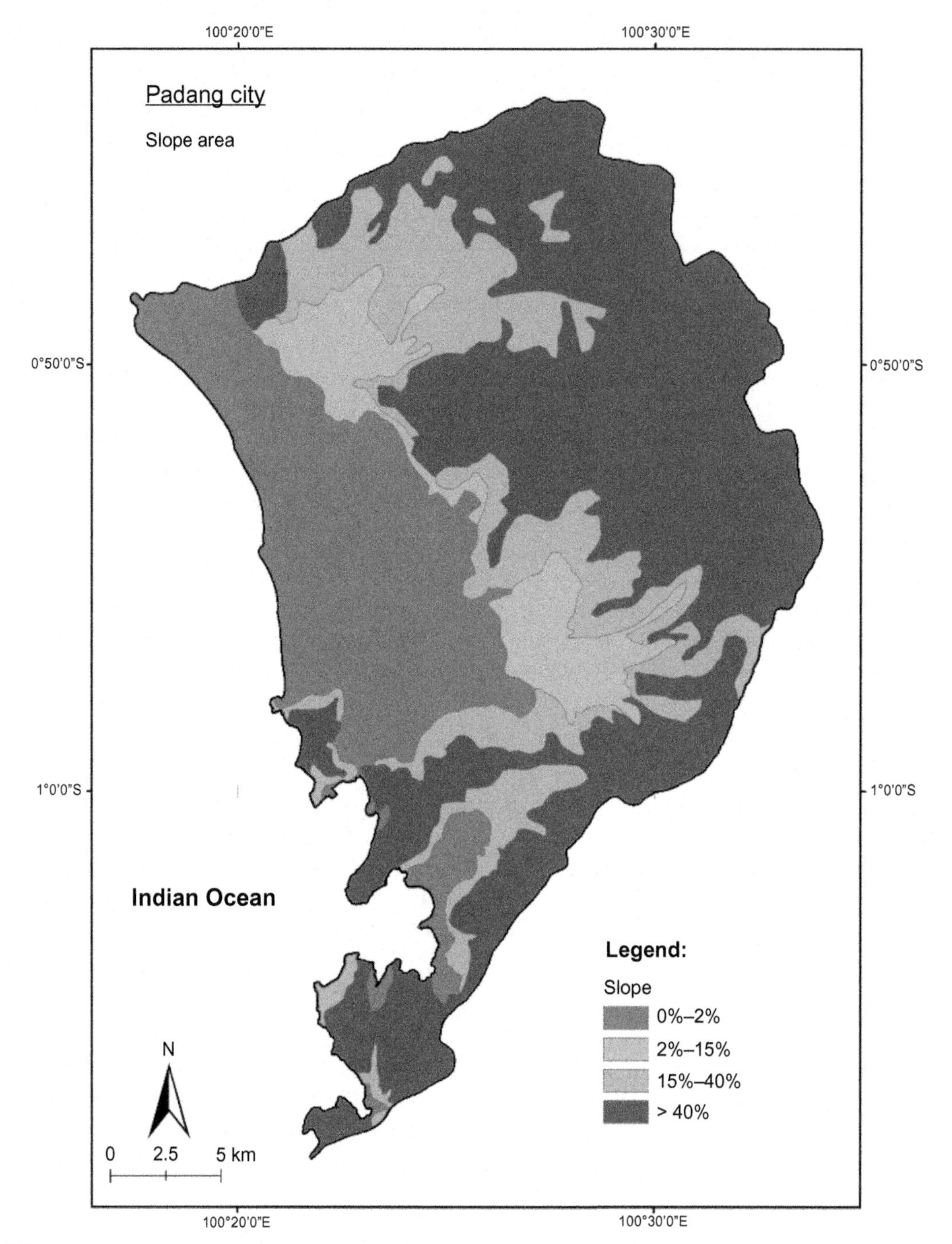

FIG. 1 Slope area in Padang. *Sources: BAPPEDA Padang and ArcGIS analysis.*

4.2 Demography

Knowledge of the population is the key to conducting development activities, planning, and evaluation. In 2015, the population of Padang reached 902,413 inhabitants, with an increase of 12,767 residents from the previous year. As a result, the density was increased from 1280 inhabitants/km^2 to 1299 inhabitants/km^2. The district with the highest population is Koto Tangah with 182,296 inhabitants. On the other hand, the land area of this district is wider than the other districts, reaching 33% of the city of Padang, with population density is low at 784.91 inhabitants/km^2. The subdistrict with the smallest population (24,137 inhabitants) and the lowest density (240 people/km^2) is Bungus Teluk Kabung. Other districts that are also less populated are Pauh, which has 468 people/km^2, and Lubuk Kilangan, which has 624 people/km^2 (Fig. 2).

According to a survey conducted by BPS (Central Bureau of Statistics (Badan Pusat Statistik)), 83.74% of the population of Padang is 15 and above, which is the labor force. Concerning gender, 63.04% of the labor force is male and 36.96% is female. Jobseekers from the population aged 15 and above amounted to 16.26% while 36.19% of the Padang population is 15 years old and not considered to be the workforce.

4.3 Disaster Conditions in Padang

Data from the Disaster Risk Index in 2013 BNPB (National Agency for Disaster Management) indicate that Padang falls in the category of high hazard, and is ranked 10th nationally and first among districts in the province of West Sumatra. Based on data from the DIBI (Data and Disaster Information Indonesia) BNPB, during 2000–16, there have been 99 occurrences of disastrous events that caused loss of life and property. Disaster events cover 10 types of disasters: flooding, flooding and landslides, tidal waves/abrasion, earthquakes, fires, land and forest fires, transport accidents, drought, cyclones, and landslides. Among these types of disasters, flooding occurs most often and accounts for as many as 34 incidents or 35% in the period between 2000 and 2016, as shown in Fig. 3.

Regarding impact, the type of disaster that causes the most casualties and property damage in Padang is the earthquake. BNPB indicates that 388 victims died as a result of earthquake disasters between 2000 and 2016. The number of heavily damaged houses was 79,016. Although the frequency of occurrence of earthquakes is small (eight times during 2000–16), the impact was extraordinary. Therefore, particular attention is needed from local government, especially regarding disaster preparedness and mitigation.

Based on the course of events from 2000 to 2012, it appears that the death toll is higher after the earthquake of September 30, 2009, that had a force of 7.6 SR. In addition to the 2009 earthquake disaster, other types of disasters—particularly floods and landslides—also caused considerable casualties with as many as 58 people dead, two missing, and 25 injured.

4.4 Earthquake and Tsunami Hazards in Padang

Threats or hazards according to Law No. 24 of 2007 Law 24/2007 (2007) are defined as events that could be catastrophic while a natural hazard is a naturally occurring event that has adverse effects on humans. For example, natural disasters caused by a geological process

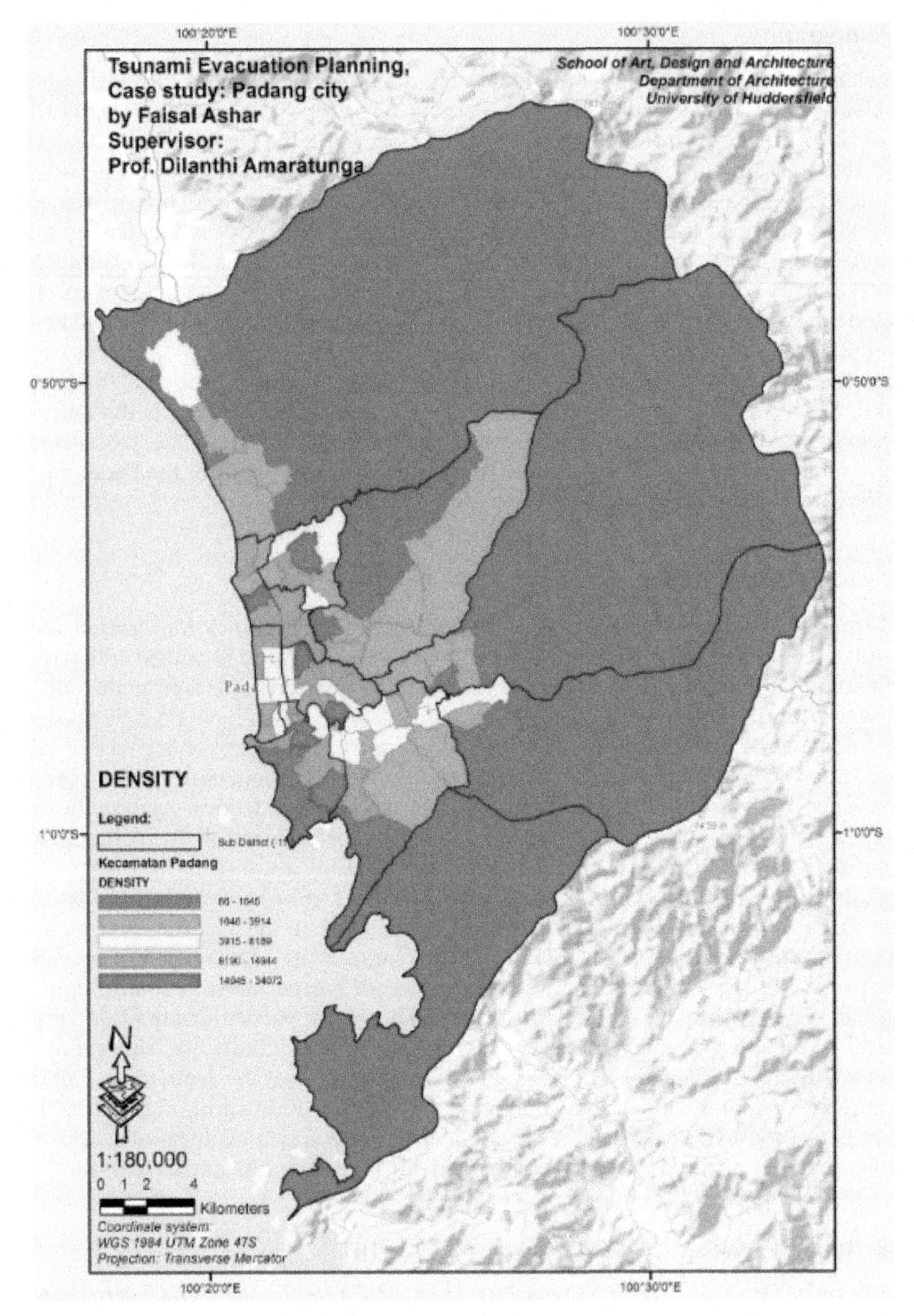

FIG. 2 Density of Padang. *Sources: BAPPEDA Padang and ArcGIS analysis.*

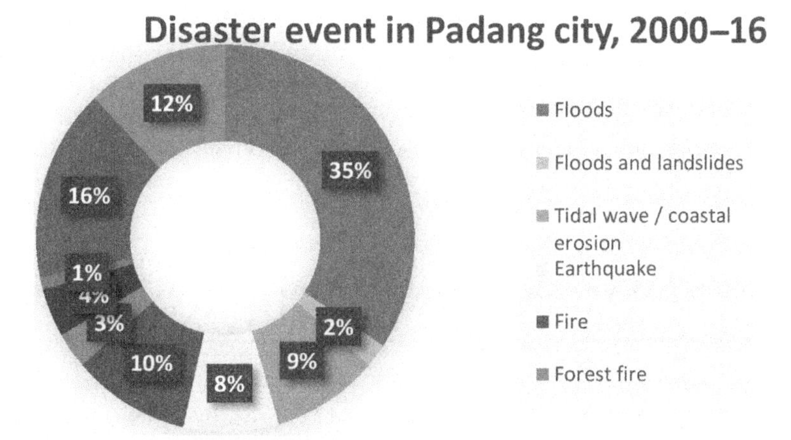

FIG. 3 Percentage of disasters in Padang, 2000–16.

are called geological disasters; these are earthquakes, tsunamis, volcanic eruptions, and landslides. Disasters caused by weather-related factors are floods, landslides, droughts, forest fires, and tornadoes. Tidal waves are known as hydrometeorological disasters.

Padang is an area that has a high level of vulnerability to natural disasters. As explained earlier, the disaster with the most significant impact in this city is the earthquake. History shows that catastrophic events in the Padang region are most often small-scale (<5 SR) earthquakes or large-scale earthquakes (>5 SR), which cause loss of life and property.

On Sept. 30, 2009, a massive earthquake (7.6 in magnitude) hit Padang, causing >300 fatalities. According to experts, the damage that occurred in West Sumatra as a result of this earthquake shows that the poor quality of construction was a hazard that was triggered by the earthquake. In Padang, schools, shops, hotels, and government buildings collapsed and buried hundreds of people. The overall official death toll was >1100 people. One of the most affected areas was the Padang Pariaman district, located in the north of Padang.

According to government data, about 200,000 homes and 2000 other buildings were damaged, where half of them are in an unusable condition. The devastating effects of the 2009 earthquake have been widely documented. Understanding the hazards, exposure, and vulnerability can help to identify the main factors of disaster risk in the community and assist in developing appropriate strategies for disaster risk reduction. What should also be remembered is that the evacuation process was not structured and many people were evacuated by private vehicles, causing heavy traffic in several points that would be very dangerous if a tsunami occurred. Further, it is essential to note that the situation was aggravated owing to the lack of an early warning system coordinated by the relevant agencies. This shows that the people still don't understand the behavior of the evacuation process. This imposes a need for solutions from the government to change behavior and increase the knowledge of community readiness when an earthquake or potential tsunami occurs.

After the 2004 earthquake and tsunami in Aceh, the threat of the earthquake and tsunami became a primary concern of all stakeholders in Padang. Earthquakes that occurred continuously in West Sumatra has been traumatic for its residents. This started in March 2005 on the island of Nias with the power of 8.5 magnitude and followed by a 6.9 magnitude earthquake in Mentawai sea in April 2005. Further, a 6.3 magnitude earthquake in Lake Batur in March 2007, SR 8.4 and 7.9 earthquakes in the seas Bengkulu and West Sumatra on 12 and 13 September 2007 and 30 September 2009, and the earthquake with a magnitude of 7.6 off the coast of Sumatra have continuously occurred. Moreover, seismologists have estimated that the next potential big earthquake would be in Mentawai.

5 TSUNAMI EVACUATION PLANNING PRACTICES AT PADANG GOVERNMENT INSTITUTIONS

The central government already has institutions associated with disaster management. In particular, BMKG is an institution that regularly (24/7) reports on the state of the weather, rain, earthquakes, and tsunami warnings (BMKG, 2016).

BMKG works with all the television stations and has an account on Twitter (@infoBMKG) in order to spread earthquake and tsunami warnings in real time. Twitter-based text and television are used because, as very often happens, the website/server of BMKG is often down during a potential tsunami or earthquake warning, owing to high traffic on the BMKG website.

Indonesia formed an agency called BNPB (*Badan Nasional Penanggulangan Bencana*) or the National Disaster Management Authority to manage disasters. The BNPB's functions are formulating and establishing disaster management policies, handling of refugees to act quickly and appropriately, and coordinating and implementing the disaster management activities in a planned, integrated, and comprehensive way.

At the regional level, Indonesia formed BPBDs to serve provincial, district, and city levels. The Regional Disaster Management Agency of West Sumatra province was established by the West Sumatra Provincial Regulation No. 9 of 2009 under the Establishment of Organization and Working Procedure of the Regional Disaster Management Agency of West Sumatra province.

On January 18, 2009, the government of Padang formed the Regional Disaster Management Agency of Padang and the Padang Local Regulation No. 18 of 2008 formed the Organization and Work of the Regional Disaster Management Agency of Padang.

The city government and DPRD Padang (legislative council) took the initiative to streamline the organizational structure (SKPD) in Padang so as not to overlap in the execution of duties. On December 4, 2012, it formed the organization and working arrangement with the Regional Disaster Management Agency and Fire Brigade Padang (BPBD-PK) in accordance with Padang Regulation No. 17 in 2012.

5.1 Tsunami Evacuation Maps

Government and related stakeholders have revised and updated tsunami evacuation maps several times. INT03 confirms that they incorporate adjustments by working on the important updates to support the process of tsunami evacuation in the future once every 2 years. These updates are carried out to improve capacity, build shelters, and open new evacuation routes.

5.1.1 Timeline Tsunami Evacuation Map Padang

KOGAMI, one of the credible NGOs that deals with disaster management, has surveyed the higher zones in order to produce a hazard zone map. The map is utilized as an initial reference to prepare Padang to face tsunami hazards. The map is designed based on the study of two famous experts: Prof. Kerry Sieh and Dr. Danny Hilman, who hypothesize that 4–5 m (MSL) of tsunami waves will hit Padang.

The hazard zone map is illustrated in three different colors:

- Red zone/tsunami hazard zone (0–5 MSL) indicates high risk of tsunami possibility.
- Yellow zone/tsunami alert zone (5–10 MSL) indicates average risk of tsunami possibility.
- Green zone/tsunami safe zone (10 MSL) indicates low risk of tsunami possibility.

Padang residents have been expected to prepare themselves by referring to the tsunami hazard map and identifying places where they live. Residents living in the red zone are expected to be able to recognize the nearest safe zones that they can reach immediately after a big earthquake (KOGAMI, 2014; Studio Mitigasi Bencana, 2016).

On August 25, 2008, at Andalas University, Padang, in cooperation with the Ministry of Marine Affairs and Fisheries (MoMAF), the Disaster Study Center and the Asia Pacific Economic Cooperation (APEC) held the International Workshop on an Official Tsunami Hazard Map for Padang (MoMAF, 2008). The workshop was attended by the local and the central government, NGOs, and scientists from Indonesia, Germany, Japan, and the United States. The purpose of this global agreement was to define what kind of hazard map can be used as an official map in Padang. The map will be applied as a reference for tsunami disaster preparedness.

After 2 years, on April 21, 2010 at the Padang Consensus II, a workshop and follow-up meeting were held to decide what type of hazard map was best for Padang. The accord was agreed upon by the head of the Municipal Disaster Management Office (BPBD) in Padang, the head of the Area Development Section—Municipal Development Planning Board (BAPPEDA) Padang, the head of the Center of Disaster Studies—Andalas University, the local advisor of the GTZ German-Indonesian Tsunami Early Warning System (GTZ-GITEWS), and the intern of the GTZ Emergency Reconstruction Project (GTZ-ERP) (BPRR, 2010). The result of the consensus is a recommendation relating to the tsunami inundation territory based on the investigation conducted by experts from Indonesia, Germany, the United States, and Japan. Most participants recommend Map 3. Validation was via the mayor's regulation (or PERWAKO).

Immediately after the consensus, on September 30, 2010, Fauzi Bahar, the mayor of Padang, released the Padang evacuation map. People from most of the related components of disaster management participated actively in the designing of the map. It was first published by Mercy Corps (NGO) through the Disaster Risk Reduction—Awareness Campaign and Government Capacity Building program. The map was printed on A2 double-sided paper and folded into its smallest size. The map contains essential information related to tsunami evacuation, and it can be freely downloaded at www.gitews.org/tsunami-kit.

Unlike previous maps, which show the hazard zones in three different colors (red, yellow, green), this map applies two colors to illustrate the perim of the tsunami hazard zone. As explained by INT05 and INT08: "*the Tsunami evacuation map released by PERWAKO/Mayor is all about the hazard zone map. It is actually derived from the early version applying three different zones, and it describes the direction of routes heading to the tsunami safe-zone, symbolized by green arrows.*" In short, this evacuation map only provides a horizontal evacuation method.

The Tsunami hazard zone map is legalized by the Padang government Act No. 4, 2012, and documented in City Spatial Planning 2010–2030 regulation (Walikota Padang, 2012). Thus, both residents and stakeholders can use the tsunami hazard zone map in the future.

Finally, in January 2014, another update was released. It was printed and distributed by Mercy Corp (NGO) with the help of PT Semen Padang Tbk. His latest version adds the location of vertical shelters and some buildings are illustrated as potential places for vertical shelters using color codes. INT05 and INT08 said: "*Previous maps were all about horizontal evacuation. Legally, both maps are similar, but by the publication of this latest version, we are able to read the potential shelters that are colored by green yellow and the siren locations.*" Similarly, the guidance to conduct horizontal evacuation is provided and marked by green arrows. This latest version, however, contains the most complete information about tsunami evacuation.

5.1.2 *The Distribution and Reprinting Process of the Latest Version of the Tsunami Evacuation Map*

The latest version of the tsunami evacuation map is printed and distributed by Mercy Corp (NGO). INT11 informs, "*We printed 2000 sheets, 900 sheets during the exhibition, and the rest were sent to village offices, the BPBD office, and we plan to print another 1000 sheets.*" Mercy Corp (NGO) has attempted to raise the funds to double print and distribute the maps to some private companies in Padang, yet the result has been quite disappointing. The only company that agreed to help the program is Semen Padang tbk. INT11 narrates it as: "*The funding raised was only from Mercy and Semen Padang. Semen Padang agreed to help for 1000 printed sheets.*" Unfortunately, other private companies were not interested. INT11 states: "*We tried to discuss it, (but) nobody was interested. Neither was the IBIS* (Hotel–red, one of the biggest companies)." Furthermore, local government is unable to reprint, and INT10 states that "*BPBD does not have any budget for it. It was not planned.*"

The current version of the tsunami evacuation map that is in use has been reprinted (3000 sheets). A total of 900 of them were distributed during the MM-Direx exhibition, and the rest were handed back to local government, in this case, BPBD expected to distribute the maps to village offices and all SKPD (= Satuan Kerja Perangkat Daerah = Local government work unit/department/agency) in Padang municipality. There were only 10 evacuation maps supplied for each village office, and the rest were given to the staff in the Padang government offices. INT10 states that "We provide 10 maps per village and I am the one who is in charge of circulating the maps to local government offices." No wonder the map only reached village offices, it did not reach residents who need them most. INT10 hoped for better distribution and stated that "*Not at all, it did not happen. We did it, not them/the KSB.*" INT04 (NGO) supports: "*It is ironic. The maps are distributed to village offices, but they are not circulated further. Those who want to read the map are given a chance to read it, come to the office, grab the map if you want.*"

5.1.3 People's Inability to Read Tsunami Evacuation Map

It is not surprising to see that many people are unable to read a map, especially the Padang evacuation map. In this case, to be able to read a map, one has to understand the content and message implied on the map. As expressed by INT02: "*Residents cannot read the map. The only thing they know is they can point out some safe zones on the map.*" INT01 said: "*Not many of them are able to read and understand how to reach safe zones. I am here, then where to go?*" Besides, the map is not only intended for Padang residents, but also for those who are new to Padang. INT01 states: "*The newcomers have to understand that this is a tsunami evacuation map that is provided for all.*"

The results of interviews reveal that there have been only a few disseminations of the map, and this lack of dissemination did not produce the expected level of awareness. INT04 emphasizes: "*First, lack of dissemination. This is important to make people recognize the existence of (the) map.*" INT01 adds: "*Once we have the maps, we have to distribute them to residents.*" INT11 suggests similar views: "*It is very true, dissemination is one of the most important things to conduct. We distribute the maps, give people an explanation. Once we have done with one village, we move to another place.*"

Some related institutions took part in the first launch and participated in the dissemination process that gives explanations to Padang residents about the map. INT01 affirms that: "*I was there in the first running. I witnessed so many volunteers joining the launch: reporters, students, and schools helped to disseminate the map.*" Large maps and big billboards were at some junctions and planned places. INT01 adds: "*Once the maps were in some places, but not any longer.*"

Some areas do not exist on the evacuation map, and some parts of the community do not have any knowledge about the map. As claimed by INT05: "*People living in the Koto Tangah to Bungus areas are not aware of the evacuation map. They have been informed that they live in a tsunami-affected area, but nobody explains the map in detail.*" INT05 is concerned about people who might misinterpret the map: "*If one is not able to read the map, he/she can interpret the color code differently.*"

Apart from the ability to understand the map, Padang residents also require a basic knowledge about the topography in order to understand the tsunami evacuation map. INT05 explains this as: "Why do rivers have the longer red mark. Why do Koto Tangah villages have more red marks than other villages? Only a few people know the answers. This is due to the contour aspect of the area; Koto Tangah is basically a swamp area." The surprising fact is that not only the general public but also some educated people are unable to read the map. INT05 states: "*The academia, students in university are confused.*"

Based on the above arguments, it can be concluded that Padang residents must be able to understand the following aspects related to the tsunami evacuation map:

1. Identify both the tsunami prone zone and the tsunami safe zone.
2. Locate the place where they live on the evacuation map.
3. Identify the location of vertical shelters.
4. Identify symbols characterizing city landmarks such as schools, markets, offices, airport, ports, bridges, etc.
5. Run along the identified evacuation routes such as the route from their own house to the safe zone or from the location where they conduct their daily activity to any nearby vertical shelter.

5.2 Tsunami Evacuation Route (Horizontal Evacuation)

There are two methods of evacuation from a tsunami threat: horizontal and vertical. Horizontal evacuation is evacuation by going away from the coast toward the tsunami safe-zones on foot (recommended) instead of driving. This method requires evacuation pathways/routes that include roads in the city, particularly major roads in residential areas.

In order to reach the safe zone, the speed of the evacuees' movements during the evacuation is the critical factor. The Japan Institute for Fire Safety and Disaster Preparedness (1987, in (Amin, 2006)) gives an overview of the walking conditions and average walking speed during a disaster evacuation.

According to Singh et al. (2010), the first tsunami wave hit Padang in 20–30 min, and it is expected that people can save themselves as soon as possible within 10 min. However, there are three groups of timing: 10 min, 20 min, and 30 min, or 600 s, 1200 s, and 1800 s. The mileage for each group can be calculated by multiplying the time by the speed of walking (0.751 m/s). Accordingly, the mileage for each group will be 450 m, 900 m, and 1350 m. The third level of this range will be applied in the analysis.

Geographic information system (GIS) is an information system that is used to analyze data related to space. GIS has several tools for analyzing spatial data, including buffer analysis, overlay analysis, and network analysis. Fig. 4 shows the level of service areas in every point of the safe zone within the range/distance of 450 m, 900 m, and 1350 m, using buffer analysis.

FIG. 4 Service area of safe zone (450 m, 900 m, and 1350 m). *Source: ArcView analysis.*

It shows that many regions are not reachable, and refugees will not be able to reach the safe zone on time, especially residents who live in the region left of the main road.

The construction of tsunami evacuation roads has been one of the significant tasks of the evacuation plan. Signposts are provided on both main routes and additional paths so that residents can use the lifelines during an emergency.

BSN (2012) identifies seven steps to determine an evacuation route: (1) data collection phase, (2) studio/data processing phase, (3) early design of the map, (4) field observation phase, (5) final design of the map, (6) the phase of evacuation map design and production, and (7) insemination phase. The roads that have been identified must be marked on the map, which is technically regarded as part of the self-evacuation procedure, in addition to the placement of signposts. Accordingly, the government of Padang has conducted all the phases, including the dissemination phase. It is described in the following map (Permana, 2013).

The evacuation route maps that are displayed in some junctions and strategic points in Padang so that residents are able to visualize them clearly. The map uses three color codes for each zone, which means it was designed long before Padang Consensus II.

The local government, in this case, the Padang government, has provided enough information about evacuation routes. Maps and big street boards are placed along some strategic routes and intersections to educate Padang residents. Unfortunately, the road surfaces are not yet physically improved. INT03 states: *"The problem is we do not have enough evacuation routes. I am now talking about numbers or quantities."* INT11 supports this by stating: *"The evacuation routes, the lifelines are not only limited in numbers, but also in quality."* INT04 also emphasizes: *"The roads are inadequate and certainly in poor condition."*

Simpang Alai-Ampang is the only road construction project that has been undertaken considering the problem mentioned above. INT10 points out: *"The Alai-Ampang routes are being constructed over and over again, but it is not taking place anywhere else. Let us mention Siteba routes".* INT10 adds: *"This Siteba route is 12 meters in width, and the government plans to widen it up to 20 meters."*

Road surface improvements for the evacuation routes are certainly crucial in some red zone areas, especially in the Koto Tangah district. There is also a shortage of evacuation routes in the red zone, specifically, in the area between Air Tawar and Koto Tangah. INT04 said: *"The fact is we do not have any evacuation routes. What if the earthquake hit us? What if it happens? There are no closer evacuation routes to use by residents."* INT10 remarks: *"One area called Brimob, it is very vulnerable, and there a lot of people living in the area while there have not yet been any lifelines arranged to accommodate them."*

The Padang government has provided a budget to build and develop some evacuation routes. INT03 confirms: *"Yes, there are some demands to add more routes. Last month, I met some staff from the public work department to discuss data regarding evacuation routes. They said more budget would be set up from national income/APBN (State Budget = Anggaran Pendapatan Belanja Negara)."* There are two proposals for the evacuation routes project: road widening and new road construction. INT03 indicates: *"If it is accepted, there will be some new evacuation routes in Padang. Road widening in the suburbs and new road construction will take place."*

This shows that enormous efforts have been taken by the Padang government to work on evacuation routes affording to its financial conditions. INT03 argues: *"They have done their best. They have money but not yet the goodwill and consistency."*

Traffic jams, which normally occur during the evacuation process, are an underlying problem associated with it. It happens because everyone uses vehicles for evacuation. Respondents evident for the traffic jams occurred during the big earthquakes that generated tsunamis. INT10 said: "*It makes me worried and doubfult about conducting horizontal evacuation as the street was usually blocked. We could not even move.*"

Generally, Padang residents understand the situation. Their understanding has improved because local government informed them through some workshops. Interviewees from NGOs claim that they have been informing people not to use their vehicles during the evacuation process. INT04 said: "*This is very important, We keep reminding them.*" INT05 hopes, "*If it works, we can slowly change their mindset.*" Some staff from government agencies/BPBDs have also taken such efforts. INT11 states: "*We communicate with Padang residents about it: do not ever use your vehicles to evacuate unless you want to be trapped in a traffic jam.*" INT10 added: "*Vehicles cause traffic jams. We repeat the same words, the same things at every chance we get.*"

The interview results show that residents' using vehicles during sevacuation is disturbing the efficiency. It is hard to explain or educate them. Another reason is that the vehicles are considered as valuable assets, and people are reluctant to lose them. INT09 states: "*There are a lot of poor people living in Padang, so when an earthquake hits them, there is always urgency to evacuate not only themselves but also their few assets, either a motorcycle or a car. Thus, traffic jams inevitably occur.*"

Another Interviewee from NGO agrees with INT09's statement and remarks: "*In fact, it is so true. I, myself would not be able to leave my own car behind when evacuating.*" INT10 adds: "*It is logical; I would not leave my motorcycle while I still have a loan.*" To sum up, Padang residents prefer to protect their assets rather than rescue themselves alive from the threat of a tsunami.

5.3 Vertical Evacuation

Vertical evacuation is the act of moving toward higher ground. Either a hill (natural shelter) or a multistory building. As the study area has no hills at all or has very flat topography, the only option remains is building more multistory buildings or utilizing the existing multistory buildings for vertical evacuation.

There are several names for vertical evacuation buildings, including tsunami evacuation buildings (TEBs) (Raskin et al., 2009), and vertical evacuation structures (FEMA, 2009). In Padang, they are called te TES (temporary evacuation shelter). They are temporary because it is expected to accommodate a displaced population only for 2–3 h before they are evacuated to a tsunami evacuation center in the safe zone.

Padang government, with the support of international donors, has constructed shelters. Some of them are schools that have a stable structure, three storys, and with a roof that serves as a tsunami evacuation area. Data from the BPBD (Disaster Management Agency) declares that there are 13 tsunami evacuation structures at this time with a total capacity of 30,550 people and the capacity for every building varies between 1000 and 3000 people (BPBDs, 2013). This amount is not sufficient for the potential loss of life of as many as 400,000 people or more. Further, the locations of the shelter buildings are not evenly distributed in tsunami-prone areas.

FIG. 5 Service area of TES, P-TES, and safe zone (450m, 900m, and 1350m). *Source: ArcView analysis.*

There is one TES in the study area, a three-story mosque with a capacity of 1500 people that has been designated by the city government through BPBD (Disaster Agency) of Padang and the NGO Mercy Corps Indonesia. One TES is not sufficient to accommodate all the displaced population of this study area. Thus, some potential shelters (P-TES) in the study area are represented in Fig. 5.

Potential TESs are existing multistory buildings that are considered to be strong and can accommodate tsunami refugees for a while. However, it is yet a necessity to determine the strength of the structure of the PTES. Some of them are structurally new buildings that have been designed to withstand strong earthquakes and tsunami waves. Several other buildings need recalculations to determine the strength of the structure.

A buffer analysis was conducted to analyze TES and P-TES with the same range that was used previously (450m, 900m, and 1350m) as the location points. As shown in Fig. 5, the distribution of TES, P-TES, and safe zone points are able to serve most of the residents in the study area. However, some spots are not covered, and people from those areas are still able to reach the safe zones.

The tsunami shelter development in Padang was started during the rehabilitation and reconstruction period after the 2009 earthquake. Damaged or collapsed school buildings were

FIG. 6 Shelter combined with elementary school building.

rebuilt with donor funds to withstand future tsunami threats. These school buildings were designed to resist large-scale earthquakes (>6 SR) and also tsunami hits.

School buildings were given precedence over other public facility buildings as tsunami evacuation sites as schoolchildren are more vulnerable to disasters. This allows the parents to safeguard themselves during a major earthquake or a tsunami without rushing to the schools to pick up their children. Schoolchildren will be able to self-evacuate directly to the roof of their school buildings. Fig. 6 shows an elementary school building that also functions as a tsunami shelter. The school itself consists of three floors, and the fourth floor is a rooftop that serves as a refuge during a tsunami. This school is also equipped with an emergency stairway to the rooftop of the building.

In addition to school buildings, mosques are also functioning as tsunami shelters. There are several reasons to encourage the mosques as tsunami shelters, including the wonder of mosques in Aceh that did not collapse from the tsunami, and many mosques served as a place of evacuation at that time. Also, the mosque is a building that is frequently used by the people of Padang almost every day. Therefore, the community is familiar with the location of the mosque and the route.

FEMA (2009) stated that tsunami-resistant buildings should incorporate walls that are not solid so that the tsunami can pass through the building smoothly. Mosques are such buildings, and their walls are open or accessible. Fig. 7 shows the Great Mosque owned by the government of West Sumatra Province. The mosque has been designed to withstand large earthquakes (>6 SR) and has large ramps to the second floor. Also in Padang, self-help communities refurbish many other mosques to serve as temporary tsunami shelters.

Fig. 8 shows a unique tsunami shelter as its functions are not combined. This unique shelter was built exclusively for evacuation by BNPB and PU (Public Work Agency). Currently, there are three similar shelter buildings, and several more buildings are expected to be built in other places.

FIG. 7 Shelter combined with grand mosque.

FIG. 8 Artificial tsunami shelter by BNPB.

Currently, Padang is more focused on improving the amount of vertical tsunami shelters rather than horizontal shelters. There were five vertical shelters at the time of data collection. INT10 said: "*Five shelters could save around 9,000 people, and the school buildings SMA 1/High School 1 around 3,000. But people have to stay standing.*" However, the number of shelters is still inadequate. INT03 said: "This *means it is not yet complete for the city of Padang. When somebody asks why, the answer is already there.*"

To anticipate the number of vertical shelters, governments and NGOs in Padang initiate the use of existing buildings that are eligible to be a vertical temporary shelter. As expressed by INT11: "*If the government were to build shelters, it would be a huge cost. High-rise buildings including schools, hotels, mosques, and government buildings that are taller than six meters have the potential to be used as evacuation sites, and we've identified those.*" Colored dots show these identified buildings or structures on the tsunami evacuation map. INT01 said: "*There is, for example, a green color here, it has been agreed upon, considered safe. Then, the yellow dot is potential shelter.*" In INT10's words: "*and then the potential shelters totalled 98.*" The next stage is to approach the owners of the buildings that have been identified as potential shelters. INT10 and INT11 said: "*Now the problem is to invite the owners to see if they are willing to use their buildings as shelters.*" If they agree, INT11 suggests the next phase: "*Because the process is long, there will be agreement, MoU will be involved in training, getting consents, and developing agreements.*" The next stage is the declaration of the government of Padang to accredit these buildings as official vertical shelters.

However, the government is reluctant to enforce it, according to INT01. The government was capable of reserving government sites, especially the buildings constructed by the government. However, according to INT01, "The government is reluctant to declare private buildings as they need to be examined." The government is concerned about labeling or certifying buildings as tsunami shelters because if the shelter collapses during the tsunami with many casualties, the government has to take the responsibility. INT03 and INT05 confirm that the fear of collapse is the basis for not labeling any private building as vertical shelters in Padang. INT06, who is an expert on earthquake-resistant building structures, also doubts the determination of these buildings as TES, as some buildings really had been designed to withstand earthquakes but were not necessarily resistant to tsunamis.

According to INT06, an assessment of a building to determine its use as a shelter includes the review of the existing location, building strength, the concrete, and the amount of steel. The foundation must be examined for the buildings specially designed to withstand tsunamis. Constraints in funding are also a barrier for assessing buildings that deserve to be named as shelters, according to INT01. INT10 says: "*It would cost 50 million rupiahs per floor to measure and to declare the building as a shelter.*"

5.4 Tsunami Sirens

The Padang government already has tsunami warning sirens and installations scattered in tsunami-prone zones. In the early stages, there were 11 sirens from foreign aid, said INT07 and INT09: "*Germany's GTZ helped (put) 11 point megaphones and tsunami sirens at each mosque in the city of Padang.*" Communication from a siren uses a wifi signal. However, according to INT07: "*Sirens (GTZ assistance) using the wifi system are no longer utilized in the city of Padang.*"

In the next phase, the sirens in Padang were replaced with a radio frequency method. According to INT07: "*The city of Padang switched to the last version of BPBDs in the provinces, using a radio frequency system, including that of BNPB.* The latest *sirens are more efficient when compared to the old, and the node (siren) can be installed everywhere, and it is simple.*"

At this time, the total number of sirens in Padang is 23. And as described in detail by INT11: "*The city of Padang installed 23 units, then BNPB also installed eight units. BMKG had already installed one unit, then the city procured two units, but this refers to models in the province.*" Sirens are placed on government offices, mosques, and community-owned shops (Fig. 9).

The siren alert runs shortly before a tsunami to warn the people to evacuate. INT05 said: "The siren was the only support of early warning systems that we have built in the

FIG. 9 Sirens on top of buildings. *Source: Own picture.*

community." Sirens only serve to inform a tsunami threat. According to INT07: "*Siren's function is to convince them to save themselves.*"

Sirens in Padang today are operated from two offices, Pusdalops Padang and Pusdalops Province. INT07 said: "*Pusdalops Padang and Pusdalops Province can run it as there are two servers. Sometimes these are turned on by BPBDs province, or BPBDs the city.*" One of these servers stays as a backup during a disaster.

Sirens in Padang are routinely tested each month at 10am. According to INT10 and INT11: "*(On the) 26th of each month the sirens are tested and fixed if there are problems. Sometimes we tested not only the siren sound but also the voice quality.*" INT07 said: "*Yesterday, the 26th, we tested only the voice and not the sirens because the voice was not very clear.*" Voice recordings are used to play during the sirens instead of live announcements as it can be played at any time (Fig. 10).

During the siren tests, officers take notes on the estimated power of the sound and any existing problems. The constraint is that there are only 23 pieces and the siren sound scope

FIG. 10 Activity Sirens test every 26th of each month.

FIG. 11 Evacuation map on busy intersection.

is not broad. INT08 said: "The *BMKG siren radius is 1 km, the simulation time in the Alai market yesterday did not sound at all, at least 500 meters.*" INT07 added that "*The siren volume is lacking, his voice is still lacking, and does not get people's attention.*"

5.5 Signs and Banners

The Padang government has installed evacuation directional signs, not only special direction signs for evacuation but also directional traffic information. Almost every major intersection in the city of Padang has a banner that shows the location and level of the tsunami hazard at the intersection (Fig. 11).

Banner placement is not only on crossroads but also at several strategic points in major roads of the tsunami-prone zones. Evacuation maps on the banner are already divided into

FIG. 12 Directional traffic information (*"Jalur Evakuasi"* = evacuation path).

zones with the installation location of the banner, and there is information indicating "You are here."

The Padang government uses a special banner to indicate the direction of evacuation lines. In addition, the government uses a traffic sign that is managed by the Department of Transportation in Padang. Evacuation path information is combined with information signposts in the city so that everyone who drives can read the directions of the evacuation path. Further, visitors also can immediately know the evacuation route from the directions indicated without having to understand the map of the city in advance. This can be seen in Fig. 12.

The Padang government with the aid of BNPB and donors has already put hundreds of evacuation signs at several points along the path of tsunami evacuation. The signs that are placed match the standard set by the Indonesian National Standard (SNI) with the numbers 7743: 2011 (BSN, 2011). In SNI signs regulations, tsunami Eevacuation is already included as a technical requirement for signposts (color, shape, size, symbol, information). Examples of signposts are shown in Fig. 13.

The problems encountered with these tsunami evacuation signs are that some of them are placed in places that are difficult to see. Also, as discussed in the previous section, some people are not able to read a map, and more importantly, the vehicle driver only has a few seconds to read and understand the map. It is suggested that further research needs to be conducted on the effectiveness of tsunami evacuation banner placement at the intersections of roads.

FIG. 13 Direction signposts for the evacuation route in Padang.

5.6 Workshops and Exhibitions

5.6.1 Workshop—Transfering Knowledge

A knowledge-sharing workshop called the Workshop Partnership and Closing Program READI was conducted at the Premier Hotel. The workshop was attended by delegates of KSB (Komunitas Siaga Bencana = Community-based Disaster Preparedness) and KSBS (School-based Disaster Preparedness) with the aim of strengthening and increasing the capacity of members in disaster risk reduction (DRR) and formulating the program, which was proposed next day (Fig. 14).

On the second day, a meeting was held to synergize ideas from the DRR community with government programs (bottom-up-top-down). On the second day, the people representing KSB proposed several programs to improve tsunami preparedness activities in their respective neighborhoods. The government explained the programs they did and government targets related to the disaster. The results of this activity show that KSB and communities have the ability to identify their needs. On the other hand, it was identified that Padang is in great need of public support (Fig. 15).

5.6.2 MM-DIREX

As a part of MM-Direx 2014 (Mentawai Megathrust Disaster Relief Exercise), a disaster management exhibition was also conducted on March 20–22, 2014. Government agencies, NGOs, and universities took part in this event (Fig. 16).

In this exhibition, government agencies from the center (BNPB), provincial (BPBDs Province of West Sumatra), and city (BPBD Padang) organized the programs related to preparedness against earthquakes and tsunamis that could threaten Padang. Likewise, the NGO, campus, and research centers also participated and shared their respective roles in tsunami

FIG. 14 Workshop READI (March 24–25, 2014) day 1.

FIG. 15 Workshop READI (March 24–25, 2014) day 2.

preparedness for the people of Padang. In this exhibition, the tsunami evacuation map (last edition) was shared as well as a live show by Radio Disaster (Classy FM) (Fig. 16).

One of the activities of MM DIREX 2014 was to perform a simulation for the control by policyholders. The activity was conducted by all government, military, NGOs (local-Int), and experts (local-Int). The activity simulated a major earthquake that occurred in Mentawai that measured 8.9 on the scale. This activated all the early warning systems and announcements until the relief was completed (Fig. 17).

FIG. 16 Disaster management exhibition was conducted on March 20–22, 2014.

FIG. 17 Command post exercise simulation.

5.7 Dissemination to the Public

5.7.1 Many People Do Not Receive the Information

Dissemination of information is not routinely performed, and the activity changes locations (villages) every year. Meanwhile, the beachside is the area where the educational activities are performed. However, the migrant population of the area changes frequently, and it makes the people unaware of the evacuation process. According to INT07: *"The turnover rate is high. For example, people move around, there are new residents, new boarding school boys who come to Padang."*

On the other hand, Padang is a great city, a capital of the province with 800,000 people, and population density at tsunami-prone zones was also very high. Further, it is difficult to reach all communities. INT11 said: *"In Padang, because of its wide scope and overcrowding nature, the villages are very dense. We see the RT (Rukun Tetangga = Neighbourhood Association/groups), approximately, 100–200 households, and several RW (Rukun Warga = Community Association/groups). This is truly exceptional."*

Another obstacle is that there are people who do not receive information. INT01 said: *"For example, when the information was disseminated, they may not be in the location of dissemination."* Communities living in areas that are remote may not get information. INT07 said *"I predict that off the coast there are only about 40–50% of people that have received the information."*

5.7.2 Dissemination by Government

The government, through the BPBDs, has been disseminating information to the public. This dissemination is executed through the levels of the organizational structure of government, from the agency to the district office, then to the village office, and finally to the RT/RW. According to INT10, BPBDs have already disseminated information to all districts in the tsunami-prone zones, but not directly to the public.

BPBD staff do not provide counseling to communities directly. Their approach is to conduct events about disasters. The obstacles in this approach are the invitees who are considered as the leader in a community but do not have the ability to transfer knowledge. INT02 mentions that *"But if it's through RT and RW, it's sometimes not forwarded to communities. That's what I experienced."* The dissemination activities are conducted by BPBD staff to communities during the Ramadan period. INT10 stated that *"Then, that was in the month of Ramadan, it was through mosques, the staff of BPBDs distributed dissemination on Pesantren Ramadan"*.

BPBDs only provide dissemination at the district level; they do not directly approach communities. Pesantren Ramadhan is the only BPBD activity where direct dissemination is undertaken. Dissemination is constrained by the population turnover rate as there is a high immigrant population.

5.8 Tsunami Evacuation Simulation

Several tsunami evacuation simulation activities have been conducted in Padang, but these activities are not routinely carried out owing to constraints on funds and organzers. Usually, these activities are organized by NGOs, yet the simulations never involve the whole population of Padang, and it usually covers one village. The participants are generally the local communities and school children. One of the simulations in which the researchers participated took place at Parupuk Tabing village, Padang on March 9, 2014, from 8 to 11 a.m. The simulation was organized by Resilient Environment Through Active DRR Initiatives (READI) of USAID, Jemari Sakato, KSB, BPBD, and Mercy Corps Indonesia (Fig. 18).

FIG. 18 Tsunami evacuation simulation activity.

After the simulation was completed, a focus group discussion (FGD) was conducted on the evaluation of the tsunami evacuation simulation. All the activities of the FGD were recorded with digital audio (Fig. 19).

5.8.1 Preparedness Measuring Tool Is a Simulation

Simulation is one of the ways to evaluate the level of success of preparedness programs. Theoretical analysis, questionnaires, interviews, or focus group discussions would not represent the real situation.

5.8.2 Budget, the Level of Frequency of Simulation

For the time being, the local government agrees to sponsor the simulation once a year. One of the staff of BPPD/INT10 affirms: "*We have the budget to hold simulation activity only once a year. Recently we performed the simulation in the district of Padang Barat and Padang Selatan. We combined two areas, and finally, we were able to collect 1,500 residents. We plan to perform it this year in Koto Tangah district.*" One of the citizens who happened to be one of the KSB members confirms this. INT09 said: "*It is better if we can conduct a simulation activity once a year to increase resident's awareness about the fact that a tsunami may happen.*" INT07 agrees: "*Local government provides its funding every year.*"

On the contrary, NGO and KSB hope to have the simulation several times a year. INT07 states: "*It should be ideally performed in some different locations once a month. It means there are 12 simulations in a year. The more often we can conduct, the easier people can memorize the procedure of simulation.*" However, INT04 states that conducting two simulations a year is more than enough.

FIG. 19 Focus group discussion (FGD).

5.8.3 Awareness to Participate

Having observed tsunami simulation activity several times in Padang, the NGOs conclude that people's awareness and willingness to participate in a simulation activity is very low. As stated by INT01: "*Only a few residents take part in some tsunami simulations.*" INT07 adds: "*We invite residents, they do not come. They even ignore the invitation to have lunch after the simulation.*"

Things would be different if the simulations were conducted at school and students were involved. According to INT01: "*If it is at school, the participants will join the program because the teachers will ask them to do so, but if it is conducted in one community or other places I cannot guarantee that the same amount of participants will join the program.*"

For this reason, more effort should be taken to raise the awareness of Padang residents to realize the importance of simulation activity. INT07 asserts: "*Yes, make them understand that they need the program. Once they realize it, they will search for it. It has not happened yet.*"

5.8.4 Incentive or Transport Cost for the Participants of a Tsunami Simulation

To involve more Padang residents in the tsunami simulations, incentives were offered for the participants such as transport cost. Some members of NGOs and staff from local government office agree that it is normal to pay the participants an incentive to get their participation. INT03 verifies: "*We want to appreciate their contribution. They ask for some money, we can fulfill it.*" Some other NGOs expressed their disappointment. INT04 asserts: "*They told me, it is disappointing, participants ask for some money.*" INT04 said: "*It is not much money. If it can replace their time for coming here, we will pay. Why not if the funding is available?*"

INT07 acknowledges the issue of providing an incentive for participants after conducting some simulations: "*BNPB provided incentives. It is a good idea. It is a trigger for participants to follow up simulation activity.*" INT07 gives one example: "*For instance, it happened in Bungo Pasang villages. Initially, it was hard to make people join the simulation, but by providing them with a transport fee, they start to take part in the activity. We just want to make them understand that this is a simulation. We want them to be familiar with it.*"

However, some argue that it would become a bad habit and affect any future simulation activity. INT10 states that this issue is strongly related to budget and expenditure. INT08 states: "*The habit, it ruins everything. People gathered. They were given 25,000 rupiahs.*" Further, INT10 explains: "*There is indeed a budget for the incentive. We provide 25,000 rupiahs for each participant, and we will allocate the money for 500 participants.*" It means we would spend 12,500,000 rupiah (around £750).

Things are worse when school children are involved. They also get the incentives from the program. INT10 asserts: "*When a simulation sponsored by MM-Direx was conducted, they paid the school children, and each student got 25,000 rupiahs. It is not at all a good way to educate children.*"

The findings suggest that simulation activities in Padang are usually held once a year. However, it would be much better if they could be conducted more than once a year to achieve a better level of community preparedness.

The fact is, people's willingness to participate in a simulation activity is very low. Pocket money or transport fees are usually given to encourage the community's participation in the simulation activity. In this case, there are two contrasting opinions regarding the transport fee given to participants, some who agree and some who disagree with this principle. On the one hand, those who agree think that a transport fee is one of the ways for the organizer to appreciate people's participation. On the other hand, the opponents think that giving transport fees will create such a bad habit for people. In the future, people would come to the simulation only for money.

5.8.5 Padang Community

Padang residents have often experienced a massive earthquake so that they have an intuition to recognize whether it was a big earthquake or not. If the earthquake was not that large, people just evacuate locally (get out of the house/building). If it is a large one, residents immediately and automatically will evacuate without waiting for the local government warning. It means that Padang residents demonstrate a natural ability to face a large earthquake and tsunami. Moreover, the community has also been educated in making buildings resistant to earthquakes since the severe shaking between 2007 and 2010 that ruined their houses.

Among the neighborhoods that live in the red zone, several KSBs (*Community-based Disaster Preparedness*) are working to establish a culture of disaster resilience within the communities. NGOs are also actively involved in preparedness activities in the form of education and dissemination to the public and other stakeholders. The private sector, such as radio broadcasting, also regularly delivers a variety of information related to disaster management.

The constraint in Padang is the coastal length of nearly 20 km, which complicates the layout design and scale in the publication of tsunami evacuation maps. Also, the government's official website does not provide information on natural disasters, especially earthquakes and tsunamis. It would be much different when compared to the government website in Oregon that has complete information. There is only one website belonging to the provincial government of West Sumatra that provides information about the disaster: http://bpbd.sumbarprov.go.id/.

Padang tsunami evacuation maps that have been printed and distributed to the public are not available on the city government's website or the website of the BPBD West Sumatra Province. The tsunami evacuation map is only available on the GITEWS website, which is a cooperation between the Indonesian government and Germany. Further, leaflets and booklets on

FIG. 20 Blue line tsunami in Padang.

disaster preparedness and tsunamis are also not available on the government website. The lack of a complete official website shows the government's reluctance on disaster preparedness.

Currently, Padang has no past tsunami data. However, the roads are marked as the boundary of the inundation area. These borders are colored in blue on the road (see Fig. 20). An inundated tsunami marking project with the blue line has also been conducted by foreign NGOs under the Blue Line Program. The blue is colored based on an inundation map area that is agreed upon by various parties.

6 CONCLUSION

It is recognized that the provision of all infrastructure for evacuation requires a long time and cost. It is further difficult to implement in Padang as it is not financially stable with low revenue. The major problem faced by the government of Padang in implementing a TEP is the time constraint for tsunami evacuation. Thus, conducting vertical evacuation is urgent compared to horizontal evacuation. Transportation facilities are far from sufficient and would cause traffic jams. Further, the number of vertical tsunami shelters is also insufficient.

Some parts of the TEP have been integrated, though not all the variables in the TEP activities have been addressed. The local government has used various efforts to develop local early warning systems and disaster management. These activities include preparing legislation (spatial planning, organization, standard operation procedure, early warning system), developing evacuation infrastructure (maps and evacuation routes as well as signposts), building shelters (vertical shelters), providing resources for government officials, and conducting community support activities.

References

Amin, B., 2006. Evacuation Shelter Building Planning for Tsunami-prone Area; a Case Study of Meulaboh City, Indonesia. (Master). ITC, The Netherlands.

Anggraeni, D.R., Artanti, W., 2013. Training manual-tsunami evacuation planning at district level. In: GITEWS (Ed.), Project for Training, Education and Consulting for Tsunami Early Warning System (PROTECTS). GTZ IS-GITEWS, Jakarta.

BMKG, 2016. Badan Meteorologi, Klimatologi, Dan Geofisika. Available from: http://www.bmkg.go.id/BMKG_Pusat/default.bmkg.

Borrero, J.C., Sieh, K., Chlieh, M., Synolakis, C.E., 2006. Tsunami inundation modeling for Western Sumatra. Proc. Natl. Acad. Sci. 103 (52), 19673–19677. https://dx.doi.org/10.1073/pnas.0604069103.

BPRR, 2010. Approval of Official Tsunami Hazard Map Based on Padang Consensus. 21 April 2010, Tsunami Kit, Padang.www.gitews.org/tsunami-kit.

BPS, 2011. Padang Dalam Angka – Padang in Numbers. Satistical Bureau (BPS) Kota Padang, Padang.

BPS, 2016. Kota Padang Dalam Angka 2016, Padang Municipality in Figure. BPS-Statistic of Padang city, Padang. https://padangkota.bps.go.id/.

BSN, 2011. SNI 7743:2011 Rambu Evakuasi Tsunami-Tsunami Evacuation Sign. BSN Indonesia, Jakarta.

BSN, 2012. SNI 7766:2012 Jalur Evakuasi Tsunami. BSN Indonesia, Jakarta.

FEMA, 2003. State and Local Mitigation Planning How-to Guide: Bringing the Plan to LIfe. Implementing the Hazard Mitigation Plan. Retrieved 18 September 2013, Available from:http://www.fema.gov/media-library-data/20130726-1521-20490-9008/fema_386_4.pdf.

FEMA, 2009. P646A-Vertical Evacuation from Tsunamis: A Guide for Community Officials.

Gede, I.S., Hoppe, M.W., 2010. Evacuation Planning in Kuta Bali, A Local Strategy for the Case of Emergency. GTZ IS-GITEWS, Jakarta.

GITEWS, 2008a. E1.1 Introduction to Risk Knowledge.

GITEWS, 2008b. E1.12 Padang Consensus on Official Tsunami Hazard Map-Protocol. Nov 2008.

Hadi, S., 2009a. Experiences in Managing Response and Preparation for Recovery: Case of Indonesia. Tsunami Global Lessons Learned. Retrieved 29 August 2016, Available from: http://kawasan.bappenas.go.id/images/data/Produk/Paparan/Experiences_In_Managing_Response_And_Preparation_For_Recovery_24_April_2009.pdf.

Hadi, S., 2009b. In: Sustainable livelihood after disaster.Case: Post Earthquake 27 may 2006 in Yogyakarta and Central Java. Building Community Centered Economies Conference. Retrieved 29 August 2016, Available from: http://kawasan.bappenas.go.id/images/data/Produk/Paparan/Sustainable_Livelihoods_After_Disaster_Australia_June_2009.pdf.

Hamilton, W.B., 1979. Tectonics of the Indonesian region, US Geological Survey Professional Paper 1078. p. 345.

Permana, H., 2013. Sosialisasi SNI Jalur Evakuasi Tsunami (SNI 7766:2012): LIPI.

Indonesia, 2016. Geografi Indonesia. Retrieved 13 Decenber 2016, Available from:http://indonesia.go.id/?page_id=479&andlang=id.

Irsyam, M., Sengara, W., Aldiamar, F., Widiyantoro, S., Triyoso, W., Hilman, D., Ridwan, M., 2010. Ringkasan Hasil Studi Tim Revisi Peta Gempa Indonesia 2010 (Summary of Study Finding of Indonesian Earthquake Map 2010 Team). Available from:http://www.preventionweb.net/files/14654_AIFDR.pdf.

Raskin, J., Wang, Y., Boyer, M.M., Fiez, T., Moncada, J., Yu, K., Yeh, H., 2009. Tsunami Evacuation Buildings (TEBs); A New Risk Management Approach to Cascadia Earthquakes and Tsunamis.

KOGAMI, 2014. Profile 'Komunitas Siaga Tsunami'. (KOGAMI) – NGO, Padang.

Undang, 2007. Undang Republik Indonesia Nomor 24 Tahun 2007 Tentang Penanggulangan Bencana, (Law of The Republic of Indonesia No. 24 Year 2007 About Disaster Management) Lembaran Negara Republik Indonesia Tahun 2007 Nomor 66.

McCloskey, J., Antonioli, A., Piatanesi, A., Sieh, K., Steacy, S., Nalbant, S., Dunlop, P., 2008. Tsunami threat in the Indian Ocean from a future megathrust earthquake west of Sumatra. Earth Planet. Sci. Lett. 265 (1–2), 61–81. https://dx.doi.org/10.1016/j.epsl.2007.09.034.

MoMAF, 2008. In: Tsunami Kit, (Ed.), Padang Consensus on Official Tsunami Hazard Map-Protocol Nov 2008. The Ministry of Marine Affairs and Fisheries (MoMAF), Republic of Indonesiawww.gitews.org/tsunami-kit.

Natawidjaja, D.H., 2007. Earthquake and tsunami sources of Indonesia: developing research-based disaster mitigation. The International Symposium on Disasters in Indonesia. Geoteknologi LIPI, Padang.

Padang, G., 2011. Materi Teknis Rencana Tata Ruang Wilayah (RTRW) Kota Padang, Tahun 2010-2010. Padang-Indonesia.

Government Regulation, 1980. Peraturan Pemerintah Nomor 17 Tahun 1980 Tanggal 21 Maret 1980, tentang Perubahan Batas Wilayah Kotamadya Daerah Tingkat II Padang.

PROTECTS, & GIZ, 2009. Evacuation Planning: Tsunami Kit. www.gitews.org/tsunami-kit.

Scheer, S., Gardi, A., Guillande, R., Eftichidis, G., Varela, V., de Vanssay, B., Colbeau-Justin, L., 2011. Handbook of Tsunami Evacuation Planning. SCHEMA (Scenarios for Hazard-induced Emergencies Management), Project n° 030963, Specific Targeted Research Project, Space Priority. Available from:http://publications.jrc.ec.europa.eu/repository/bitstream/111111111/15978/1/lbna24707enc.pdf.

Scheer, S.J., Varela, V., Eftychidis, G., 2012. A generic framework for tsunami evacuation planning. Phys. Chem. Earth 49, 79–91. https://dx.doi.org/10.1016/j.pce.2011.12.001.

Seymour, A., Posadas, B.C., Coker, C.E., Langlois, S.A., Coker, R.Y., 2010. In: Community disaster preparedness: An index designed to measure the disaster preparedness of rural communities.Paper Presented at the Coastal Research and Extension Center, Mississippi.

Simpson, D.M., 2008. Disaster preparedness measures: a test case development and application. Disaster Prev Manag 17 (5), 645–661. https://dx.doi.org/10.1108/09653560810918658.

Singh, S.C., Hananto, N.D., Chauhan, A.P.S., Permana, H., Denolle, M., Hendriyana, A., Natawidjaja, D., 2010. Evidence of active backthrusting at the NE margin of Mentawai Islands, SW Sumatra. Geophys. J. Int. 180 (2), 703–714. https://dx.doi.org/10.1111/j.1365-246X.2009.04458.x.

Socialresearchmethods, 2006. The Planning-Evaluation Cycle. Retrieved 18 September 2013, Available from: http://www.socialresearchmethods.net/kb/pecycle.php.

Spahn, H., Hoppe, M., Usdianto, B., Vidiarina, H., 2010. GTZ IS-GITEWS, (Ed.), Planning for Tsunami Evacuations, A Guidebook for Local Authorities and other Stakeholders in Indonesian Communities. Tsunami Kit, Jakarta, Indonesia. www.gitews.org/tsunami-kit.

Spahn, H., Kesper, A., 2010. Evacuation Planning in Tanjung Benoa, Successful Cooperation between Communities and the Private Sector. GTZ IS-GITEWS, Jakarta.

Studio Mitigasi Bencana, 2016. Konsepsi Risiko dan Sistem Penanggulangan Bencana di Kota Padang. In: Studio Mitigasi Bencana dan Adaptasi Perubahan Iklim. SAPPK ITB.

Taubenböck, H., Goseberg, N., Setiadi, N., Lämmel, G., Moder, F., Oczipka, M., Klein, R., 2009. "Last-Mile" preparation for a potential disaster–interdisciplinary approach toward tsunami early warning and an evacuation information system for the coastal city of Padang, Indonesia. Nat. Hazards Earth Syst. Sci. 9 (4), 1509–1528. https://dx.doi.org/10.5194/nhess-9-1509-2009.

UNESCO-IOC, 2009. Tsunami Risk Assessment and Mitigation for the Indian Ocean; Knowing your Tsunami Risk–and What to Do About It. IOC Manuals and Guides No. 52, UNESCO, Paris.

UNESCO-IOC, 2015. Tsunami Risk Assessment and Mitigation for the Indian Ocean; Knowing your Tsunami Risk—and What to Do About It, second ed. IOC, Perth Manuals and Guides 52.

Padang, W., 2012. Rencana Tata Ruang Wilayah Kota Padang Tahun 2010–2030. Padang.

Further Reading

InfoIndonesia, 2014. Peta Indonesia. https://infoindonesia.files.wordpress.com/2014/05/peta-indonesia-bagus-besar.jpg.

Ricos, M., 2008. Community Based Disaster Management. IDEP Foundation.

CHAPTER

19

Numerical Analysis of Changes in Ocean Currents and Density Structures in Miyako Bay, Japan, Before and After the Great East Japan Earthquake

*Tomokazu Murakami**, *Shinya Shimokawa**, *Toshinori Ogasawara*†

*National Research Institute for Earth Science and Disaster Resilience, Tsukuba, Japan
†Iwate University, Morioka, Japan

1 INTRODUCTION

Miyako Bay, Japan, the target area of this study, is located at 39.58°N, 141.95°E and 39.65°N, 142.00°E. It is 10km in length and 6km in width. Two rivers, the Hei and Tsugaruishi, flow into the middle and inner parts of the bay, respectively. The bay contains a wide variety of fisheries resources, including herring, flounder, and *Zostera marina*. Recently, it has been reported that the present distribution of *Z. marina* in the bay is different to that before the Great East Japan Earthquake of March 11, 2011 (Space.geocities, 2014). A possible reason for this is that the topographic changes (mainly to the sea floor) resulting from the earthquake-induced tsunami could affect the ocean currents and density structures in the bay. For conducting the quantitative evaluations of the effect and sustainable use of fisheries resources in the bay, the clarification of the influence of topographic changes caused by the tsunami on the current and density structures in the bay is necessary.

Observations of water current velocity, salinity, and temperature were conducted before the earthquake using an Aanderaa current meter and a salinity-temperature-depth profiler at five points in the bay (Okazaki, 1994). A multifunction water quality meter was used to take measurements of water quality in the inner part of the bay (Yaname, 2010). However, we

https://doi.org/10.1016/B978-0-12-810473-6.00020-0

© 2019 Elsevier Inc. All rights reserved.

cannot evaluate the influence of topographic changes caused by the tsunami on the current and density structures because no observations were conducted after the Great East Japan Earthquake. Moreover, a detailed numerical simulation has not been conducted in the bay; therefore, the spatial and temporal changes in the ocean currents and density structures in the bay are still unknown.

The aim of this study was to investigate the effects of topographic changes resulting from the tsunami caused by the Great East Japan Earthquake on the ocean current and density structures in Miyako Bay, Japan, using numerical analysis. We reproduced the ocean current and density structures in the bay in February, May, August, and December 2010 using our original ocean model (Murakami et al., 2011) with preearthquake topography and clarified the seasonal changes in the structures over the entire bay. Then, we repeated the numerical simulation for the postearthquake topography and evaluated the effect of topographic changes on the ocean currents and density structures.

2 SIMULATION METHOD

We used the Coastal Ocean Current Model (CCM) developed by Murakami et al. (2011). Although it is a primitive ocean model, the CCM is our original ocean model that utilizes a multisigma coordinate system. The multisigma coordinate system supports the accurate calculation of inflow rates of seawater from offshore, which greatly improves the reproducibility of ocean currents in the inner bays. The details of the model can be found in Murakami et al. (2011).

To include all four seasons, the calculation periods were set as February 10–24, May 13–27, August 16–30, and December 1–15, 2010. Atmospheric data from the Automated Meteorological Data Acquisition System (AMeDAS), sea level data from the Japan Meteorological Agency (JMA), and information on the temperature, salinity, and flow rate of the Hei and Tsugaruishi rivers into the bay from the Japan Fisheries Agency (JFA) in the bay were used as the initial and boundary values of the CCM. The sea-floor topography of Miyako Bay before the Great East Japan Earthquake used in this study is shown in Fig. 1. The topography has 50 m grid spacing and is constructed from the JODC (Japan Oceanographic Data Center)—Expert Grid Data for Geography-500 m (J-EGG500) and the JODC-Bathymetry Integrated Random Dataset (J-BIRD) of the Japan Oceanographic Data Center (JPDC), and marine cadaster data from the Hydrographic and Oceanographic Department of the Japan Coast Guard (HOD, JCG). The postearthquake sea-floor topography of the bay used in this study is shown in Fig. 2. This is based on airplane laser measurements taken over 11 days (June 11–21) in 2011 conducted by the HOD and the JCG.

The method used here is the same as that in (Murakami et al., 2013). However, the topographies used in this study were modified to correctly represent the coast lines of the bay, including seawalls and structures, using a nautical chart (Japan Coast Guard, 2012) and recent aerial photographs (taken in June 2013). The nautical chart was published in August 2012, but indicates data from both before and after the earthquake. The numerical simulations in this study were performed using these new topographies.

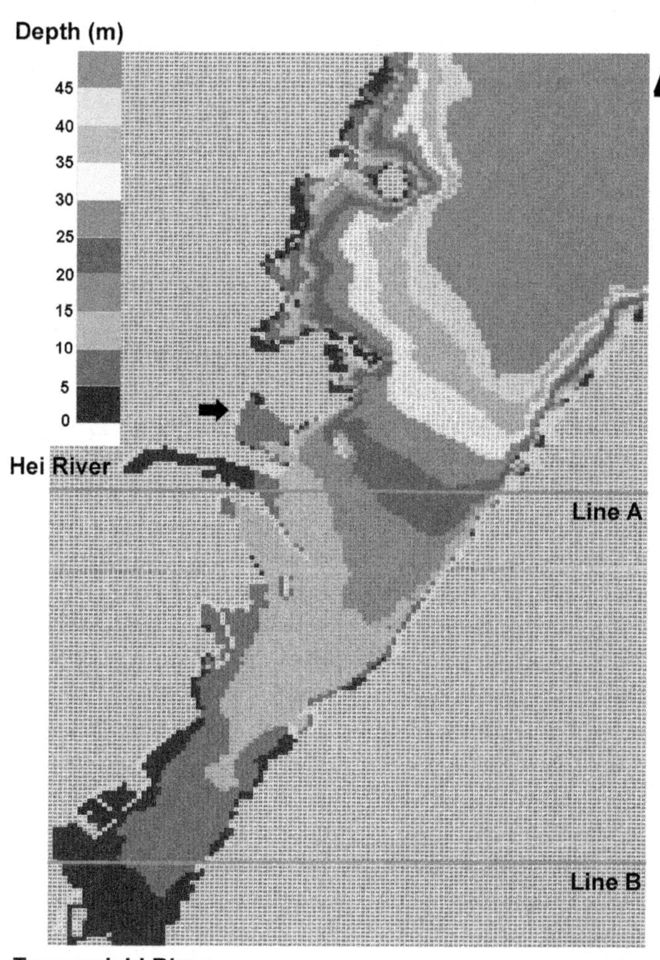

FIG. 1 Sea-floor topography of Miyako Bay before the Great East Japan Earthquake. Small bay indicated by the *arrow* is referred in Section 3.2.

Fig. 3 shows the amount of topographic change after the earthquake. Erosion of more than 2m caused by the tsunami occurred in broad areas in the inner and middle parts of the bay. The maximum value was 5m in the middle part of the bay.

3 RESULTS

3.1 Seasonal Changes in Oceanic Currents and Density Structures in Miyako Bay Before the Great East Japan Earthquake

Figs. 4 and 5 show the observed wind speeds and wind directions, respectively, in Miyako Bay for February, May, August, and December during the calculation period. The average

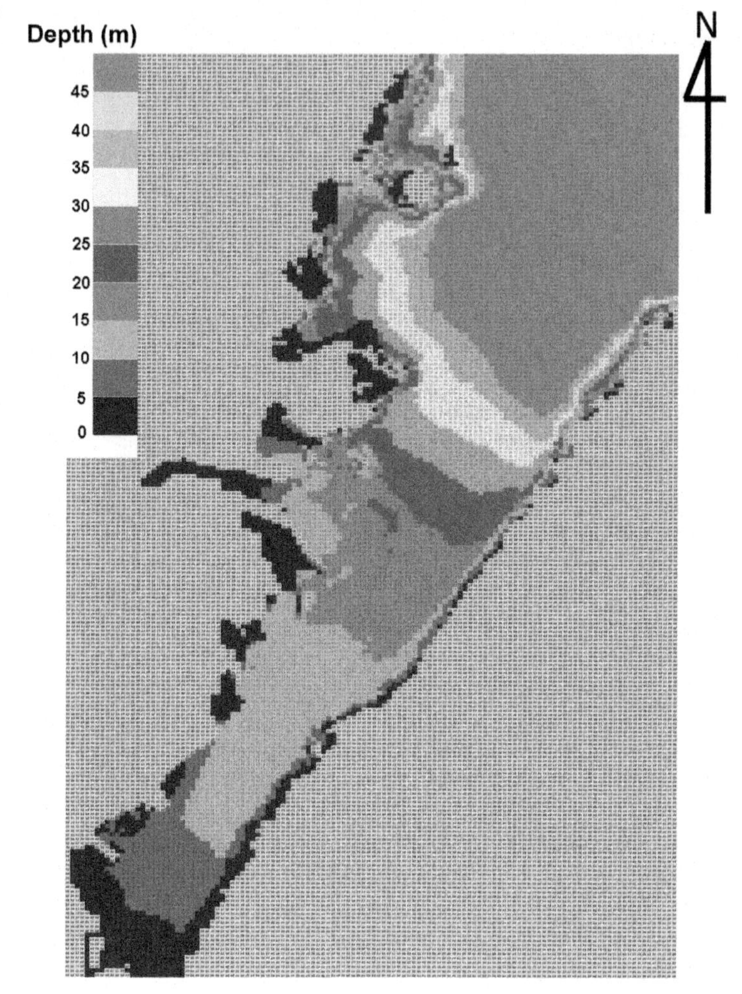

FIG. 2 Sea-floor topography of Miyako Bay after the Great East Japan Earthquake.

wind speeds in February, May, August, and December were 2.3 m/s, 2.3 m/s, 1.9 m/s, and 2.6 m/s, respectively. Seasonal differences were small, and the maximum wind speed of 20 m/s on December 3 was caused by the approach of a "bomb" cyclone (Fig. 4D). Although daily changes in the wind direction were observed in all the seasons, the main wind direction was southwest. Therefore, these results suggest that the average wind speeds and wind directions in the bay were almost constant throughout the year.

Fig. 6 shows the average distribution of sea surface density in the bay during February, May, August, and December, calculated using the topography before the Great East Japan Earthquake. The sea surface density is expressed by σt, which is defined as the density deviation from 1000 kg/m^3. This indicates that the sea surface densities in winter (Fig. 6A and D) were smaller than those in summer (Fig. 6B and C) in the entire bay. As stated above, the average wind speed and wind direction conditions in the bay were almost constant throughout

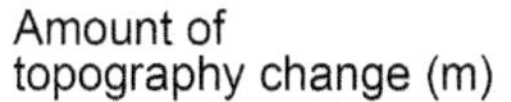

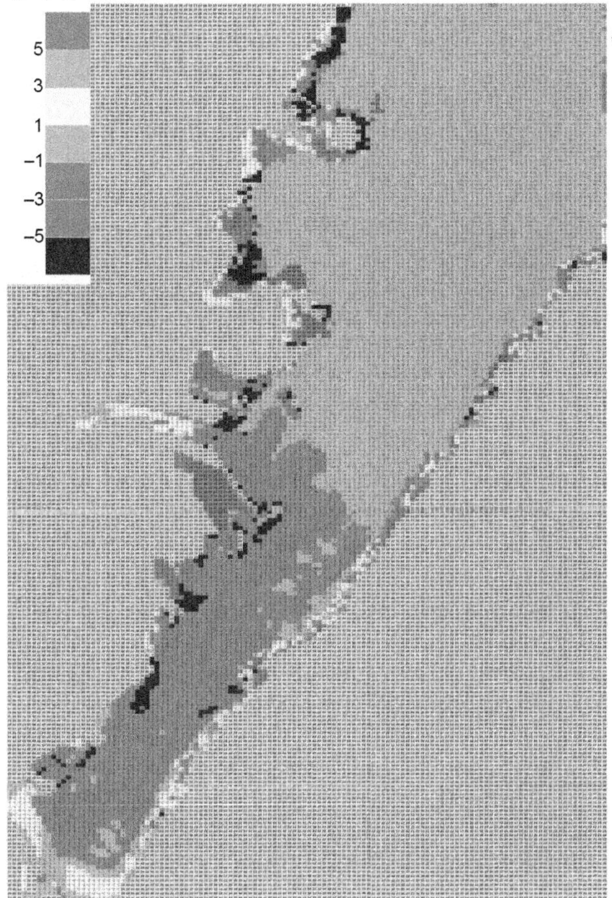

FIG. 3 Amount of topographic change before and after the Great East Japan Earthquake. *Positive* and *negative* values indicate accumulation and erosion, respectively, after the earthquake.

the year. The flow of the Hei River, which is located in the central part of the bay (Fig. 1), in summer and winter was $14.5\,m^3/s$ and $21.0\,m^3/s$, respectively (Japan Fisheries Agency (Ministry of Agriculture, Forestry and Fisheries), Japan Forest Agency (Ministry of Agriculture, Forestry and Fisheries) and Water and Disaster Management Bureau (Ministry of Land, Infrastructure, Transport and Tourism), 2004). The flow of the Tsugaruishi River in the inner part of the bay (Fig. 1) in summer and winter was $1.9\,m^3/s$ and $7.0\,m^3/s$, respectively (Japan Fisheries Agency (Ministry of Agriculture, Forestry and Fisheries) et al., 2004). The river flow in winter was larger than that in summer for both the rivers. Therefore, these results suggest that decreases of sea surface density in winter season were caused by increase of river flows of Hei and Tsugaruishi rivers. In addition, the results indicate that the changes in the sea surface density between May and August and also between February and December were small. Thus, hereafter we focus only on changes for August and December.

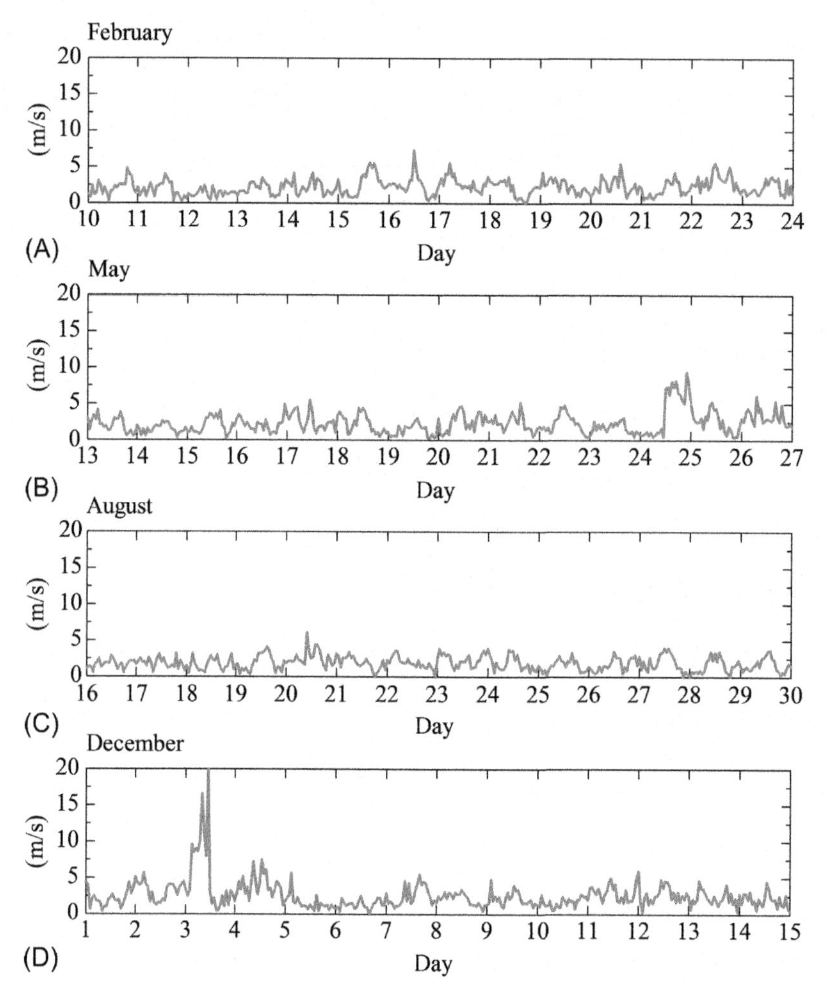

FIG. 4 Observed wind speeds in Miyako Bay for (A) February, (B) May, (C) August, and (D) December in the calculation periods.

Fig. 7 shows the cross sections along lines A and B (Fig. 1) of average current velocity in the bay for August and December, calculated using the preearthquake topography. The cross sections along line A (Fig. 7A and B) show that northward currents from the inner part to the mouth of the bay were present at depths shallower than 5m in both August and December. In depths below 5m on the east side of the bay, southward currents can be distinguished for both August and December. In depths below 5m on the west side of the bay, a northward current occurred in August, but a southward current occurred in December. The cross-sections along line B (Fig. 7C and D) show that the northward currents from the inner part to the mouth of the bay were present in depths above 2m in both August and December. In depths below 2m, southward currents were

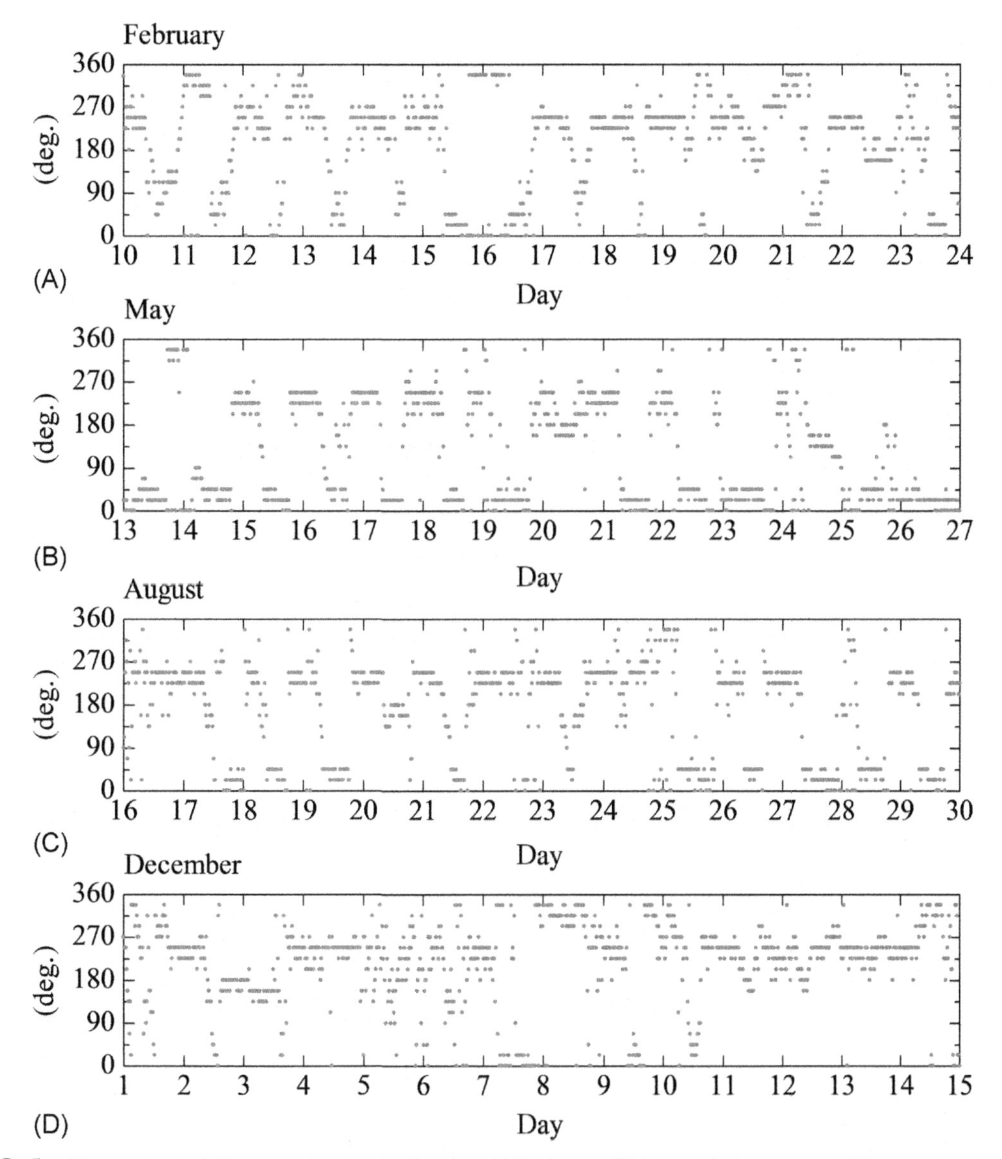

FIG. 5 Observed wind directions in Miyako Bay for (A) February, (B) May, (C) August, and (D) December in the calculation periods. Wind directions are expressed as degrees, for example, 0 degrees and 90 degrees mean north wind and south wind, respectively.

distinguished for both August and December, but the velocities in August were smaller than those in December. These results indicate that the current directions changed with the season.

As stated above, the maximum wind speed of 20 m/s on December 3 was caused by the approach of a "bomb" cyclone (Fig. 4D). Fig. 8 shows the distributions of sea surface

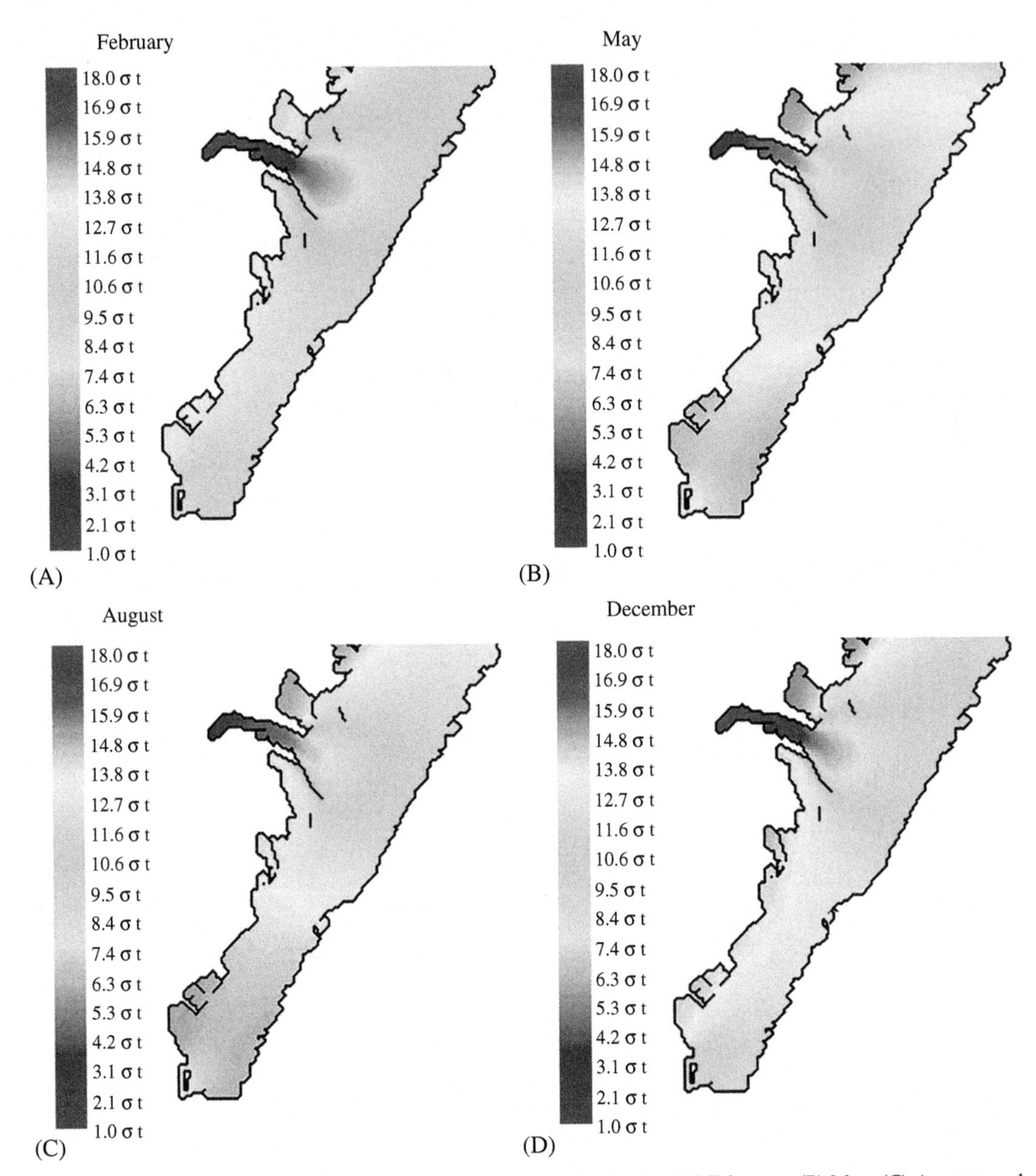

FIG. 6 Average distributions of sea surface density in Miyako Bay for (A) February, (B) May, (C) August, and (D) December, calculated for the topography before the Great East Japan Earthquake. The sea surface density is expressed by σt, which is defined as the density deviation from $1000\,kg/m^3$.

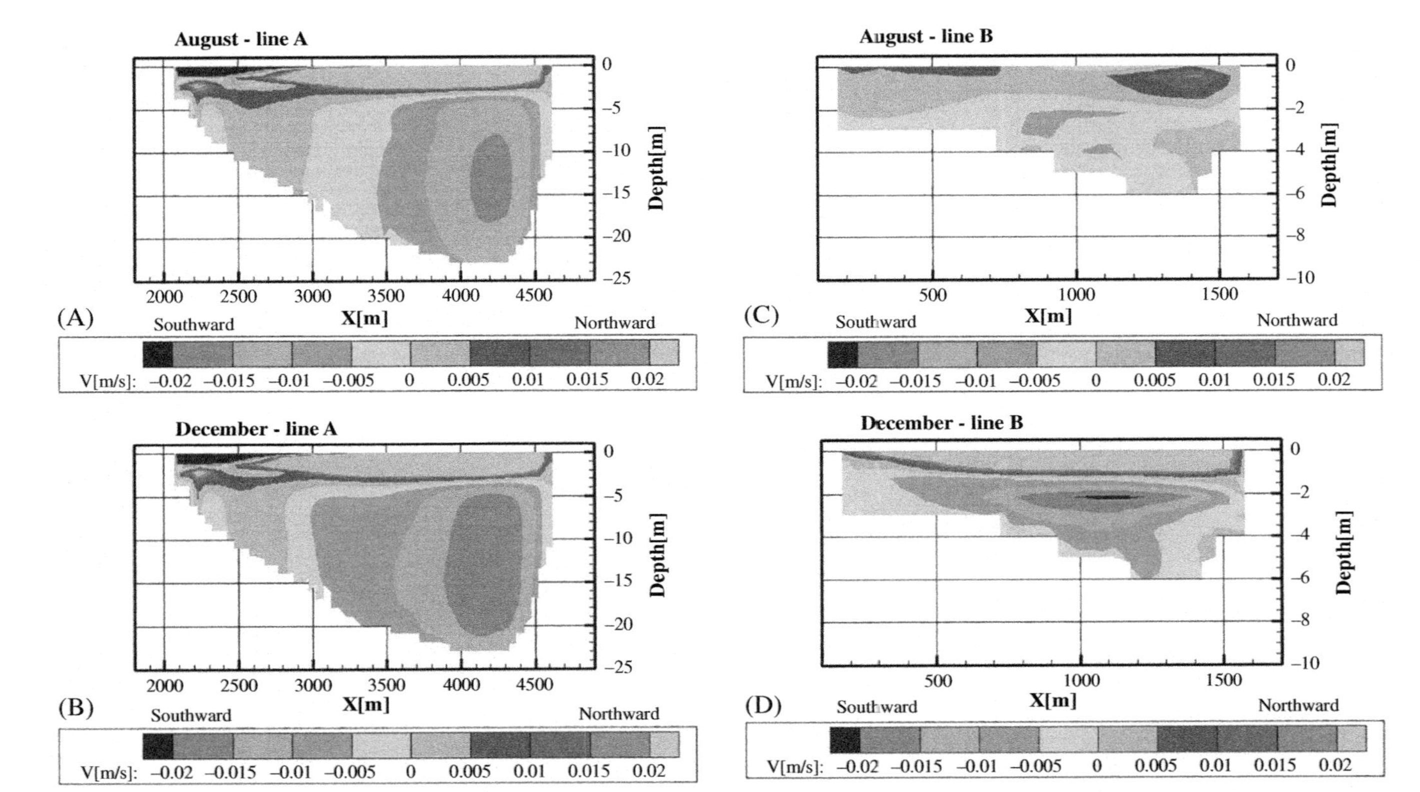

FIG. 7 Cross-sections of average current velocity in Miyako Bay for (A) August (line A), (B) December (line A), (C) August (line B), and (D) December (line B), calculated for the topography before the Great East Japan Earthquake. Lines A and B are shown in Fig. 1. *Positive* and *negative* values indicate northward and southward flows, respectively.

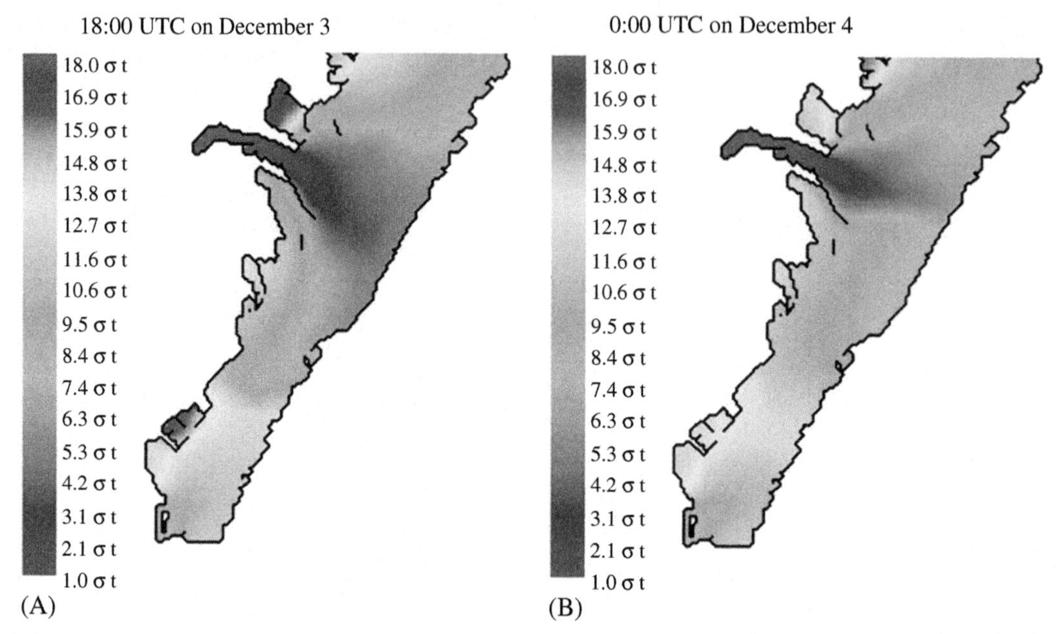

FIG. 8 Distributions of sea surface density in Miyako Bay at (A) 18:00 UTC on December 3, 2010, and (B) 00:00 UTC on December 4, 2010, calculated for the topography before the Great East Japan Earthquake.

density in the bay at 18:00 UTC on December 3, 2010, and 00:00 UTC on December 4, 2010, calculated for the preearthquake topography. By comparing the sea surface density at 18:00 UTC on December 3 (Fig. 8A) with the average for December (Fig. 6D), we can see that the sea surface density decreased with the approach of the bomb low pressure. By comparing the sea surface density at 00:00 UTC on December 4 (Fig. 8B) with the average for December (Fig. 6D), we can see that the decrease of sea surface density was a temporary phenomenon.

Fig. 9 shows the cross sections of current velocity along line A on December 3, 2010, calculated for the preearthquake topography. At 11:30 UTC (Fig. 9B), when the bomb low pressure was approaching, the northward and southward currents could be distinguished in the upper and lower layers, respectively, from the effect of a south wind at 20 m/s (Figs. 4D and 5D). At 14:00 UTC (Fig. 9C), when the wind direction changed from south to north, southward and northward currents could be distinguished in the upper and lower layers, respectively, which were in opposite directions to the currents at 11:30 UTC. At 20:00 UTC (Fig. 9D), when the wind speed was low (approximately 2 m/s), the distribution of current velocity was similar to the average distribution (Fig. 9A). Therefore, these results suggest that the distribution of current velocity in the bay changed over a short time because of the effect of meteorological disturbances such as the strong winds of a "bomb" cyclone.

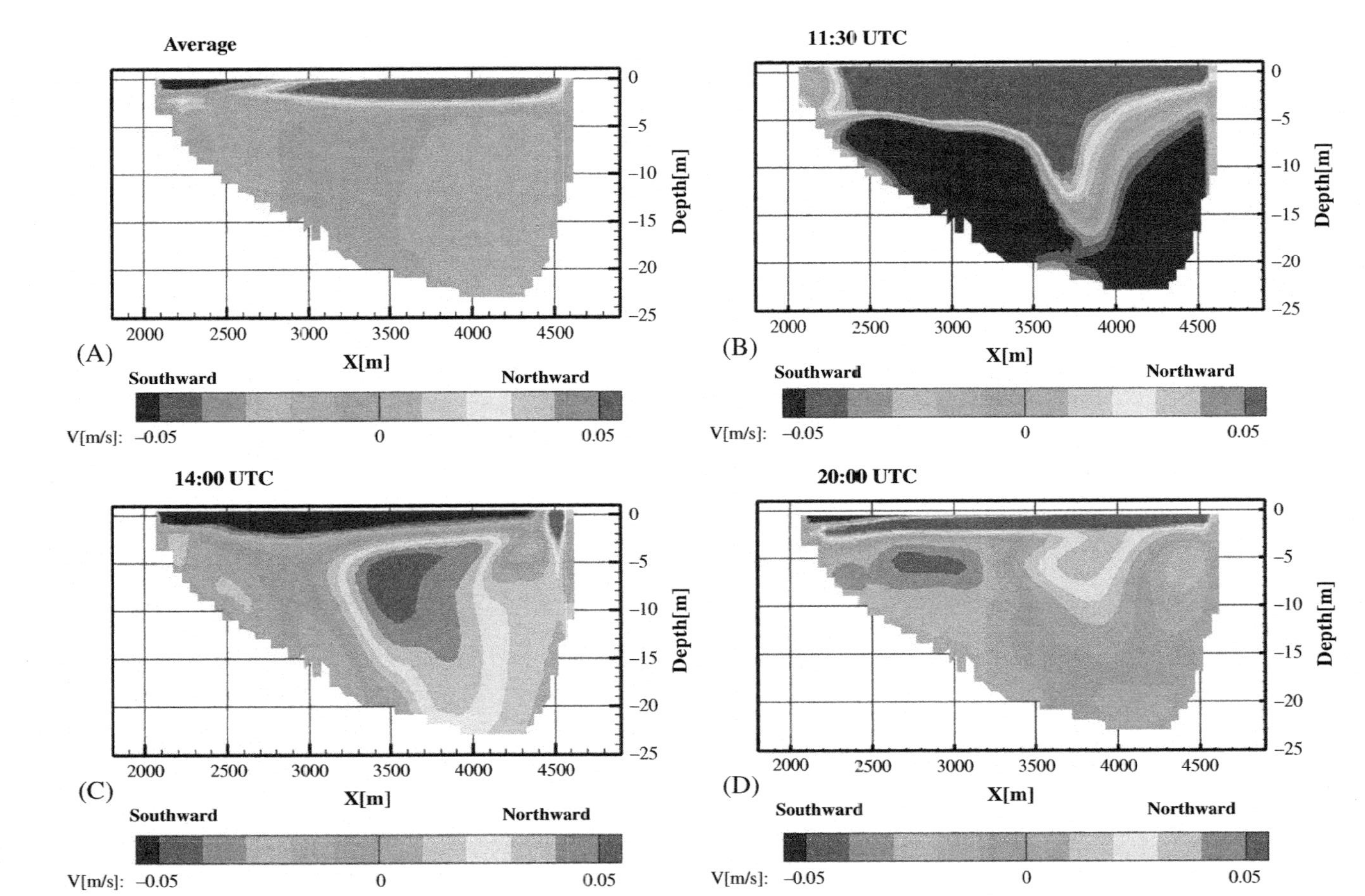

FIG. 9 Crosssections along line A of current velocity in Miyako Bay (A) average on December 3, 2010, and at (B) 11:30 UTC, (C) 14:00 UTC, and (D) 20:00 UTC on December 3, 2010, calculated using the topography before the Great East Japan Earthquake. Line A is indicated in Fig. 1. *Positive* and *negative* values indicate northward and southward flows, respectively.

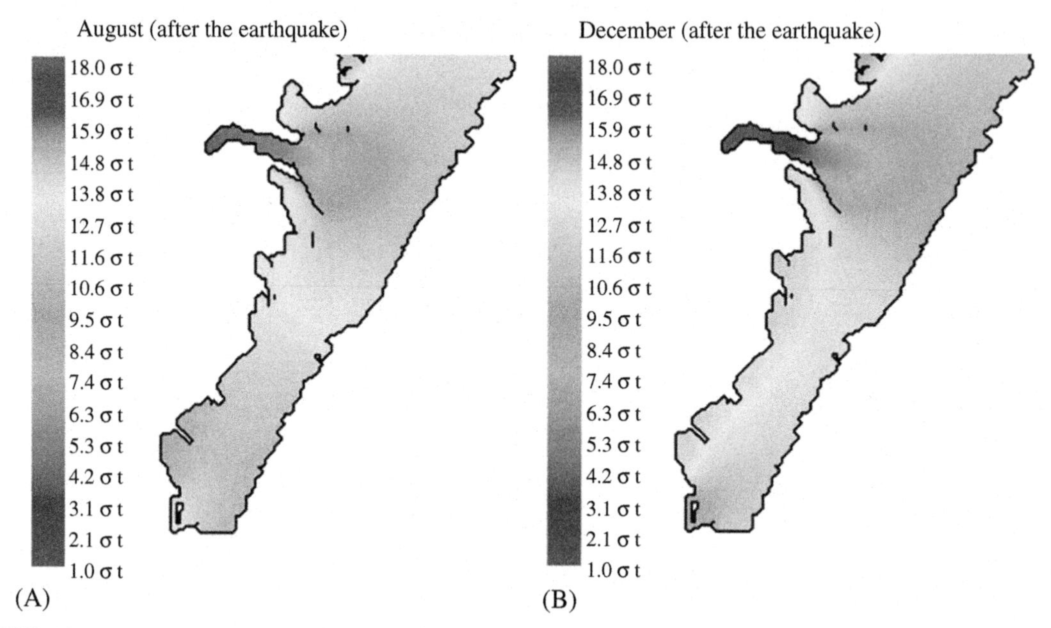

FIG. 10 Distributions of average sea surface density in Miyako Bay in (A) August and (B) December, calculated for the topography after the Great East Japan Earthquake.

3.2 Influence of Tsunami-Caused Topographic Changes on the Ocean Current and Density Structures in Miyako Bay

Fig. 10 shows the distributions of average sea surface density in the bay in August and December, calculated for the postearthquake topography. The sea surface density in a small bay located at the north of the Hei River (indicated by the arrow in Fig. 1) in August after the earthquake (Fig.10A) was smaller than that before the earthquake. This is speculated to be because sea wall destruction by the earthquake allowed the flow of low-density water from the Hei River into the small bay. This tendency could also be observed in December (Figs. 10B and 6D). Therefore, we concluded that the topographic changes in the bay caused by the earthquake-induced tsunami locally influenced sea surface density.

Fig. 11 shows the cross sections along lines A and B (Fig. 1) of average current velocity in the bay in August and December, calculated for the postearthquake topography. In cross-sections along line A in August (Figs. 7A and 11A) and along line B in December (Figs. 7D and 11D), the distributions of current velocity showed few differences. In contrast, in cross-sections along line A in December (Figs. 7B and 11B), the southward currents at depths below 5 m on the east side of the bay after the earthquake were stronger than those before the earthquake. In cross-sections along line B in August (Figs. 7C and 11C), the current directions in the west side of the bay changed from southward before the earthquake to northward after the earthquake. Therefore, we conclude that the topographic changes in the bay resulting from the tsunami caused by the earthquake had little influence on ocean current structure over the entire bay, but had a strong effect on the structure locally, such as reversal of current directions.

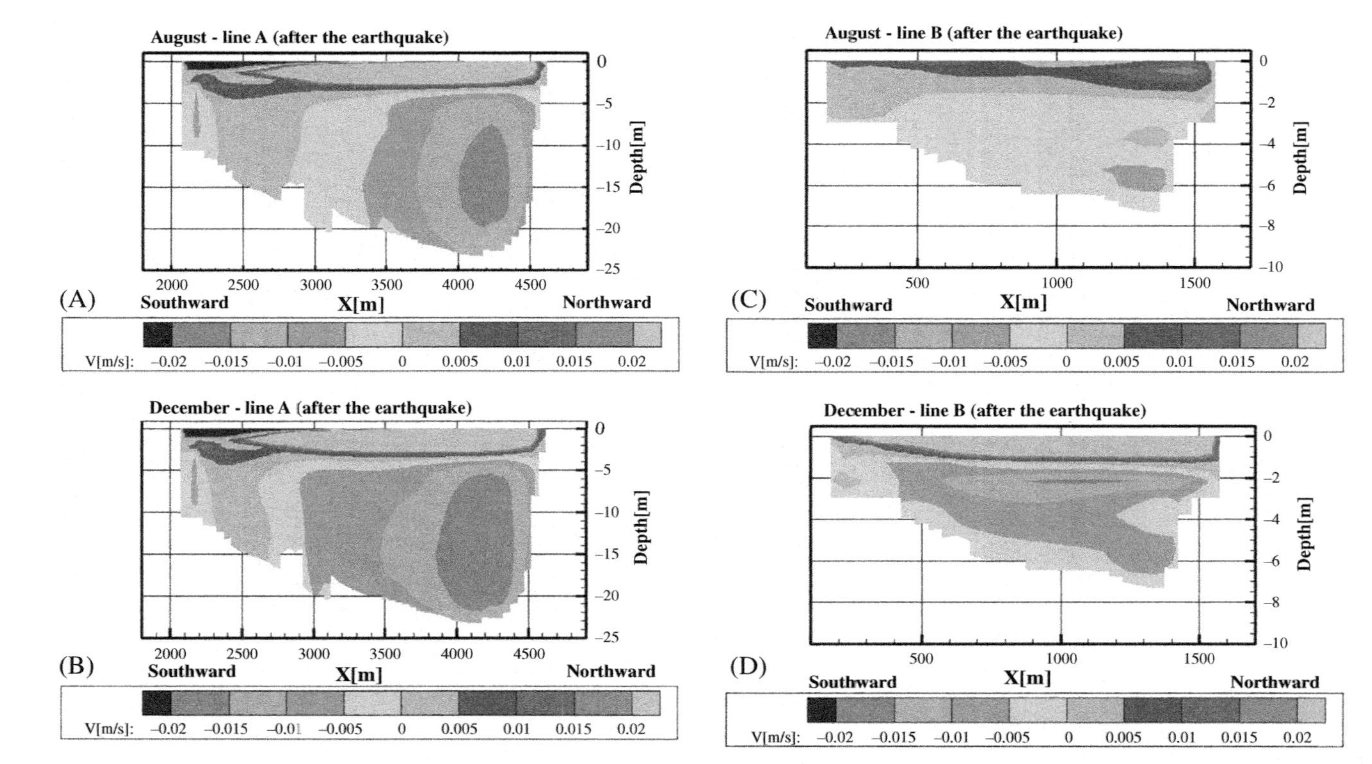

FIG. 11 Cross-sections of average current velocity in Miyako Bay for (A) August (line A), (B) December (line A), (C) August (line B), and (D) December (line B), calculated using the topography after the Great East Japan Earthquake. Lines A and B are indicated in Fig. 1. *Positive* and *negative* values indicate northward and southward flows, respectively.

4 SUMMARY

In this study, we conducted numerical simulations to evaluate the influence of topographic changes resulting from the tsunami caused by the Great East Japan Earthquake on the ocean current and density structures in Miyako Bay, Japan, using our original ocean model (which uses a multisigma coordinate system) for the topographies before and after the earthquake. The results show that the topographic changes caused by the earthquake-induced tsunami have had little influence on ocean current and density structures over the entire bay, but locally affected the sea surface density and directions of the ocean currents.

The numerical analysis in this study clarified the magnitude of the influences of tsunami-caused topographic changes on the ocean current and density structures in Miyako Bay. However, the models of ecosystem performance based on the ocean current and density structures in the bay are necessary to elucidate the effects of the changes in the sea surface density and the direction of the ocean currents on the use of fisheries resources. In the near future, we intend to conduct such simulations and clarify the effect of the topographic changes caused by the earthquake-induced tsunami on ecosystems in the bay.

Acknowledgments

This research was supported by Grant-in-Aid for Scientific Research 16K01292 and 16K13876 from JSPS (Japan Society for the Promotion of Science) and a research project of the National Research Institute for Earth Science and Disaster Prevention.

References

Japan Coast Guard, 2012. Nautical Chart, W54 (Ishinomaki Wan to MiyakoKo). (written in both Japanese and English), Available from: http://www.jha.or.jp/shop/index.php?main_page=product_info&products_id=518.

Japan Fisheries Agency (Ministry of Agriculture, Forestry and Fisheries), Japan Forest Agency (Ministry of Agriculture, Forestry and Fisheries) and Water and Disaster Management Bureau (Ministry of Land, Infrastructure, Transport and Tourism), 2004. Survey Report of Investigation of Creation Method of Fertile Fishing Ground Environment Considering Connections Among Forest, River and Ocean. pp. 298–312. (in Japanese), Available from: http://www.mlit.go.jp/river/press_blog/past_press/press/200401_06/040427/.

Murakami, T., Kawaguchi, C., Ogasawara, T., 2013. Numerical analysis of coastal currents and density structures in Miyako Bay affected by meteorological disturbances and topographic changes due to tsunami. Ann. J. Civil Eng. Ocean B3-69, I_718–I_723 (in Japanese with English abstract).

Murakami, T., Yoshino, J., Yasuda, T., Iizuka, S., Shimokawa, S., 2011. Atmosphere–ocean–wave coupled model performing 4DDA with a tropical cyclone bogussing scheme to calculate storm surges in an inner bay. AJEDM 3, 217–228.

Okazaki, M., 1994. The circulation of sea water and the variation of the properties in some bays of Sanriku coast, eastern coast of northern Japan. Bull. Coast. Oceanogr. 32, 15–28 (in Japanese with English abstract).

Space.geocities, 2014. Panel for Study of the Seaweed Bed and Land Reclamation in Miyako Bay. http://space.geocities.jp/miyakowannomoba/index.html. (in Japanese).

Yaname, K., 2010. Study on physical environment factors affecting inhabitation and growth of Herring's larvae and juveniles in Miyako Bay. Res. Rep. Int. Coast. Res. Cent. 35, 9–11 (in Japanese).

CHAPTER

20

Assessment of Potential Oil Spill Risk Along Vishakhaptnam Coast, India: Integrated Approach for Coastal Management

S. Arockiaraj, R.S. Kankara

National Centre for Coastal Research (NCCR), Ministry of Earth Sciences, Government of India, Chennai, India

1 INTRODUCTION

Of all the pollutants in various forms entering the sea, oil very often appears to be the substance that draws the most attention. Similarly, of all sources of marine pollution, oil-tanker accidents very often appear to be a significant source of marine oil pollution from operational spills. The ensembles of oil spills are the model-based simulations of oil trajectories that help in the statistical characterization of the risk in a study area (Olita et al., 2012). The environmental risk assessment always helps in the spill response strategies, with the best professional judgment on possible impact on the coast and water column organisms (Bejarano and Michel, 2016; Kirby and Law, 2010). National preparedness is necessary to combat an oil spill, which includes prevention, response, cleanup, and remediation of the affected coast. Prior knowledge of the sensitivity of the coastal resources is essential to assess the damage that might be caused by the oil spills (Adler and Inbar, 2007). Identification of coastal resources in the longer coastal stretch can be achieved through multispectral satellite images and the application of classification techniques (Hossain et al., 2009). The geographical information systems (GIS) modeler can be used for risk assessment to explore the possibility of oil spill combat techniques while interactive tools and settings can be employed to quantify the impacts on coastal resources (Kankara et al., 2016; Kankara and Subramanian, 2007). Considering the

Coastal Management
https://doi.org/10.1016/B978-0-12-810473-6.00021-2

© 2019 Elsevier Inc. All rights reserved.

importance of coastal protection in this study, a seasonal oil spill modeling and risk assessment has been carried out for selected locations along the Visakhapatnam coast.

2 STUDY AREA

The ideal location of Visakhapatnam is almost midway between Calcutta and Chennai at a latitude 17°41′N and a longitude of 83°18′E, serving a vast and rich hinterland (Fig. 1). The port has three: the outer harbor, the inner harbor, and the fishing harbor. The port caters to key industries such as petroleum, steel, power, and fertilizer besides other manufacturing industries. It also plays a catalyst role for the agricultural and industrial development of the hinterland spreading from the south to the north.

The East Indian coast has faced nearly 20 oil spills (Blue Waters, 2012), and many of them were not formally addressed for the potential for major oil or hazardous substance spills. Table 1 shows the number of oil spills along the Vishakhapatnam coast. Even though the number of spills is small, the possibility of occurrence in the future is alarming as the transport of petroleum through waterways has considerably increased in recent years (OSMPRMC, 2003). The spills have negative impacts on beaches, wildlife, fishing, and

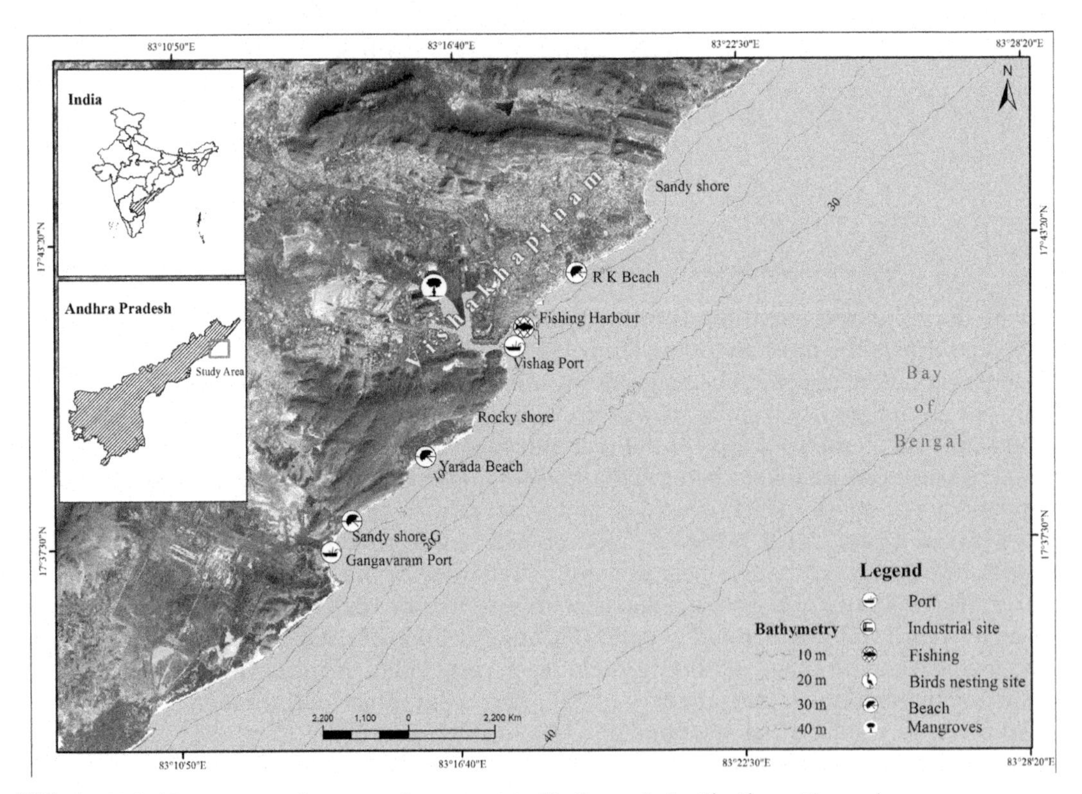

FIG. 1 Vishakhapatnam study area on the east coast of India marked with oil sensitive region.

TABLE 1 Major Oil Spills on the Vishakhapatnam Coast of India From 1990

S. No	Date	Quantity Spilled (t)	Position	Vessel/Other Incidents
1	March 26, 1995	200/diesel	Off Vizag, Andhra Pradesh	Dredger Mandovi-2
2	September 14, 1997	Naptha, diesel gas	Vizag, Andhra Pradesh	HPC refinery
3	June 2, 2005	49,537/cargo and 640/FO	Vishakhapatnam Port	MV Jinan VRWD-5
4	September 1 2005	100 tons HO	Off Visakhapatnam	MV Royal Ocean 2

FO, fuel oil; *HO*, heavy oil.
Source: Blue waters (Newsletters on Marine Environmental Protection).

tourism. Considering the major oil transport, the study locations were selected to carry out the oil spill trajectory modeling, environmental sensitivity mapping, and risk assessment (Fig. 1).

3 METHODOLOGY

A coastal resource information system was developed in a GIS platform for the Vishakhapatnam coast. Oil spill fate and trajectory models were designed to identify areas likely to be affected by oil spills. A GIS modeler tool was developed to integrate the results of oil spill simulations to identify the resources at risk in the GIS environment. The authors discussed with experts in the relevant areas to arrive at the qualitative scaling of risk assessment and sensitivity analysis for the resources of the Vishakhapatnam coast.

3.1 Oil Spill Trajectory Modeling

An oil spill model was used to generate the oil trajectory for various scenarios. The simulations of the hydrodynamic model (HD) were calibrated and validated with the observed data and were used in the oil spill trajectory. The movement of oil was based on the HD model. The oil characteristics, bathymetry, oceanographic parameters such as sea surface temperature, tides and currents, meteorological parameters such as wind speed and direction, etc., are given as inputs to the model based on the actual data provided. The oil spill simulation of a hypothetical accidental oil spill shows that the movement of oil mainly relies on the combined effects of tidal current and wind.

The trajectory models for the northeast and southwest monsoons were run for predicting the fate and trajectory of oil spilled with the actual wind data (wind speed 6m/s and different wind direction for two seasons); the simulation was carried out for 72h of the oil spill with about $100\,m^3$ of crude oil. The two different spill locations and their origins have been specified S1. The spill location S1 (single point mooring) is placed 3km from the shore.

3.2 Weathering Model ADIOS2

Automated data inquiry for oil spills (ADIOS2) is an oil spill response tool to assist oil spill responders and contingency planners in making decisions on potential response strategies.

ADIOS2 helps to assess the oil weathering and clean-up model to help develop clean-up strategies based on estimates of the amount of time that spilled oil will remain in the marine environment. This model requires the appropriate oil type, the meteorological condition of the study area, and the amount of product spilled. In order to calculate the oil budget for all the scenarios, the seasonal meteorological conditions were given as input.

The northeast monsoon conditions were as follows: water temperature: 26°C, wind speed: 3.5 m/s, wind direction: 45 degrees, current speed and direction: 0.10 m/s and 220 degrees, wave height: 0.6 m, and salinity: 32 ppt. A volume of 100 m^3 of crude oil was selected from the model and the result is shown in Fig. 3A. The oil mass balance for the southwest monsoon is shown in Fig. 3B. The environmental conditions for that monsoon were as follows: water temperature: 26°C, wind speed: 3 m/s, wind direction: 225 degrees, current speed and direction: 0.45 m/s and 140 degrees, wave height: 1.0 m, and salinity: 32ppt.

3.3 Coastal Resources and Sensitive Groups

3.3.1 *Biological Resources*

The coast of Vishakhapatnam consists of various biological resources that are sensitive to oil spills and are presented in the maps. The importance of these resources is considered based on their concentration, nesting or breading grounds, commercially important species, and rare, threatened species (Sriganesh et al., 2015). Biological resources are indicated with symbols in the maps (Fig. 5). Biological and related information such as plankton concentration (ICMAM-PD, COMAPS 2007–2011),which is highly vulnerable to oil spills, were taken into consideration. Samples of these microorganisms were collected for the concerned study area at different intervals of 0, 0.5, 1, 3, and 5 m depth during the particular seasons. The impact on biological resources such as the benthos, phytoplankton, and zooplankton was calculated based on the spill area of that particular spill and the amount of planktonic population within the area.

3.3.2 *Human Use Resources*

Human use resources are under threat due to oil spills causing economic loss to the users. They are grouped based on their uses, such as (i) high recreational purpose, (ii) highly protected natural resources, (iii) resource extraction sites (shell mining), and (iv) cultural resources, etc. Multispectral images (Landsat-7 ETM+) were used for coastal resource mapping with classification techniques. The atmospheric corrections were carried out in order to reduce the haziness of the images. The histogram minimum method discussed (Hadjimitsis et al., 2010) here is known as the dark pixel (DP) atmospheric correction method. The Landsat-7 ETM+ was subjected to the DP method, which produced reasonable corrections in the cloud-free image. The atmospherically corrected images were then georeferenced with the field-collected ground control points.

3.4 Environmental Sensitivity Indexing

The map includes information on classified (biological and human-use) resources, allowing shoreline habitats to be ranked according to sensitivity to oil spill impact, natural persistence of oil, and ease of cleanup. Information on biological resources includes oil-sensitive species population on shore habitats and pelagic organisms, which was received from the Coastal Ocean Monitoring and Prediction System database (ICMAM-PD, COMAPS 2007–2011). The coastal resource sensitivity values were based on information developed with the Delphi Method, a structured process of collecting and using qualitative knowledge from a group of experts by means of a series of questionnaires. This was used as the base to arrive at the risk assessment and sensitivity analysis of the classified resources of Vishakhapatnam.

A numerical modeling technique was applied for simulating spatial and temporal patterns of an oil spill under given environmental conditions. The bathymetry of the study area was prepared from Indian Naval Hydrographic Chart no. 313, which has a scale of 1: 50,000. The results of these simulations were used to identify the resources at risk. Coastal resource information such as estuaries, lagoons, natural beaches, fishing zones, shell mining, brackish water marshes, and saltpans was obtained from satellite images and field verification. Personal discussions with experts in relevant areas helped us to arrive at the qualitative scaling of risk assessment and sensitivity analysis for the resources of the Vishakhapatnam coast.

The sensitivity criteria (1–10) and priority index were assigned to the coastal resources using GIS based oil spill trajectory modeling approach as described by Kankara and Subramanian (2007). The river/creek mouth, sandy beaches are more sensitive to oil spills in comparison to rocky shore, harbor etc. due to their scientific, environmental and economic importance. 30% weightage was given to sensitivity of resources to oil pollution followed by 10% for scientific values, 30% for environmental importance and 10% for economic importance to assess relative response sensitivity and priority index. The relative response of sensitivity was derived from the summation of the sensitivity criteria and the weight factor of each resource (Tables 2A and 2B).

3.5 Oil Spill Risk Assessment

The Oil Spill Risk Assessment Model (OSRAM) discussed here does take into account the likelihood of the event occurring and the measures that can be taken to mitigate the potential impacts. This OSRAM (Fig. 2) generated in GIS is to assess the risk on the basis of the direction of the spill movement and the proximity of the spill patch to coastal resources. The ESI ranking adopted the following steps

- Relative Response of Sensitivity (a) ($Wi \times Si$)

Where Wi—weighting factor (sensitivity for oil pollution 30%, cultural and social values 10%, scientific value 20%, environmental importance 30%, and economic consideration 10%), and Si—Sensitivity Index.

- Risk (b) = Nearest Distance of Oil Patch (VB conditional command).
- Priority Index (PI) = ($a \times b$).
- Priority Order = (PI) %.
- Environmental Sensitivity Index ESI (VB conditional command).
- Impact on Biological Resources = (No/m^2) × Spill Area.

The relative response of sensitivity of the coastal resources was derived from the weightage and sensitivity factors as mentioned. The risk of the resource at that particular spill was calculated on the basis of the distance of the oil patch on the shore. The priority index is the multiplication of the (a) and (b), which shows the high risk of the resource. The affected quantity

TABLE 2A Oil Spill Risk Assessment for Scenario 3/14 (Northeast Monsoon)

Sensitivity and priority index (including sensitivity criteria [Si] [1–0], where 1 is least sensitive and 10 is most sensitive; weighting factor (wi): 1—low, 2—medium, 3—high; and risk factor: 1—low, 2—medium, 3—high) (Priority Order: A—Most Priority, C—Least priority)

Resources	Shoreline (km)	Sensitivity for Oil Pollution (1–10) Weight (30%)	Cultural and Social Values (1–10) Weight (10%)	Scientific Value (1–10) Weight (20%)	Environmental Importance (1–10) Weight (30%)	Economic Consideration (1–10) Weight (10%)	Total Relative Response of Sensitivity (a) = (wi × si) (%)	Oil Patch Distance to the Resources (km)	Risk (b)	Priority Index (a × b)	Priority Order	Environmental Sensitivity Index
Sandy shore	2.9	8	4	5	5	2	5.5	2.8	1	5.5	5.2%	C
RK beach	5.6	10	9	5	8	6	7.9	1.1	2	15.8	15.1%	B
Fishing harbor	6.3	2	8	2	5	8	4.1	0.5	3	12.3	11.7%	B
Port	14.0	10	4	1	5	5	5.6	0.9	3	16.8	16.0%	B
Break water	3.2	1	2	4	2	5	2.4	0.2	3	7.2	6.9%	C
Rockyshore	2.8	3	1	4	2	1	2.5	0.6	3	7.5	7.2%	C
Sandy shore G	3.0	8	4	5	5	2	5.5	1.4	2	11.0	10.5%	B
Yarada beach	3.5	10	9	5	8	6	7.9	0.6	3	23.7	22.6%	A
Rockyshore_G	1.7	3	1	4	2	1	2.5	1.1	2	5.0	4.8%	C

TABLE 2B Oil Spill Risk Assessment for Scenario 9/14 (Southwest Monsoon)

Sensitivity and priority index (including sensitivity criteria [Si] [1–0], where 1 is least sensitive and 10 is most sensitive; weighting factor (wi): 1—low, 2—medium, 3—high; and risk factor: 1—low, 2—medium, 3—high) (Priority Order: A–Most Priority, C–Least priority)

Resources	Shoreline (km)	Sensitivity for Oil Pollution (1–10) Weight (30%)	Cultural and Social Values (1–10) Weight (10%)	Scientific Value (1–10) Weight (20%)	Environmental Importance (1–10) Weight (30%)	Economic Consideration (1–10) Weight (10%)	Total Relative Response of Sensitivity (a) = (wi × si) (%)	Oil Patch Distance to the Resources (km)	Risk (b)	Priority Index (a × b)	Priority Order	Environmental Sensitivity Index
Sandy shore	2.9	8	4	5	5	2	5.5	2.9	1	5.5	4.6%	C
RK beach	5.6	10	9	5	8	6	7.9	1.0	3	23.7	19.6%	B
Fishing harbor	6.3	2	8	2	5	8	4.1	0.4	3	12.3	10.2%	B
Port	14.0	10	4	1	5	5	5.6	0.7	3	16.8	13.9%	C
Break water	3.2	1	2	4	2	5	2.4	0.2	3	7.2	6.0%	C
Rockyshore	2.8	3	1	4	2	1	2.5	0.4	3	7.5	6.2%	C
Sandy shore G	3.0	8	4	5	5	2	5.5	0.9	3	16.5	13.7%	C
Yarada beach	3.5	10	9	5	8	6	7.9	0.4	3	23.7	19.6%	A
Rockyshore_G	1.7	3	1	4	2	1	2.5	0.7	3	7.5	6.2%	C

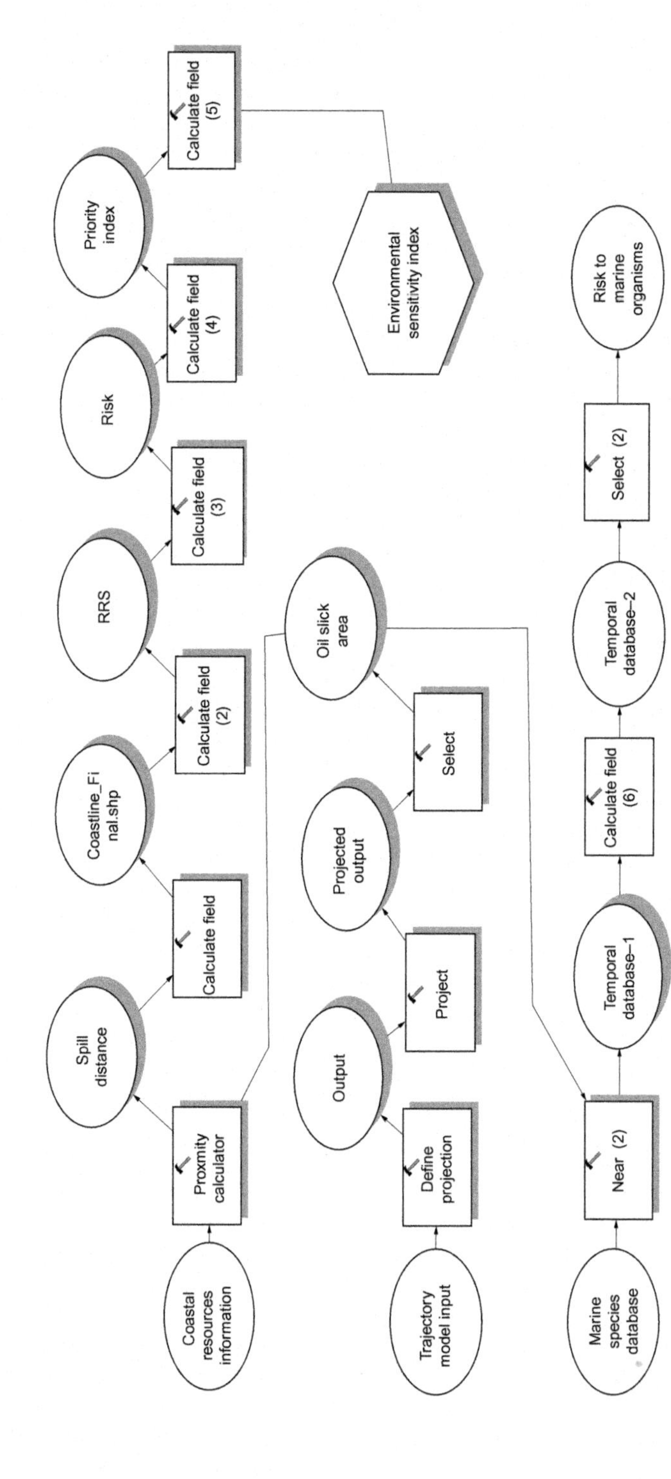

FIG. 2 GIS modeler flowchart showing different components in the workflow and derivation of ESI output.

of biological resources in the water is derived from the sampled number of species and multiplied with the spill area. The analysis of the priority index for Vishakhapatnam suggests the highest ranking to the beaches and sandy shore, which are biologically rich. The ESI for the resources was denoted with the letters A, B, and C, which stand for high priority (A) to low priority (C) for the timely protection of resources during the response.

The modeling output result was incorporated into the OSRAM to find the impact of oil at that particular season. The continuous oil spill happening from the ship anchorage location was used in the model with seasonal hydrodynamic meteorological conditions. Their trajectory output was fed into the OSRAM for possible risk assessment. The maps were prepared for the 14 scenarios using the integrated output from OSRAM.

4 RESULTS AND DISCUSSION

4.1 Oil Spill Model

The spill scenarios were generated for two seasonal hydrodynamic regimes i.e. southwest and northeast monsoon. The net current direction was northerly during southwest monsoon. The oil that continuously spilled moved toward the northern side for 15 km from the ship anchorage location. The behavior of the slick movement is more or less similar in various scenarios, irrespective of the quantities of oil spilled in the same season. The area of oil spread differs depending on the source quantities. In the second scenario, the northeast monsoon period, the oil slick moves toward the southern direction from location S1 as a continuous spill. The oil thickness formed for a continuous spill varies from 0 to 6 mm and 0 to 0.04 mm in the continuous spill for the northeast monsoon and southwest monsoon, respectively. The movement of the oil spill and its thickness varied with time due to its weathering property. The modeling results revealed that under the under calm conditions, oil moved as per the path of the streamlines of tidal flow (Fig. 3). However, the wind played a major role for the transport of the oil slick at the surface.

4.2 Weathering Model

In the weathering model scenario, the oil mass balance was computed for crude oil No. 2. It was found that about 30%–40% of spilled oil was lost due to evaporation, and 5%–8% was vertically dispersed in the water column for 7 h out of 24 h, as shown in Fig. 4A for the northeast monsoon. The weathering model for the southwest monsoon shows a difference in the dispersion, ranging from 8% to 10% vertically, and so the remaining volume decreased to be 50% in 24 h of the spill. It was evident that the evaporation process was very fast at the initial stage of the spill and slowed down subsequently. The oil dispersion in the water column increased considerably with time due to the increase in density of the oil and the variation of viscosity and density of the oil. However, the dispersion of diesel suddenly increased two hours after the spill occurred. This analysis provides valuable information that remediation using dispersants or the application of booms will be effective in the initial phase of the spill (Mohan et al., 2014).

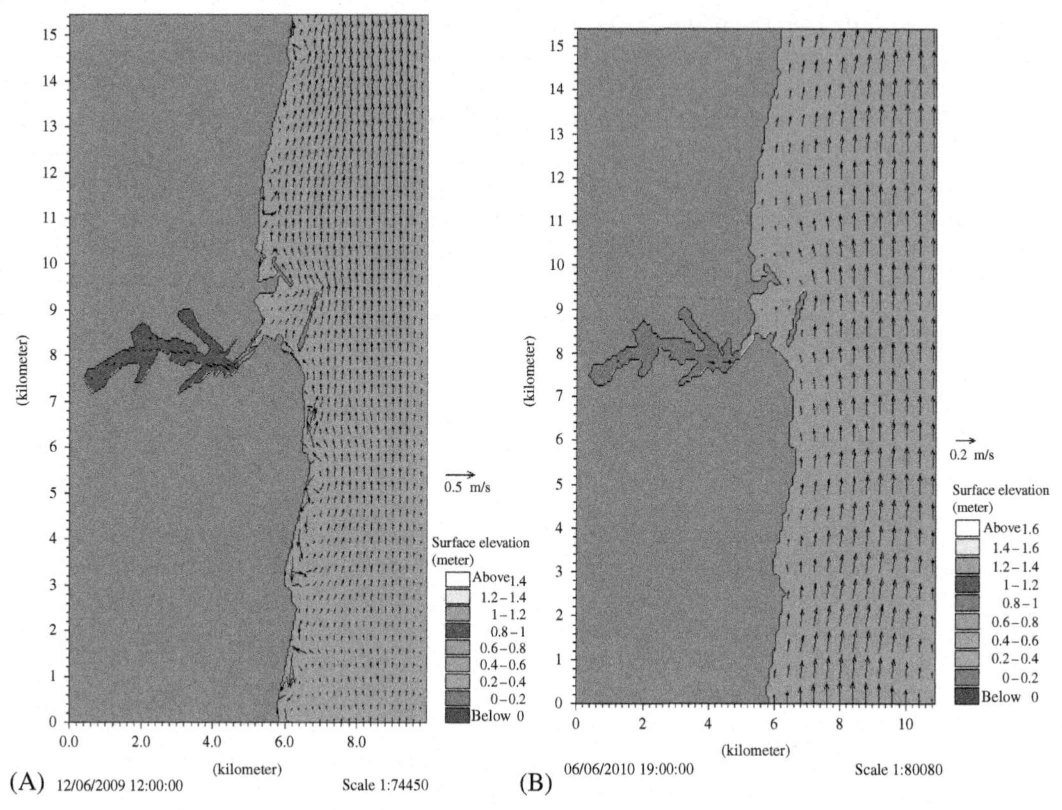

FIG. 3 (A) Surface elevation, current speed, and direction during the northeast monsoon. Mean current speed was 0.5 m/s during this season. (B) Surface elevation, current speed, and direction during the southwest monsoon. The current speed was recorded to be 0.2 m/s during this season.

4.3 Scenario-Based Risk Assessment

The spill model in the seasonal scenarios considered the seasonal atmospheric and hydrodynamic parameters. The seasonal scenarios considered the properties of the mean of the seasonal current, the wind direction, and the speed for the oil-spill model. The possible coastal impact was found during the northeast monsoon whereas during the southwest monsoon, the coastal impact is caused when the spill occurs at the single point mooring (SPM) location. The biological resources include species-rich areas that are ecologically important such as fish nursery areas and bird nesting sites. The natural sandy shore and fishing locations are the ecologically sensitive areas found along the Vishakhapatnam coast. Because pollutants such as oil occur in surface waters, they affect plankton functions. Observations showed the changes in phytoplankton population composition and the reduction of zooplankton populations during an oil spill as well as an inhibition in algal primary production (Parsons et al., 2015).

The results of the seasonal scenario show that when oil spills from location 1 during the northeast monsoon, it most likely affects the Vizag port and the Erada Beach (Fig. 5A). As per the modeler results, it is found that most of the northern part of the Vishakhapatnam

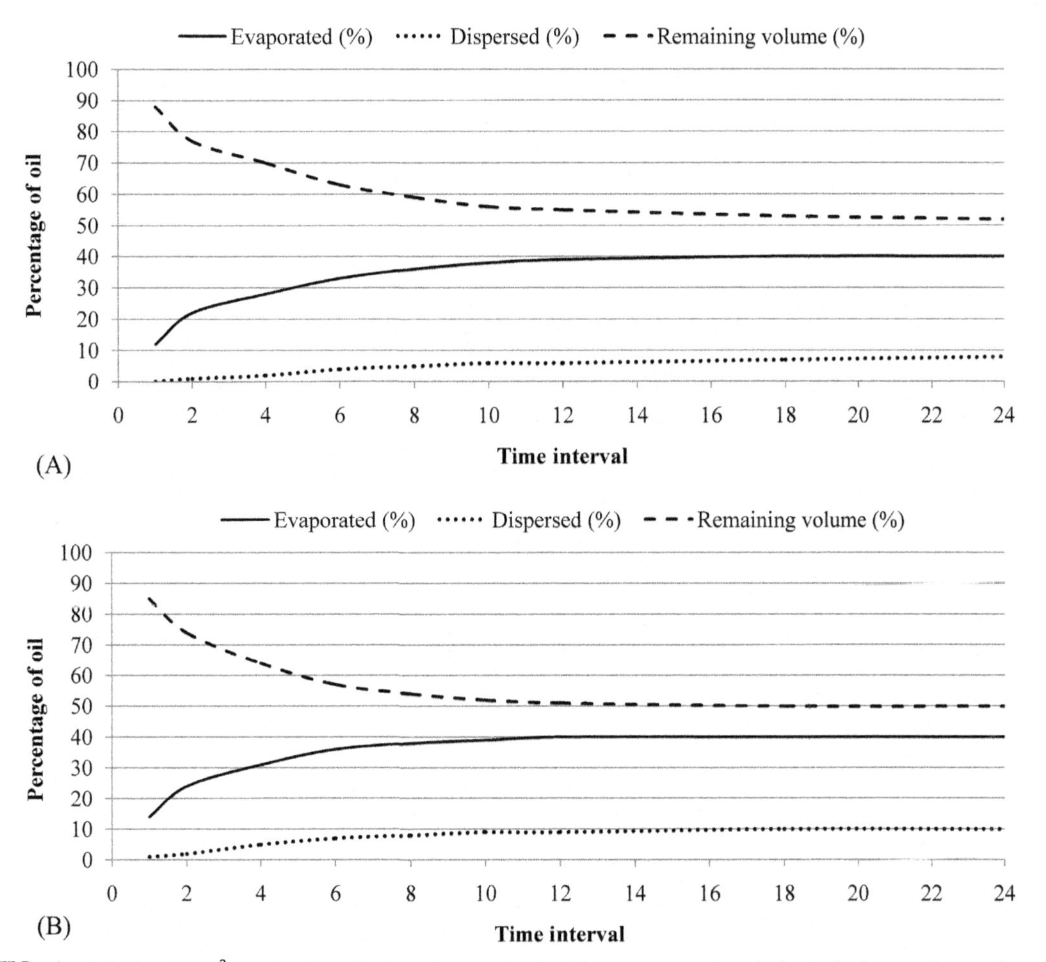

FIG. 4 (A) The $100\,m^3$ crude oil spilled continuously at different time intervals for 24h during the northeast monsoon. It is showing low dispersion of 9% with increased remaining volume. (B) The $100\,m^3$ crude oil spilled continuously at different time intervals for 24h during the southwest monsoon. There is high dispersion of 10% with less volume of oil remaining.

coast, specifically the fishing harbor and port, is at a high risk of oil spill (Tables 2A, 2B, and 3). The southern portion is less affected as per the study.

4.4 Risk to Marine Organisms

Planktonic organisms are the most affected and experience the heaviest toxic impact from n oil spill (Patin, 2013). These planktonic organisms have spatial and temporal patterns that decide their exposure to pollution in the water column. Considering the possible mortality of these organisms, a quantitative assessment has been made in a scenario within the GIS platform. The quantity of dissolved and dispersed oil in the water column is used to assess

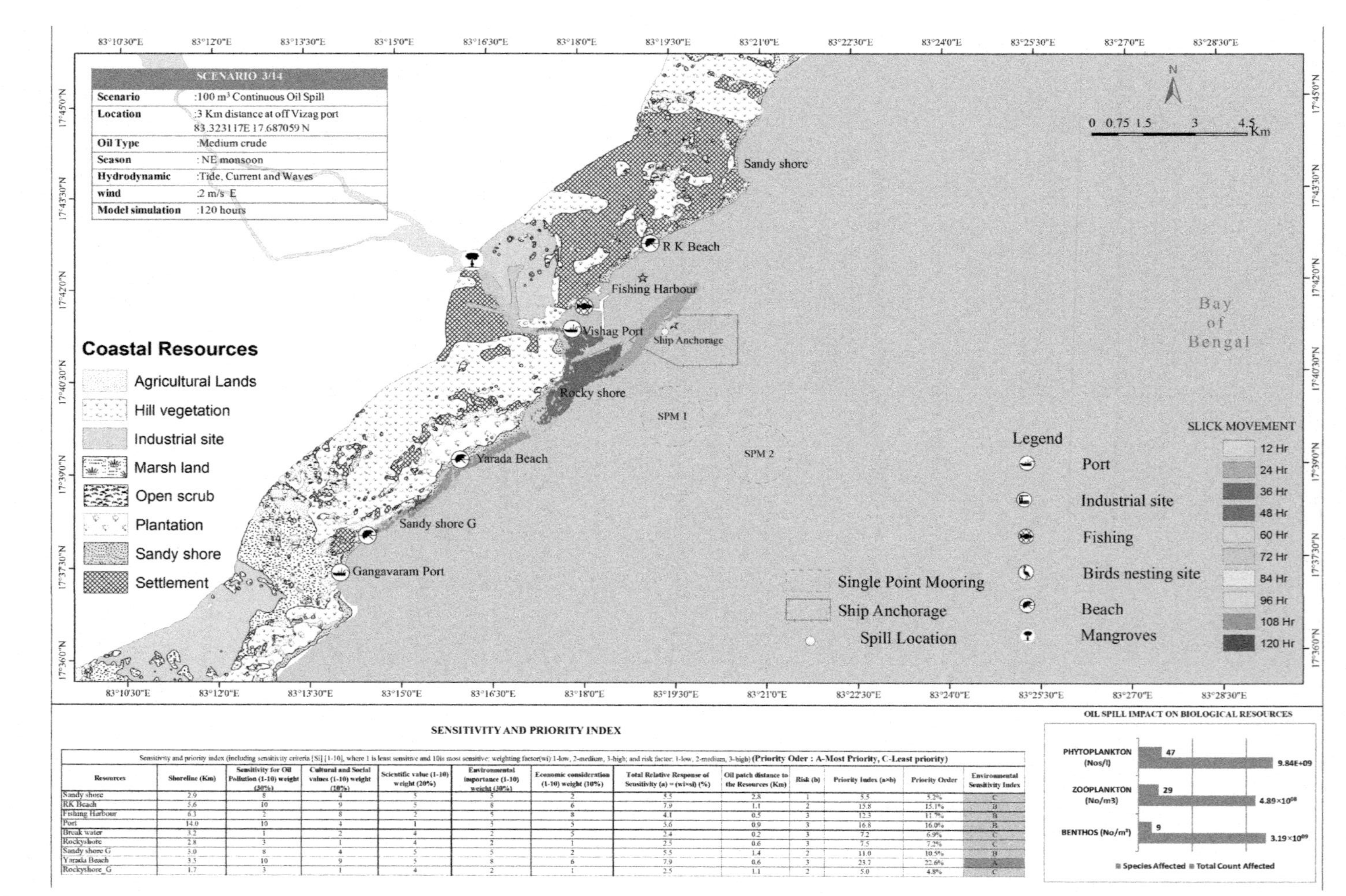

SENSITIVITY AND PRIORITY INDEX

Sensitivity and priority index (including sensitivity criteria [Si] [1-10], where 1 is least sensitive and 10is most sensitive; weighting factor(wi) 1-low, 2-medium, 3-high; and risk factor: 1-low, 2-medium, 3-high) (Priority Oder : A-Most Priority, C-Least priority)

Resources	Shoreline (Km)	Sensitivity for Oil Pollution (1-10) weight (30%)	Cultural and Social values (1-10) weight (10%)	Scientific value (1-10) weight (20%)	Environmental importance (1-10) weight (30%)	Economic consideration (1-10) weight (10%)	Total Relative Response of Sensitivity (a) = (wi×si) (%)	Oil patch distance to the Resources (Km)	Risk (b)	Priority Index (a×b)	Priority Order	Environmental Sensitivity Index
Sandy shore	2.9	8	4	5	5	2	5.5	2.8	1	5.5	5.2%	C
RK Beach	5.6	10	9	5	8	6	7.9	1.1	2	15.8	15.1%	B
Fishing Harbour	6.3	2	8	2	5	8	4.1	0.5	3	12.3	11.7%	B
Port	14.0	10	4	1	5	5	5.6	0.9	3	16.8	16.0%	B
Break water	3.2	1	2	4	2	5	2.4	0.2	3	7.2	6.9%	C
Rockyshote	2.8	3	1	4	2	1	2.5	0.6	3	7.5	7.2%	C
Sandy shore G	3.0	8	4	5	5	2	5.5	1.4	2	11.0	10.5%	B
Yarada Beach	3.5	10	9	5	8	6	7.9	0.6	3	23.7	22.6%	A
Rockyshore_G	1.7	3	1	4	2	1	2.5	1.1	2	5.0	4.8%	C

FIG. 5 (A) Oil spill risk assessment of Scenario 3/14 showing a high sensitivity index for Yarada Beach and high mortality for marine microorganisms during the spill. *(Continued)*

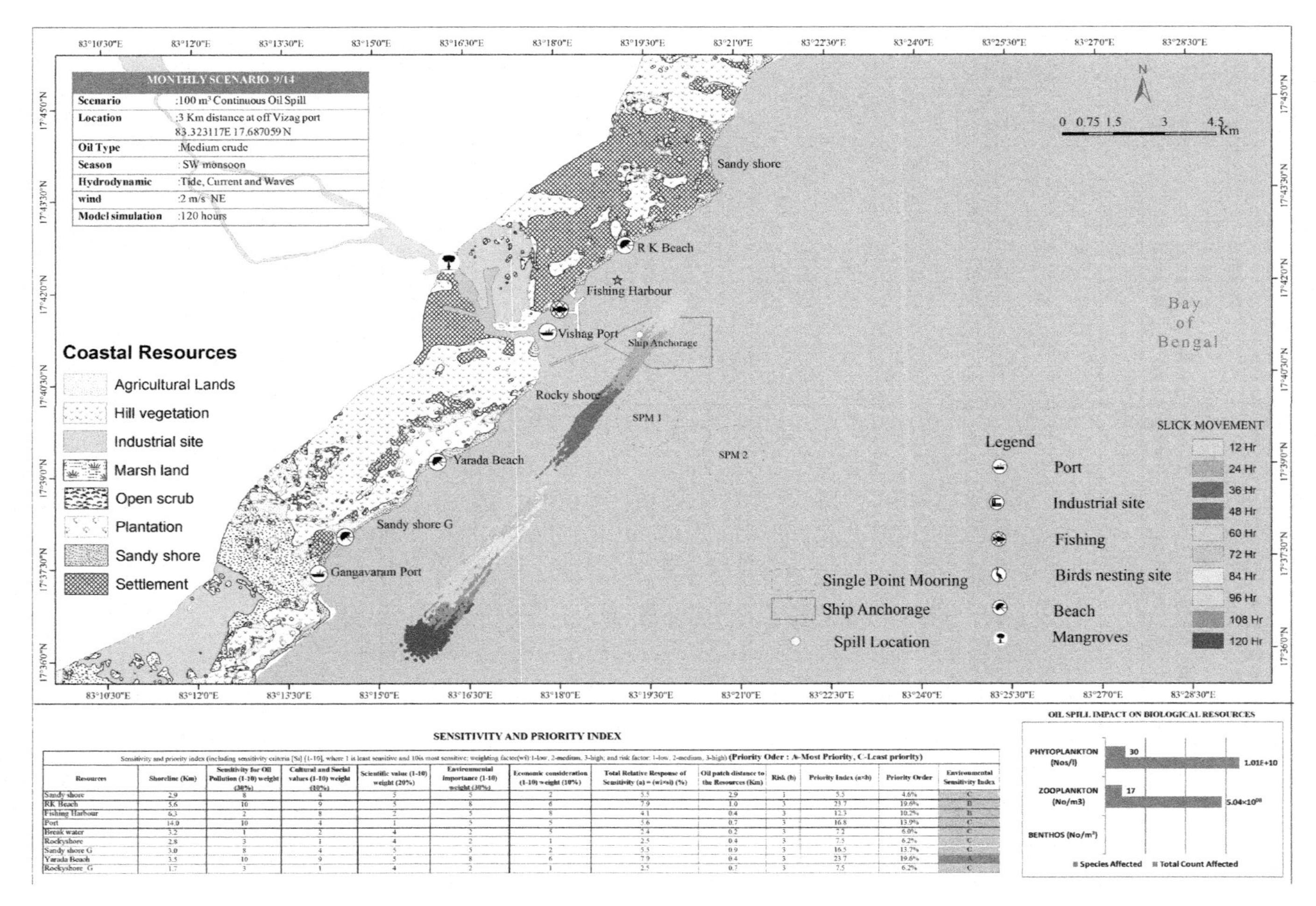

FIG. 5—CONT'D (B) Oil spill risk assessment of Scenario 9/14 showing high sensitivity index for Yarada Beach and low mortality of marine microorganisms during the spill.

TABLE 3 Model Parameters of Scenarios and the Impact Assessment of the Seasonal Scenario Oil Spill

Model Set-Up Parameters			Model Prediction Output			
Scenarios	**Wind Speed (mean) m/s**	**Wind Direction (Deg)**	**Spill Direction in Model Prediction**	**Minimum Time to Reach Shoreline (h)**	**Possible Impact on Coastal Resources**	**Area Affected on the Coast (km^2)**
NORTHEAST MONSOON						
Scenario 1/14	2	N	S	–	In water	0
Scenario 2/14	2	NE	SW	–	In water	0
Scenario 3/14	2	E	W	36	Vizag Port, Yarada Beach	5.2
Scenario 4/14	2	SE	NW	36	Fishing Harbor, RK Beach	1
Scenario 5/14	2	SW	NE	96	RK Beach	3.2
Scenario 6/14	2	W	NE	–	In water	0
Scenario 7/14	2	NW	E	–	In water	0
SOUTHWEST MONSOON						
Scenario 8/14	2	N	E	–	In water	0
Scenario 9/14	2	NE	SW	–	In water	0
Scenario 10/14	2	E	W	72	Yarada Beach	0.8
Scenario 11/14	2	SE	W	60	Yarada Beach	3.8
Scenario 12/14	2	S	NW	60	Vizag Port, Yarada Beach	0
Scenario 13/14	2	SW	N	108	RK Beach, Fishing Harbor	1.6
Scenario 14/14	2	W	E	–	In water	0

the risk to the marine organisms. The Vishakhapatnam coast would face high mortality of these marine organisms during the northeast monsoon based on our risk assessment (Fig. 5A and B).

5 CONCLUSION AND RECOMMENDATION

The aim of the present study was to identify the potential locations that are at high risk due to accidental oil spills, using modeling and mapping techniques. The ecologically sensitive areas, beaches of commercial and recreational importance, and historical sites along the coastline were mapped to carry the sensitivity analysis against oil spills. The degree of sensitivity of these areas in terms of oil spill risks, integrated with the oil trajectory and the likely impact on flora and fauna, have also been assessed.

The model study proves very essential to regulate the sea- and shore-based activities in the study area. The modeling results integrated with GIS produced an analysis-oriented output of the coastal resources. The risk assessment reveals (Tables 2A and 2B) that during both the northeast and southwest monsoons, the impact on the coast occurs only when the wind direction is from east to southwest. Also, the possibility of the oil reaching the shore depends on hydrodynamic and meteorological conditions (Table 3).

References

Adler, E., Inbar, M., 2007. Shoreline sensitivity to oil spills, the Mediterranean coast of Israel: assessment and analysis. Ocean Coast. Manag. 50, 24–34.

Bejarano, A.C., Michel, J., 2016. Oil spills and their impacts on sand beach invertebrate communities: a literature review. Environ. Pollut. 218, 709–722.

Blue Waters, 2012. Jan, V.X. Issue, A Publication of the Indian Coast Guard From the Director General' s Desk, XIII.

Hadjimitsis, D.G., Papadavid, G., Agapiou, A., Themistocleous, K., Hadjimitsis, M.G., Retalis, A., Michaelides, S., Chrysoulakis, N., Toulios, L., Clayton, C.R.I., 2010. Atmospheric correction for satellite remotely sensed data intended for agricultural applications: impact on vegetation indices. Nat. Hazards Earth Syst. Sci. 10, 89–95.

Hossain, M.Z., Tripathi, N.K., Gallardo, W.G., 2009. Land use dynamics in a marine protected area system in lower andaman coast of Thailand, 1990–2005. J. Coast. Res. 255, 1082–1095.

Kankara, R.S., Arockiaraj, S., Prabhu, K., 2016. Environmental sensitivity mapping and risk assessment for oil spill along the Chennai Coast in India. Mar. Pollut. Bull. 106 (1-2), 95–103.

Kankara, R.S., Subramanian, B.R., 2007. Oil spill sensitivity analysis and risk assessment for Gulf of Kachchh, India, using integrated modeling. J. Coast. Res. 235, 1251–1258.

Kirby, M.F., Law, R.J., 2010. Accidental spills at sea–risk, impact, mitigation and the need for co-ordinated post-incident monitoring. Mar. Pollut. Bull. 60 (6), 797–803.

Mohan, R., Kankara, R.S., Venkatachalapathy, R., 2014. Oil spill trajectory modelling of Chennai coast, east coast of India.Proceedings of 5th Indian National Conference on Harbour and Ocean Engineering 5–7 February 2014, CSIR-NIO Goa, India, 2014, pp. 94–99.

Olita, A., Cucco, A., Simeone, S., Ribotti, A., Fazioli, L., Sorgente, B., Sorgente, R., 2012. Oil spill hazard and risk assessment for the shorelines of a Mediterranean coastal archipelago. Ocean Coast. Manag. 57, 44–52.

OSMPRMC (Oil Spill Management Project Review Management Committee), 2003. Road Map for Oil Spill Management for India.

Parsons, M.L., Morrison, W., Rabalais, N.N., Turner, R.E., Tyre, K.N., 2015. Phytoplankton and the Macondo oil spill: a comparison of the 2010 phytoplankton assemblage to baseline conditions on the Louisiana shelf. Environ. Pollut. 207, 152–160.

Patin, S., 2013. Environmental impact of crude oil spills. In: Reference Module in Earth Systems and Environmental Sciences. Elsevier Inc.

Sriganesh, J., Kankara, R.S., Venkatachalapathy, R., 2015. Environmental Sensitivity Index (ESI) mapping for oil s pill hazard – a case study. Int. J. Remote Sens. Geosci. 4, 8–13.

CHAPTER

21

Using Low-Cost UAVs for Environmental Monitoring, Mapping, and Modelling: Examples From the Coastal Zone

David R. Green, Jason J. Hagon*,†, Cristina Gómez*,‡, Billy J. Gregory*,§*

*AICSM/UCEMM, Department of Geography and Environment, School of Geosciences, University of Aberdeen, Aberdeen, Scotland, United Kingdom

†GeoDrone Survey Ltd., United Kingdom

‡INIA-Forest Research Centre, Department of Forest Dynamics and Management, Madrid, Spain

§DroneLite, Forres, Moray, United Kingdom

1 INTRODUCTION

Aerial photography and eventually video imagery have been acquired using small aerial platforms for more than 30 years. Small-scale, radio-controlled (RC) model aircraft and helicopters using small 35 mm SLR and video cameras have been used to acquire panchromatic, color, color infrared (CIR), and multispectral aerial photography for a wide range of environmental applications (Green, 2016). While this technology was not initially viewed as a serious source of aerial photography by many, developments in miniaturized sensors, camera and battery technology, data storage, and small multirotor and fixed-wing aerial platforms, known as unmanned aerial vehicles (UAVs), over the past 5 years have served to reinvent the potential that such small platforms and sensors have for the low-cost acquisition of a wide range of aerial data and imagery. Coupled with the development of ready to fly (RTF) technology, low-cost digital cameras, GPS, image processing, soft-copy photogrammetry software, and multispectral, hyperspectral, thermal, and LiDAR sensors, UAVs now offer a sophisticated means of acquiring many different high-resolution photographic and 4K video

Coastal Management
https://doi.org/10.1016/B978-0-12-810473-6.00022-4

© 2019 Elsevier Inc. All rights reserved.

datasets for small-area coverage studies. With the aid of low-cost image processing and soft-copy photogrammetric software, photographic stills can easily be mosaiced, and three-dimensional models of the terrain and features constructed.

Beginning with a *reflective* examination of the early small-scale aerial radio-controlled (RC) platforms and cameras as a contextual setting for today's applications, this chapter sets out to explore the current potential of these small-scale platforms, the sensors, and some of the software that now provides the means to capture high-resolution imagery of the environment and to process it into spatial information. The chapter considers some of the current aerial platforms, sensors, technological developments, and the means by which the data and imagery can now be processed into information. Focusing specifically on coastal applications, the chapter is illustrated with three examples of this airborne technology spanning some 30 years of small-scale aircraft systems. This serves to show just how rapidly the technology has developed in recent years, evolving into the sophisticated systems now in use.

2 IN THE BEGINNING

The use of small aircraft to acquire aerial photography is not by any means new. Accounts in the literature, admittedly scattered across many academic journals, magazines, books, and the Internet, much of it not of an academic or scientific origin, reveal that a wide range of different aerial platforms have been successfully used over the years as a means to acquire aerial photography of varying types and formats for many different environmental applications.

Long before the modern-day RTF UAVs became widely available and so popular, there were already many examples of small-scale aerial platforms in existence. These were known as RC model aircraft (both fixed-wing and helicopter) that were usually free ranging, although some were tethered, with most of these largely powered by fuel engines. These were generally known as RC aircraft or helicopters while others were referred to as small or model aircraft (Green, 2016). Similar in concept, aside from hobby and aerobatic uses, these platforms were also occasionally flown for aerial photography and were able to carry small cameras that could be triggered remotely to obtain aerial photographic coverage (Wedler, 1984). Compared to the aerial platforms available today, model aircraft generally required specialist operation and experience (as they were quite difficult to fly) and DIY skills while none had any of the modern attributes such as GPS, directional lights, and RTF capability.

Over the past 30 years, numerous studies have been reported in the literature citing the use of small-scale aircraft for field and environmental research where remotely sensed data needed to be collected. For example: Gregory et al. (1974), Anon (1975), Hoer (1976), Lapp and Shea (1976), Kerr (1977), Taylor and Munson (1977), Lemeunier (1978), Miller (1979), Jonsson et al. (1980), Bowie (1981), Ellis et al. (1981), Clark (1982), North East River Purification Board (1982), Syms and Turner (1982), Velligan and Gossett (1982), Tomlins and Lee (1983), Canas and Irwin (1986), NASA (1996), Roustabout (n.d.).

A great deal of commercial interest also existed with many small companies being established at this time to take advantage of a wide range of small-scale RC platforms to acquire low-altitude aerial photography, for example, High-Spy in the United Kingdom (www.highspy.co.uk).

According to Harding (1989), the earliest model aircraft photography was reportedly taken in 1939, although other researchers have suggested an even earlier date of 1917. Significantly these are both during wartime periods when experimentation with unmanned aerial reconnaissance was already being undertaken. Wester-Ebbinghaus (1980) used a model helicopter for photography suitable for engineering applications. Tomlins (1983) used such platforms to acquire aerial photography for a variety of applications, including forestry, pollution detection, wildlife habitat assessment, site mapping, publicity, wildlife inventories, and shoreline mapping. Tomlins and Manore (1984) report the use of aerial photography from model airborne platforms for assessing landfill operations, fisheries habitats, coastal studies, and environmental impact assessments (EIA). Wedler (1984) used a wide range of small platforms to monitor the ambient sediment load at pipeline crossings and to acquire vertical atmospheric temperature profiles. Thurling et al. (n.d.) used model aircraft to acquire crop biomass data and to investigate the extent of weed infestation in agricultural crops. Harding (1989) used model aircraft at Wessex Water to photograph sites for water quality surveys of reservoirs as well as publicity photographs of sewage treatment works. De Wulf and Goossens (1994) reported on the use of remotely piloted aircraft to acquire color infrared (CIR) photography to study winter wheat yield estimation. Fouche and Booysen (1994) wrote a number of papers about the application of remotely piloted aircraft (RPA) for the detection of nitrogen deficiency in crops and stress detection in tree crops. Mullins (1997) reported on the future development of so-called "palmtop" planes or micro air vehicles (MAVs) for remote sensing.

Wedler (1984) lists a wider range of environmental applications (Table 1). Some commercial scientific and nonscientific applications of the time are also shown in Table 2.

3 PLATFORM NAMES

The available literature reveals a wide range of different types, sizes, and specifications for small aerial platforms with many different designs, including gliders and helicopters. Other

TABLE 1 A Range of Environmental Applications (Wedler, 1984)

- Monitoring ice conditions
- Observing coastal erosion and deposition features
- Recording ports and harbor sites
- Monitoring flood conditions at river ice jam sites
- Measuring storm-induced beach erosion/deposition
- Recording of engineering structures and facilities
- Detection of underground utilities
- Quality of rural highway decks
- Older dam sites and river embankment protection
- Preliminary site mapping
- Observations of domesticated farm animals and their environs
- Observations over tree nurseries and experimental agricultural farms
- Tree species identification and tree spacing
- Snowmelt behavior
- Field drainage patterns

TABLE 2 Commercial Applications of SSAs

- Commercial sites
- Real estate site survey
- Company presentation and advertising brochures
- Local government municipal projects
- Local trades
- Legal work and law enforcement
- Construction industry
- Private properties
- Golf courses
- Plane to plane
- Insurance companies, accident sites
- Landscaping companies
- Mortgage companies
- Appraisal companies
- Land management
- Environmental Impact Assessment (EIA)
- Gifts for example, aerial photographs of homes

airborne platforms are also mentioned, such as airfoils, kites, model rockets, and balloons (e.g., Johnson, 1978; Kendall and Clark, 1979; Levanon, 1979; Deuze et al., 1989; Marks, 1989). While these platforms have been successfully used for many different applications over the years, they will not, however, be discussed further in this chapter.

The names given to some of these small platforms vary quite considerably in the literature. Some are clearly just different names for essentially the same thing while others serve to distinguish between the platforms on the basis of their size, for example, wingspan, or certain other characteristics such as the payload. The most common names used are shown in Table 3. A useful definition provided by Wedler (1984, p. 44) also states:

> Radio-controlled aircraft (RCA) ... are small, retrievable aircraft, are remotely controlled directly and/or indirectly, are built to carry a camera pay-load, and operate under specific photo mission requirements. They are not reduced-scale models, as in "model aircraft", but are full-size 1:1 scale aircraft (albeit small) with their own design and construction and flight characteristics.

TABLE 3 Common Names Given to SSA Platforms

- Toy aircraft
- Model aircraft
- Small-scale aircraft
- Small-scale model aircraft
- RCA (remotely controlled aircraft)
- RPV (remotely piloted vehicle)
- RPA (remotely piloted aircraft)
- Photo drones
- Reconnaissance minidrone
- Surveillance minidrone
- Radio-controlled aircraft (RCA)
- Unmanned aerial vehicle (UAV)

On the whole, the terms RPA (remotely piloted aircraft), RPC (remotely piloted craft), RPV (remotely piloted vehicle), and photo-drone appear to correspond to the larger examples of these small aircraft. Photo-drones are generally the largest with very large wingspans. They generally include more sophisticated and more expensive radio-controlled platforms designed to carry large cameras and payloads. According to Wedler (1984), these examples also belong to one of four classes of drones: target drones; harassment or decoy drones; weapons-carrier drones; and reconnaissance or surveillance drones. They have largely been associated solely with military applications.

Examination of some example specifications also seems to suggest that, on the whole, the word "model" (as in model aircraft) applies to the very small versions of these aircraft with a small wingspan, the sort that are typically flown by model-aircraft enthusiasts or hobbyists. These platforms are capable of carrying a small camera strapped to the underside of the aircraft. The word *model*, however, is perhaps an unfortunate label to have become associated with many of these aircraft. While they are a *model* in the sense that they are sometimes, but not always, small-scale replicas, they are far from being a model in many other respects. The authors of this chapter therefore consider the names "toy" and "model" inappropriate for use because they first provide an incorrect impression of the potential of these platforms, tending to conjure up an amateurish rather than a professional role, and second, they are usually associated with hobbyists where the aircraft is only capable of carrying a relatively small payload. Furthermore, by using such a heading, these platforms are often not considered to be a serious contender for aerial photographic data acquisition.

The alternative terminology often used is "small-scale aircraft" (SSA). This is perhaps the most appropriate because it helps to dispel the idea of the plane being a model, and is one that is generally equally applicable to almost all these aircraft, referring only to the size. SSA therefore seems to be the most appropriate terminology because it suggests that they are scaled-down versions of a larger aircraft; they will therefore be referred to as such in the remaining relevant parts of this chapter.

4 TYPES OF PLATFORMS

SSAs fall into two distinct categories: the fixed wing and the rotary wing, with the latter more commonly known as helicopters.

4.1 Fixed Wing

Fixed-wing platforms have been subdivided into high-wing monoplanes or biplanes, with both powered and power/powerless categories; the latter are better known as gliders. In practice, fixed-wing SSAs have tended to find more favor for applications than rotary wings in the past. The reason for this is that monoplanes have good stability and a slow flying capability. Biplanes, however, although more complicated to construct and more susceptible to damage, offer advantages of an increased wing area and therefore an increased capacity to carry equipment and to fly slowly. Seldom mentioned in the literature, gliders have also found uses where there is a need for quiet operation, for example, wildlife monitoring and habitat surveys, but they generally offer less of a practical solution than a powered plane.

4.2 Rotary Wing

Helicopters appear to have been far less popular in the past because they were often more costly to buy, more difficult to fly, suffered from vibration, were unstable, had low-payload capacities and, if they suffered from engine failure, could easily fall from the sky with the inevitable resultant loss of both the platform and the camera equipment. However, a quick search using *model and aircraft* as the search words revealed a surprising number of commercial companies using helicopters for aerial work. Benefits often cited were the ability to hover and the different photographic perspectives possible. The problem of vibration from the main rotor, tail rotor, and engine also seemed to be something of the past, being solved through the use of damping systems. Also, the larger ones can be flown in higher wind conditions and are more stable (personal communication and experience—Borich Aircams—www.borichaircams.co.uk).

5 CONSTRUCTION

SSAs for aerial photography often originated as a standard model kit purchased from a model hobby shop, or alternatively were purpose built. Some kit aircraft had modifications to accommodate a camera mounting with strengthened landing gear and more powerful engines. More commonly, model enthusiasts were often responsible for custom-built designs specifically to carry out the aerial tasks.

Typically, wingspans for fixed-wing platforms used in the United Kingdom were between 1.8 and 3.7 m (Thurling, 1987). According to Thurling (1987), approximately one-third of the designs available included ailerons (a flap hinged to the trailing edge of the aircraft wing to provide lateral control as in a bank or roll) in addition to an elevator (control surface on the tail-plane of an aircraft to allow it to climb or descend), and rudder controls (a vertical control surface attached to the rear of the aircraft to steer it in conjunction with ailerons), and most included flaps (a movable surface attached to the trailing edge of an aircraft wing that increases lift during take-off and drag during landing).

Undercarriage configurations were usually of the tricycle type with two rubber wheels plus either a nose wheel (fixed or movable) or a tail dragger. On the whole, the undercarriage had to be both flexible and forgiving while also being very robust. Getting a fixed-wing SSA airborne was much easier to achieve than getting it down again on a level surface using an undercarriage that is strong enough to take the impact.

Most fixed-wing SSAs used removable wings, tail-planes, and undercarriages for transportability. Elastic bands or nylon bolts were also used for fixing, for example, the wings because they give or break on heavy impact. In the event of an accident, this could minimize the damage to the rest of the aircraft and to the camera. Many of the SSAs were constructed of an open balsa wood frame with plywood reinforcement and covered with lightweight nylon fabric. Some also used expanded polystyrene wings, which had the advantage of being easily repaired, but the disadvantage of being heavier and easier to break. GRP (glass-reinforced plastic) construction was also used, as it had to be both strong and provide protection against damage but with the added penalty of weight.

While there were a number of kit-based helicopters also available for purchase, in practice most of those used for flying aerial photography appear to have been specialist equipment. Hi-Cam (www.hicam.com) in Australia cited a 1.5 m long platform flying at 0–300 ft, complete with a film camera and a microvideo camera with a 2.5 GHz microwave downlink for monitoring and flying. In addition, this platform also carried a mini-DV digital camcorder. Most helicopters available were made of lightweight materials using a combination of plastic, carbon fiber, and glass fiber.

6 ENGINES

Engine requirements varied markedly according to the size of the plane, the wing area, and the operating circumstances. Originally, most were either 2-stroke or 4-stroke engines. For small planes, the 2-stroke examples were powered by a methanol/castor oil mixture while the larger planes usually required a petrol engine. The position of the engine mounting on the plane also varied with options to be mounted at the front, the rear, or on top of the wings. Placement on the wing top or at the rear of the aircraft avoided the problem of oily exhaust fumes being blown back onto the camera lens. Mounting the engine at the front, however, required a pipe to take the oily mixture from the engine exhaust behind the camera. Most early helicopters also utilized gas engines. In all cases, to help avoid vibrations from the engine being transmitted to the camera, it was usual to rubber-mount the engine and pack the camera with plenty of latex foam.

7 OPERATIONAL CONSIDERATIONS

Practical operation of RC aircraft was well documented by Buckle (1985). Launching of SSAs depended very much on their size and the environment in which they were to be used. Systems were designed to be launched by hand whereby the operator or operator's assistant supported the aircraft while it was prepared for take-off. At the appropriate time, it was thrown gently upward and forward. Others were power launched from flat ground, for example, roads, tracks or a grass surface where available; the roof of a car or van; from water using small pontoons attached to the undercarriage; and from snow or ice with the aid of skis (Wedler, 1984). Some were even launched with the aid of a catapult and ramp. Providing the undercarriage was fairly robust and the wheels large enough, it was also possible to use ground that was not too densely vegetated and undulating as a runway.

Various opinions exist as to how difficult these aircraft were to launch and subsequently to fly and land. Discussions with experienced pilots suggested that considerable specialist experience and skills were required for overall control and operation; in many ways flying a small-scale aircraft was actually considered much harder than flying the real thing! Certainly the fixed-wing aircraft were easier to work with relative to the rotary wing (helicopter), but both needed plenty of practice and experience. Given the cost of camera equipment and the platform then, it was advisable, perhaps even vital, to seek the advice and assistance of an expert.

Getting the aircraft off the ground often proved to be far less of a problem than getting it down again safely, especially in one piece. It was often advisable when launching the aircraft to ensure that the runway was level and had sufficient length for take-off. It was also advisable to take off into the wind and to avoid strong sidewinds. Landing the aircraft was sometimes more difficult and it was usually best achieved in tall grass or in a bush to cushion the landing and to protect the camera that should in any case be well recessed into the underside of the fuselage. A hard landing, not necessarily in the exact location planned, often led to the aircraft and camera being badly damaged and even being written off, although a carefully designed aircraft would usually survive a crash landing quite well in practice. Windy conditions can also be quite hazardous for landing and take-off as well as attempting to keep the camera vertical during the flight.

Some enthusiasts (e.g., Wedler, 1984) have also reported the use of waterborne aircraft mounted on floats. These presumably posed a completely different set of operational problems, mostly relating to ensuring that the aircraft can be retrieved and also that the camera and film remained dry.

Maintenance is also an essential part of successful operation and in this context it was frequently noted that inspection of the airframe regularly was required to maintain the aircraft and engine carefully and to avoid any problems. Nearly all operators also carried many spares, including a second aircraft and camera. Some operators also reported the inclusion of a parachute or alternative recovery system to prevent possible damage prior to landing or should a failure occur.

8 POSITIONING

While the addition of a camera to an SSA's normal payload could prove a handful even for the most competent SSA pilot as particular care was needed when launching, landing, and during low-level slow flying of the aircraft, judging distances, flight altitude, and attitude, especially when flying over long distances, was equally difficult.

One of the problems identified when flying an SSA is positioning it at the desired height and judging its flight attitude as well as the yaw, pitch, and roll. For the experienced flyer, this is far less of a problem. Commercial operators suggested that image acquisition required both a pilot and a photographer to ensure correct locational positioning of the airborne platform and ultimately to aid in the acquisition of good aerial photography. Some made use of a GPS on helicopter platforms as well as a video downlink to provide a "virtual experience" of being in the air. A number of alternatives for height estimation were also available, including the use of an altimeter (unfortunately not always cost-effective for low-cost operations), painting a series of black bars or stripes on the underside of the wings that could be resolved at various distances by sighting and guesswork, and using optical range finders, or using trigonometry.

The problems associated with the attitude of the platform in relation to the position of the camera lens were often compensated for by mounting the camera on gimbals, which ensured that the lens was always pointing directly downward. Determining the field of view (FOV), a related problem, was also equally difficult. Once again, the experienced operator was able to ascertain fairly accurately when the plane was correctly positioned. Alternatives were to mount a video camera on the aircraft together with a downlink to a monitor (Fagerlund and Gunnershed, 1975) to provide the pilot with a visual positioning of the aircraft. However, this also was not always practical because of the cost of the additional equipment, the requirement for the video downlink, the size of the aircraft, and its overall potential payload limitations.

A similar problem was that of being able to judge how far away the aircraft was from the operator and whether the camera was covering the desired area. Various solutions were proposed. One was to fly the area in stages, using colleagues positioned in the field to signal when the aircraft was overhead. Another alternative, suggested by Wedler (1984), was to allow long-distance operation up to a distance of 15 km by placing the operator in a truck traveling along the flight line. This was only practical, however, in terrain that was accessible to a vehicle by road and not too rough.

Particular care needed to be exercised in some areas to avoid other low-flying aircraft. In the highland areas of Britain, it is vital to contact the relevant authorities to establish the flight plans of military jets prior to planning a flight. Part of the problem for SSA pilots is that it is difficult to gain an idea of the spatial position (X, Y, Z) of the SSA relative to other airborne vehicles at a distance from the operator. To improve the potential visibility of the SSA, solutions such as painting the SSA in a bright color, for example, fluorescent green/orange, have been used.

Considerable care also has to be exercised when flying these aircraft to avoid other people and power cables, as even aircraft of this size can be lethal or cause quite a lot of damage in a collision. Therefore, a comprehensive insurance policy is important. In some areas, it is not permitted to fly near airports, for example, but other areas are also banned because of the noise factor. To ensure complete and successful control of the aircraft, radio transmitters must be free from interference as loss of control can also be a possibility because of the dependency on a radio signal for control. Similar issues can arise when flying near model-aircraft clubs if other people are flying on the same frequency.

9 FLYING REGULATIONS

In most countries, there have been flight regulations in place for SSA for some time. In the United Kingdom, for example, the Civil Aviation Authority (CAA) regulations restricted the total flying weight of an SSA to 7 kg and below (various communications with the CAA, 1988/1989). Most if not all countries also have some legislation in place.

10 PHOTOGRAPHY

Photographic scales acquired, using flying altitudes of between 30 and 300 m, have been used to provide contact scales of between 1:800 and 1:70,000 (Wedler, 1984). Both panchromatic and color film were often used for either oblique or vertical photographs, most frequently taken with a film speed of 100 ASA (American Standards Association). Most of the cameras used were 35mm SLR (single lens reflex) and Polaroid cameras, although some researchers have reported the use of 8mm movie cameras and small video systems. A typical camera system would have been a Contax 137D 35mm camera with an electric film wind and a 50mm f1.7 Zeiss Planar lens (Boddington, 1989). Another would have been a Konica FS-1/FT-1 fitted with a Konica 22mm or 40mm lens (Boddington, 1989; Harding, 1989). Shutter speeds of 1/500s or faster were generally used (Boddington, 1989). Wester-Ebbinghaus (1980) used both Hasselblad and Rolleiflex 60 × 60 format cameras. Others report the use of Kodak 110 cameras (Harding, 1989; Wedler, 1984). Researchers have also mentioned film speeds of 400 ISO using a 28mm lens at f/5.6, with the use of a Kodak Portra 400VC color negative film because of the fine grain structure. By being closer to the subject, a telephoto lens was not needed, which allowed slower shutter speeds and more light to the film as well as the use of slow, fine grain, high contrast film.

11 ADVANTAGES

SSAs were soon shown by a number of researchers, for example, Wedler (1984), to have a number of distinct advantages over more conventional light aircraft, autogyros, and microlights, the platforms most often used for acquiring small-format aerial photography. Beyond the low cost of these small platforms and the imagery, the operation and maintenance was comparatively cheap, all of which was deemed important for small-area studies. Another advantage was that the photography could be acquired at virtually any time (except under extreme weather conditions such as high wind and rain). The availability of a small, easily transported aircraft, also provided relative freedom for image acquisition; no booking was needed for flying time, and the aircraft, or the risk of flight cancellation due to poor weather conditions. Flying regulations for light aircraft also prevent low-altitude flying (<200m) for the acquisition of large-scale photography, something that was easily achievable with an SSA. Furthermore, when special types of aerial photography taken at one or more specific times of the day/year are needed, the SSA platform was ideal as it allowed complete flexibility. In addition, taking aerial photography at a lower altitude reduced the effects of atmospheric haze and improved the clarity of the imagery.

12 TECHNOLOGICAL DEVELOPMENTS

Interestingly, toward the end of the initial popularity of SSAs for aerial work, Warner et al. (1996) mentioned rapid technological advances in lightweight optics, electronic equipment, strong composite materials, high performance engines, and digital cameras as some of the later developments in SSA that were considered to greatly improve their future potential for aerial imagery. In addition, some operational examples started to use video cameras to aid in SSA navigation, and there was the growing use of small videocams on helicopter platforms.

13 TODAY

Reflecting briefly on the SSA platforms and sensors used in the past for aerial photography provides a very useful and important contextual setting for the current platforms that have become so popular today. While SSAs were commonly seen in the past as "toys for the boys" with limited professional and commercial roles in serious aerial data acquisition, it is clear from the relatively few published examples cited in the literature that many of the original SSAs were very sophisticated and practical airborne data acquisition platforms that utilized—what were at the time—very up-to-date technologies to fly the aircraft and acquire the aerial imagery. Although all these technologies have all clearly evolved considerably since then, providing considerable improvements in the opportunities to acquire and process imagery, many of the practical, operational, and flying constraints today are still remarkably similar to those faced in the past. They are now, though perhaps more relevant than ever because of the rapid growth in the end-user community now able to access and use RTF platforms, the lower costs associated with these technologies, and the ease with which the data and imagery acquired can now be processed into information.

Today, unmanned systems are still associated with a host of terms in the literature and the media: unmanned aerial systems (UAS), drones, remotely piloted aircraft (RPA), unmanned vehicle systems (UVS), and unmanned airborne or aerial vehicles (UAV) all reflecting the variety of system configurations and fields of application in use. Different sources use UAV or UAS as the preferred term, and although UAV is the term adopted by the UK Civil Aviation Authority (CAA), others suggest that UAS is more correct.

An unmanned aerial vehicle (UAV) is flown without a pilot on board and is either remotely and fully controlled from another place (e.g., ground, another aircraft, space) or programmed and fully autonomous (ICAO, 2011). A UAV comprises the flying platform—an aircraft designed to operate without a human pilot on board; the elements necessary to enable and control its navigation, including taxiing, take-off, launch, flight, and recovery/landing; and the elements needed to accomplish mission objectives, such as sensors and equipment for data acquisition and transfer of data, including devices for a precise location when necessary.

As noted earlier, aerial and remotely controlled systems for surveillance and the acquisition of Earth surface data have a relatively long history, typically originating with military activities. Photogrammetry and remote-sensing technologies identified the potential of UAV-sourced imagery acquired at low altitudes with high spatial resolution more than 30 years ago (Colomina and Molina, 2014). However, civilian research on UAVs only began in the 1990s (Skrypietz, 2012). Currently, the rapid emergence of UAVs in many civilian applications has once again raised awareness of the vast potential of these aerial systems.

In the United Kingdom, UAVs are usually classified by their size and weight, from small and lightweight (<2.7kg) with a relatively short distance range up to systems with more than a 20,000km range and a weight of approximately 12,000kg. To date, only small platforms (<20kg) can be used for civilian applications in the United Kingdom.

13.1 Multirotor UAV

One category of UAV currently available, and perhaps the most popular for recreational flying, aerial photography, and video work, has been the multirotor UAV platform (Fig. 1).

FIG. 1 A multirotor UAV.

Multirotors usually have four, six, or eight rotors powered by electric motors. Similar in many ways to single-rotor helicopter platforms that are still also widely available, multirotors now provide a more stable aerial platform for a range of cameras, including cameras on mobile phones, the GoPro Hero series, and many DSLR (digital single lens reflex) cameras, for example, Panasonic GH4. Popular examples of multirotors include the DJI Phantom and DJI Inspire (Quadcopter—4R), the DJI S900 (Hexacopter—6R), and the DJI S1000 (Octacopter—8R) (www.dji.com).

As this technology has developed, so too has the range of platforms from toy to specialist, and from small to large. Developments in technology are now also providing nanodrones, miniature UAVs able to carry small still and video cameras. The palm-sized Micro Drone 2 (http://www.micro-drone.co.uk) weighs 0.034 kg, has a flying range of 120 m, and endurance of 6–8 min. Other small drones can now be flown as tethered aerial vehicles to circumvent the risks associated with free flying; for example, the Pocket Flyer by CyPhy Works (http://cyphyworks.com/pocket-flyer) is an 0.080 kg tethered platform that can fly continuously for 2 h or more.

Compared to early SSAs, modern rotary-wing aircraft have very complex mechanics to allow them to be flown at low speeds. Among their main strengths, rotary-wing UAVs can fly vertically, take off and land in a very small space, and hover over a fixed position and at a given height. This makes rotary-wing UAVs well suited for applications that require maneuvering in tight spaces and the ability to focus on a single target for extended periods (e.g., inspections). Disadvantages of rotary-wing UAVs are that they can be less stable than fixed-wing aircraft under some conditions and also are more difficult to control during flight. Single-rotor and coaxial rotor platforms (with two counter-rotating rotors on the same axis) are very similar to conventional helicopters, with a single lifting rotor and two or more blades. These platforms maintain directional control by varying blade pitch via a servo-actuated mechanical linkage. Single-rotor and coaxial rotor UAVs are typically radio-controlled and powered by electric motors, although some of the heaviest examples use petrol engines.

Multicopters have an even number of rotors and utilize differential thrust management of the independent motor units to provide lift and directional control. As a general rule, the more rotors, the higher the payload they can take; they are also more functional in strong wind conditions, as the redundant lift capacity provides for increased safety and more control in the event of a rotor malfunction or failure. Rotary-wing UAVs are commonly used to capture oblique aerial photographs and video, and may be used for mapping. Some commercially available vehicles are also equipped with GPS/IMU subsystems. This makes them capable of autonomous flights that significantly improve the capability to undertake repeat aerial video and photography to cover the ground in a systematic manner for mapping applications.

13.2 Fixed-Wing UAVs

Fixed-wing UAVs are an alternative to multirotor platforms and have some distinct advantages and disadvantages. While larger and with a capacity to carry a large payload, fixed-wing aircraft are able to carry out aerial sorties for a longer duration of time and can cover larger areas under autonomous flight. Disadvantages include complicated launch and retrieval requirements and limitations to linear and areal coverage. Fixed-wing UAVs are characterized by a relatively simple structure, making them reasonably stable platforms that are relatively easy to control during autonomous flights. Their efficient aerodynamics enable longer flight duration and higher speeds. This makes fixed-wing UAVs ideal for applications such as aerial surveys that require the capture of georeferenced imagery over large areas. On the downside, fixed-wing UAVs need to fly forward continuously and need space to both turn and land. These platforms are also dependent on a launcher (person or mechanical) or a runway to facilitate take off and landing, which can have implications on the type of payloads they carry. Typical lightweight fixed-wing current commercial platforms have a flying wing design (Fig. 2) with wings spanning 0.8–1.2m, and a very small fin at both ends of the wing. In-house vehicles tend to have slightly longer wings to enable carrying the required heavier sensors (Petrie, 2013). A second type of design is the conventional fuselage.

FIG. 2 A fixed-wing UAV.

The dimensions are 1.2–1.4 m length for the fuselage and 1.6–2.8 m for the wing. In the United Kingdom, there are around 20 companies operating commercial airborne imaging services using fixed-wing UAVs (Petrie, 2013).

13.3 Combined Multirotor and Fixed-Wing Platforms

A relatively recent development has been that of an aerial platform that combines the best characteristics of a rotary-wing (take off and landing, hovering) and a fixed-wing platform (long distance flight). An example of this technology is the VTOL Flying Wing (http://www.vtol-technologies.com).

14 SENSORS

Rapid developments in digital and microprocessor technology have led to a wide range of low-cost and miniaturized digital cameras and sensors that are suitable for mounting on UAVs. These include panchromatic, color and color Infrared (CIR), multispectral, hyperspectral, thermal, and LiDAR sensors. This also includes the possibility to make use of digital cameras and attachments on mobile phones. While not all sensors have been developed specifically for UAVs, as UAV technology has evolved so too there has been a demand for specialist sensors for various applications. Different sensors are limited to different platforms by the payload and lift capacity of the aerial platform. A growing range of both passive and active sensors measuring naturally occurring radiation reflected or emitted by the target objects is becoming available. Some are off the shelf while others are now being developed specifically for UAVs.

14.1 Passive Sensors

Passive optical sensors measure radiation in the visible (0.4–0.7 μm) and infrared (0.7–14 μm) part of the electromagnetic spectrum. They rely on the sun as the illumination source, which makes them only suitable in daylight conditions. They are also limited by atmospheric effects such as clouds, haze, or smoke. There are an increasing number of optical sensors now available for UAVs, ranging from small cameras capable of still photography and video (e.g., GoPro series and iLook (http://www.walkera.com/index.php/Goods/info/id/37.html) cameras), to both small and large DSLR cameras, stereo cameras, multispectral and hyperspectral cameras, high resolution cameras on smartphones, or low-cost developments such as the HackHD camera (www.hackhd.com). The range of opportunities provided by optical sensors is constrained by various issues concerning the digital frame cameras that can be deployed on lightweight UAVs. These include the camera weight relative to the available UAV payload; the very small format of the camera images; the numerous nonmetric characteristics of many of the lower-cost cameras; the lenses and resolution; the photographic intervals; the need for very short exposure times to help combat the effects of platform instability (i.e., due to speed, roll, pitch, and yaw); the requirements for high framing rates arising from the speed of the UAV platform over the ground from a very low altitude; and the very large longitudinal and lateral overlaps (i.e., percentage endlap and sidelap) that need to be employed for mapping purposes.

14.1.1 Multispectral and Near Infrared

Multispectral imagery is produced by sensors that measure reflected energy within several specific bands of the electromagnetic spectrum. They usually have three or more different band measurements in each pixel of the images they produce. Examples of bands in these sensors typically include visible green, visible red, and near infrared (NIR). Simultaneous measurement of multiple spectral wavelengths provides information that can be visually or automatically interpreted. For a given location, algebraic combinations of values in various spectral wave bands can be very useful in the detection of environmental features; for example, multispectral imaging of vegetation is very useful in the identification of plant stress, disease, and nutrient or water status. Marcus UAV (http://www.marcusuav.com) has a prototype of a custom payload manifold for the housing of a Tetracam ADC Lite MS (http://www.tetracam.com/Products-ADC_Lite.htm) camera system on a UAV. Using infrared camera images collected at specific time intervals, overlapped and aligned, and some spectral filtering software (e.g., Pixel Wrench (http://www.pixelwrench.co.uk)) normalized difference vegetation index (NDVI) images can be derived to provide information about plant condition and status.

14.1.2 Shortwave Infrared

Radiation in the shortwave infrared (SWIR) (typically 0.9–1.7 μm) is not visible to the human eye but can be sensed by dedicated indium gallium arsenide (InGaAs) sensors. Images from an InGaAs camera are comparable to visible images in their resolution and detail, making objects easily recognisable (as opposed to thermal imagery). One of the main benefits of SWIR imaging is its low power consumption, as it uses a thermoelectric cooler or no cooler if the dark current is low enough while still providing good-enough imagery in low-light conditions. InGaAs sensors can be made extremely sensitive, literally counting individual photons. Thus, when built as focal plane arrays with thousands or millions of tiny sensor pixels, SWIR cameras will work in low light conditions.

Only a few commercial companies are making SWIR cameras and even fewer are making the detector material, indium gallium arsenide (InGaAs). Sensors Unlimited (www.sensorsinc.com) and Teledyne Judson (www.judsontechnologies.com) are the only two US developers of SWIR technology, subject to strongly controlled exporting regulations. Xenics (www.stemmer-imaging.co.uk) in Belgium, Allied Vision Technologies (www.alliedvision.com) in Germany, and Chunghwa (www.leadinglight.com.tw) in Taiwan are other providers. Sensors Unlimited—UTC Aerospace Systems (utcaerospacesystems.com)—has introduced the smallest SWaP (size weight and power) SWIR camera for unmanned vehicles. The 640×512 pixel (25 μm pitch) camera weighs 27 g. The $25.4 \times 25.4 \times 25.3\,mm^3$ total volume allows it to easily fit onboard most unmanned aerial or ground vehicle systems (UAS or UGS).

14.1.3 Hyperspectral

Hyperspectral imaging samples a wide variety of bandwidths in the light spectrum to provide a rich dataset and detect objects of interest not visible to single-bandwidth imaging sensors. With a larger number of fine spectral bandwidths, the identification of specific conditions and characteristics is greater. Sensors with hundreds of bands (e.g., 255) are increasingly being used for many applications. Developers of hyperspectral sensors provide flexible and customizable options for the number and resolution of spectral bands in the visible and infrared (e.g., Rikola Ltd. (www.rikola.fi)). Headwall Photonics (www.headwallphotonics.com), for example, now specializes in hyperspectral imaging sensors that are small and rugged enough to fit on relatively small UAVs. Adding hyperspectral imaging as a standard element in an electro-optical and infrared (EO/IR) sensor suite for UAVs has, however, until recently presented a difficult engineering challenge.

14.1.4 Thermal

Some surfaces and features have been found to show up well in thermal infrared (TIR) imagery because of temperature differences between the different surfaces. Lightweight thermal cameras have been adapted or specifically developed for use in UAVs. The FLIR Quark 640 (www.flir-tau-buy.com/product/flir-quark-2-thermal-camera/) long-wave infrared (7.5–13.5 μm) thermal sensor was incorporated by Sky-Watch (Denmark) into one of its drones. FLIR Quark 640 is a very small ($22 \times 22 \times 12\,mm^3$ without lenses) sensor with a flexible configuration of lenses (6–35 mm), scene range from −40°C to 160°C, and sensitivity of 50 mK. Its weight depends on configuration but is <30 g. Tamarisk 320 developed by DRS Technology (www.drsinfrared.com) is a long-wave infrared (8–14 μm) thermal sensor of similar size ($\sim 30 \times 30 \times 30\,mm^3$). It can also be configured with lenses (7–35 mm) and has a scene range from −40°C to 67°C and sensitivity of 50 mK. Its weight depends on the configuration, ranging from 30 to 135 g.

14.1.5 Fluorescence

Fluorescence spectroscopy has also proven to be useful for some applications. Estimates of fluorescence (F) can be derived from multispectral and hyperspectral radiance sensors, exploiting the Fraunhofer line and decoupling F from the reflected flux. Furthermore, optical indices related to F can be derived from reflectance sensed by multispectral sensors. However, the quantitative estimation of F from the air is complicated by the absorption of the atmosphere en route to the sensor, and approaches to deal with atmospheric effects have yet to be developed (Meroni et al., 2009). For the estimation of F, very high spectral resolution (0.05–0.1 nm) sensors are recommended (Meroni et al., 2009) to allow resampling and application of estimation methods, from multispectral to hyperspectral.

14.2 Active Sensors

Active sensors emit radiation and measure the fraction reflected by the target objects as well as the difference in the time between emission and reception. Active sensors require power supplied by a source that inevitably adds some considerable weight to the aerial system. For this reason, active equipment is less versatile for use on UAVs when compared with passive equipment. Radar synthetic aperture radar (SAR) is a type of radar using relative motion between an antenna and its target region to provide distinctive coherent-signal variations. SAR pulses radio waves at wavelengths of 0.002–1 m repeatedly toward a target region. The many echo waveforms received successively at the different antenna positions are coherently detected and stored, and then postprocessed together to resolve elements in an image of the target region. SAR is used for a wide variety of environmental applications. Oil spills in the ocean or waterbodies can be detected using SAR imagery because the oil changes the backscatter characteristics of the water. As with thermal imagery, differential imaging is necessary for the detection of oil leaks. All weather, night, and day capacity for data collection makes radar technology appealing and convenient for surveillance in difficult environments.

14.2.1 Radar

Radar systems have been integrated into large UAV systems by the military, but there is still a need for small, low-cost, high-resolution radar systems specifically designed for operation on small UAVs. Brigham Young University (BYU) in the United States has developed some compact, low-cost, low-power SAR systems, including a series of microSAR systems designed for operation on small UAVs. The microSAR design represents a trade off between coverage and precision versus cost and size. It is an ultra low-power system (16 W) designed for operation on a UAV with ~2 m wingspan. The system records data continuously for more than an hour on a pair of compact flash disks, which can then be loaded onto a laptop for processing data into images using SAR image formation and autofocusing software. Image downlink capacity and real-time processing are being developed. The microSAR system consists of a stack of circuit boards ($7 \times 8.5 \times 7\,cm^3$) and two flat microstrip antennas ($0.1 \times 0.5\,m^2$). Minimal enclosures reduce the flight weight to <1 kg. Unlike conventional SAR in which short pulses are transmitted and received separated by an interval, microSAR transmission and reception occur simultaneously via continuous-wave linear-frequency modulation, enabling low-power operation. To optimize performance, microSAR uses bistatic operation in which transmission and receipt occur via different antennas. Designed for operation at 130–800 m height and speeds of 20–50 m/s, microSAR has a swath width of 200–900 m with a nominal one-look spatial resolution of $0.1 \times 0.6\,m^2$, which is multilook averaged to $1 \times 1\,m^2$ in processed imagery. The averaging reduces the *speckle noise* inherent in SAR images. The first microSAR operated in the C-band (5.56 GHz), but microSAR systems using other bands have also been built.

ImSAR (Utah, United States) has developed a NanoSAR (http://www.imsar.com/pages/products.php?name=nanosar) series improving from the first NanoSAR-A (0.5 m resolution operating at 500 m height) to NanoSAR-C (0.3 m resolution at 2000 m height). The NanoSAR-C weighs <1 kg. ImSAR radar has printed circuit board technology in place of the heavy metal tubes that serve as radio wave guides in standard synthetic aperture radars. Etched on fiberglass boards, the radar circuits are similar to the lightweight circuits used in laptop computers and cell phones. NanoSAR-C has a range of 1–16 km and power consumption between 25 and 70 W.

14.2.2 LiDAR

Over the last few years, airborne light detection and ranging (LiDAR) surveys have become popular for many applications. Typically, LiDAR surveys are carried out using manned aircraft flying at 1500 m or above. At this height they are susceptible to atmospheric conditions and have poor image resolution or a small image footprint. In general, the main constraint to the use of LiDAR on a UAV has been the size and weight of the sensor. UAV imaging systems are generally much lighter than an equivalent LiDAR system. This is due in part to two factors: LiDAR sensors are not nearly as small and light as cameras and a LiDAR system depends entirely upon an accurate inertial navigation system (INS). The imaging system in contrast uses traditional photogrammetry and does not require any INS. As a result, UAV imaging systems can cover a larger area in less time at less initial expense.

Riegl (www.riegl.com/products/unmanned-scanning/ricopter) has developed a LiDAR system for UAVs, the VUX-1, beaming 500k shots per second of NIR radiation; it has an estimated accuracy of 10mm. Its dimensions are $22.7 \times 18.0 \times 12.5\,cm^3$ and it weighs 3.6kg. The RIEGL VUX-1 is designed to be mounted in any orientation and has a 330 degrees field of view (FOV). Its maximum range is 900m. Yellowscan has developed a LiDAR sensor for UAVs with 2.2kg beaming 800k shots per second with multiecho technology (three echoes per shot) in the 905nm wavelength. It has power for 2h of autonomy with a maximum range of 100–150m depending on conditions and a 100 degrees FOV. Velodyne (velodynelidar.com/vlp-16.html) has developed a new HDL-32E LiDAR sensor with a scanning rate of 700k shots per second. At <1.5kg and smaller than $15 \times 9\,cm^2$, this LiDAR scanner is therefore ideal for UAV applications. Velodyne has also announced a new LiDAR sensor with a scanning rate of 300k shots per second, 0.6kg of weight, and dimensions of $10 \times 6.5\,cm^2$. Both LiDAR systems work with radiation of 905nm and have a range of ~100m. Hokuyo (www.hokuyo-aut.jp/02sensor/07scanner/utm_30lx.html) has developed a number of small lightweight LiDAR devices for UAVs, for example, UTM-30LX. The UTM-30LX is $60 \times 60 \times 87\,mm^3$ size and 0.37kg, has a scanning rate of 40k shots per second, and a range up to 30m. Interface to the sensor is through USB 2.0 with an additional synchronous data line to indicate a full sweep. The FOV is 270 degrees and a 12V power source is required. Characteristics such as range, flying height, and power consumption are trade offs to consider for specific applications and platforms systems.

To accommodate LiDAR sensors on UAVs, Phoenix Aerial has recently developed the Scout, designed to host Velodyne LiDAR sensors and a combination of LiDAR and photogrammetric equipment. The Scout has dimensions of $12.5 \times 22.4 \times 18.5$ cm and weighs 2.5kg.

Most recently, Riegl has developed a specialized UAV platform and bathymetric sensor—Bathycopter—as a means to survey water depths (bathymetry) in shallow water areas (http://www.riegl.com/uploads/tx_pxpriegldownloads/BathyCopter_at_a_glance_2015-09-11.pdf).

15 VIDEO AND STILL CAMERAS

The GoPro Hero series of cameras and competing ultracompact and lightweight cameras (e.g., Walkera iLook) have proven themselves to professionals and amateurs alike as ideal sensors for recording high-resolution digital still photographs and video on UAV platforms such as the DJI Phantom series. Some UAV systems now also take advantage of smartphones and computer tablets instead of traditional screen viewers to monitor the aerial view of the surface. With the addition of video downlink hardware and view screens (first person view (FPV)), aerial flights on UAVs can be monitored in real time (although it is necessary to acquire the help of an additional person to the UAV pilot to monitor the flight using the FPV screen), even on smart devices such as phones and computer tablets.

High-resolution imagery (e.g., 12+ MP) provides detailed color and filtered imagery from low to medium altitude flights on platforms such as the DJI Phantom 1, Phantom 2, and 3D Robotics IRIS (3dr.com). Using traditional tools and techniques and digital image processing software (e.g., AgiSoft, MosaicMill, 2d3 (insitu.com), or Pix4D) the imagery acquired can then be interpreted and analyzed manually or on screen, or the information can be extracted using semiautomated approaches. Three-dimensional (3D) orthophoto imagery and DTMs can also be generated.

Larger platforms can carry larger payloads and even multiple cameras. This means that much better cameras, for example, the Panasonic GH4 DSLR, can be mounted on a UAV with the end result being much higher-resolution digital stills and video (e.g., 4 K) that moves the imagery from amateur to the professional quality required for commercial applications. In addition, the capability to carry a larger payload means that a better gimbal can be used as well as numerous other sensors to provide complementary imagery.

16 STEREO CAMERAS

Stereo cameras have been utilized for a number of roles on UAVs. First, they have been used for capturing stereoimagery: taking two photographs simultaneously from two slightly different viewpoints provides a basis for generating 3D imagery or views. Cameras such as the Fuji Finepix stereo camera have been successfully mounted on small N-copters for this purpose (Haubeck and Prinz, 2013). Second, stereo cameras can be used as the basis for UAV navigation systems. Stereoimagery can be used to work out the distances to any obstacles, such as planes, buildings, or mountains ahead; to detect the nearest object in the field of view so that a warning can be issued if the UAV is about to collide with the obstacle; and to enable the UAV to use the distance-to-obstacle feature to calculate a flight path to avoid the obstacle. Stereoimaging can be accomplished through the use of two imaging CCD cameras and suitable software, such as the Stanford Research Institute (SRI) Small Vision System software. One image can be fitted with a vertical polarizer, the other with a horizontal polarizer. The difference between the images from the two imagers can be used to detect the presence of water because only light with a horizontal polarization is reflected from a water surface.

17 COASTAL AND MARINE APPLICATIONS

UAVs have been used for many different applications around the world in the last 5 years. As their ease of use improves, so the number and range of applications have increased rapidly. In particular, coastal and marine environments requiring high-resolution local data and information or multitemporal monitoring data and imagery for change detection are well-suited to the use of UAVs carrying a range of different sensors, including RGB cameras and thermal and hyperspectral sensors. Klemas (2015) notes that *UAVs (now ....) have the capability to effectively fill current observation gaps in environmental remote sensing and provide critical information needed for coastal change research* .... (p. 1265).

In recent years, with the widespread availability of small, off-the-shelf, RTF UAVs and accompanying photographic and video sensors as well as the rapidly evolving GPS, navigational, and sensor technologies, a number of coastal/marine applications have emerged that take advantage and demonstrate the value of this technology for use by coastal researchers and managers. There are now many different suppliers of UAV platforms or frames, some ready built, others available as kits for construction. These are along with some very specialist custom frames costing considerably more. Depending on the requirements of a coastal study, there are many examples of small off-the shelf sport cameras, for example, GoPro and DSLR cameras, all of which can be mounted on gimbals that are in use alongside more expensive sensors, such as thermal cameras, hyperspectral and LiDAR sensors, and most recently the bathymetric LiDAR (Riegl) mentioned earlier.

Most of the journal papers on coastal applications of UAVs referenced in this chapter are from 2014 to 2016, which coincides with the rapidly growing recognition of the technology, its affordability, and its ease of use for many different applications ranging from relatively simple photographic and video acquisition, to more sophisticated monitoring and small area surveying exercises. While each paper addresses different coastal applications, the large majority of these papers not only highlight the context behind the growth in applications, for example, technology, battery power, GPS accuracy, and survey capability, but also the sophistication of the soft-photogrammetry software now available to process the imagery as well as the importance of operational safety and flight regulations. Additional considerations are the low costs of data acquisition (Ryan, 2012), large-scale mapping, and spatial resolution requirements. Special UAV environmental considerations are also cited by a number of authors in relation to coastal work. For example, Guillot and Pouget (2015) mention the challenging nature of coastal environments for flying UAVs and the need to take into account parameters such as wind, water, temperature, and climate. Mancini et al. (2013) recorded a number of important observations in relation to the use of UAVs in dune systems, which include planning take off and landing sites to avoid setting the sand in motion by the UAV rotors, and avoiding damage to the functioning of the rotors and camera lens. Klemas (2015) observes the value of UAVs to increase operational flexibility and provide greater versatility. In addition, Klemas cites a number of studies that make use of different types of UAV platforms and sensors. Hugenholtz et al. (2012) also note that improvements in the design of flight control systems have transformed these platforms into research-grade tools capable of acquiring high-quality images and geophysical/biological measurements. Klemas (2015) mentions studies that have mapped tidal wetlands, coastal vegetation, algal blooms, sand bar morphology, the locations of rip channels, the dimensions of surf/swash zones, coastal hazards, fishing surveillance, coastal erosion studies, and combined aerial views from a UAV (drone) with measurements from autonomous underwater vehicles (AUVs) to get an unprecedented look at coastal waters off the coast of southern Portugal. Others have used UAVs in combination with other platforms and sensors to monitor and track marine mammals.

The role of fixed-wing versus multirotor platforms is also discussed in many papers, serving to provide a useful current overview of the status of these platforms and sensors while also highlighting the complementary value of these platforms and their sensors for monitoring, mapping, and surveying different aspects of the coastal environment. In addition, reference is still commonly made to other small airborne data acquisition platforms such as tethered kites and balloons of blimps (Ryan, 2012), which also have potential for some applications and can be both cost effective and easy to use, even if they fall outside the UAV domain. As with all such small platforms and systems, however, they also have their limitations.

Some of the applications of UAVs to coastal environments have included monitoring coastal and intertidal habitats such as mangroves, saltmarsh, and sea grass (Ryan, 2012) to gather orthophotos as well as the use of video transects for intertidal bathymetric habitat mapping, to assess inaccessible areas, facilitate repeat surveys, measure the extent and coverage of vegetation, and for monitoring dredging activity and plume extent. Also mentioned are applications that include the generation of 3D point clouds through stereophotography or LiDAR, routine marine fauna observation (MFO) including cetaceans and turtle nesting activity using fixed-wing aircraft, minor oil spill contingency tracking, hyperspectral vegetation classification, detailed engineering inspections of pipelines, offshore structures, thermal imaging, and health and safety intervention. While some of the platforms used for environmental applications have been specialized custom examples, the rapid growth in small low-cost platforms has led to more of these platforms being standard off-the-shelf UAVs, for example, the DJI Phantom has become very popular as an RTF. Inevitably, the opportunity to monitor changes to the coastal environment has been one of the applications where UAVs have considerable potential given the ease with which it is now possible to acquire multitemporal photographic imagery to see changes to the coastline in terms of erosion and deposition, but also to be able to measure the change for example, in cutback and to determine the rate of coastal erosion. Appeaning Addo et al. (2016) have used a DJI Phantom 3 to monitor and map coastal changes due to erosion in relation to coastal protection structures in Ghana along the Volta Delta shoreline. Using a relatively standard simple platform and sensor configuration, they were able to generate aerial photographs of the protection structures and surrounds as well the generation of high-resolution orthophotos and DEMs for the detection and analyses of both planimetric and volumetric changes. The practicality of the method for repeated monitoring and surveying is also highlighted. Anderson et al. (2015) have utilized UAVs to gather data about spatial ecology and coastal water quality, citing the key advantages as being that *they can be launched, operated, and the data accessed and analyzed within hours* as well as systems that can be cheaply and easily used by aquaculture businesses (e.g., pelagic fish and shellfish farms) and by coastal environmental managers (e.g., UK Environment Agency) to easily and repeatedly monitor near-shore water quality.

In another study, Guillot and Pouget (2015) have used UAVs specially adapted for coastal environments (e.g., salt, wind, and moisture) for monitoring and precision mapping of the dunes and dikes of Oleron Island in France before and after tides and storms. The UAV—known also as a Coastal UAV—is a modified DJI F550 and is capable of flying in high wind conditions. It is designed to be resistant to moisture and sand particles. Interestingly, and perhaps sensibly, they also made use of an off-the-shelf rugged sports camera with a standard fisheye lens. Postprocessing of the imagery acquired was carried out to remove the fisheye distortion from the imagery and to geocorrect the imagery using the ground control points (GCPs). AgiSoft soft-copy photogrammetric software was used to generate an orthomosaic, a digital surface model (DSM) and a 3D PDF (portable document file) for input to a geographic information system (GIS). A DDVM (difference of digital volumetric model) was generated for the different dates of imagery to determine change. DSM and DEM files provided the means to generate two-dimensional (2D) topographic profiles in the GIS software. Benefits of the research that were identified included the ease with which it is possible to design a low-cost, coastal-specific platform using off-the-shelf technology for use in a relatively inhospitable environment to gather repetitive high resolution data and imagery from which it is possible to quantify coastal change at a cm resolution from the UAV-derived products. Further projects for coastal monitoring include the study of sand and pebble movement, birds, waste, and seaweed, which can provide valuable information to improve ecosystem knowledge in the future.

Mancini et al. (2013) describe how UAVs have the potential to generate data and information to understand coastal processes using low-altitude UAV aerial photographs and structure from motion (SfM) to generate DSMs comparable to ground surveys of the morphometry of a dune and beach system in Italy. Comparison of ground and aerial surveys using UAVs reveals the considerable potential of the latter to complete accurate surveys of dunes with fewer cost, time, and manpower requirements. The research used a nonstandard UAV with a DSLR, RTK GPS, and autonomous flight control.

Turner et al. (2016) consider UAV technology to be a tried and tested technology for surveying work, specifically engineering applications, and to be a cost-effective solution for coastal zone applications. They also note that no more step changes in UAV technology or ease of usability are required. With the aid of RTK GPS, Turner et al. (2016) now consider this technology to be a practical, effective, and routine poststorm survey tool for coastal monitoring in Australia at spatial and temporal resolutions not previously possible. They also provide a brief overview of some of the recent UAV applications to coastal engineering and management.

Pereira et al. (2009) focused on the benefits of recent evolutions in UAV technology that include autonomous take off and landing capabilities for aerial gravimetry, aerial photography, surveillance and control of maritime traffic, fishing surveillance, and detection and control of coastal hazards. Such developments include advances in the distance and duration drones can fly for maritime surveillance in harsh coastal environments.

Goncalves and Renato (2015) have used UAV imagery and photogrammetry to derive topographic information for coastal areas while Long et al. (2016) also used UAV imagery to monitor the topography of a tidal inlet.

18 SOME FURTHER EXAMPLES

In this chapter, three different examples are used to illustrate the potential of these of small airborne platforms and sensors to gather data on various aspects of the coastal environment. Spanning some 30 years, these provide a good indication of how the aerial platform and sensor technologies have advanced over a relatively short period of time to the present day, empowering a wide range of people with the potential to acquire and process high resolution environmental data.

BOX 1. MONITORING AND MAPPING MACROALGAL WEEDMATS IN THE YTHAN ESTUARY, SCOTLAND, UNITED KINGDOM, WITH AN SSA

As part of a long-term monitoring and mapping project to map macroalgal weedmats in the Ythan Estuary, north of Aberdeen in Scotland, (Fig. 3), using different sources of remote sensing, for example, satellite imagery and aerial photography, part of the work involved the acquisition of large-scale color and filtered aerial photography from a large-scale model aircraft (Fig. 4). The Centre for Remote Sensing and Mapping Science (CRSMS) at the University of Aberdeen used the combination of an SSA, a global positioning system (GPS), digital image processing software (Erdas Imagine), a geographic information system (GIS) (ESRI's ArcView), and 35mm SLR color and filtered photography to capture multitemporal, high-resolution imagery from which to map the extent of macroalgal weedmats in the Ythan Estuary over a number of years (Green and Morton, 1994; Green, 1995).

Prior to the aerial overflights, ground control points (GCPs) were placed in the flight area. Each GCP made use of a white plastic fertilizer bag with a wooden stake (Fig. 5) the center of which was surveyed with the aid of a Trimble GPS. Each overflight was undertaken using a different film/filter combination.

The aerial photographs (Fig. 6) were printed and then scanned using a desktop scanner. Erdas Imagine was used to mosaic and geocorrect the aerial photographs. The mosaicked imagery was then input to the GIS and with the aid of the onscreen digitizing tools, the pan and zoom functionality, and the seven factors of aerial photointerpretation, the macro-algal weedmats units were identified and delineated as map layers to create weedmat maps for a number of years. With the aid of GIS, it was then possible to calculate the area of weedmat coverage for each year and to determine the changes in location and areal extent from 1 year to the next. Combined with other layers of map information, for example, point pollution sources and estuary substrate, it was then possible to establish some correlation between the locations of the weedmats from year to year.

BOX 2. SPEY BAY—MONITORING A DYNAMIC COASTLINE WITH A MULTIROTOR UAV

With the assistance of Borich Aircams (www.borichaircams.co.uk), color aerial photographs and video were captured using both an Align S690 Hexacopter and an Align S800 Helicopter (Fig. 7) along the waterfront at Spey Bay in Scotland (Fig. 8), as part of a funded experimental study to try two different small-scale UAV platforms carrying digital SLR cameras as the basis for coastal monitoring and mapping. Aerial overflights were conducted in the summer of 2015 under clear skies with a Panasonic GH4 DSLR camera. The resulting aerial photographs were input to the AgiSoft soft-copy photogrammetry software (www.agisoft.com) to generate photographic mosaics as well as 3D models of the beach at Spey Bay (Fig. 9). The project demonstrated the ease with which high-resolution aerial imagery can easily be acquired with small-scale, battery-operated UAV platforms and then processed into various products for subsequent analysis and interpretation.

The project also provided the basis to test the potential of a UAV platform for vertical and oblique aerial photography and video imagery. The data and imagery were used to support a coastal monitoring and mapping application, and demonstrated the stability of a large UAV platform and camera system. In addition, there was the additional possibility to test new sensors, for example, LiDAR and hyperspectral, either on their own or in conjunction with other sensors. Demonstration of this UAV capability provided a number of new aerial acquisition opportunities, with further potential to test the platform and sensors for a number of other marine and coastal tasks.

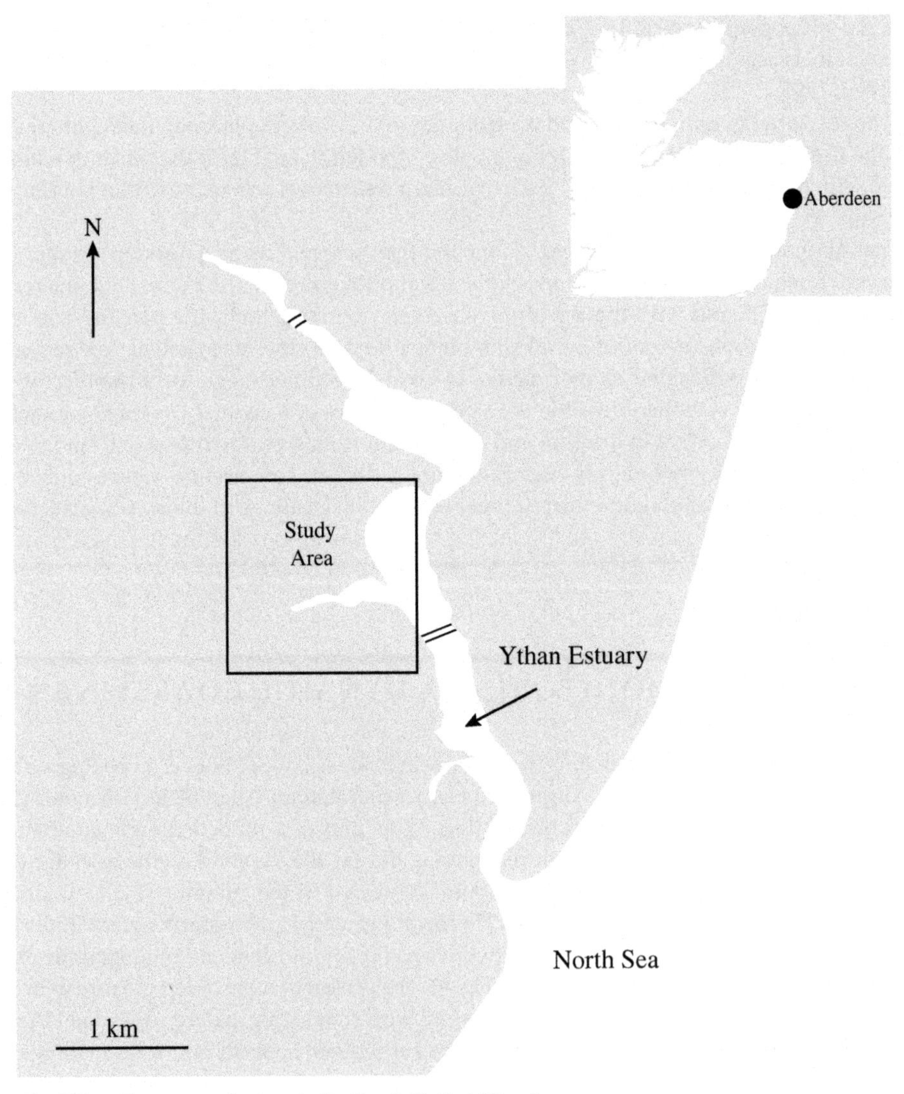

FIG. 3 The Ythan Estuary study area in Scotland, United Kingdom.

FIG. 4 Fixed-wing SSA used for aerial overflights of the Ythan Estuary.

FIG. 5 A ground control point (GCP) marker.

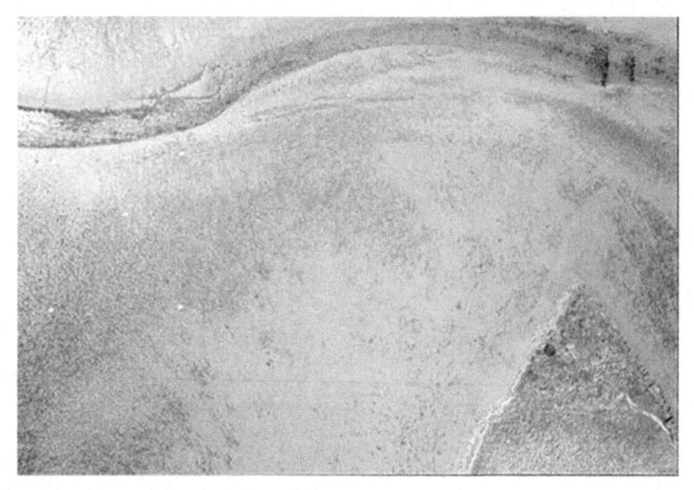

FIG. 6 Aerial photograph of macroalgal weedmats in the Sleek of Tarty, Ythan Estuary.

FIG. 7 Align 800 Helicopter (Borich Aircams).

BOX 3. CHANGE DETECTION AND RATES OF EROSION ON THE NORFOLK COASTLINE USING A MULTIROTOR UAV

Introduction

The poorly consolidated glacial till deposits, separated by silt, clay, and sand, leave sections of the north Norfolk coastline particularly vulnerable to coastal erosion (Thomalla and Vincent, 2003; Poulton et al., 2006). In recent years, aggressive erosion along the coastline has caused the collapse of several houses and roads in the most vulnerable areas.

Accurately quantifying rates of erosion is important for a variety of coastal management decisions, including creating vulnerability indices and shoreline management plans (SMPs).

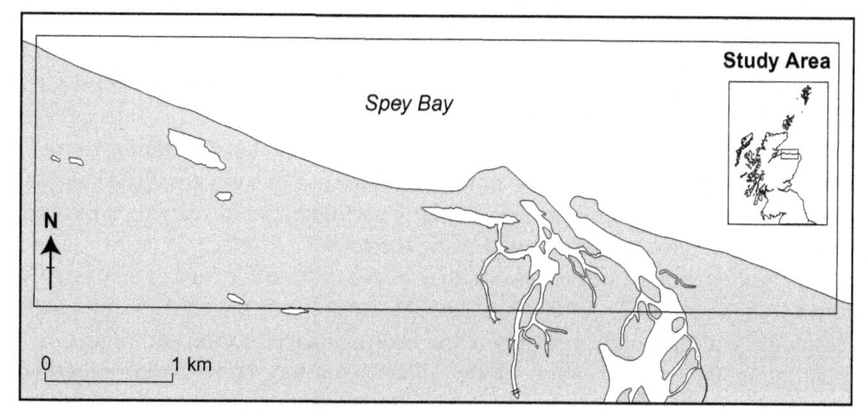

FIG. 8 Spey Bay, Moray, Scotland, UK study area.

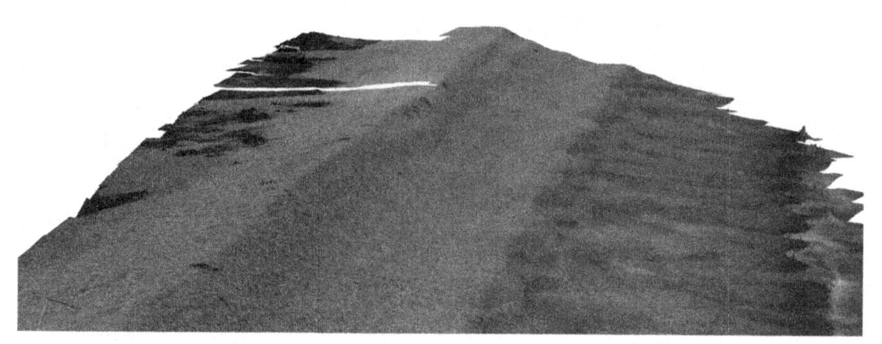

FIG. 9 3D model of part of Spey Bay beach using AgiSoft.

In this example, a combination of Environment Agency (EA) airborne LiDAR data and UAV data was used to identify erosion hotspots along the Norfolk coastline. Airborne LiDAR data was used to identify areas suffering from high levels of erosion along 12 km of coastline while UAV imagery was used to provide up-to-date rates of erosion and high-resolution 3D models, DSMs, and orthomosaics at Happisburgh—one of the worst affected areas.

LiDAR Data

Airborne LiDAR data, collected by the UK's Environment Agency (EA), were obtained for 1999, 2009, and 2013. The data were processed using ESRI's ArcGIS software from which a baseline for the cliff toe from 1999 to 2013 was then derived. With the use of the Digital Shoreline Analysis System (DSAS) tool (coast.noaa.gov/digitalcoast/tools/dsas; Thieler et al., 2009), total and average rates of erosion were calculated along 12 km of coastline. The results revealed erosion *hotspot* areas at various locations along the coastline.

UAV Data

In May 2016, a DJI Phantom 3 was flown at the Happisburgh study site to obtain aerial imagery. The aim was to use UAV imagery in combination with the historic LiDAR data to investigate coastal changes in the area. Ninety-five images were collected using Drone Deploy—an autonomous flight planning application—to acquire images of the site. The images were then uploaded from the UAV and imported into Pix4D, a soft-copy photogrammetry software package, where they were processed (Fig. 10).

The final processed results included a 3D point cloud, a TIN model, an orthomosaic, a DSM, and a reflectance image. In addition to users exporting data for use in software, Pix4D allows users to calculate volumes of any area of interest, a particularly useful tool for identifying the volume of material that may have been eroded or accreted along the coast.

Rectified and georeferenced orthomosaics were also exported for further analysis. The DSAS tool was used to quantify levels of erosion throughout the area covered by the UAV-derived data. A commonly cited problem with shoreline change quantification techniques is the difficulty in identifying the exact location of the cliff top or toe. However, the high-resolution UAV orthomosaics can help to address this problem by allowing the user to zoom in to very large scales to identify the cliff top or toe–ultimately reducing statistical error margins. UAV data collected and processed in this way can provide orthomosaics and 3D models with an accuracy of below 1 cm per pixel, something that has traditionally only been possible with very costly LiDAR systems (Fig. 11).

Fig. 12 shows how UAV-derived data can be used in conjunction with other data sources–LiDAR and 1:1000 Ordnance Survey (OS) maps–to accurately quantify spatial change at the coastline over time. The UAV-derived data allows more accurate 3D models and orthomosaics

FIG. 10 Pix4D—*solid black circles* indicate where the images were captured and the *gray line* shows the UAV flight path.

FIG. 11 A 3D TIN model produced in Pix4D. The location of where each photograph was captured is indicated by the spheres/squares.

to be produced than is otherwise available and at relatively low cost. Furthermore, accurate 3D models can be useful when explaining complex environmental processes in stakeholder engagement meetings, where audiences may be from a wide variety of nongeographical backgrounds (Brown et al., 2006).

Although it is unlikely that large stretches of the coastline could be mapped using multirotor UAVs under current Civil Aviation Authority regulations, it is possible that selective inspection of specific areas using UAVs in combination with other data sources could be advantageous over conventional mapping and monitoring approaches.

Conclusions

1. Small off-the-shelf UAV platforms can be used to collect high-quality spatial information.
2. The outputs from soft-copy photogrammetric software such as Pix4D can be analyzed in industry-standard GIS programs to aid in coastal change detection analysis, the results of which can aid in the development of coastal management plans.

19 ADVANCES IN THE TECHNOLOGY

Advances in UAV and related technologies have been very rapid in recent years. Hazel and Aoude (2015), for example, list the following areas that are in various stages at present: *advanced manufacturing techniques; batteries and other power; communication systems; detect, sense, and avoid capabilities; GPS; lightweight structures; microprocessors; motors; engines; and sensors.* Within a very short space of time, many of these areas of research are already finding their way into off-the-shelf products, for example, the DJI Phantom now has Flight Limit and

Change detection along the coast: Happisburgh, Norfolk—1999 – 2016

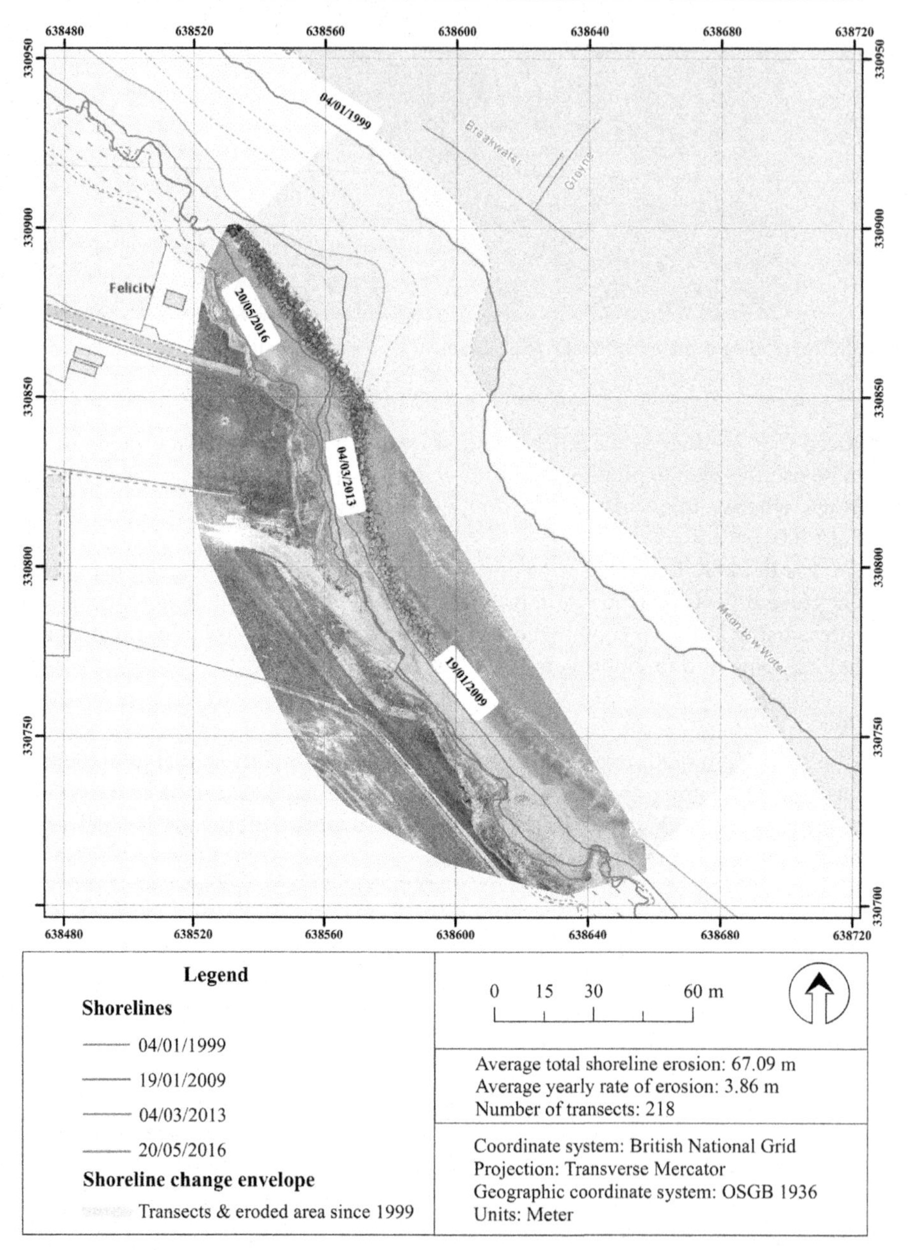

FIG. 12 UAV-derived data used in conjunction with other data sources—LiDAR and 1:1000 Ordnance Survey (OS) maps—to accurately quantify spatial change at the coastline over time.

No-Fly Zone technology built in, and for the Phantom 4 this includes a collision avoidance system, the Obstacle Sensing System (Fisher, 2016).

19.1 Platforms

There are now many manufacturers of both multirotor and fixed-wing aerial platforms of varying types and configurations. These include quadcopters, hexacopters, and octa/okta-copters of various different sizes, types, and costs, from toys to professional kits. Alongside the standard platforms, there are now also a number of waterproof platforms that not only allow the UAV to land and take off from waterbodies, but also to prevent loss of the platform should it ditch in water owing to a power or physical malfunction for example, Aquacopter Bullfrog (www.aquacopters.com) and Splashdrones (www.fpvfactory.com). These developments now provide operators with a wide choice of platforms that range from the simple and easy to maintain to the professional platform that requires a fair amount of expert maintenance. Today, the technology is still evolving very rapidly and there are a number of ongoing areas of development that will significantly enhance the future potential of UAV technology.

19.2 Multiple Drone Configurations

Although most applications of drones to date have involved only a single platform, there has been a lot of very sophisticated ongoing research into the development and deployment of multiple platforms or so-called *swarms* flying in configuration. According to Madey (2013), *Swarms offer numerous advantages over single UAVs, such as higher coverage, redundancy in numbers, and reduced long-range bandwidth requirements.* Already, SenseFly, for example, has developed software for controlling multiple drones for mapping remote areas (eijournal.com/news/industry-insights-trends/swarm-technology-for-drone-mapping-released).

19.3 Ready to Fly

The growing number of RTF drones now available is allowing virtually anyone to fly and capture airborne data and imagery as the basis for environmental monitoring, mapping, and modeling applications. Although some of the larger platforms require more skills, knowledge, and understanding of the platform and sensor combination as well as a certain level of training required to fly safely and within the current legislation (e.g., CAA), the freedom to acquire high-resolution data and imagery when needed is really unprecedented. The growing ease of use is one of the factors that is driving the popularity of these platforms, especially when compared to many earlier SSA platforms.

19.4 Batteries

Because most of the platforms and their attached systems are now mostly reliant upon batteries, for example, LiPo (Lithium Polymer), developments in battery technology have become essential to allow for longer flight duration and reliability. There is currently considerable effort being put into developing battery technology with rapid advancements; for instance, the ratio capacity to size is increasing, which facilitates more power consuming activities such as longer flights and heavier payloads (Gomez and Green, 2015). Various new battery technologies are also emerging that are more compact and more lightweight alongside their extended life. Metal-air battery technologies (e.g., zinc-air, aluminium-air, lithium-air) are being developed. These batteries have valuable qualities that will benefit the UAV industry, for example, high-density power. Graphene cells, although not yet commercially available, also have potential for UAVs. For the moment, though, LiPo batteries of varying sizes and capacities seem to dominate the market, although lithium sulfur technology (Li-STM) is viewed as the next development (www.barnardmicrosystems.com/UAV/engines/batteries.html). Within a very short space of time, battery life, even for smaller UAVs (e.g., the DJI Phantom) has increased the flight time from 5–8 min to 15–20 min, and even longer with some larger and enhanced LiPo batteries. Naturally, larger UAVs with more payload and lift capacity can extend this flight time further using multiple onboard battery packs. Other sources of energy used by a few UAVs are fuel cells and solar power. Some even use new technologies such as microgenerators, microturbines, and chemically powered systems that may replace batteries in the future. Nevertheless, battery technology is far from becoming obsolete and new designs introduced by battery manufacturers now offer relatively light, high capacity and reliable sources of power (Gomez and Green, 2015).

19.5 Autonomous Navigation, GPS, and Collision Avoidance

Autonomous flight capability is well developed for the larger and more sophisticated drone platforms and covers the ability of the platform to navigate among other obstacles in the flight path. This is clearly advantageous for systems that may be expected to fly in enclosed spaces (Gomez and Green, 2015). Although many small UAVs are still flown manually, with each new release of UAV platform model (e.g., DJI Phantom 3 and 4) new technology has made it increasingly easy for a pilot to fly the aircraft. With the aid of improved GPS units and mobile phone and tablet Apps (e.g., Drone Deploy (www.dronedeploy.com), Ground Station (www.dji.com/product/pc-ground-station); Litchi (flylitchi.com)), it is now possible to preplan a UAV flight—either before going onsite or onsite—and let the software and UAV do the work, for example, flying coverage of a field or specified area. This may include auto take off and landing, operation of the camera to acquire stereoimagery, altitude settings, and so on. Low-cost RTK GPS units are now also becoming available for UAVs to provide more accurate positional locations to aid in, for example, aerial survey applications; for example, Piksi (www.swiftnav.com/piksi.html).

With the unfortunate number of incidents involving *drones* covered by the media and the press as well as the desire to accomplish more complicated flight paths and commercial uses of drones for surveying and product delivery, there has been a growing emphasis on including as standard built-in collision avoidance systems and software that prevents pilots from flying in certain areas or within a certain distance of, for example, an airport (Fisher, 2016).

19.6 Sensors

Many of the early aerial sensors were off-the-shelf digital cameras of varying makes and sizes used without much modification for use on UAVs. One of the most popular has been the

GoPro Hero (gopro.com) series of small, waterproof, and physically robust digital sports cameras capable of taking photographic stills and video footage. Although many applications have used these cameras *as is*, recent developments have seen a number of modifications to the camera sensor wavelengths and the lens types. Manufacturers have developed low-cost specialist modifications for digital cameras for off-the-shelf platforms that extend the capabilities for environmental monitoring for example, MapIR (www.mapir.camera) and HackHD (www.hackhd.com). Further developments lay with the higher end digital SLR and video cameras that can provide very high-resolution imagery from slightly larger airborne platforms with higher payloads. The development of UAV gimbals by companies such as DJI, for example, their ZenMuse (www.dji.com/product/zenmuse-z15) has provided the means to carry off-the-shelf digital cameras that are recognized as highly suited to aerial photography from small-aircraft platforms.

A growing demand for additional aerial remote sensors to extend the wavelength-sensing capabilities beyond the visible and the NIR includes thermal cameras. One example is the range of modifications offered by IRPro (www.ir-pro.com) in the United States to extend the sensitivity of GoPro cameras to include multispectral images and NDVI as well as a range of flat lenses that remove the oft-cited disadvantages of the GoPro, namely the semifisheye lens. Another advantage of these customized popular sensors is the end product that, in the case of the GoPro NDVI modification, offers the end user, client, or customer a service with a product that is presented in the form of information that can then be used in in situ planning and decision-making. The Flir Tau 2 (www.flir.com/) is a small thermal sensor designed specifically for the small UAV platform, for example, DJI Phantom and DJI Inspire. Likewise, hyperspectral cameras from Headwall Photonics (www.headwallphotonics.com) and Lidar sensors, for example, LidarPod (www.routescene.com) and Riegl's Bathycopter LiDar (www.riegl.com), are all products specifically designed to fit the payload and size constraints of SSAs. The miniaturization of these sensors offers the potential to gather new, unique data and environmental information from small airborne platforms. The major constraint for many operators at present, though, is the price associated with these particular sensors, and while the costs are dropping they are still generally outside the budget of most organizations who may only use the equipment infrequently.

19.7 Software

An increasing demand to be able to process the digital still and video imagery in various different ways has also grown rapidly. Initially, many hobbyists made use of commercially available or open source (opensource.org) digital image processing (DIP) software, and occasionally the soft-copy photogrammetric modules within these. While some people made use of specialist environmental remote-sensing software, for example, Erdas Imagine (www.hexagongeospatial.com), ENVI (www.harrisgeospatial.com), others utilized graphic software such as Adobe Photoshop to undertake lens and light corrections and to stitch together multiple photographic stills into mosaics. Not surprisingly, within the past 5 years various software vendors have seen the opportunity to develop a number of relatively low-cost commercial and educational soft-photogrammetry and image-processing software specifically tailored to imagery acquired from UAV platforms and common digital cameras such as the GoPro. Commercially available products such as AgiSoft (http://www.agisoft.com),

Pix4D (https://pix4d.com), and, most recently, ENVI-UAV (http://www.harrisgeospatial.com), have emerged to allow rapid processing of UAV imagery into a range of products such as mosaics and 3D visual surface models. With each new release of the software, these products are rapidly improving and, like RTF drones, the interfaces are becoming easier to use for the nonspecialist. Higher-end softcopy photogrammetric software solutions are also offered by MosaicMill (http://www.mosaicmill.com), Pix4D, and others. Another product is GeoApp UAS from Hexagon Geospatial (www.hexagongeospatial.com). Open source software such as AirPhotoSE (http://www.uni-koeln.de/~al001/airphotose.html) also provides the basis for researchers to utilize UAV imagery, in this case specifically targeting Archaeology.

20 SUMMARY AND CONCLUSIONS

Small-scale aerial platforms and sensors have clearly come a considerable way in a relatively short period of time, going from model aircraft with fuel-based engines and SLR cameras that needed to be flown by a specialist to the current plug-and-play RTF platforms with custom-designed UAV sensors. In a relatively short period of time, UAV technology has advanced from being a hard-to-use novelty toy to a very serious aerial platform that offers considerable research and commercial potential, in whatever form, to gather unique data and imagery from a range of altitudes and coverages for small area, large-scale requirements for a very wide range of environmental applications.

The coastal examples used to illustrate this chapter are evidence of the considerable potential these platforms have always had, but which have now entered a different league in terms of what can be achieved. The reduced costs and ease of image acquisition, processing, analysis, and generation of usable products has grown phenomenally in the last 5 years. Recent developments also show that the technologies are still evolving very quickly and with the ongoing developments in the technology, this is set to continue, grow, and expand in the next few years. The significance of these small aerial platforms is perhaps best summed up by Wyman (2015) who observes, *By 2035, the number of unmanned aerial vehicle operations per year will surpass that of manned aircraft.*

Acknowledgments

I would like to extend special acknowledgements to two people who have contributed to my long-term interest in RC models, and now UAVs, over the years:

(1) Ed Wedler, whose office I shared in the Ontario Centre for Remote Sensing (OCRS) in the mid/late 1980s when I was a postgraduate student studying at the University of Toronto in Canada. He was the person who was first responsible for cultivating my interest in SSAs all those years ago.

(2) Norrie Kerr of Aberdeen, Scotland, for his considerable practical knowledge and expertise and for willingly spending a lot of his free time flying the various model aircraft he designed and built over the Ythan Estuary to the north of Aberdeen.

An interest that has endured to the present day, leading finally to the establishment of my UAV Centre for Environmental Monitoring and Mapping (UCEMM) at the University of Aberdeen in the Department of Geography and Environment, Scotland, with a squadron of multirotor UAV hardware, accompanying software, and applications.

References

Anderson, K., et al., 2015. 'Coastal-eye'—Monitoring Coastal Waters Using a Lightweight UAV. Unpublished document. Available from: http://nercgw4plus.ac.uk/files/2015/07/Coastal-eye-monitoring-coastal-waters-using-a-lightweight-UAV.pdf.

Anon, 1975. Remote aerial photography from a model helicopter. Br. J. Photogr. 122.

Appeaning Addo, K., Jayson-Quashigah, P.-N., Rovere, A., Mann, T., Caella, E., 2016. Monitoring Coastal Protection Structures Along the Volta Delta Shoreline Using Unmanned Aerial Vehicle (UAV). Unpublished Poster.

Boddington, D., 1989. Cyclops: aerial photography of field experiments using an RPV. Radio Control Models and Electronics, 448–450. June 1989.

Bowie, P., 1981. Eye in the Sky. RC Model Builder, pp. 38–40. June 1981.

Brown, I., Jude, S., Koukoulas, S., Nicholls, R., Dickson, M., Walkden, M., 2006. Dynamic simulation and visualisation of coastal erosion. Comput. Environ. Urban. Syst. 30 (6), 840–860.

Buckle, B., 1985. Practical safe operation of very small RPVs for civil users.Proceedings of the 5th International Conference on RPVs—Remotely Piloted Vehicles. Bristol, UK, 9–11 September 1985, pp. 4.1–4.9.

Canas, A.A.D., Irwin, D.A., 1986. Airborne remote sensing from remotely piloted aircraft. Int. J. Remote Sens. 7 (12), 1623–1635.

Clark, A.S., 1982. Canadair rotary-wing RPV technology development—PART II.Proceedings of the 3rd International Conference on Remotely Piloted Vehicles. Bristol, UK. 13th–15th September 1982. Paper II.

Colomina, I., Molina, P., 2014. Unmanned aerial systems for photogrammetry and remote sensing: a review. ISPRS J. Photogramm. Remote Sens. 92, 79–97.

De Wulf, R.E., Goossens, R.E., 1994. In: Winter wheat yield estimation with CIR imagery from a remotely piloted aircraft.Proceedings 1st International Airborne Remote Sensing Conference and Exhibition. Applications, Technology, and Science: Today's Progress for Tomorrow's Needs. Strasbourg, France, 12–15 September 1994.

Deuze, J.L., Devaux, C., Herman, M., Santer, R., Balois, J.Y., Gonzalez, L., Lecomte, P., Verwaerde, C., 1989. Photopolarmetric observations of aerosols and clouds from balloon. Remote Sens. Environ. 29, 93–109.

Ellis, R.M., Totten, J.A., Fuller, A.R., 1981. In: The use of radio controlled aircraft in pollution studies.Proceedings 2nd International Conference on Remotely Piloted Vehicles. Bristol, UK. Paper 11.

Fagerlund, E., Gunnershed, N., 1975. Systems analysis and development of a mini-RPV for reconnaissance. The 'Skatan' Project. FDA Report D30021-t1. National Defence Research Institute, Stockholm. 50 p.

Fisher, J., 2016. DJI Adds Collision Avoidance System to Phantom 4 Drone. PC Mag UK. March 2016, http://uk.pcmag.com/dji-phantom-3-professional/75655/news/dji-adds-collision-avoidance-system-to-phantom-4-drone.

Fouche, P.S., Booysen, N., 1994. Low altitude surveillance of agricultural crops using inexpensive remotely piloted aircraft.Proceedings 1st International Airborne Remote Sensing Conference and Exhibition, Strasbourg, France, 11th–15th September 1994, pp. III-315–III-326.

Gomez, C., Green, D.R., 2015. Small-scale airborne platforms for oil and gas pipeline monitoring and mapping. Unpublished AICSM Interface Report for Redwing Ltd., 54 p.

Goncalves, J.A., Renato, H., 2015. UAV photogrammetry for topographic monitoring of coastal areas. ISPRS J. Photogramm. Remote Sens. 104, 101–111.

Green, D.R., 1995. Preserving a fragile environment: integrating technology to study the Ythan estuary. Mapp. Aware. 9, 28–30.

Green, D.R., 2016. Acquiring environmental remotely sensed data from small-scale aircraft for input to geographic information systems (GIS). Unpublished Paper, 34 p.

Green, D.R., Morton, D.C., 1994. In: Acquiring environmental remotely sensed data for input to geographic information systems.Proceedings of the AGI'94 Conference: Broadening Your Horizons, Birmingham, UK, 15th–17th November 1994, pp. 15.3.1–15.3.27.

Gregory, T.J., Bailey, R.O., Nehms, W.P., 1974. RPVs—exploring civilian applications. Astronaut. Aeronaut., 38–47 (September).

Guillot, B., Pouget, F., 2015. UAV application in coastal environment, example of the Oleron Island for dunes and dikes survey. Int. Arch. Photogramm. Remote. Sens. Spat. Inf. Sci. XL-3/W3, 321–326. ISPRS Geospatial Week 2015, 28 Sep–03 Oct 2015, La Grande Motte, France.

Harding, B., 1989. Model aircraft as survey platforms. Photogramm. Rec. 13 (74), 237–240.

Haubeck, K., Prinz, T., 2013. A UAV-based low-cost stereo camera system for archaeological surveys—experiences from Doliche (Turkey). Int. Arch. Photogramm. Remote. Sens. Spat. Inf. Sci. XL-1/W2, 195–200. UAV-g2013, 4–6 September 2013, Rostock, Germany.

Hazel, B., Aoude, G., 2015. In Commercial Drones, the Race Is On: Aviation's Fastest-Growing Sector Outpaces US Regulators. Oliver Wyman Aviation, Aerospace and Defense, Marsh and MacLennan Companies. 16 p.
Hoer, J., 1976. Aerial photography from model aircraft. Br. J. Photogr. 123 (8), 156.
Hugenholtz, C.H., Moorman, B.J., Riddell, K., Whitehead, K., 2012. Small unmanned aircraft systems for remote sensing and earth science research. EOS Trans. Am. Geophys. Union 93 (25), 236.
ICAO, 2011. ICAO circular 328, Unmanned Aircraft Systems (UAS). Technical Report, International Civil Aviation Authority, Montreal.
Johnson, G.W., 1978. In: Balloon photography for archaelogical exploration and mapping.ASP Proceedings. 43rd Annual Meeting. February 27th–March 5th, 1978.
Jonsson, I., Mattsson, J.O., Okla, L., Stridsberg, S., 1980. Photography and temperature measurements from a remotely piloted vehicle. Oikos 35, 120–125.
Kendall, D.J.W., Clark, T.A., 1979. Balloon borne for infrared Michelson interferometer for atmospheric emission studies. Appl. Opt. 18 (8), 346–353.
Kerr, N., 1977. Aerial Surveying in Miniature. A Magazine for People in the North Sea Oil Industry, Roustabout, p. 35.
Klemas, V.V., 2015. Coastal and environmental remote sensing from unmanned aerial vehicles: an overview. J. Coast. Res. 31 (5), 1260–1267.
Lapp, H.S., Shea, E., 1976. In: A night photo system for remotely piloted vehicles.Proceedings Society of Photo-Optical Instrumentation Engineers, Reston, VA, 24th–25th March 1976, pp. 174–178.
Lemeunier, P., 1978. Electronic reconnaissance with pilotless aircraft. Electro. Appl. Ind. 247, 41–43.
Levanon, N., 1979. Ice elevation map of Queen Maud Land, Antarctica from balloon altimetry. Nature 278 (507), 842–845.
Long, N., Millescamps, A., Benoît, G., Bertn, X., 2016. Monitoring the topography of a dynamic tidal inlet using UAV imagery. Remote Sens. 8 (5), 387.
Madey, A.G., 2013. Unmanned aerial vehicle swarms: the design and evaluation of command and control strategies using agent-based modeling. Int. J. Agent Technol. Syst. 5 (3), 1–13.
Mancini, F., Dubbini, M., Gattelli, M., Stecchi, F., Fabbri, S., Gabbianelli, G., 2013. Using unmanned aerial vehicles (UAV) for high-resolution reconstruction of topography: the structure from motion approach on coastal environments. Remote Sens. 5, 6880–6898.
Marks, A.R., 1989. Aerial photography from a tethered helium filled balloon. Photogramm. Rec. 13 (74), 257–261.
Meroni, M., Rossini, M., Guanter, L., Alonso, L., Rascher, U., Colombo, R., Moreno, J., 2009. Remote sensing of solar-induced chlorophyll fluorescence: review of methods and applications. Remote Sens. Environ. 113, 2037–2051.
Miller, P., 1979. Aerial photography from radio-controlled aircraft. Aerial Archaeol. 4, 11–15.
Mullins, J., 1997. Palmtop planes. New Sci. (2076). 9 p.
NASA, 1996. Workshop on Remotely-Piloted Aircraft for U.S. Global Change Research. November 12th–15th 1996, Williamsburg, VA.
North East River Purification Board, 1982. Aerial photography using a model aircraft—an evaluation. Unpublished Report. 16 p.
Pereira, E., Bencatel, R., Correia, J., Félix, L., Gonçalves, G., Morgado, J., Sousa, J., 2009. Unmanned air vehicles for coastal and environmental research. J. Coast. Res. (56), 1557–1561. Proceedings of ICS2009.
Petrie, G., 2013. Commercial operation of lightweight UAVs for aerial imaging and mapping. GEOInformatics 1, 28–38.
Poulton, C.V.L., Lee, J.R., Hobbs, P.R.N., Jones, L., Hall, M., 2006. Preliminary investigation into monitoring coastal erosion using terrestrial laser scanning: case study at Happisburgh, Norfolk. Bull. R. Geol. Soc. 56, 45–64.
Roustabout. n.d. A magazine for people in the north sea oil industry. Aerial Surv. Miniat.. Issue No. 133. p.35.
Ryan, D., 2012. In: Unmanned aerial vehicles: a new approach for coastal habitat assessment.Presentation, Worley Parsons Western Operations, OGeo. September 2012. 27 slides.
Skrypietz, T., 2012. Unmanned Aircraft Systems for Civilian Missions. Brandenburg Institute for Society and Security (BIGS), Potsdam. Policy Paper No. 1, 28 p.
Syms, P., Turner, P.S., 1982. ASAT: the UK's first turbojet RPV.Proceedings 3rd International Conference Remotely Piloted Vehicles. Bristol, UK. 13th–15th September 1982. Paper 12.
Taylor, J.W., Munson, K., 1977. Jane's Pocket Book of Remotely Piloted Vehicles: Robot Aircraft Today. Collier Books, London. 239 p.
Thieler, E.R., Himmelstoss, E.A., Zichichi, J.L., Ergul, A., 2009. Digital Shoreline Analysis System (DSAS) version 4.0—an ArcGIS extension for calculating shoreline change. U.S. Geological Survey, p. 1278. Open-File Report 2008.

Thomalla, F., Vincent, E.E., 2003. Beach response to shore-parallel breakwaters at Sea Palling, Norfolk, UK. Estuar. Coast. Shelf Sci. 56 (2), 203–212.

Thurling, D.J., 1987. In: Design and operation of low-cost remotely-piloted aircraft for scientific field research. Proceedings of 6th International Conference on Remotely Piloted Vehicles, Bristol, UK, 6th–8th April 1987.

Thurling, D.J., Harvey, R.N., and Butler, N.J., n.d. Aerial Photography of Field Experiments Using Remotely Piloted Aircraft. Unpublished Manuscript, 7 p.

Tomlins, G.F., 1983. Some considerations in the design of low-cost remotely-piloted aircraft for civil remote sensing applications. Can. Surv. 37 (3), 157–167.

Tomlins, G.F., Lee, Y.J., 1983. Remotely piloted aircraft—an inexpensive option for large-scale aerial photography in forestry applications. Can. J. Remote. Sens. 76–85. BC Regional Issue.

Tomlins, G.F., Manore, M.J., 1984. In: Remotely piloted aircraft for small format aerial photography. Proceedings 8th Canadian Symposium on Remote Sensing, pp. 127–135.

Turner, I.L., Harley, M.D., Drummond, C.D., 2016. UAVs for coastal surveying. Coast. Eng. 114 (2016), 19–24.

Velligan, F.A., Gossett, T.D., 1982. In: Extended use of the aquila RPV system. Proceedings 3rd International Conference on Remotely Piloted Vehicles. Bristol, UK. 13th–15th September 1982. Paper 6.

Warner, W.S., Graham, R.W., Read, R.E., 1996. Small Format Aerial Photography. Whittles Publishing, Scotland. 347 p.

Wedler, E., 1984. Experience with radio-controlled aircraft in remote sensing applications. Proceedings 1984 ASP-ARSM Convention, Washington DC, March 11th–16th 1984, pp. 44–54.

Wester-Ebbinghaus, W., 1980. Aerial photography by radio-controlled model helicopter. Photogramm. Rec. 10 (55), 85–92.

Wyman, O., 2015. http://www.oliverwyman.com/our-expertise/insights/2015/apr/in-commercial-drones–the-race-is-on.html.

Further Reading

Kosowsky, L.H., Graziano, R.S., Wagner, R., Dunlap, D., 1977. In: A millimeter wave surveillance radar for RPVs. Proceedings Southeastcon'77, Imaginative Engineering through Education and Experience, Williamsburg, VA, 4th–6th April 1977, pp. 238–243.

The Way Forward

R.R. Krishnamurthy, M.P. Jonathan, Bernhard Glaeser

Analysis of the case study experiences in Bangladesh, India, Indonesia, Japan, Mexico, Philippines, Sri Lanka, and the United Kingdom on various aspects of coastal management approaches, policies, practices, and technology applications has proved that global coastal communities have gained numerous experiences that have facilitated the derivation of appropriate innovations in the field of coastal management. Based on the synthesis of case studies across the world, it is expected that the global coast may face the following challenges, which are most certainly likely to pose important research questions to our researchers:

1. Providing equitable access to coastal resources and coastal governance.
2. Mitigating coastal hazards and disaster risk reduction (DRR).
3. Enhancing the resilience of vulnerable communities.
4. Integrating indigenous knowledge with technology applications.
5. Training and capacity building on Integrated Coastal Zone Management, including higher education in coastal management.

1 COASTAL GOVERNANCE

Population explosion and poverty are the main causes for the challenges in providing equitable access to resources, which have already been impacted due to the overexploitation, pollution, and deterioration of environmental quality in major parts of underdeveloped and developing countries. In Asia, India's coastal regulatory mechanism, introduced by the Federal Ministry of Environment and Forests in 1991, has faced serious implementation issues. Also, being a regulatory approach, it did not incorporate the perceptions of local communities or the scientific basis, thereby leading to several amendments and modifications. Finally, the ministry has changed the Coastal Regulation Zone (CRZ) implementation to a management approach by incorporating stakeholder views as well as scientific data after two decades. Similarly, Sri Lanka has introduced the concept of a Special Coastal Management Area (SCMA) to manage about 27 designated coastal "hot spots" all over its coast. As an island nation, Sri Lanka relies on tourism as an important source for its economy and the issues from coastal tourism are still being considered as a challenge in the country. The lack of public policies and programs at the local level leads to sectoral conflicts and enhances the risk of environmental degradation in many coastal regions. On the Central American Mexican coast over the past two decades, there have been several changes in coastal policies related to the development of smart cities and tourism sectors. This is due to the overexploitation of the coastal cities as well as the natural calamities that often hit the regions from both the Pacific and Atlantic weather systems. Several countries have, therefore, taken up initiatives to introduce and implement new policies and programs to attain a balance between development and environmental quality. Coastal governance is, however, going to be one of the biggest challenges in the coming decades, without which environmental governance will not be a reality. Additionally, governance includes transparency, accountability, a corruption-free administration, and the participation-cum-sharing of intentions of all stakeholders. Conflicts and contradictions of interest among the stakeholders pose serious challenges to coastal governance.

2 COASTAL HAZARDS MITIGATION AND DISASTER RISK REDUCTION

The global coast has witnessed the increased intensity and frequency of climate-related disasters over the last three decades. Several countries have improved their warning systems for cyclones and coastal floods but serious challenges remain, especially to mitigate the damages due to such episodic events as tsunamis. Countries such as Indonesia, Japan, and Mexico have limited time to implement effective evacuation, as witnessed during the 2011 Tohohu earthquake and tsunami. Much more localized plans and the participation of local authorities are found to be very vital in a tsunami evacuation plan (TEP) as experimented in Padang City, as this involves phased activities. The major lacunae in an effective TEP are (1) lack of shelters, and (2) the acceptance of the local community as well as local authorities, apart from trained volunteers, to carry out effective evacuation. This is the area where several underdeveloped and developing countries need expertise and technological support from developed countries. Likewise, hurricanes, which often hit the Mexican Coast, teach a lesson during every season in one way or the other. The establishment of model sites is very vital in spreading the importance of mitigation and DRR through multinational and multiinstitutional collaboration to carry out field-based research in this domain.

3 COMMUNITY RESILIENCE

Coastal erosion is posing a serious threat to traditional communities, and this is being further aggravated by accelerating sea level changes. Coastal inhabitants in the thousands or even more are forced to move from their original place of living, making them "climate refugees." The stability of the coastline depends on various factors, including upland activities such as construction of dams and land use/land cover changes, which ultimately have an impact on the sediment load supplied to the near-shore regions. Countries with more deltaic regions such as Bangladesh are experiencing the chronic disaster of coastal erosion and inundation, posing a serious threat to traditional communities. Coastal erosion and inundation often result in saline water incursion into coastal aquifers and hence, coastal groundwater resources need to be managed effectively. Various controlling measures such as breakwater construction, groynes, or geotubes are being experimented with all over the world to mitigate the impact of coastal erosion. Artificial recharge methods are being experimented with for mitigating saline water intrusion, apart from controlling the overexploitation of groundwater in coastal areas. The changes in suspended sediments concentration (SSC) in coastal waters have an impact on the fishery potential and, in certain cases, lead to conflicts between farmers and fishermen. Continuous monitoring of the ecological parameters of coastal waters on a long-term basis is required, which is being carried out in India, because both coastal biodiversity and livelihood security are at risk. Hence, there is a need to enhance the resilience of vulnerable communities through appropriate scientific measures, as discussed in the case studies of Bangladesh, India, Malaysia, and Mexico.

4 FUSION OF MODERN TECHNOLOGY AND INDIGENOUS KNOWLEDGE

Spatial information technology tools have proved to be very vital for producing scientific databases for the end users and planners in coastal management. Thanks to their affordability and reliability, these tools are spreading to every nook and corner and are producing user-friendly products. Critical coastal habitats such as mangroves, coral reefs, and seagrass meadows are being monitored more precisely using multispectral, multisensor remote sensing data with the support of GIS and DGPS tools. India is one of the pioneering countries in utilizing the satellite-derived sea surface temperature (SST) data together with marine water productivity to demarcate Potential Fishery Zone (PFZ) information and translate it into vernacular languages for dissemination to several fishing centers all over the country. India has achieved considerable success in science-based, people-centered coastal ecosystem management with the participation of stakeholders in which the above technology tools have played vital roles. The indigenous knowledge of traditional communities has, however, not been given due recognition and importance, especially in managing coastal resources. For example, the fishing community knows where to fish and when to fish based on sea water color and other parameters, which they have traditionally followed. While the PFZ information helped them to locate potential areas for catching fish, it failed to provide information on the type of fish in the catch. In the Mexican coastal regions, the development of coastal cities for ecofriendly tourism activities always needs the special involvement of not only the local government, but also the dedication of the local community to conserve the region for tourism through sustainable development. Hence, there is a need to look for fusion information derived from modern technology with indigenous knowledge to effectively manage coastal resources.

5 INTEGRATED COASTAL ZONE MANAGEMENT (ICZM) TRAINING AND OUTREACH ACTIVITIES

Outreach is one of the important domains in ICZM, which needs periodic and site-specific training for various target groups. In the late 1990s, ICZM training for key decision-makers in the Indian government, under the umbrella of the UK DFID funded Indo-British program, helped not only to enrich knowledge but also to accept the importance of intersectoral approaches toward coastal management. In parallel, higher education in coastal management is also gaining importance in various developing countries. Conservation programs on protecting marine mammals, turtles, and beach quality as well as proper waste management are some of the areas focusing on nature protection in the coastal areas of Mexico. More attention is required in offering quality higher education in coastal management through international collaboration. A global consortium of institutions with provisions for the exchange of students and faculty is the current need, especially for producing trained and skilled human power in coastal management.

Index

Note: Page numbers followed by *f* indicate figures, *t* indicate tables, *b* indicate boxes and *p* indicate plates.

C

D

N

T

U